Environmental Science

Series Editors: R. Allan • U. Förstner • W. Salomons

Springer
Berlin
Heidelberg
New York
Hong Kong
London
Milan
Paris
Tokyo

Gerhard Gerold
Michael Fremerey
Edi Guhardja (Eds.)

Land Use, Nature Conservation and the Stability of Rainforest Margins in Southeast Asia

With 87 Figures

Springer

Prof. Dr. Gerhard Gerold
Department of Landscape Ecology
Institute of Geography,
Georg-August-University
Goldschmidtstr. 5
37077 Göttingen, Germany
ggerold@gwdg.de

Prof. Dr. Michael Fremerey
Institute for Socio-cultural Studies
Kassel University
Steinstr. 19
37213 Witzenhausen, Germany
fremerey@wiz.uni-kassel.de

Prof. Dr. Edi Guhardja
Kantor Kerjasama Jerman
Gunung Gede, Bogor, Indonesia
ctfm-ipb@indo.net.id

ISBN 3-540-00603-6 Springer-Verlag Berlin Heidelberg New York

Cataloging-in-Publication Data applied for

A catalog record for this book is available from the Library of Congress.

Bibliographic information published by Die Deutsche Bibliothek
Die Deutsche Bibliothek lists this publication in the Deutsche Nationalbibliografie; detailed bibliographic data is available in the Internet at <http://dnb.ddb.de>.

Springer-Verlag Berlin Heidelberg New York
a part of Springer Science+Business Media GmbH

springeronline.com

Camera ready by Jürgen Übener
Cover design: Struve & Partner, Heidelberg
Printed on acid-free paper 30/2132/AO 5 4 3 2 1 0

Preface

Southeast Asia constitutes one of the world's most extended rainforest regions. It is characterized by a high degree of biodiversity and contains a large variety of endemic species. Moreover, these forests provide a number of important and singular ecosystem services, like erosion protection and provision of high quality water, which cannot be replaced by alternative ecosystems. However, various forms of encroachment, mostly those made by human interventions, seriously threaten the continuance of rainforests in this area.

There is ample evidence that the rainforest resources, apart from large scale commercial logging, are exposed to danger particularly from its margin areas. These areas, which are characterized by intensive man-nature interaction, have been identified as extremely fragile systems. The dynamic equilibrium that balances human needs and interventions on the one hand, and natural regeneration capacity on the other, is at stake. The decrease of rainforest resources is, to a substantial degree, connected with the destabilization of these systems. Accordingly, the search for measures and processes, which prevent destabilization and promote stability is regarded as imperative. This refers to both the human and the natural part of the forest margin ecosystem.

Applying the concept of stability to a system which, by its very nature, is in a permanent process of change, deserves some explanations. Stability, in this context, does not imply any standstill of conditions or processes. It rather denotes a state of resilience which is required to keep the basic functions of the system intact. This refers, on the one hand, to the persistence of human livelihood which, to some extent, depends on continuous interventions into the natural environment. On the other hand, it applies to the capacity of the natural ecosystem to cope with such interventions, i.e. to regenerate and to provide a basis for human livelihood in the long run. Hence it follows that stability of forest margin areas has to be seen in terms of space and time. With regard to the first, a stable margin area remains spatially unchanged, preventing adjacent ecosystems of the rainforest from further conversion and fragmentation. With regard to the latter, the stability concept implies that the natural ecosystems provide long-term resilience to balance anthropogenic as well as natural disturbances.

In view of the multitude of unanswered questions concerning the issue described, the promotion and sharing of discipline-specific expertise still seem to be indispensable. However, the investigation of factors supporting stability or cause destabilization ultimately requires a holistic approach which combines socio-economic, biological and technological views and knowledge. Moving towards such an integrated perspective is one of the major objectives of the long term German-Indonesian research program on the "Stability of Rainforest Margins" (STORMA), jointly conducted by the universities of Göttingen, Kassel, Bogor (IPB) and Palu (UNTAD) and funded by the German Research Association (DFG). This collaborative program started in 2000 and focuses on the margin area of the Lore Lindu National Park in Central Sulawesi.

This reader reflects recent research findings on "Land Use, Nature Conservation and Stability of Rainforest Margins in Southeast Asia", discussed in an international Symposium held at Bogor from September 29 to October 3, 2002. This event, jointly organized by the four above-mentioned universities, was meant to represent STORMA's state of the art and, at the same time, match it with relevant research in other parts of the world, with special emphasis on Southeast Asia. This includes the identification of gaps of knowledge which may lead to the definition of meaningful research issues in the future.

The reader starts with a discussion of the *socio-economic dimension* of forest margin destabilization (Chapter 1). Deforestation and degradation of forests raise legitimate environmental concerns and can create high social costs for current and future generations. If unchecked by policy and institutional arrangements among stakeholders, private benefits (be they driven by despair or greed) from use of forest resources will continue to trigger chain-saw or slash-and-burn action in the remaining tropical forest margins of our world. Policies have therefore a pivotal role to play in the sustainable use of natural resources. Pluralistic policy frameworks and suitable governance structures are preconditions for appropriate policy design and effective policy enforcement. Moreover, policy design requires an understanding of the determinants, actors, stakeholders, and processes of deforestation and forest degradation. On the other hand, local practices of sustainable forest use and resource management need to be analyzed and documented in order to guide supporting policy measures. This part of the book provides social science research in various disciplines (cultural geography, sociology, agricultural and institutional economics, and political science) that seeks to offer some insights useful for better understanding the problem as a precondition for the design of improved policies. Case studies are presented from forest margins in Indonesia, Thailand and Vietnam, and feature topics such as land tenure, migration, informal institutions of resource management, as well as socio-cultural patterns and economic models of land use. A paper synthesizing the current knowledge on the causes of deforestation at the global level and related policy options and a paper reviewing the developments and policies in Indonesia precede the micro-level analysis in the three Southeast-Asian countries.

The second chapter deals with aspects of *biodiversity and conservation*. Major questions in recent research related to these themes refer, among others, to strategies of land use which maintain a high degree of biodiversity, the response of plant and animal groups to land use changes, and the relationship between biodiversity and ecological functions. Answers to such questions could provide relevant input to the design of adequate conservation concepts. The discussion on this matter in the present reader begins with a survey of a holistic approach to study resource use patterns in an old forest landscape in Ethopia. It is followed by an account of resource utilization in a protected area complex in India, showing that appropriate natural resource management is absolutely necessary for warranting the livelihood of the rural population. Complementary to these contributions, which particularly highlight the interplay between man and natural environment, the third article deals with the long-term impact of rainforest conversion by shifting cultivation on tree diversity, species composition and soil nutrients. It is based on research in West Kalimantan. The other five contributions in this chapter concentrate on innovative biological findings in the context of the STORMA project in Central Sulawesi. These include the patterns of tree diversity in secondary forests, a study which has been accompanied by the establishment of the first herbarium in Sulawesi, as well as the diversity of butterflies that shows how an important biodiversity indicator taxon responds to increasing forest modification and land use. Two contributions are autoecological accounts, dealing with the ability of selected vertebrates to persist in disturbed habitats, i.e. the Dian's tarsier *(Tarsius dianae)* and the Tonkean macaque *(Macaca tonkeana)*. These two primate species have never before been thoroughly studied in their natural environment. Finally, a model on bird species loss through deforestation is presented. It shows that relatively minor changes in forest cover within protected areas may have important consequences for the regional and global biodiversity.

Chapter 3 gives a comprehensive view on the interrelated aspects of *water and nutrient cycles, land use systems and the modelling of land use*. The first part deals with particular effects of forest conversion and land use on water and nutrient fluxes. Research results indicate the specific role of the regional context and the outstanding impact of human intervention in terms of forest conversion and land use patterns. In this context, the first two contributions relate to the hydrological consequences of forest conversion at two different stages: (1) recent conversion, and (2) older land use development with concomitant conflicts on water tenure. The ensuing two papers concentrate on field experiments with annual crops, aiming at optimizing the use of soil nutrients. Three contributions on the impact of different forest use strategies on soil nutrients round off this section. Corresponding data refer to modern agroforestry systems (rubber plantations) and to the traditional forest gardens. A comparative analysis investigates nutrient cycles and biodiversity in a secondary forest and in a mature forest plantation.

The authors of the two concluding papers take a perspective that includes a number of socio-economic and environmental factors to explain land use changes in Southeast Asia. They demonstrate the feasibility of spatially explicit regression models as tools for explaining land use changes as well as for scenario analysis.

The reader closes with a postscript which highlights the experience of two years of *research in an intercultural setting*. This experience was considered to be worth sharing, as many schemes nowadays involved in research on global environmental issues tend to be not only interdisciplinary but also intercultural in composition. This raises the problem of minimizing potential frictions and, at the same time, maximizing the benefits emerging from the diversity of expertise, knowledge systems and methodological approaches. At this point it is appropriate to convey special thanks to the German Academic Exchange Service (DAAD) which, by its generous co-funding, has particularly facilitated the inclusion of the intercultural issue into the proceedings of the Symposium.

The editors

Contents

List of Contributing Authors

Abate, A., M.Sc.
Lehrstuhl Biogeographie, Universität Bayreuth
Universitätsstr. 30, 95447 Bayreuth
Germany

Abdulkadir-Sunito, M., M.Sc.
Dept. of Agricultural Socio-Economics, Bogor Agricultural University
Duta Pakuan E/I no.1, Bogor 16127
Indonesia

Anthofer, J., Dr.
Inst. of Plant Production and Agroecology in the Tropics and Subtropics, University of Hohenheim
Garbenstr. 13, 70599 Stuttgart
Germany

Arnaud, M., Statistician
Cirad-Tera
73 avenue JF Breton, 34398 Montpellier cedex
France

Bechstedt, H.-D., Dr.
Fac. of Agriculture, Chiang Mai University
Hohenheim Office, New Building, 2nd floor, Chiang Mai 50200
Thailand

Birner, R., Dr.
Inst. for Rural Development, University of Göttingen
Waldweg 26, D-37073 Göttingen
Germany

Brodbeck, F, M.Sc. trop.for.
Inst. of Tropical Silviculture, University of Göttingen
Büsgenweg 1, 37077 Göttingen
Germany

Chamsai, L., M.Sc.
The Uplands Program (SFB 564), University of Hohenheim
New Building, Second Floor, Fac. of Agriculture, Chiang Mai University
Chiang Mai 50200
Thailand

Claassen, N., Prof. Dr.
Inst. of Agricultural Chemistry, University of Göttingen
Carl Sprengel Weg 6, 37075 Göttingen
Germany

Dao Chau Thu, Assoc. Prof. Dr.
Sustainable Agriculture Research and Development Center, Hanoi Agricultural University Gia Lam
Hanoi
Vietnam

Diaz-Novellon, S., Engineer in tropical agronomy
CNEARC
1101 avenue agropolis, 34090 Montpellier
France

El Dessougi, H., Dr.
Inst. of Agricultural Chemistry, University of Göttingen
Carl Sprengel Weg 6, 37075 Göttingen
Germany

Elbel, C., Dipl.-Ing. agr.
Inst. of Rural Development, University of Göttingen
Waldweg 26, 37073 Göttingen
Germany

Engel, S., Dr.
Centre for development Research (ZEF), University of Bonn
Walter-Flex- Str.3, 53113 Bonn
Germany

Engelsman, W., Ir.
Laboratory of Soil Science and Geology, Wageningen University
P.O. Box 37, 6700 AA Wageningen
The Netherlands

Erasmi, S., Dr. rer.nat.
Geographisches Institut, University of Göttingen
Goldschmidtstr. 5, 37077 Göttingen
Germany

Faust, H., Dr. phil.
Inst. of Geography, Dept. of Social and Cultural Geography, University of Göttingen
Goldschmidtstr. 5, 37077 Göttingen
Germany

Fremerey, M., Prof. Dr.
Inst. for Socio-cultural Studies, University of Kassel
Steinstr. 19, 37213 Witzenhausen
Germany

Gerold, G., Prof. Dr.
Inst. of Geography, Dept. of Landscape Ecology, University of Göttingen
Goldschmidtstr. 5, 37077 Göttingen
Germany

Gradstein, S. R., Prof. Dr.
Inst. of Plant Sciences, University of Göttingen
Untere Karspüle 2, 37073 Göttingen
Germany

Guhardja, E., Prof. Dr.
Dept. of Biology, Fac.of Basic Sciences, Bogor Agricultural University
IPB Kampus Baranang Siang, Bogor
Indonesia

Hammer, M., B.S.
Inst. of Agricultural Engineering in the Tropics and Subtropics
University of Hohenheim (495)
70593 Stuttgart
Germany

Härtel, M., M.Sc.
Inst. of Geography, University of Göttingen
Goldschmidtstrasse 5, 37077 Göttingen
Germany

Keenan, R. J., Dr. trop.for.
Bureau of Rural Sciences, Kingston
Australia

Kessler, P., J.A. Dr.
National Herbarium of The Netherlands, University of Leiden Branch
P.O. Box 9514, 2300 RA Leiden
The Netherlands

Kleinhans, A., M.Sc.
Inst. of Geography, Dept. of Landscape Ecology, University of Göttingen
Goldschmidtstr. 5, 37077 Göttingen
Germany

Kreisel, W., Prof. Dr. phil.
Inst. of Geography, Dept. of Social and Cultural Geography, University of Göttingen
Goldschmidtstr. 5, 37077 Göttingen
Germany

Kroschel, J., PD Dr.
Inst. of Plant Production and Agroecology in the Tropics and Subtropics, University of Hohenheim
Garbenstr. 13, 70599 Stuttgart
Germany

Lamb, D. Assoc., Prof. trop.for.
University of Queensland
Brisbane Qld 4072
Australia

Langi, M.A., Dr. trop.for.
Sam Ratulangi University
Manado 95115 North Sulawesi
Indonesia

Langkau, M.
Centre for Nature Conservation (Dept. I), University of Göttingen
Von-Siebold-Strasse 2, 37075 Göttingen
Germany

Lawrence, D., Dr.
Dept. of Environmental Sciences, University of Virginia
P O Box 400123, Charlottesville, Virginia 22902-4123
USA

Maertens, M., M.Sc.
Institut of Rural Development, Georg-August-University Göttingen
Waldweg 26, 37073 Göttingen
Germany

Mansjoer, S., Dr.
Primate Research Center, Bogor Agricultural University
Jl. Lodaya 3 Bogor 16151, West Java
Indonesia

Mardiastuti, A., Dr.
Primate Research Center, Bogor Agricultural University
Jl. Lodaya 3 Bogor 16151, West Java
Indonesia

Merker, S., Dr.
Centre for Nature Conservation, University of Göttingen
Von-Siebold-Str. 2, 37079 Göttingen
Germany

Mitlöhner, R., PD Dr.
Inst. of Tropical Silviculture, University of Göttingen
Büsgenweg 1, 37077 Göttingen
Germany

Mühlenberg, M., Prof. Dr.
Centre for Nature Conservation (Dept. I), University of Göttingen
Von-Siebold-Strasse 2, 37075 Göttingen
Germany

Müller, D., Dr.
Inst. of Rural Development, University of Göttingen
Waldweg 26, 37073 Göttingen
Germany

Müller-Hohenstein, K., o. Prof. Dr.
Lehrstuhl Biogeographie, Universität Bayreuth
Universitätsstr. 30, 95447 Bayreuth
Germany

Nawichai , P., M.Sc.
The Uplands Program, Chiang Mai University
80 Moo.7, Soi Ton Sang, Sansai Luang Subdistrict, Sansai District, Chiang Mai, 50210
Thailand

Neef, A., Dr.
The Uplands Program (SFB 564), University of Hohenheim
New Building, Second Floor, Fac. of Agriculture, Chiang Mai University
Chiang Mai 50200
Thailand

Nuryartono, N., Ir. M.Sc.
Inst. of Rural Development, Universität of Göttingen
Waldweg 26, 37073 Göttingen
Germany

Overmars, K.P., Ir.
Laboratory of Soil Science and Geology, Wageningen University
P.O. Box 37, 6700 AA Wageningen
The Netherlands
and:
Centre of Environmental Science, Leiden University
P.O. Box 9518, 2300 RA Leiden
The Netherlands

Panggabean, H., Dr.
Fakultas Psychologi, Universitas Atmajaya
Jl. Jenderal Sudirman 51, Jakarta 12930
Indonesia

Penot, E., Dr. Economics
Cirad-Tera
73 avenue JF Breton, 34398 Montpellier cedex
France

Pitopang, Ramadhanil, M.Sc.
Dept. of Forest Management, Fac. of Agriculture, Tadulako University
Kampus Bumi Tadulako, Tondo Palu
Indonesia

Pombo, R., M.Sc.
Primate Research Center, Bogor Agricultural University
Jl. Amirullah No 10/20 Makassar 95131, South Sulawesi
Indonesia

Rosyadi, S., Dr.
Jenderal Soedirman University
Jalan HR Boenjamin Purwokerto 53122 Central Java
Indonesia

Sangkapitux, C., Dr.
Dept. of Agricultural Economics, Fac. of Agriculture, Chiang Mai University
Chiang Mai 50200
Thailand

Sanim, B., Prof. Dr.
Dept. of Economics, Inst. Pertanian Bogor
Jl. Raya Gunung Gede, Bogor 16153
Indonesia

Sawhney, P., Dr.
Centre for development Research (ZEF), University of Bonn
Walter-Flex- Str.3, 53113 Bonn
Germany

Schlosser, G., M.Sc.
Inst. of Plant Production and Agroecology in the Tropics and Subtropics, University of Hohenheim
70593 Stuttgart
Germany

Schwarze, S., M.Sc. agr.
Inst. of Rural Development, University of Göttingen
Waldweg 26, 37073 Göttingen
Germany

Sirisupluxuna, P., Dr.
Dept. of Agricultural and Resource Economics, Fac. of Economics, Kasetsart University
Bangkok 10900
Thailand

Sitorus, MT Felix, Dr.
Rural Sociology Program, Bogor Agriculture University
Jl. Bangka IXA/84, Jakarta Selatan
Indonesia

Spreer, W., M.Sc.
Inst. of Agricultural Engineering in the Tropics and Subtropics, University of Hohenheim
70593 Stuttgart
Germany

Sunito, S., Drs.
Dept. of Agricultural Socio-Economics, Bogor Agricultural University
Duta Pakuan E/I no. 1, Bogor 16127
Indonesia

Tjitrosoedirdjo, S. S., Dr.
Dept. of Biology, Fac. of Basic Sciences, Bogor Agricultural University
Bogor
Indonesia

Tjondronegoro, S. M. P., Prof. Dr.
Inst. Pertanian Bogor
Jl. Pakuan Indah No. 3, Bogor 16143
Indonesia

van Mensvoort, M. E. F., Dr. Ir.
Laboratory of Soil Science and Geology, Wageningen University
P.O. Box 37, 6700 AA Wageningen
The Netherlands

van Rheenen, T., Dr. Ir.
Inst. of Rural Development, University of Göttingen
Waldweg 26, 37073 Göttingen
Germany

van Zalinge, R., Ir.
Laboratory of Soil Science and Geology, Wageningen University
P.O. Box 37, 6700 AA Wageningen
The Netherlands

Veldkamp, A., Prof. Dr. Ir.
Laboratory of Soil Science and Geology, Wageningen University
P.O. Box 37, 6700 AA Wageningen
The Netherlands

Verburg, P.H., Dr. Ir.
Laboratory of Soil Science and Geology, Wageningen University
P.O. Box 37, 6700 AA Wageningen
The Netherlands

Waltert, M., Dr.
Centre for Nature Conservation (Dept. I), University of Göttingen
Von-Siebold-Strasse 2, 37075 Göttingen
Germany

Wannitpradit, A., M.Sc.
Dept. of Social Anthropology, Fac. of Social Sciences, Chiang Mai University
Chiang Mai 50200
Thailand

Weber, R., M.Sc.
Inst. of Geography, Dept. of Social and Cultural Geography, University of Göttingen
Goldschmidtstr. 5, 37077 Göttingen
Germany

Weidelt, H.-J., Prof. Dr.
Inst. of Tropical Silviculture, University of Göttingen
Büsgenweg 1, 37077 Göttingen
Germany

Wiriadinata, H., Dr.
Herbarium Bogoriense, Research Center for Biology, LIPI
Bogor
Indonesia

Wirth, T., Dipl. Forsting. ETH
The Uplands Program (SFB 564), Vietnam Office, University of Hohenheim
1 Dai Co Viet, Hanoi
Vietnam

Wunder, S., Ph.D., D.Sc.
Center for International Forestry Research
Indonesia

Xyooj, Y., B.S.
The Uplands Program (SFB 564), University of Hohenheim
New Building, Second Floor, Fac. of Agriculture, Chiang Mai University
Chiang Mai 50200
Thailand

Yustian, I., M.Sc.
Dept. of Biology, University of Sriwijaya
Jl. Palembang-Prabumulih, km 32, Inderalaya, OKI, South Sumatra 30662
Indonesia

Zeller, M., Prof. Dr.
Inst. of Rural Development, University of Göttingen
Waldweg 26, 37073 Göttingen
Germany

zu Dreele, A., M.Sc.
Inst. of Agricultural Chemistry, University of Göttingen
Carl Sprengel Weg. 6, 37075 Göttingen
Germany

List of Editors

Gerold, G., Prof. Dr.
Inst. of Geography, Dept. of Landscape Ecology, University of Göttingen
Goldschmidtstr. 5, 37077 Göttingen
Germany

Fremerey, M., Prof. Dr.
Inst. for Socio-cultural Studies, University of Kassel
Steinstr. 19, 37213 Witzenhausen
Germany

Guhardja, E., Prof. Dr.
Dept. of Biology, Fac.of Basic Sciences, Bogor Agricultural University
IPB Kampus Baranang Siang, Bogor
Indonesia

Claassen, N., Prof. Dr.
Inst. of Agricultural Chemistry, University of Göttingen
Carl Sprengel Weg 6, 37075 Göttingen
Germany

Priess, J., Dr.
Center for Environmental Systems Research, University of Kassel
Kurt-Wolters-Straße 3, 34109 Kassel
Germany

van Rheenen, T., Dr. Ir.
Inst. of Rural Development, University of Göttingen
Waldweg 26, 37073 Göttingen
Germany

Waltert, M., Dr.
Centre for Nature Conservation (Dept. I), University of Göttingen
Von-Siebold-Strasse 2, 37075 Göttingen
Germany

Zeller, M., Prof. Dr.
Inst. of Rural Development, University of Göttingen
Waldweg 26, 37073 Göttingen
Germany

Chapter 1:

Social and Economic Development and Change

Policy Options for Stabilising the Forest Frontier: A Global Perspective

Sven Wunder

1 Introduction

The purpose of this paper is to summarise different research results about the impact of macro-level factors and "extra-sectoral" policies on tropical forest cover. Specifically, we are interested in the forest margins – i.e. the spatial transition zone between tropical forests and converted land uses. What are the policy factors that accelerate frontier expansion, and which ones tend to slow it down? The term "extra-sectoral" refers to all the things that happen outside of forests and forestry, yet nevertheless have a significant effect on forests. For instance, how do changes in international trade and a country's balance of payment affect deforestation? What does it mean for pressures on forests that a country drastically devalues its currency? What is the role of population growth?

In answering these and other questions, we will mainly draw on published and ongoing research carried out by the Center for International Forestry Research (CIFOR), which the author of this paper is affiliated to. For most of a decade, CIFOR has carried out a tropics-wide research programme on the "underlying causes of deforestation". The empirical results from this programme are the main source of knowledge for this paper. We will supplement this knowledge with other selected empirical studies that demonstrate how these macro factors and policies eventually "trickle down" to the forest. But the main objective is to synthesise the "big picture". Readers interested in the specific case studies that shape this "big picture" are referred to publications describing the underlying studies.

A key hypothesis is that what happens to tropical forests is more determined by events outside the forest arena than by what happens inside the forest sector. In other words, the extra-sectoral impacts will often be more important than, say, the new forest law, the participatory tree-planting project or the environmental education programme that is implemented in the forest margins. That does not mean that the latter type of intervention is useless. What it does mean is that some macroeconomic and extra-forestry factors tend to set the scene for success or failure of the projects and strategies of forest-margin stabilisation strategies, so that the promoters of these strategies need to have a realistic vision about the direction and proportions of impacts. In some cases, the macro-decision makers should also explicitly take into account how forests are affected before they make their "extra-sectoral", macro-level choices.

2 Definitions: Deforestation and Forest Degradation

2.1 Deforestation

Many different deforestation definitions exist, but in this paper we employ the terminology used by the United Nations Food and Agricultural Organization (FAO 2000a). According to FAO, a forest is an area of a minimum 0.5 ha size, covered by a tree canopy of at least 10%, with trees that can reach more than 5m height, subject to the constraint that the area should not be under an alternative (e.g. agricultural or urban) use. Deforestation would thus be any change in conditions that means the area no longer qualifies as a forest. In the majority of cases, deforestation occurs because the area's tree canopy-cover is reduced to less than 10% by converted land uses. This conversion can be permanent (e.g. urban expansion) or temporary (e.g. shifting cultivation). This means that we identify deforestation with a radical removal of tree cover - in most cases a conversion to other land uses.

Note that this definition does not say anything normative about whether deforestation is good or bad. Although much deforestation research is driven by a legitimate concern about the rapid loss of tropical forests, the desirability of these land-use change processes has to be assessed separately, based on a subsequent analysis of the costs and benefits of forest loss to different stakeholders at variable levels of aggregation.

2.2 Frontier Deforestation

Obviously, there are different means and ways to get rid of a forest. In this paper, we are particularly interested in frontier deforestation - the process of moving into large blocks of previously continuous forests. This transition zone is also often referred to as "forest margins". The process of frontier deforestation and advancing forest margins has to be distinguished from the clearing of forest remnants in preestablished agricultural or in peri-urban areas. Both types of forest loss have important implications, but for two reasons we have a special interest in the forest-frontier margins. First, the conservation of frontier forests has been given special weight from a biodiversity point of view (Bryant et al. 1997). Second, there is evidence that once forests are fragmented, they disappear more rapidly in incremental processes that are harder to stop (Mertens and Lambin 1997; Rudel with Horowitz 1993). In terms of attacking root causes of tropical forest loss, it thus makes sense to have a special interest in forest frontiers and rainforest margins.

What does frontier deforestation look like in spatial terms? The top row of Figure 1 shows three different forms of frontier deforestation. First, large clearing for commercial purposes can appear as a geometric shape, for instance in the case of the expansion of soybean production in lowland Bolivia and Brazil (Kaimowitz and Smith 2001). A second type is the corridor shape, which is often found from settlement and agricultural activities around new roads, such as the logging roads being built into the humid forest zone of Cameroon (Mertens and Lambin 2000).

Finally, a third frontier-clearing prototype is the fishbone pattern, known from directed settlement programmes (such as Indonesia's *Transmigrasi*), where land is allocated to settlers in strips along a road or a settlement nucleus.

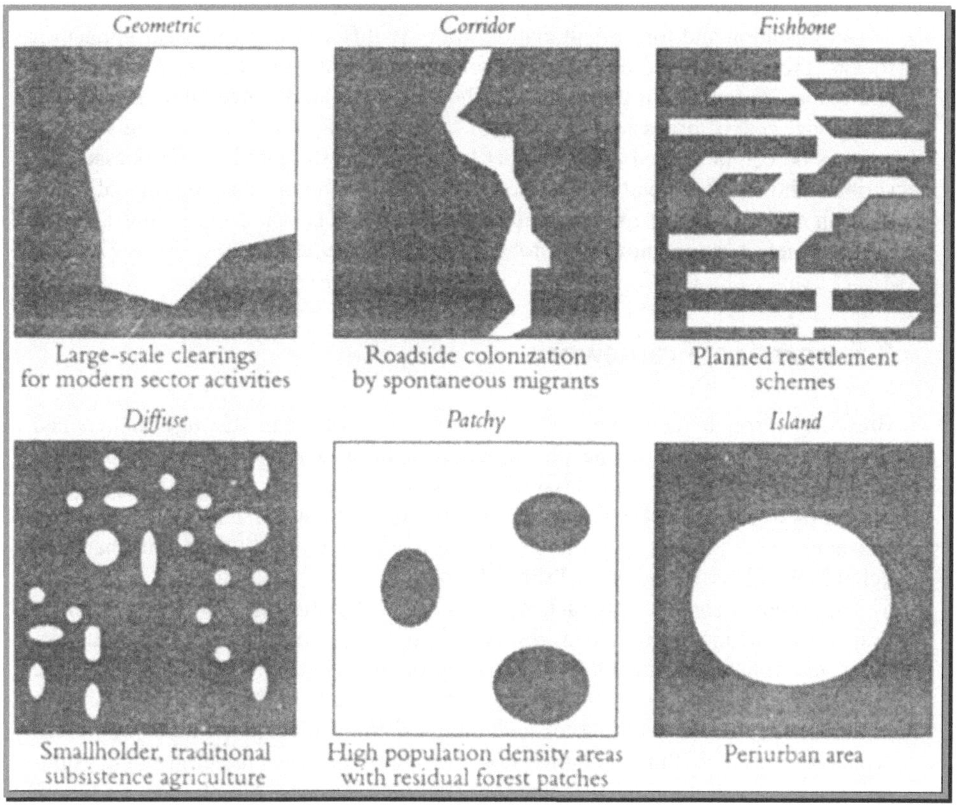

Fig. 1. Six spatial forest--non-forest patterns (Mertens and Lambin 1997)

2.3 Forest Degradation

Besides deforestation, there are also numerous forest degradation processes – a residual category of interventions that significantly affect forest quality and structure, but do not deprive an area its status as a forest. Notably, this includes selective logging, which reduces forest canopy-cover, but normally not below the 10% minimum threshold. On the other hand, clear-cut harvesting for pulp harvesting would usually been seen as deforestation, to the extent that they fully eliminate the canopy cover.[1] Other examples of forest degradation include repeated exposure to fire, over-extraction of firewood or over-grazing (both mostly in dry forests), or over-harvesting of bush meat -- also called *defaunation*, and especially important in Central Africa.

Deforestation and forest degradation not only differ in their physical impact on forests; they also tend to be dissimilar processes in economic terms. Deforestation is often an investment in future uses of the converted land, since there usually is a non-trivial cost of preparing the land for alternative uses before the benefits from conversion can be reaped. On the other hand, the forest-degradation processes described above often tend to be more associated with a "cashing in" of rents through the over-harvesting of various forest products – i.e. producing an economic benefit here and now, but probably less in the future.

3 Theoretical Framework

Browsing through the literature on deforestation, one can distinguish between three main approaches to the phenomenon of deforestation: 1. The impoverishment approach; 2. The neo-classical approach, and 3. The political ecology approach (see Wunder 2000: Chapter 2 for further discussion). These schools differ substantially as to what they identify as the main drivers, agents and mechanisms behind forest loss, as shown in Table 1 below.

The impoverishment approach points to a combination of poverty and demographics as the main mechanism responsible for forest loss, creating a vicious circle of environmental degradation driven by the growing number of smallholders. Obviously, population growth plays a prominent role here; low labour absorption at the frontier and a low pace of technological innovation mean that Malthusian scenarios dominate. Shifting cultivation for subsistence uses under growing population pressures is an applied scenario of this type, leading to both reduced fallow periods and dwindling forest resources.

Neo-classical analysts rather see the ill-defined forest property rights as the main evil: an open or quasi-open access to forestland at the frontier encourages smallholders and large investors alike to open up the forest and claim land rights afterwards. Agents are not so much forced by deterministic and vicious circles; they rather react to opportunities in a rational and optimising way, even when they happen to be poor. Labour supply tends to be flexible; if there are good rewards to forest colonisation and conversion, people will have more children and, notably, migrants will come in from outside to fill job opportunities.

[1] In FAO's use of the terms, that would only hold if the area is not *intended* to be reforested after the clear cut. In a critique of the FAO concept, it is argued that intentions and predictions about post-clearing land uses in the tropics are extremely uncertain, making the FAO definition highly speculative (Wunder 2003).

Table 1. Three Deforestation Schools

School / Questions	Impoverishment Approach	Neo-classical Approach	Political ecology Approach
What main, single factor is responsible for deforestation?	'The growing number of poor'	Open-access property rights	Capitalist investors crowd out peasants
Who is the principal deforestation agent?	Smallholders	Various agents	Capitalist entrepreneurs
What is driving the dynamics of deforestation?	A gradual push with deterministic, vicious circles	Optimising agents react to pull incentives	Capitalist pull, land expulsion and smallholder push
What impact have demographics and labour absorption	Labour absorption is low. Labour abundance boosts deforestation	Labour mobility is high and labour supply is very elastic	General labour scarcity at the frontier causes deforestation
What effect has a rise in the peasant's farm output prices?	It causes lower farm production and less deforestation	It causes higher farm production and more deforestation	It causes lower farm production and less deforestation

Source: Wunder (2000)

The political ecology approach points specifically to externally driven processes, e.g. large capitalist farmers or ranchers, as the main agents of deforestation. This can either be directly through their additional land demand, or by disrupting local land-use systems and by crowding out small farmers, who hence are pushed further into the forest. In other words, according to this school, the rich deforest for greed while the poor mostly do so for need. Normally, population growth is a subordinate factor in this picture.

Obviously, the three schools take competing approaches to the explanation of deforestation, but this does not necessarily mean that any one of them is universally more correct than the others. Forest-loss processes differ across the tropics, and one will find examples from different parts of the world that fit any of the three schools, as we will see below. On the other hand, as shown in the last row of Table 1, some of the predictions by the three schools are directly opposed, allowing us to test their relevance directly.

Consider that a small forest-margin farmer producing cocoa as his main cash crop is suddenly facing higher cocoa prices that substantially increase his revenues. What would be the impact on deforestation? In the political ecology and especially the immiseration approach, smallholders that are better off would need to produce less to make the same money - or they would be able to feed more mouths without having to push into new forest areas for cultivation. The assumption is that they only produce a certain "target revenue", which is sometimes also called a "full belly" economy – so to say, you only work until your stomach can be filled up with food. The opposite reaction occurs under the neo-classical standard economic assumptions of profit maximisation and unlimited wants. Farmers faced with a higher profitability in cocoa will allocate more labour, capital and land to cocoa to take maximum advantage of the price boom. This means normally that they will deforest more, rather than less.

4 Economic Models of Tropical Deforestation

The book “Economic models of tropical deforestation” (Kaimowitz and Angelsen 1998) was a state-of-the-art review of a range of different types of models explaining forest loss in the tropics. Most of the literature in this field is from the 1990s. The 133 models reviewed included *analytical models* (25), *household empirical models* (17), *spatial regression models* (9), *regional regression models* (20), *computable general equilibrium models* (14) and *global regression models* (38). The advantage of this synthesis is that it gives us a global snapshot of "what matters" in terms of economic incentives for land-use changes.

The countries under analysis ranged from the larger forest countries to some with limited forest coverage. Most frequently represented were Brazil (12), Costa Rica (6), Ecuador (5), Mexico (5), Indonesia (7), Philippines (4), Thailand (5), Cameroon (3) and Tanzania (4). In general the quality of data is lower the larger the coverage. Household models tend to have good-quality data as the scientist is in control of the data collected, just like spatial regression models where data often come from remote sensing imagery. On the other hand, global regression models used national deforestation estimates, generally based on FAO forest assessment or yearbook data that exhibit a number of insecurities and serious problems (Rudel and Roper 1997; Grainger 1996; Matthews 2001).

In analytical terms we should generally make a distinction between:

- Sources and agents of deforestation – who deforests, for what purpose?
- Immediate causes of deforestation - the agents' decision parameters.
- Underlying causes of deforestation - broader contextual changes.

4.1 Effects of Price Changes

As a main result, the synthesis shows that higher agricultural prices in most cases stimulate more forest clearing. Farmers react positively to the opportunity of more profitable cultivation, compared with other alternatives. Hence, they increase their income by cultivating more land themselves, or newcomers will be attracted. This picture thus favours the neo-classical approach, at the expense of the political-ecology and impoverishment school.

Second, according to the model results changes in relative prices between agricultural products can also alter the balance between land uses, which affects deforestation. In particular, if farmers produce both *land-extensive*[2] food crops and *land-intensive* cash crops, and choose mainly between these two livelihood options in their land use, than a rising relative price of food crops over cash crops will tend to cause higher deforestation.

Third, higher timber prices can also stimulate deforestation, although the evidence is weaker than for agricultural commodities. This happens because better prices tend to stimulate a more rapid harvesting rate, which indirectly opens up forested areas for conversion, mainly through road building (see below).

4.2 Factors Affecting Costs

Policies and other interventions that favour agriculture will in most cases cause higher deforestation. Higher agricultural productivity, lower input prices, lower land prices, and lower transport costs are among the most important factors identified (Kaimowitz and Angelsen 1998). In most places, deforestation is thus fairly well explained by expanding agriculture (Andersen et al. 2002 ; Barbier 2001).

Road building near or into forest areas is the single most important factor causing deforestation. It lowers transport costs for both timber and agricultural products, so that these commodities can "pay their way out" to the marketplace. By making viable a series of economic activities and enabling more intensive human

[2] "Land-extensive" here means with a *high* input of land per output unit, "land-intensive" the reverse.

settlement, roads are thus often the first but decisive step towards forest conversion.

Higher rural wages, or higher labour opportunity costs in terms of new employment options, will reduce deforestation. This is because forest clearing is a particularly labour-intensive activity. The effect will be particularly strong when forest-based options constitute an "employment of last resort", which people turn to in periods of economic crisis when alternative, better remunerated employment options become scarce.

Fertiliser subsidies that make purchased fertilisers cheaper can in some cases reduce deforestation, and their withdrawal can increase it. This is an exception to the general pattern of agricultural subsidies promoting forest loss. It applies especially in contexts where slash-and-burn is a prime cause of deforestation, and where the main importance of forests is as a source of nutrient inputs into agriculture. Cheaper alternative fertilisers will then tend to reduce conversion for that purpose.

4.3 Land Tenure

We remember from last section that the neo-classical approach pointed to insecure land tenure as a key factor behind deforestation. But Kaimowitz and Angelsen (1998), as well as other empirical studies (e.g. Wunder 2000), find land-tenure security to be an ambiguous factor vis-à-vis the determination of forest loss. This is a controversial issue, where probably more research is needed. People who have insecure tenure and access rights can only plan for limited time periods; the more long-term the benefits, the less secure is it that the land user with insecure rights can reap them, and the less (s)he will be inclined to invest in the land.

In general, (more) secure tenure will help the land user adopt long-term profitable solutions. In some circumstances, that will favour forest management, but in many cases it will not. It depends on whether forestry is actually the long-run most profitable option - or whether that is cattle ranching, oil-palm estates, soybean fields - or even to sell the land to other parties. Depending on the socio-economic context, secure tenure seems to have a more positive effect on tree planting and agro-forestry than on natural forest management. Trees take time to grow, so the decision to allocate land almost per definition requires control over the land until harvest. But there is nothing in and off itself that makes sure that more secure tenure leads to more forests in the landscape; especially examples from Latin America show the opposite because pastures for cattle ranching often are the most rewarding and convenient land-use option in the long run.

One factor to consider is thus what land uses are favoured by secure tenure, but another one is the process by which this secure tenure is established in the first place. Forest clearing is often seen as a sign of active occupation - "the land is being worked" - that discourages others from taking possession. On the other hand, forestlands are often seen as "idle" territory inviting invasion. This means that deforestation often helps establish property rights (“homesteading”) – whether by informal tenure recognition among a group of land-colonising settlers, or by the process of getting formal land tenure through a state agency controlling that land has actively been cleared. Homesteading thus promotes "excessive", speculative deforestation - beyond what can be explained by a pure economic rationale. People may in some cases clear forests simply to obtain control over the land, regardless of what is the most profitable land-use option.

5 The Role of Agricultural Technology

Forests and agriculture are generally the most extensive land uses in the tropics, and they tend to compete for land with forestry. Hence, higher agricultural land demand becomes the main driver of forest loss. Nonetheless, the balance between recipient sub-sectors of new agricultural land is quite different between tropical continents:

- Cattle-ranching heavily dominates land-use change in Latin America: most deforested lands end up as pastures in land-extensive ranching systems.
- In Central Africa and in South Asia, extensive swidden systems for food crops (plantains, tubers, etc.) require large land areas for crops and fallows.
- In Southeast Asia and West Africa, logging of highly priced timbers has played a larger role in opening up forest frontiers. In Southeast Asia, much land has subsequently been converted to cash and estate crops (oil palm, cocoa, coffee, etc.).

It has been argued that the Green Revolution with its drastic increase in the productivity of staple crop production has saved a lot of forests, and that further rises in yields will be necessary if the remaining wildlands in the tropics are not to be sacrificed (Borlaug 2002). The logic of the Borlaug hypothesis would seem similar to that of the impoverishment political ecology approach: if prime agricultural areas can produce higher yields, then production need not expand into marginal lands. But how precisely is the relationship between technological innovation and forest loss at different scales and under variable scenarios? A workshop held in Costa Rica in 1999 brought together a range of case studies around this topic, published later in a book (Angelsen and Kaimowitz 2001). This section will present some main results, and compare them to the model outputs from last section.

As a general observation, technological advances in agriculture, as well as the introduction of new profitable crops, will tend to make agriculture more profitable. So, in a given location, region or country, technological progress will usually cause higher deforestation, just as higher output prices do. This is what one would expect from neo-classical reasoning, with upward sloping producer supply curves (see above). However, several specific scenarios can change the picture, depending on factors such as farmers' production functions and output markets (see Angelsen and Kaimowitz 2001 for technical details). Let us look broadly at the factors separating the two cases:

Q.1. Under what circumstances do new technologies or products *reduce* forest clearing?

1. **When labour-intensive techniques/ products are being introduced in forest-scarce regions**. In these cases, some labour that might otherwise have deforested forest fragments will be absorbed by the use of new techniques that are applied to relatively large cultivated areas.
2. **Shifts in dual systems towards the sedentary, more land-intensive type**. Imagine a dual production system, with on the one hand productive, fertile, irrigated prime agricultural areas in the lowlands, and on the other marginal, rainfed uplands with low yields. If new technologies are only applicable to the prime areas with best conditions, this will reduce output prices and diminish forest pressures in marginal zones (Jayasurya 2001). This process has driven forest regrowth in many marginal zones of developed countries.[3]
3. **Introduction of high yield varieties (HYVs)** of cereals and other staples, which have an inelastic demand, will lower food prices. So, if the output market is limited and/or demand is highly price-elastic, then an increased production will trigger lower agricultural prices, which tends to reduce deforestation.

Obviously, the second and third cases represent the Borlaug hypothesis, underlining the role of scale. Let us return for a moment to our cocoa farmer from above, and assume that (s)he successfully introduces a new cocoa high-yield variety, which has been developed in that particular region. Consider three scenarios:

- **A:** The farmer sells his cocoa to the world market, and no other producer regions adopt similar yield-improving techniques. Hence, our farmer and his colleague innovators can sell unlimited additional cocoa at the same price. That provides farmers with good extra earnings, and they would be inclined to clear forest to plant more of the new variety.
- **B:** Assume now, alternatively, that sales go via middlemen with large accumulated stocks, so that the latter would only buy the additional cocoa at a reduced price. In that case, producer gains and incentives for new cocoa-led deforestation would be lower.
- **C:** Finally, suppose that all cocoa farmers in the world adopt the new yield-enhancing technique at the same time. That would flood the market with cocoa supplies and, depending on the demand elasticity of chocolate consumers, lead to a fall in world-market cocoa prices. This price fall would ultimately also reduce the incentives to expand cocoa production into the forest margins.

In other words, the Borlaug hypothesis remains valid at the aggregate world-market level - or when markets are restricted by policy or by transport costs. Yet, when innovations occur at a lower scale, with access to external markets that fix output prices, then it is likely that technological progress raises local land demand and increases local pressures on forests.

---

[3] See e.g. Mather and Needle (1998) and Rudel (2001)

Q.2. In which cases would new technologies or products *accelerate* forest clearing?

The first general answer is that, more often than not, new technologies will increase deforestation. The second is that this is in particular likely to happen when one finds:

1. Labour-saving or -displacing products or techniques (e.g. mechanisation of crop cultivation, ranching, soybean introduction), combined with a flexible supply of capital. Part of the redundant labour will here be "set free" to expand into the forest margins;
2. Eradication of plant and animal diseases is a powerful tool to make production across-the-board more profitable (just like a price increase does), and thus also stimulates land demand and forest conversion;
3. Export booms with products that demand large initial immigration of labour, which subsequently is "set free" under bust periods to expand into the forest;
4. Forest margins with a high population density, high population growth and/or flexible immigration of labour.

Note that in all the cases illustrated above, forestland is mainly to be considered as an available reservoir of land, which will accommodate fluctuations in the demand for new agricultural lands. Perhaps the strongest result is the fourth observation. It implies that if one has an agricultural frontier with a flexible labour supply - probably a condition valid for most tropical frontiers - almost no matter what type of technologies you introduce and safeguards you take, higher profitability will go hand in hand with higher deforestation. That is a somewhat uneasy message to send to the managers of Integrated Conservation and Development Projects (ICDPs) who aim to make both the environment and local people better off simultaneously by means of improved agricultural systems. Yet, this picture is consistent with the problematic practical results of most ICDPs (Gilmour 1994), and hence a revision of the overly optimistic Brundtland-report view on "win-win" options related to commodity production in tropical forests (Angelsen 1997; Fisher 2001; Wunder 2001a).

6 Comparing Macro-Economic Links

After a short general introduction, this section will highlight how factors and policies at the national level "trickle down" to the forest level. The section will draw mainly on CIFOR country-comparative work specifically on long-run land-use changes in eight tropical oil countries (Wunder 2003). The primary cases here were Cameroon, Venezuela, Gabon, Ecuador and Papua New Guinea, with secondary studies on Indonesia, Nigeria and Mexico. We will also draw on comparative CIFOR research about the forest implications of policy responses to macroeconomic crisis and to structural adjustment in Indonesia, Bolivia and Cameroon (Kaimowitz et al. 1998; Ndoye and Kaimowitz 2000; Sunderlin et al. 2001).

6.1 The General Picture: Economic Growth and Poverty

In the 1950, Simon Kuznets found that income inequality was rising in the early stages of an economic development process, while being reduced again in the later stages (Kuznets 1955). This pattern of an inverted U-curve over development phases has come to be known as the "Kuznets curve". More recently, scholars have also tried to look out for an Environmental Kuznets Curve (EKC), testing whether also the environment "has to get worse before it can get better" by means of economic development. While the pattern seems to fit some "brown" environmental problems (such as industrial emissions) fairly well, there has only been meagre support for an EKC on deforestation. In developing countries, economic growth is correlated with multiple sources of absolutely higher land demand, and even the *rate of* deforestation among developing countries does not seem to go down systematically in later stages of economic development (Culas and Dutta 2002). Consequently, it is unlikely that tropical countries can economically 'grow their way out' of high deforestation scenarios - except in particular cases where urban sectors and the service economy have a very dynamic role (see below). In that special case, they come to resemble more the case of developed economies at very high income levels; countries that grow crops in specialised high-yield systems and can afford to import the bulk of the most land-demanding commodities from other countries.

By the same token, poverty and its reduction over time have an ambiguous effect on deforestation (Reardon and Vosti 1995; Angelsen 1997). On one hand, poverty alleviation typically is associated with higher labour (opportunity) costs, which tends to reduce both forest clearing and degradation. On the other hand, when people get less poor they also start to consume more protein-rich foodstuff like meat and dairies, which has an impact on forests. They may also save more money, which alleviates their capital constraints vis-a-vis investments requiring forest clearing. As we will see below, the aggregate impact of poverty alleviation on forests depends on the relative weight of these different factors.

6.2 The Role of Trade and Foreign-Exchange Inflows

Both the forthcoming book on eight specialised oil and mineral exporters (Wunder 2003) and aggregate-level statistical comparisons of this group with tropical non-mineral exporters (Mainardi 1998; Sunderlin and Wunder 2000) confirm two basic facts. First, oil- and mineral-rich countries in the tropics on average retain a greater share of forest cover and, second, they tend to lose these remaining forests at a slower pace.

The core reason is that they have an abundant inflow of foreign exchange from mineral exports, which allows for higher government spending levels that attract people to the cities. At the same time, a more appreciated real exchange rate makes both agriculture and timber extraction less competitive than in non-mineral countries. This under-development of agriculture and forestry has a protective impact on forests, especially if the accompanying policies also come to have a favourable impact on forest conservation. Hence, the development path of these countries becomes more urban-based. Road building into forested areas and other rural development policies become widely neglected. The urban population still consumes resources that leave an "ecological footprint" on forests, but peri-urban cultivation systems tend to be more land-intensive than those practised by a rural-based population. In most cases, at the national level, urbanisation is unambiguously good for forest conservation.

Conversely, from the research on crisis and structural adjustment we know that the opposite scenarios of foreign-exchange scarcity and currency devaluation often lead to an increased emphasis on land- and forest-based resources and a "re-ruralisation" of the economy, which eventually also increase pressures on the forest margins. One factor is that relative prices make farming and logging more profitable, hence land users expand these activities to additional land. Another one is that urban employment declines, making low-remunerative rural-based activities the default option to secure livelihoods. Finally, a third general pattern is that crisis and sharp price fluctuations induce risk-reducing diversification strategies, e.g. in rural areas a larger portfolio of crops is grown by farmers so as to be prepared for unexpected income shortfalls. All these three effects increase pressures on forests.

6.3 What Policies Hurt Forests?

Not only the external condition created by trade and foreign exchange inflows matter; the domestic policy responses are also crucial in determining the net deforestation outcome. The forthcoming book on tropical oil and mineral exporters (Wunder 2003) identified the following ten major fields where national policies would *accelerate* deforestation.

1. Rural road building (or improvement) through/ near forests

Those countries that had strong rural road-building programmes (e.g. Ecuador and Indonesia) also had high deforestation, confirming the micro impact of roads from above.

2. Large gasoline subsidies

Not only roads reduce transport costs; cheap fuel has similar (though reversible and non-spatial) effects of enabling agriculture or timber harvesting from remote areas. Fuel subsidies thus accelerate forest clearing.

3. Large government spending at the frontier

Providing social infrastructure (schools, health services) in frontier areas helps to attract migrants and strengthens colonisation, and is thus conducive to deforestation.

4. Currency devaluation

In the macroeconomic sphere, devaluation is a powerful tool to change relative prices and production incentives. If agriculture and timber harvesting are a tropical country's main trade-exposed sectors, then making them more competitive through sharp and repeated devaluation will accelerate deforestation.

5. Generous forest concessions

A government that generously allocates land to concessionaires on favourable terms in an aggressive attempt to attract investors will tend to face more rapid extraction rates. Thus, forest areas are also being opened up more rapidly for conversion.

6. Import protection of land-extensive sectors

Generally, protectionism has ambiguous impacts on deforestation. Yet, we can safely say that import protection of certain land-extensive sectors like cattle ranching and dairy farming in parts of Latin America or slash-and-burn produced food crops in Central Africa hurts forests. These protected sectors then over-extend into marginal soils with very low returns. Protected domestic timber sectors can also be highly wasteful in their use of wood resources when lack of import competition induces them to become inefficient.

7. Subsidised credits for these land-extensive sectors

If the government provides specific subsidised credits for the mentioned sectors, this will further over-expand them, at the expense of forests.

8. Resettlement into forested areas

'Transmigration'-type programmes (like in Indonesia) where people are resettled from densely populated areas out into the forest, under the slogan of "bringing people with no land to a land with no people", will obviously accelerate deforestation.

9. "Homesteading" land-tenure rules

Land-tenure agencies often allocate property rights to settlers only if they can prove that they open up and convert 'unproductive' forestland. As explained above, this fosters speculative forest clearing beyond of what is mandated by production motives.

10. Abandon all family-planning programmes in favor of a pro-natalist strategy

There is no doubt that population growth tends to accelerate forest loss, since more people need more land to satisfy their needs. Population growth is a "slow driver", working indirectly, time-lagged and correlated more clearly at aggregated scales. It also does not open up the forest margins on its own,[4] but it is very important as a "fuel" to empower and to upscale the deforestation triggers.

6.4 What Policies Protect Forests?

Conversely, what policy package has *de facto* worked in the eight tropical oil countries as an effective protection of the forest margins? Most of the points listed here are a direct reversal of the above-mentioned factors that accelerate forest loss. The applied "macro-policy conservation recipe" looks like this:

[4] Rudel with Horowitz (1993) assert that, rather than causing frontier deforestation, rural population growth is more instrumental in eliminating forest fragments in pre-established agricultural zones.

1. Neglect the rural road network
2. Spend all the oil money in the cities
3. **Sell gasoline at its 'normal' price**
4. Keep over-valued exchange rates
5. **Tax logging companies heavily**
6. Heavily tax export agriculture
7. **Liberalise food imports**
8. Resettle people out of the forest to near roads
9. Waste budgets on agro-industrial 'white elephants' and ignore rural smallholders
10. Create a business environment where few people find it worthwhile to produce

First, compared to the previous "reverse" list, some factors merit additional explanation. Regarding (5), the implication is that if governments are able to capture the bulk of stumpage values, the rate of timber extraction will be slower, which will also reduce some of the conversion that is enabled by logging roads and other "opening-up" effects. For (7), more food imports will reduce the size of domestic land-extensive cultivation, and possibly increase overall efficiency in the use of resources. (8) indicates that some resettlement programmes, notably in Central Africa, have actually curbed deforestation, because they have moved people *out of* remote forested areas and into roadside settlements with typically more land-intensive agricultural production. (9) refers to the fact that an inefficient use of public funds in agricultural parastatals and misguided mega-projects has come to benefit forests because these activities never accomplished to clear the land they originally had planned to. Likewise, (10) notes that if a general rent-seeking mentality surges, as occurred in the oil countries, then the lack of entrepreneurial spirit will become a serious obstacle to any type of commodity production, which obviously also relieves pressures on forests.

Second, we should note that of these *de facto* effective conservation policies, only one - taxation of logging operations (5) - originates in the forestry sector itself, whereas another one - resettlement (8) - affects forested area directly. All the other measures are "extra-sectoral" - they were "blind" strategies of "conservation by chance", with measures originally designed to achieve completely different goals.

Third, we could ask the question how this set of policy recommendations likely would be received by development decision-makers - say, the Minister of Planning or a World-Bank team helping to design a structural adjustment programme and a poverty-reduction strategy. Probably, only three out of the ten components (the ones in **bold** in the Box) would earn positive marks vis-à-vis a list of "good development policies". These would be the elimination of gasoline subsidies (2), an effective taxation of logging rents (5) and a more liberal food-import regime (7). All the other measures, from urban policy biases to semi-corruptive practices and excessive interventions, would be perceived as having negative impact on economic development and poverty alleviation - some of them in a decisive way. This indicates that the hard trade-offs between tropical forest conservation and economic development do not only occur at the micro-intervention level (see above), but also when we look at the big policy decisions at the macro level.

7 What Does This Mean for Indonesia?

7.1 Screening the Main Deforestation Drivers

We will now turn to a brief examination of policies and deforestation causes in a single country, Indonesia. We will do that by comparing the pattern of policies and macroeconomic development over the last decades with that of land-use change and forest loss. Initially, as mentioned above the model synthesis by Kaimowitz and Angelsen (1998) also included seven models for Indonesia. By looking at the main deforestation factors identified in these models, we can obtain at least a preliminary idea about what factors have driven forest loss in that country. The main factors found to cause higher forest loss over time or space in these models fall into the three domains of agriculture, logging and infrastructure:

Agriculture
- high output prices and/or low input costs
- high productivity and/or good soil quality

Logging
- high timber prices and/or low timber extraction costs
- the type (and terms of operation) of logging concessions

Infrastructure
- low transport costs (e.g. rural road density, type, maintenance)

From this set of factors, we can already make some observations vis-à-vis the three schools of deforestation from Section 2, Table 1:

- Commercial incentives have been the dominant drivers. Although some factors (such as soil quality) can also be subsistence-driven, most relate to markets.
- "Neo-classical" (and possibly "political ecology") explanations are more relevant than "immiseration". The deforestation-accelerating effect of higher productivity/ output prices/ lower input costs clearly shows that farmers react to "pull" incentives.
- The impoverishment mechanism of a vicious circle whereby a poor and increasing population is pushed to convert new forests to grow food crops is less relevant for the last three decades in Indonesia. If one increases agricultural profitability at the forest margin, one should thus expect forest loss to go up in Indonesia, not down.
- Logging has had a larger deforestation role than in most other countries. It has helped to open up forest frontiers. The sector has probably also quite often pro-

vided the basic capital for alternative land uses, e.g. oil-palm companies that have depended on harvesting timber profits first to then finance investments in the estate crops.

7.2 New Order Policies and Development Strategies

In spite of the financial crisis and political turmoil accompanying President Suharto's downfall, for his three decades of "New Order" period as a whole, Indonesia's economic development record has been impressive. Policies have been characterised by:

1. Macroeconomic policies widely praised for their prudence, continuity and timeliness (Gelb and Glassburner 1988; Bevan 1999b).
2. An economic strategy of openness to 'mobile' capital both of foreign and Chinese-Indonesian origin. Relatively liberal rules for capital movements and other measures to attract investment (Winters 1996).
3. A competitive real exchange rate - including the active use of currency devaluations (Warr 2000), favouring the exports of urban labour-intensive industries (textiles, electronics, etc.), cash and estate crops, and forest-based industries - with the consecutive rise of timber in the 1970s, plywood in the 1980s, and pulp & paper in the 1990s (Barr 2001: Chapter 2).
4. As a result of 1., 2. and 3., a remarkably high per-capita growth in national income and private consumption, turning Indonesia from an extremely poor country in the late 1960s to a middle-income economy in the 1990s.
5. Significant policy attention to rural development, agriculture and food security - including massive increases in rice productivity (Scherr 1989).
6. As a result of 4. and 5., major long-run progress in rural and urban poverty alleviation and also in non-income welfare (higher life expectancy and primary education enrolment, reduced child mortality, etc.)(Hill 1992; Sunderlin 1993; World Bank 1999b).
7. Aggressive land-use policies opening up forest margins through generous timber concessions, transmigration programmes and rural road construction.

7.3 Forest Lost and Converted

Turning now to the forest sphere, Indonesian deforestation figures are notoriously uncertain (Sunderlin and Resosudarmo 1996). A handful of remote-sensing based studies exist, but variable forest definitions and coverage make the estimates very difficult to compare. Nonetheless, in the following we give some rough numbers on both measured forest loss and its converted uses over the last two decades. These are based on estimates from Forest Watch Indonesia/ Global Forest Watch (FWI/GFW 2002), a consultancy report comparing a variety of sources (Muhamad 2002) and a synthesis of additional statistics on cropped areas (Wunder 2003: Chapter 9):

Deforestation between 1980-2000:	**about**	**30**	**million ha**
Conversion 1980-2000:			
Estate-crop and perennials expansion		12+	million ha
Forestry expansion (pulp harvest and plantations)		3÷	million ha
Food-crop expansion (incl. swidden)		7-8	million ha
Total converted uses 1980-2000:	**about**	**22-23**	**million ha**

As indicated by the ranges and +/÷ signs, these figures are subject to great uncertainties. Deforestation figures depend on the assumption regarding the differences in definitions and coverage (see above). The expansion of estate and cash crops is underestimated, as some of the minor crops are not accounted for. Forestry expansion is probably over-estimated, as some of the areas harvested by clear cuts have been put into cash and estate crops, and thus are double-counted. In addition, both natural forests and forestry plantations are "forests" in FAO terms, so converting one to the other should not count as deforestation.

Still, it seems worthwhile to get the guesstimates down on paper, allowing for an explicit discussion of the proportions. Even the rough figures show that perennial and estate crops make up more than half of converted land use. They also seem to indicate an inconsistency between deforestation and alternative uses. As by far most of Indonesia's "default" vegetation cover is tropical forest, we would expect deforestation and converted land use to approximately match. But the total conversion of about 22-23 million ha falls about 25% short off the alleged deforestation figure. The numbers don't add up in Indonesia!

One possible explanation is that forest clearing 'runs ahead' of conversion because of the economic attraction to harvest timber and pulp resources, combined with a lack of capital to put the cleared land under alternative uses. Ecological processes could also play a role, like *alang-alang* (imperata) grassland invasions and repeated fires that come to convert forests into "wastelands". But part of the discrepancy could also be explained by outright errors in land classification leading to overestimated deforestation figures.

Returning to the policy aspects, the tentative figures confirm the suspicion from above that commercial "neo-classical" motives have clearly been dominating in the Indonesian deforestation case: cash- and estate-crop expansion has been much more important in quantitative terms than the increase in food-crop areas.[5]

What role have macroeconomic policies played for the outcome of accelerated forest loss? Did forest loss contribute to economic development - or was it even a necessary condition? This is not the place to analyse these complex questions in detail. One can cast doubt on whether the policies of generous large-scale timber concessions contributed much to the positive macroeconomic outcome. The wood-

[5] Even many so-called "food crops" are actually major cash generators for their producers, reflecting the growing importance of national markets. This further reinforces the vital role of commercial processes.

based industries generated much foreign exchange and accumulated capital for re-investments in other sectors, but the direct employment effects were limited and often local people in rural areas were expelled from their land, in the way it is described in the "political ecology" literature. On the other hand, much of the expanding cash and food crops were owned by smallholders, helping to appreciably consolidate the rural economy and alleviate poverty - much more than in countries with marked urban policy biases (see above). The development of labour-intensive agriculture was an effective motor for poverty alleviation, but it had a deforestation cost. Many of the policy measures were good for the macro-economy - and even for the majority of poor Indonesian people, but most elements were also bad news for forest conservation. This underscores the notion from last section of important policy trade-offs at the macroeconomic level.

8 Conclusion and Policy Recommendations

Agriculture is the great land-use competitor of tropical forests, and by far most deforestation occurs in order to increase farmlands. Hence, most policies and interventions that favour the expansion of agricultural production also come to decrease forest area – at least that is true for the type of forest-abundant agricultural frontier areas and forest margins that we have been concerned with in this paper. In some cases, agricultural expansion is driven by an increasing poor population growing food crops with land-extensive swidden cultivation (impoverishment approach). In others, it is the emergence of new market opportunities that drives the process (neoclassical approach). In a third set of cases, it is the clashes between these two processes that provide the main impetus (political ecology approach). Independent of what the dynamics are, there are normally strong underlying factors, most of them outside the forest sector, that enable and empower the conversion of forests to alternative uses.

Notably, even agricultural 'intensification' that increases per-hectare yields can accelerate forest loss. Intensification is still often seen as an area- and forest-saving factor, but that effect is highly context- and scale-dependent. The assumptions certainly hold for widespread intensifying innovations that reduce the total market price through their general supply-boosting effect. Yet, where adoption is limited, innovators increase production but prices remain high, so they will in most cases scale-up their now more profitable production. Hence, they will tend to deforest more, rather than less. Almost any agricultural investment in frontier areas with flexible labour supply promotes deforestation. So, it is hard to design agricultural programmes in these regions without a negative effect on remaining forest. In providing policy recommendations as to what could be done *strictly with the aim to stabilise the forest margins*, we were able distinguish between those factors that directly affect land extensification through a spatial effect, and those that work through the macro-level context.

8.1 Reduce Land Extensification

- Stop building or improving roads in remote rural areas near tropical forests!
- Don't give subsidised credits and inputs to "land-hungry" production sectors!
- Don't give out overly generous forest concessions to the timber companies!
- Stop rewarding deforesting squatters with secure land rights!
- Stop moving and/or directing people into forests!
- Stop financing development projects in the forest margins!
- Use instead resources and incentives in favour of other areas (e.g. pre-established "prime" agricultural zones, urban areas, peri-urban agricultural systems)!

The last two recommendations will appear controversial, and need qualification. They do certainly not imply that it is *impossible* to design forest-margin projects that through institutional fine-tuning and micro-adjustment of interventions will be successful in stabilising land demand at the forest margin. But it is very difficult to do so, and among the many projects that have tried to achieve it, the majority have failed. One does not have to be an Economics Nobel Laureate to predict that when you spend money in the forest margins – be it on health, education, R&D, value added activities and especially on agriculture – in the medium run these investments will have spin-offs that tend to attract more people and foster economic development. But more people and more development both mean in most cases more local land demand. That land is usually made available by converting forests. This is a serious risk that even the most well-intentioned forest-margin project will face.

8.2 Create a Conservation-Conducive Macro-Level Context

- Promote high urban labour absorption to keep people from migrating to the frontier!
- Avoid excessive economic fluctuations through careful adjustment policies!
- Avoid that currency devaluation makes agriculture and logging overly attractive!
- Don't provide subsidies to make fuel cheaper!
- Liberalise food and timber imports!
- Reduce population growth as an important long-run driver!

As for the spatially explicit recommendations above, many of them are deeply problematic from a development perspective. Unfortunately, many 'good' development policies (for economic growth and poverty reduction) are bad for forest conservation. Conversely, some 'bad' development policies come to protect forests. These *de facto* conservation successes are the result of 'blind' strategies and unintentional side-effects from macro policies. In particular, non-forestry (extra-sectoral) policies prove to be much more important for forests than forest policies proper. This part of the picture is not very encouraging.

Still, among the above-mentioned policy tools and interventions, there are some 'win-win' options that are promising for both forest conservation and economic development. Generally, the removal of subsidies with 'perverse' forest impacts (fuel, cheap agricultural inputs) has such a potential. Forestry-sector reform in developing countries can potentially help to capture (and distribute more fairly) timber stumpage values while also slowing down the "opening up" of forest frontiers. Speculative land-tenure arrangements caused by "homesteading" rules could be eliminated, and provide some social benefits at the same time. Import liberalisation in the timber and food sectors could reduce forest loss while increasing economic efficiency and (arguably) fostering national development in the long run.

Yet, it becomes clear from the above that the interface between forest conservation and (local or national) development in the tropics exhibits more trade-offs than synergies. A logical consequence from that diagnosis is that one should experiment much more with direct compensations for environmental services, i.e. rewarding local land users for forest conservation yielding benefits to outsiders (related to watershed, tourism, carbon-storage and biodiversity). Only if they are compensated in a *quid pro quo* for their opportunity costs of conserving the forest will they take these external benefits into account in their land-use decisions. Although experiences in the tropics with these schemes are incipient, they are certainly expanding (Landell-Mills and Porras 2002; Pagiola et al. 2002)– and they are badly needed as applied conservation tools in a world where tropical forests continue to recede.

References

Andersen LE et al. (2000) *The dynamics of deforestation in the Brazilian Amazon*. Cambridge University Press, Cambridge (UK)

Angelsen A (1997) The poverty-environment thesis: was Brundtland wrong? *Forum for Development Studies* No. 1

Angelsen A, Kaimowitz D, eds (2001) *Agricultural technologies and tropical deforestation*. CABI Publishing, Wallingford (UK)

Barbier EB (2001) The economics of tropical deforestation and land use. *Land Economics* 77(2): 155-72

Barr C (2001) Banking on sustainability: Structural adjustment and forestry reform in post-Suharto Indonesia: CIFOR, Bogor and WWF, Washington DC

Bevan D et al. (1999b) The political economy of poverty, equity, and growth: Nigeria and Indonesia: A World Bank comparative study. World Bank, Washington DC

Borlaug N, (2002) We can feed the world: Here's how. *The Wall Street Journal, Monday, 13 May*

Bryant D et al. (1997) The last frontier forests: ecosystems and economies on the edge: what is the status of the world's remaining large, natural forest ecosystems? Forest Frontiers Initiative. World Resources Institute, Washington DC

Culas R, Dutta D (2002) The underlying causes of deforestation and the environmental Kuznets curve: a cross-country analysis. University of Sydney, Sydney

FAO (2000a) FRA 2000. On definitions of forest and forest change. *FRA Working Paper* 33. FAO, Rome

Fisher RJ (2001) Poverty alleviation and forests: experiences from Asia, Occasional Paper. RECOFTC Kasetsart University, Bangkok. Available at www.RECOFTC.org/publications_at_recoftc.htm

FWI/GFW (2002) The state of the forest: Indonesia. Forest Watch Indonesia, Bogor and Global Forest Watch, Washington DC

Gelb A with Glassburner, B (1988) Indonesia: Windfalls in a poor rural economy. In: Gelb A, Associates (eds) Oil Windfalls: Blessings or Curse? Oxford University Press, New York, pp 197-226

Gilmour DA (1994) Conservation and development - seeking the linkages: Paper read at International Symposium on Management of Rainforests in Asia, at University of Oslo, 23-26 March

Grainger A (1996) An evaluation of the FAO tropical forest resource assessment, 1990. The Geographical Journal 162,1

Hill H (1992) Regional development in a boom and bust petroleum economy: Indonesia since 1970. Economic Development and Cultural Change 40(2): 351-79

Jayasurya S (2001) Agriculture and deforestation in tropical Asia: an analytical framework. In: Angelsen A, Kaimowitz D (eds) Agricultural technologies and tropical deforestation. CABI, Wallingford

Kaimowitz D, Angelsen A (1998) Economic models of tropical deforestation: A review. Center for International Forestry Research, Bogor

Kaimowitz D et al. (1998) Considering the impact of structural adjustment policies on forests in Bolivia, Cameroon, and Indonesia. *Unasylva* 49(149): 57-64

Kaimowitz D, and Smith J (2001) Soybean technology and loss of natural vegetation in Brazil and Bolivia. In: Angelsen A, Kaimowitz D (eds) Agricultural technologies and tropical deforestation. CABI International with CIFOR, Wallingford (UK), pp 195-211

Kuznets S (1955) Economic growth and income inequality. *American Economic Review* XLV (1):1-28

Landell-Mills N, Porras IT (2002) Silver bullet or fools' gold? A global review of markets for forest environmental services and their impacts on the poor, Instruments for sustainable private sector forestry series. International Institute for Environment and Development (IIED), London

Mainardi S (1998) An econometric analysis of factors affecting tropical and subtropical deforestation. *Agrekon* 37(1): 23-45

Mather A, Needle CL (1998) Trends in global forest cover: issues in explanation and modelling. In: Himiyama Y, Crissman L Information bases for land use/cover change research. Proceedings of IGU-LUCC'97, Brisbane

Matthews E (2001) Understanding the FRA 2000: Forest Briefing No. 1. World Resources Institute, Washington DC

Mertens B, Lambin E (1997) Spatial modelling of deforestation in southern Cameroon: spatial disaggregation of diverse deforestation processes. Applied Geography 17(2): 143-162

Mertens B, Lambin EF (1997) Spatial modeling of deforestation in Southern Cameroon: spatial disaggregation of diverse deforestation processes. Applied Geography 17: 143-68

Mertens B, Lambin EF (2000) Land-cover-change trajectories in southern Cameroon. Annals of the Association of American Geographers 90(3): 467-494

Muhamad NZ (2002) Rates of deforestation in Indonesia. An annotated bibliography: Unpublished consultancy report. CIFOR, Bogor (Indonesia)

Ndoye O, Kaimowitz D (2000) Macro-economics, markets, and the humid forests of Cameroon, 1967-1997. Journal of Modern African Studies 38(2): 225-53

Pagiola, S. et al. eds (2002) Selling forest environmental services: Market-based mechanisms for conservation and development. Earthscan, London

Reardon T, Vosti SA (1995) Links between rural poverty and the environment in developing countries: asset categories and investment poverty. World Development 23(9)

Rudel T, Roper J (1997) The paths to rain forest destruction: cross-national patterns of tropical deforestation, 1975-1990. World Development 25(1): 53-65.

Rudel T, with Horowitz B (1993) Tropical deforestation: small farmers and land clearing in the Ecuadorian Amazon. Columbia University Press, New York

Rudel TK (2001) Did a Green Revolution restore the forests of the American South? In: Angelsen A, Kaimowitz D (eds) Agricultural technologies and tropical deforestation. CABI, Wallingford

Scherr SJ (1989) Agriculture in an export boom economy: a comparative analysis of policy and performance in Indonesia, Mexico and Nigeria. World Development 17(4): 543-560

Sunderlin WD (1993) Environment, equity and the state: a perspective through the Java Social Forestry Program. Ph.D. thesis, Graduate School Cornell University

Sunderlin WD, Resosudarmo IAP (1996) Rates and causes of deforestation in Indonesia: towards a resolution of the ambiguities: CIFOR Occasional Paper No. 9. Bogor, CIFOR

Sunderlin WD et al. (2001) The effect of economic crises on small farmers and forest cover: a comparison of Cameroon and Indonesia. In: Palo M, Uusivuori J (eds) World Forests, Society, and Environment. Kluwer, Dordrecht, pp 115-125

Sunderlin WD, Wunder S (2000) The influence of mineral exports on the variability of tropical deforestation. Environmental and Development Economics 5: 309-332

Reardon T, and Vosti SA (1995) Links between rural poverty and the environment in developing countries: asset categories and investment poverty. World Development 23(9)

Warr PG (2000) Indonesia: Trade shocks and construction booms. In: Collier P, Gunning JW (eds) Trade shocks in developing countries, Volume 2: Asia and Latin America. Oxford University Press, pp 326-354

Winters JA (1996) Power in motion: capital mobility and the Indonesian state. Cornell University Press

World Bank (1999b) World Development Report 1998/99. World Bank, Washington DC

Wunder S (2000) The economics of deforestation: the example of Ecuador: Macmillan Press St. Martin's Press, Houndmills New York

Wunder S (2001a) Poverty alleviation and tropical forests - what scopes for synergies? World Development 29(11): 1817-33

Wunder S (2003) Oil wealth and the fate of the forest. a comparative study of eight tropical developing countries (forthcoming). Routledge, London

Forest Margin Protection and Community Involvement

Sediono M.P. Tjondronegoro

1 Introduction

If Nancy Peluso's "Rich Forests, Poor People" may have been a valid observation until 1992, now ten years later (2002) the title is definitely no longer appropriate, not even for larger Outer Islands such as Sumatera, Kalimantan or Sulawesi and Papua. The situation is becoming extremely cumbersome for millions of shifting cultivators living in or around forest areas. A more recent mapping of the forest cover of Indonesia by the Ministry of Forestry (MOF) has revealed that the rate of deforestation in Indonesia approximately doubled between 1985 and 1997, from less than 1.0 million ha to at least 1.7 million ha each year (Holmes 2002). The mapping predates the widespread forest fires of 1997-98 and the extensive illegal logging that followed the political crisis of 1998. The analysis of the MOF data concludes that over 20 million ha of forest cover have been lost over a twelve year period including 6.7 million ha in Sumatera and 8.5 million ha in Kalimantan. This amounts to an average annual rate of 1.67 million ha, nation wide-roughly 600 ha per day or 190 ha per hour. The rate in the three islands of Sumatera, Kalimantan and Sulawesi is 1.45 million ha per year. The rate of forest loss between 1985 – 1997, which has been quoted from the Holmes' report shows the disastrous decline of Indonesia's tropical rain forest (Annex.1 & 2). In central Sulawesi where also the Lore Lindu is located, the forest area constituted 72.3% of total area in 1985. However, the forest area declined to 56.7% in 1997, which means 15.6% in 12 years or more than one percent per annum, equivalent to an average of 35.000 ha. The forest status areas are the latest preliminary figures available from the MOF in October 1999, following integration with provincial spatial plans (paduserasi = harmonious integration)

On balance, however, it should also be reported that the latest revision of forest status compared with the former agreed forest use categories (Tata Guna Hutan Kesepakatan, TGHK) has resulted in an increase of conservation area by 1.82 million ha, an increase in the area of Protection Forestry by 3.22 million ha, an increase in production forest by 2.32 million ha, but a decrease in limited Production Forest by 8.13 million Ha. The increase in conservation areas reflects the designation of new national parks. However, it remains a question to what extent damage has been inflicted upon the older existing conservation areas.

Forest coverage in the conservation areas differs between the provinces and no information has been reported for Central Sulawesi, but for the conservation areas of North Sulawesi are covered by 90% with forests. Since the Reformation in the country in 1998 there seems to be less control and supervision over all categories of forest areas, and there is a deep concern among NGOs that deforestation will accelerate.

Greater Regional Autonomy, following the promulgation of law no. 22 & 25/1999 on the same subject, is suspected to be one reason to increase the region's revenue (PAD). One serious problem is also the often conflicting tasks performance of the National Land Agency (B.P.N) on the one hand and the provincial and district government on the other. The former has been assigned to issue land certificates, and is to some extent valued by the number of certificates that it has issued, while the latter has been assigned with spatial planning. The two assignments have often contradicted one another, and it is not clear which institution should mediate to solve such problems in good spirit.

2 Conversion of Forest Areas

The question of who bears the greatest responsibility for conversion of forest areas is indeed a difficult question to answer, since there are the irresponsible logging companies, which after tree felling often do not care about reforesting the area that has been logged, there are the illegal logging organizations as well as individuals; and there are the shifting cultivators who clear relatively small forest plots for food crop production. The shifting cultivators cause relatively less destruction, because the size of land converted for the purpose of food crop production will often range between 1 to 2 ha. At a later stage rubber trees re-cover the converted forest land.

The three-stage recovery is a standard practice in all shifting cultivation areas. Forest concessions to domestic and foreign companies have without doubt increasingly limited this extensive agricultural practice. Foreign forest concessionaires have issued stringent limitations to prevent shifting cultivators from practicing shifting cultivation which often causes conflicts related to tenurial rights. Shifting cultivators often claim forest concessions agreed upon by the government, as ***Tanah Adat***. (land under customary law).

Though forest mapping has been completed in most of the provinces in the Outer Islands there seems to be limited distribution of the maps. In the absence of a more detailed land-use map, it is also difficult for regional administrators to compose a spatial plan. Though district officers may limit their competence to confine spatial planning to land surface, and not include riches underneath the land surface, if mining is more profitable to the central government, disputes easily arise.

The author of this paper observed that in South Kalimantan several projects had to be terminated because coal was found. In such cases the central government has the last say on the basis of Chapter 33 of the 1945 Constitution saying that all natural resources which can be exploited for general welfare should be given first priority. The legal status of converted forest areas is frequently not quite clear, but the MoF tends to maintain its claim upon them.

3 Illegal Logging

The general trend in this risky, but very lucrative business, is frightening when considered from a deforestation point of view. Illegal logging, necessary to maintain the operations of the country's timber mills was producing 20 million m^3 annually. The total raw material shortfall supplied by illegal logging was 37 million m^3 in 1997, rising to as high as 56.6 million m^3 in one year only (1998). Though the figures may not be quite accurate there is an abundance of anecdotal evidence to support the stated figures. Apparently illegal logging and forest area conversion for other uses of land occurred initially on the low lands, however, as population increased and concessionary forest projects as well as estate land broadened, or restrictions are imposed to allow mining concessions, the shifting cultivators are unwillingly driven up hill and into the mountainous regions, denuding the uplands and subsequently giving way to soil erosion and floods in the lower plains. Therefore, it is easy to understand how also national parks are in increasing jeopardy of losing their initial functions i.e. to preserve the richness of biodiversity as well as wild life.

In principle the Biodiversity Protection Area (Kawasan Perlindungan Plasma Nutfah, KPPN) should be managed by the Forest Concessionaires (HPH). However, in several provinces it is known that concessionaires have displayed serious neglect. The trend as described above can be illustrated by e.g. Leuser National Park in Aceh, Bukit Dua Belas National Park in Jambi. The Director General of Forest Protection & Nature Preservation (Perlindungan Hutan dan Konservasi Alam, PH & KA) has earlier admitted his deep concern about forest destruction both by concessionaires who commit so-called "Corporate crimes" by disregarding the rules in the agreement.

The decision made by the Director General of Forest Management no. 208/1993, stating criteria for good management have been neglected. Among the criteria are: (1) production continuity; (2) preservation of the forest resources, (3) environmental conservation, and (4) preservation of cultural and social institution (Romimohtarto cs. ed. 1997). As a consequence there are a number of NGOs which try to spread the message, and empower local *adat* (customary) communities, with regard to their existence and survival. One of them is WALHI (Wahana Linkungan Hidup / Vehicle for Living Environment), but there are others, among which is ELSAM (Safitri 2000).

Village studies have been completed in Sangeh, Bali; Sumber Agung Lampung, and many other locations in East Kalimantan (Thamrin & Gunawan 1997). After mainly looking at the lucrative aspects of managing forests, it has now become extremely urgent also to rediscover the social relevance. There is an urgent need not only to look at nature and the world as mainly a source of human prosperity and welfare, but to submit ourselves to a cosmological balance, which we now often describe as earth's ecology.

4 Government Initiatives

The Government of Indonesia is very well aware of the serious trend of destruction. There is definitely a legal basis of environmental management as the promulgated laws and acts show. Twenty years ago a law no.4/1982 on The Basic Provision for Environmental Management was promulgated. In chronological order several laws and conventions have been promulgated as follows:

- Law no.5/1990 on Conservation of Natural Resources.
- Government Regulation no.20/1990 regarding Water Pollution Control.
- Law on Spatial Planning no.24/1992
- Government Regulation no. 51/1993 EIA
- Convention on Biodiversity, law no. 5/1994
- Convention on Climate Change, law no. 6/1994
- A Government Regulation was issued on Management of Hazardous Waste no.19/1994.

More recently the Basic Forestry law no.5/1967 has been reviewed for a second time and in its latest version promulgated as law no.41/1999. Meanwhile, however, the law on Mining (no.11/1967) has also been readjusted and rephrased (no.8/2001), while more Departments and Directorate Generals are in the process of reviewing their respective laws chq. D.G. of Estates, Ministry of the Environment, Department of Marine and Fisheries. All, however, has not been achieved in a fully integrative manner, since the Basic Agrarian law no.5/1960, which is still valid, is not clearly referred to. For many government officers in the region a very confusing situation has been created, since they do not have a sound legal basis to implement instructions in coordination. This has been the experience of many district heads, such that one way out is to act on the ground of law on Regional Autonomy no.22 & 25/1999, by allowing joint committees at district level to design spatial planning. A bill on the Management of Natural Resources is under discussion.

5 NGOs Role

In a sphere of ineffective law enforcement by government agencies, NGOs, are deeply concerned about environmental management and their role has become increasingly important. Most likely also due to solidarity feelings with foreign NGOs Indonesia's Lembaga Swadaya Masyarakat or LSMs have taken over functions which government agencies have failed to perform. Participatory approaches at field level with the local communities are the key to more effective implementation of programs in the broad context of sustainable development and environmental management.

There are NGOs active at National level to lobby governmental institutions and policy makers including members of parliament (DPR) and People's Consultative

Assembly (MPR). But also at provincial level, NGOs have emerged as correcting factors to the regional notables. One of the more established NGOs at this level is WALHI. More recently WARSI (Wahana Informasi Konservasi = Conservation Information (vehicle) Group) has strongly supported a System of Peoples Forest. Some 39 smaller NGOs in the region have joined and completed case studies in villages on the forest margins (Darusman 2000). International foundations such as the Ford Foundation have warmly supported many NGOs. Attention was not only limited to environmental management, but included the closely related problem of poverty to which people living in forest margin areas are subjected.

6 Social Organization

Broadly speaking one might say that people living in communal entities in or around forest margins are shifting cultivators and forest product collectors. Daily family life sustenance is usually provided by way of cultivating small plots for food crop growing around their little huts. These huts are rather distant from the *kampungs* where family life is more stable and permanent, particularly in the post-harvest period. Economic institutions such as markets, financial or credit providing organizations are operating in such kampungs, which therefore often provide the link with and services characteristic of the modern economic system. Shifting cultivators and forest product gatherers, people live in relative isolation from their kampung neighbors, at least for certain seasons of the year. Individuality, however, does not influence their consciousness about land ownership, which remains communal, or the right of usufruct if it concerns forest products.

Yet there seems to be a common understanding and acceptance that one's usufruct rights are delineated in and confined to certain limited areas. These areas are vast enough to meet the needs of the family. Understandably as population densities increase, and government projects such as HPH increase, such "usufruct forest areas" decline in size. Since private ownership of land has been denied in customary law, boundaries are also vaguely marked by natural borders such as rivers, ravines, or indicated by trees such as palm (enau) or durian. The first man who clears primary forest has the right to pass on the right to his eldest son. Genuine independence in land clearing that shifting cultivators have been accustomed to make it extra difficult to bring them under regular administrative control. There is one way, however, to make the kampung more attractive and useful to both indigenous people and new comers. Kampungs can be made more attractive by, for example, providing services through government established or sponsored institutions, such as markets, schools, health posts/clinics, village banks, post office and so forth. The problem has often been where to locate those service providing institutions. In the case of transmigration it is customary to build them in the center of the villages; social envy among the community has to be avoided, i.e. indigenous people are not to be treated as second citizens in their own territory.

Ideas have been developed to establish a tripartite sector village (indigenous, new settlers and future industry), in the center of which the services are built.

This idea emerged in the Department of Transmigration and Cooperatives in the mid nineteen seventies, but it has not been consistently implemented.

Here again we are confronted with a problem of integrated spatial planning. Hopefully similar plans may become more effectively realized after the promulgation of the Regional Autonomy law no.22 & 25 / 1999 at district level. So far, however, many district officers found chapters on delegation of authorities by the central government confusing, particularly those regarding agrarian issues.

7 Forest Fires

According to periodic monitoring large scale forest fires have occurred in Indonesia every 3 to 4 years since 1982, but the most destructive one occurred in 1997 due to vast land clearing operations by large estates expanding them for oil palm cultivation. The spread of forest fires worsened in 1997 by the very dry season as a result of the El-Nino Southern Oscillation (ENSO) phenomenon. Only part of the forest fires was caused by shifting cultivators (10%), but the largest part was caused by land clearing by HPH and HTI concessionaires. Not only did neighboring countries such as Singapore and Malaysia suffer from the smoke due to the forest fires, but Indonesia lost a large area of forest, not counting the impact on living conditions of the local population. Large areas of forest have been cleared for the cultivation of oil palm (see Table 1).

Table 1. Statistics from the Directorate General of Plantations show the following figures

No.	Year of land clearing	Area cleared Ha	Crude oil production targets X 1.000.000
1.	1994	1.792.413	4.1
2.	1997	2.048.891	7.0
3.	2000	2.103.620	10.1

8 Community Forestry

A fact is that indigenous communities living in or close to forest areas in developing countries as observed on all continents belong to the lower economic stratum of society. The arrival of forest concessionaires to start their logging and timber operations may employ some of the needed wage labourers, but most shifting cultivators who live in those communities are not accustomed to get involved in such technical production processes. Workers will therefore be recruited from elsewhere resulting in a more heterogeneous community around the forest area under production. Moreover, the shifting cultivators are losing their territory for shifting agriculture and forest product gathering. Community forestry as a concept is to prevent indigenous shifting cultivators from becoming a menace to the logging and timber companies, by offering them alternative jobs.

Simultaneous, food-and-tree crop cultivation (tumpang sari) is offered as one way out, but the time-span within which it can be practiced becomes a problem when the tree canopy reaches such a height such that it over-shadows the food crops. The problem in efforts to develop social forestry has often been that of "property rights" both on land as well as forest, and low educational level of the local community as a results of which employment opportunities are also limited. Relative poverty will consequently emerge in the newly established heterogeneous community. Social envy and conflicts often find their roots in the above process of marginalization of indigenous (adat) communities.

A fundamental change in the joint management of natural resources will have to be conceptualized, including: (1) a redefinition of state/public-and adat land; (2) clear and enforceable rules regarding rights and duties of the companies and local communities; (3) democratic procedures relating to the allocation of natural resources.

Meanwhile the Basic Forestry Law No. 5/1967 has been revised and promulgated as Forestry Law No. 41/1999, which became a major reference for Government Regulation No. 34/2002 on Drafting a Plan for Forest Management. However, so far law enforcement has been too weak to sustain effective environmental management.

9 Conclusion

For the past few decades the MoF and PERHUTANI have developed social and community forestry units in rather scattered locations in the country. So far, there has been relatively few information on the inclusion or integration of such projects into larger administrative units such as Kecamatans (sub-districts).

Therefore, in closing, the question is to what extent the Lore Lindu case will be able to make any contribution in the above direction, such that forest preservation within the National Park and rural development around it will be sustained in harmonious balance.

References and Other Literature that was Consulted

Boedi H (2002) Menuju Penyempurnaan Hukum Tanah Nasional. Jakarta, Penerbit Universitas TRISAKTI

Calkins R (1994) Indonesia Environment and Development; Challenges for the Future. Wahington D.C. World Bank Report No.12083 - IND

Darusman D (ed. 2000) Ketika Rakyat Mengelola Hutan: Pengalaman dari Jambi (When People get to Manage the Forest; Experiences from Jambi Province), KPSHK, Bogor

Darusman D (cs; eds. 2001) Resiliensi Kehutanan Masyarakat di Yogyakarta, Debut Press

Holmes, DA (2002) Where have all the Forest Gone ? Environment and Social Development East Asia * Pacific Region, Discussion Paper. The World Bank, Washington D.C.

Ladejinski W (1964) "Agrarian Reform as Unfinished Business" dalam Wakinsky LJ (ed), Oxford Union Press

Peluso NL (1992) Rich forest, poor people; resources control and resistance in Java. University of California Press, Berkeley.

Romimohtarto K cs (eds. 1997) Peran Pelestarian Hidupan Liar dan Ekosistemnya dalam Pembangunan Nasional yang berkelanjutan. (Translation: Role of Wild Life Preservation and its Ecosystem in the Context of Sustainable National Development). YSI (Indonesia Wildlife Fund, IWF), Jakarta

Rostow WW (1960) The Stages of Economic Growth: A non-communist manifest. Cambridge University Press, London. Hal. 40-43

Safitri, MA (2000) Desa, Institusi Lokal dan Pengelolaan Hutan; Refleksi Kebijakan dan Praktik. (Desa, Local Institution and Forest Management; Reflection on Policy and Practice) Elsam, Jakarta

Sastrapradja, D. (1976) Compilation of Papers submitted on Occasion of a Workshop Nature Protection and Preservation in the Man and the Biosphere (MAB) Program

Sunito S cs (1999) Wuasa; Case Study of a Village in Lore-Lindu Region. IPB, Bogor

Soedargo (1962) Perundang-undangan Agraria Indonesia. N.V. Eresco. Jilid I, Bandung.

Stark W (1948) The Ideal Foundations of Economic Thought; Three Essays on the Philosophy of Economic. Routlege & Kegan Poul Ltd, Norwich & London

Sumardjono MSW (2002) "Undang-undang Agraria; Menyelesaikan Pekerjaan Rumah" dalam E. Suhendar. Ibid. Hal. 85-94

Suhendar E dkk. (eds, 2002) Menuju Keadilan Agraria; 70 Tahun Gunawan Wiradi. AKATIGA, Bandung. Hal. 27-39

Thamrin J, Gunawan R (1997) Industrialisasi. Kehutanan dan Dampaknya Terhadap Masyarakat Adat. Yayasan AKATIGA, Bandung

Tjondronegoro SMP, G. Wiradi G (2002) "Menelusuri Pengertian Istilah Agraria "dalam

White B (2002) Agrarian Debates and Agrarian Research in Java, Past dan Present; dalam

Sitorus F dkk. (eds, 2002) Menuju Keadilan Agraria. Bandung, AKATI

Annex 1: Forest Status, Forest Cover and Forest use

		Forest status[2]						Forest cover		Forest use[3]		
Province	Total land Area	Conserv. Forest	Protection Forest	Limited Production Forest	Production Forest	Permanent Forest Status	Conversion Forest	1997 Forest Cover[1]	Total cover as percentage of permanent Forest	Logging concessions	Timber eatates allocated	Total allocated forest industry
Aceh	5,674,800	852,421	1,884,500	37,392	601,392	3,335,613	0	3,611,953	108%	1,087,500	376,564	1,464,064
N.Sumatra	7,250,100	253,885	1,924,535	760,958	871,183	3,810,561	37,797	1,891,819	50%	710,600	486,640	1,197,240
W.Sumatra	4,169,000	846,175	910,533	246,383	407,849	2,410,940	189,346	1,944,015	81%	152,830	0	152,830
Riau	9,859,700	560,237	1,323,801	0	2,649,608	4,533,646	334,521	5,071,891	112%	2,719,603	684,312	3,403,915
Jambi	4,873,900	676,120	191,130	340,700	971,490	2,179,440	0	1,603,079	74%	651,350	189,941	841,291
S.Sumatra	10,226,300	822,300	879,390	298,600	2,269,690	4,269,690	774,100	1,248,209	29%	1,231,850	590,069	1,821,919
Bengkulu	2,090,400	444,882	252,042	182,210	41,830	920,964	70,360	899,858	98%	198,900	5,000	203,900
Lampung	3,386,700	442,500	331,531	44,120	192,902	991,053	153,459	361,319	36%	0	282,835	282,835
SUMATRA	47,530,900	4,878,520	7,657,462	1,910,271	8,005,654	22,451,907	1,559,853	16,632,143	74%	6,752,633	2,615,361	9,367,994
W.Kalimantan	14,753,000	1,435,480	2,335,045	2,421,950	2,235,700	8,448,175	582,320	6,713,026	79%	3,139,810	876,749	4,016,559
C.Kalimantan	15,360,400	680,580	1,014,130	4,593,003	4,448,222	10,735,935	0	9,900,000	92%	4,085,000	391,843	4,476,843
S.Kalimantan	3,749,000	176,615	554,139	155,268	687,834	1,573,856	265,638	999,182	63%	174,000?	549,474	723,473
E.Kalimantan	19,721,000	2,166,212	2,935,478	4,755,494	4,727,488	14,584,672	0	13,900,000	95%	4,602,000	1,290,113	5,892113
Kalimantan	53,583,400	4,458,887	6,858,792	11,925,715	12,099,244	35,342,638	847,958	31,512,208	89%	11,826,810	3,108,179	14,934,989
N.Sulawesi	2,655,500	429,065	341,447	552,573	168,108	1,491,193	34,812	1,300,000	87%	408,650	0	408,650
C.Sulawesi	6,032,900	676,248	1,489,923	1,476,316	483,034	4,125,521	269,911	3,400,000	82%	1,440,925	31,392	1,472,317
S.Sulawesi	6,245,100	843,966	1,928,597	828,255	186,666	3,787,484	102,073	2,300,000	61%	352,000	135,706	487,706
SE.Sulawesi	3,681,000	274,069	1,061,270	419,244	633,431	2,388,014	212,123	2,000,000	84%	491,500	61,594	553,094
SULAWESI	18,614,500	2,223,348	4,821,237	3,276,388	1,471,239	11,792,212	618,419	9,000,000	76%	2,693,075	228,692	2,921,767
3 Island Total	119,728,800	11,560,755	19,337,491	17,112,374	21,576,137	69,586,757	3,025,960	57,144,351	82%	21,272,518	5,925,232	27,224,750

Java & Bali	13,820,400	468,233	728,651	394,316	1,633,383	3,224,583	0	?	?	0	0	0
Nusatenggara	8,074,000	567,714	1,571,418	651,257	676,326	3,446,715	352,667	?	/	60,500	170,307	230,807
Maluku	7,801,900	443,345	1,809,634	1,653,625	1,053,171	4,959,775	2,034,932	5,538,506	112%	2,547,425	188,689	2,736,114
Irian jaya	41,480,000	7,539,300	11,452,990	3,365,475	10,379,684	32,737,449	2,671,275	33,160,231	101%	11,582,673	?	11,582,673
Indonesia	190,905,100	20,579,347	34,900,184	23,177,047	35,318,701	111,975,279	8,084,834	95,843,088	84%	35,739,486	6,311,228	42,050,774

[1] Source: Table 1 in Holmes 2002

[2] Source: advance draft based on *pemaduserasian TGHK dan RTRWP, October 1999, Sekretaris Badan Planologi Hutbun.*

[3] Source: MoFEC website for concessions, *Statistik Perusahaan Hutan Tanaman Industri* 1996 for HTI (BPS).

[4] The Riau data list Protection Forest as only 361,967 ha, but then a fotnote states that the total does not include 961,834 ha of peat and mangrove forests.

The source table notes that the on forest status areas are preliminary for Aceh, North Sumatra, Riau, and West Central and East Kalimantan.

See footnotes to Table 1 Concerning forest area data.

Annex 2: Forest Cover within Forest Status Boundaries

	Conservation Forest			Protection Forest			Total Protected Area			Production Forest			Permanent forest estate		
	TGHK	Forest	%	TGHK	Forest	%	TGHK	Forest	%	TGHK	Forest	%	TGHK	Forest	%
Aceh	832,453	811,498	978	972,290	904,813	93	1,804,743	1,716,311	95	1,812,050	1,425,684	79	3,616,793	3,141,995	87
N.Sumatra	253,557	213,219	4	1,543,337	689,618	45	1,796,894	902,837	50	2,251,312	894,438	40	4,048,206	1,797,275	44
W.Sumatra	539,915	376,098	70	1,242,256	901,199	73	1,782,171	1,277,297	72	1,098,571	699,623	64	2,880,742	1,976,920	69
Riau	378,437	335,531	89	426,017	308,875	73	804,454	644,406	80	4,125,048	2,907,930	70	4,929,502	3,552,336	72
Jambi	645,720	509,910	79	204,642	170,677	83	850,362	680,587	80	1,492,791	802,905	54	2,343,153	1,483,492	63
S.Sumatra	605,836	278,644	46	775,680	376,867	49	1,381,516	655,511	47	2,615,506	648,930	25	3,997,022	1,304,441	33
Bengkulu	307,721	267,907	87	459,612	382,616	83	767,333	650,523	85	249,235	172,633	69	1,016,568	823,156	81
Lampung	388,743	264,035	68	341,243	117,139	34	729,986	381,174	52	309,448	18,383	6	1,039,434	399,557	38
SUMATRA	3,952,382	3,056,842	77	5,965,077	3,851,804	65	9,917,459	6,908,646	70	13,953,961	7,570,526	54	23,871,420	14,479,172	61
W.Kalimantan	1,279,467	1,129,627	88	2,296,136	1,860,959	81	3,575,603	2,990,586	84	4,814,921	2,447,850	51	8,390,524	5,438,166	65
C.Kalimantan	632,708	584,769	92	840,176	804,564	96	1,472,884	1,389,333	94	9,397,431	7,181,448	76	10,870,315	8,570,821	79
S.Kalimantan	127,962	65,174	51	441,148	263,794	60	569,110	328,968	58	1,375,640	793,315	58	1,944,750	1,122,283	58
E.Kalimantan	1,783,624	1,525,213	86	2,866,921	2,741,102	96	4,650,545	4,266,315	92	9,887,197	7,643,158	77	14,537,742	11,909,473	82
KALIMANTAN	3,823,761	3,304,783	86	6,444,381	5,670,419	88	10,268,142	8,975,202	87	25,475,189	18,065,541	71	35,743,331	27,040,743	76
N.Sulawesi	416,727	383,530	92	327,609	287,167	88	744,336	670,697	90	832,523	832,523	78	1,576,859	1,319,608	84
C.Sulawesi	633,160	549,035	87	1,322,124	1,143,232	86	1,995,284	1,692,267	87	2,206,506	2,206,506	75	4,161,790	3,350,611	81
S.Sulawesi	125,019	97,914	78	2,029,571	1,391,288	69	2,154,590	1,489,202	69	1,209,838	1,209,838	59	3,364,428	2,206,731	66
SE.Sulawesi	291,016	223,497	77	529,478	457,826	86	820,494	681,323	83	1,551,231	1,551,231	77	2,371,725	1,875.432	79
SULAWESI	1,465,922	1,253,976	86	4,208,782	3,279,513	78	5,674,704	4,553,489	80	5,800,098	5,800,098	73	11,474,802	8,752,382	76
TOTAL	9,242,065	7,615,601	82	16,618,240	12,801,736	77	25,860,305	20,417,337	79	45,229,248	45,229,248	66	71,089,553	50,272,297	71

Notes : areas having cloud, or without data, are assumed to have forest cover.
Source: Badan Planologi, 1999

Historical Impacts on Use and Management of Natural Resources in the Rainforest Margins of Central Sulawesi

Werner Kreisel, Robert Weber & Heiko Faust

1 Introduction

The use and management of natural resources in the rainforest margins of Central Sulawesi has become a focus of national and international research very recently. That is because the area was always located in a political and economic periphery and the pressure on the rainforest resources had been relatively low. But nowadays the awareness of resource management of tropical rainforests has risen. Deforestation has got worldwide attention mainly because of its climatic relevance. Also a growing population pressure in Indonesia accompanied by an actual political and economic transformation process has intensified the impacts on the resources in Central Sulawesi. Especially during the last two decades the migration into the region has increased and strengthened the pressure on the access to land. But the foundations for the cultural landscape happened much earlier. These historically developed patterns of population and settlement structure lead to different attitudes of today's resource management and effects on land use in the region. The reflection of the processes, which show the way to a local differentiation of resource management around the Lore Lindu National Park, is stressed in this paper and will be exemplified with micro studies at the village level. The paper focuses on the external and internal impacts on the shape of the cultural landscape during different historical periods of the last century and reflects the demographic development as well as the land use changes in the region.

Following Mitchell (2000), over time, the cultural landscape becomes incredibly complex with each "introduction of a different culture" in an area, a rejuvenation of the cultural landscape sets in, or a new landscape is superimposed on remnants of an older one. The accreted layers of the cultural landscape could therefore be peeled back by an attentive geographer to determine exactly "cultural history in its regional articulation" (ibid.). So, the first step is to trace back the historical development of settlements and the ethnic structure of the local people in the region. The aim is to reconstruct phases of continuity and phases of change and to identify the external and internal cultural influences for different time phases in cross-sections. Therefore, political impacts caused by different phases of Indonesian history (i.e. colonial rule, independence) are reflected. Against this background the actual patterns of population density, settlement and population structure of the region are examined. As one of the most important parts of the demographic analysis, migration flows and its motives for different historical phases are compiled and differentiated for the ethnic groups. In general, the local population, which is highly divergent, can be divided into local and migrant groups. For these groups traditional values and attitudes as well as mobility patterns are worked out in case studies.

The paper is structured from macro level down to micro level. After the description of theoretical orientation of this paper and methodology of the study, a demographic and historical overview of the whole study area will provide the macro level settings important to understand the further analysis. Thereafter, a characteristic of two sub-regions and two villages will provide the necessary empirical data for the subsequent discussion of deviations between the different levels from the common regional patterns of cultural landscape.

2 The Theoretical Approach on Cultural Landscape

The basis for the concept of cultural geography used here to explain the shape of the landscape around the Lore Lindu National Park relies on the classical approach on "The Morphology of Landscape" by Carl Oscar Sauer (1925, 1963). In this approach, "the natural landscape is being subject to transformation at the hands of man, the last and for us [geographers] the most important morphological factor. By his culture he makes use of the natural forms, in many cases alters them, in some destroys them" (Sauer 1963: 341). Sauer underlines the importance of the "[...] agency of man on earth". He writes "the cultural landscape is fashioned from a natural landscape by a culture group. Culture is the agent, the natural area is the medium, and the cultural landscape the result. The natural landscape is very important for it supplies the materials out of which the cultural landscape is formed. The shaping force, however, lies in the culture itself" (Sauer 1963: 343). Sauer forcefully reverses environmental determinism of the cultural landscape, because it is "[...] derived from the mind of man, not imposed by nature, and hence are cultural expressions". The cultural landscape is therefore an *effect*, and culture (working in, with, and against nature) is a cause of what we see in any scene upon which we look.

Cultural development and transformation, including conquest of indigenous peoples by imperial power like in Central Sulawesi, constantly created and recreated the places and landscapes in which people lived. The problem of the approach is, that by placing culture as an agent, it tends to abstract cultural constructions and it tends to separate them from their social and material environment. Those constructions are usually regarded as entities with independent existence and causative power. Its explanation is based on the transcendental foundation, and by tending to essentialise it, the "culturalism" approach frequently fails to take into account a wider social context in which varying different constructions are produced and stated.

In Germany the ideas of Schlüter (1928) on cultural landscape are similar to the concept of Sauer, but he has an even more physiological focus of the morphology of the landscape and the basis are the visible materialistic objectives. The classical theoretical formalistic and genetic approaches of landscape analysis were extended by Bobek (1948) and Bobek and Schmithüsen (1949) with the concept in principle of man's determination of landscape by functions. Haggett´s "Locational Analysis in Human Geography" (1972) and his assumptions on conventional geography in "A Modern Synthesis" (1991) try to put physical settings and anthropological behaviour together, thus systematically putting together the various disciplines of Geography.

The actual discussion on a critical cultural landscape is formulated by Jackson (1989) in his "Maps of Meaning". In contrast to classical and conservative approaches, which see a physical setting as a context or a "container" for society, in this theory it is seen as a structure created by society itself. This is because human ideas are expressed in behaviour, which then creates cultural landscape. In return, these cultural landscapes reflect social relations and institutions, and they shape subsequent social relations. A characteristic of this approach is to emphasise the so-called socio-spatial dialectical model. Social and cultural approaches on landscape are enlarged by Werlen´s (1993) theories on regionalisation of "ordinary culture" and his theories of action. He argues that the relation of culture and space in the contemporary conditions of life needs to be defined in a new way because today a variety of cultural worlds and lifestyles meet in small areas (Werlen and Wälty 1995). Werlen (1995) adds that social and cultural universes have no fixed spatial existence. Therefore action-centred categories and not only space-centred categories are needed for the geographical social and cultural research. In addition Crang (1998) gives an approach on locating culture, regional personalities, cultural areas and landscape as well as Peet (1998) adds his modern geographical thoughts and critics on the relations between society and the natural environment. Mitchell (2000) focuses more on the cultural aspects determining spaces and demonstrates an urgent need for a revised conception of culture in geography.

Based on this theoretical context the cultural landscape emerges from different layers of cultural impact over time on the region, until today on the acting people who use the natural resources for their needs. In this understanding, the traditions and rituals of the local communities decide on the use of the land and process the social and spatial change. To explain these changes it is necessary to interpret the landscape as shaped by the energies and practices of peoples to accord with their cultures over time. On the other hand, the landscape should be seen as a symbolic system. That is, how it is shaped according to the politics, the beliefs of the inhabitants and the meanings invested in the landscape. The extent of expressing cultural traditions in terms of value for the community life and the family status can be seen as an indicator of regional cultural obligations. Both interpretations are to be seen and explained in a historical perspective either in local beliefs, or migrants transported it to the region. In some villages traditions still have high importance for the people themselves, in other villages the phenomenon does not occur. Therefore the authors try to combine the classical approaches on cultural landscape with the modern ideas to reach a deeper interpretation of the research area.

3 Methodology

The regional historical and cultural overall analysis is based on the evaluation of contemporary and recent literature and statistics, historical plans and cartographic material in archives of Indonesia and the Netherlands. It focuses on historical background – development and structure of population and settlements. For example, the *memories van overgave*, reports prepared by the regional colonial officials, offer an abundance of data and information for the reconstruction of cultural landscape development in Central Sulawesi. However, they represent a one-dimensional view, namely that of the colonial ruling power. Moreover, these reports are very clearly structured. But the appearance of erased passages in the reports reflects the censorship that limits the use of the sources. Nevertheless, the *memories van overgave* are, besides published sources, the only pool of relatively micro-level data that delivers information about the impacts of Dutch colonial rule on population and settlement growth as well as on natural resource management. They have been analysed for the Dutch colonial administrative unit *onderafdeeling* Palu[1] under the special research approach of STORMA[2] sub-project A1.

The official statistics give an overview of the population development and the distribution pattern, but the results give little information for the years before 1980, in particularly at micro level. That is why selected micro studies are an integrated part of the regional overall analysis. Quantitative empirical data on demography have been taken from the statistical sampling frame of STORMA sub-project A4 in which twelve out of 115 villages of the research area were randomly selected and questioned in 326 households (Zeller et al. 2002). These data give a large amount of information, especially for the micro level, but they do not deeply explain processes, influences, and backgrounds of resource management over time. The standardised information of these sample studies for instance cannot compare local groups and migrant groups with respect to their cultural attitudes to land use systems and to potential conflicts. Therefore, the processes of tradition and cultural change in the framework of the above mentioned aspects are exemplified in detail for four villages located in the surrounding valleys of Lore Lindu National Park by in-depth qualitative research. Mainly semi-structured interviews were conducted in selected households, with formal and informal leaders of the villages and further key persons concerning the research questions. The synthesis

---

1 Under Dutch colonial rule the Indonesian archipelago was divided into the following administrative units: *Gewesten* (e.g. Groote Oost), *Residencies* (e.g. Manado), *Afdeelingen* (e.g. Donggala), *Onderafdeelingen* (e.g. Palu) and *Landschappen* (e.g. Sigi-Biromaru).

2 STORMA (Stability of Rainforest Margins in Indonesia) is a DFG-financed collaborative research programme of the Universities of Goettingen, Kassel (both Germany) and Palu (Central Sulawesi, Indonesia), and IPB Bogor (Indonesia). Since July 2000, researchers of these institutions study interdisciplinary the processes of stabilisation and destabilisation in the vicinity of Lore Lindu National Park, Central Sulawesi. Sub-project A1 examines these processes in terms of cultural geography.

of the regional analysis and the empirical case studies gives an overview of the development of the cultural landscape.

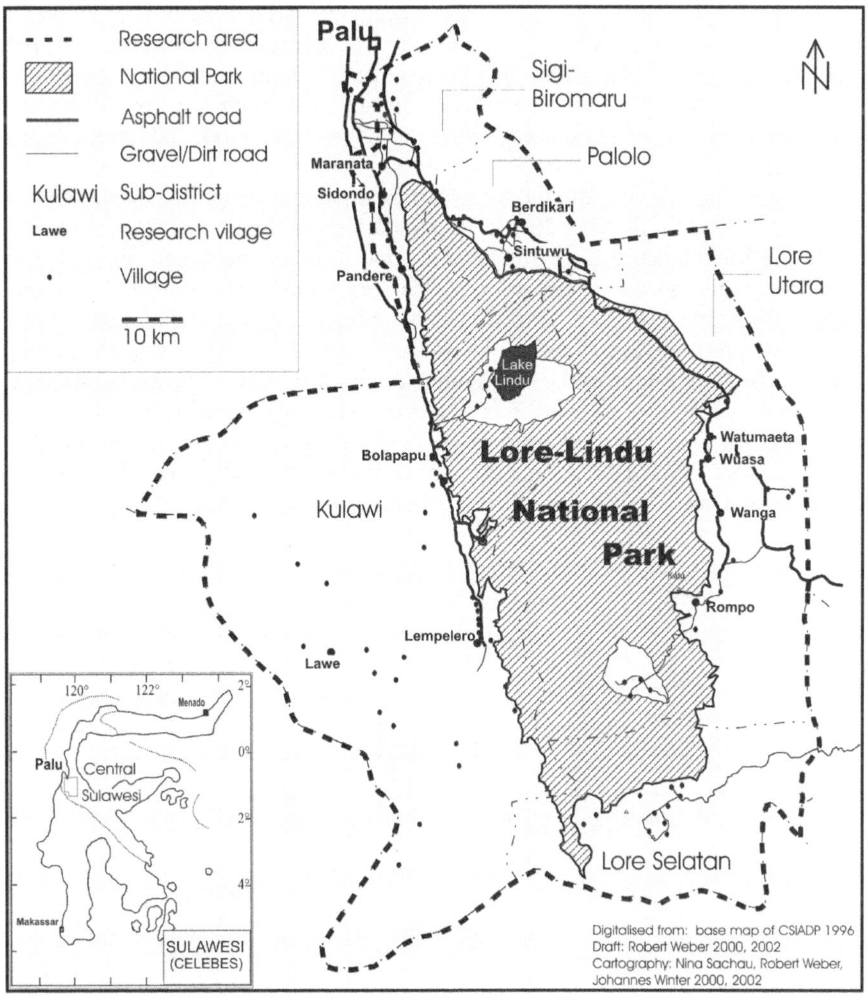

Fig. 1. Location of study area and research villages

4 Regional Demographic Overview

The research area (see fig. 1) is administratively divided into five sub districts (*Kecamatan*)[3] and 119 villages (*Desa*): Kecamatan Kulawi, with the sample villages of Bolapapu, Lawe and Lempelero, Kecamatan Sigi-Biromaru, with the sample villages of Maranatha and Pandere and Sidondo II, Kecamatan Palolo, with the sample villages of Sintuwu and Berdikari, Kecamatan Lore Utara, with the sample villages of Watumaeta, Wuasa, Wanga and Rompo, and the Kecamatan Lore Selatan.

Using village survey data the estimated total population in the five districts lies at 132 thousand for the year 2001. The population grew from 62.000 inhabitants in 1970 to 83.000 inhabitants in 1980 and to about 120.000 inhabitants in 1996. Over the past two decades the population in this area grew by 60 %, resp. 2.4 % annually. In this time span, the population size doubled in Palolo and almost tripled in Lore Utara (see fig. 2.). In those two districts population growth was already high in the 1980s and increased considerably during the 1990's. In Sigi-Biromaru a slowdown in population growth is observed. The population density is 18.7 inhabitants per km². The northern districts, Sigi-Biromaru and Palolo, are populated more densely: 86 and 43 people per km² respectively. In the other three districts population density is very low: around 10 people per km² (Faust et al.).

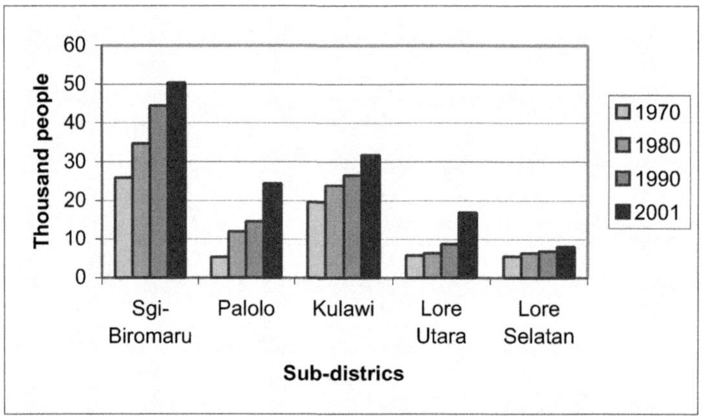

Fig. 2. Variation of population increase in the different sub-districts (Source Sensus Propinsi Sulawesi Tengah 1970, 1980, 1996, 2001. BPS Sulawesi Tengah)

[3] The policy of regional autonomy however, which can be regarded as one outcome of the nation's political development of the post-Soeharto era (since 1998), led to a restructuring of administrative units. Thus, since 1991 the Lore region consists of three subdistricts: Lore Utara, Lore Tengah and Lore Selatan. Since January 2002 the sub-region of Kulawi, Pipikoro, became a sub-district of its own as well. In this report the former administrative units are used.

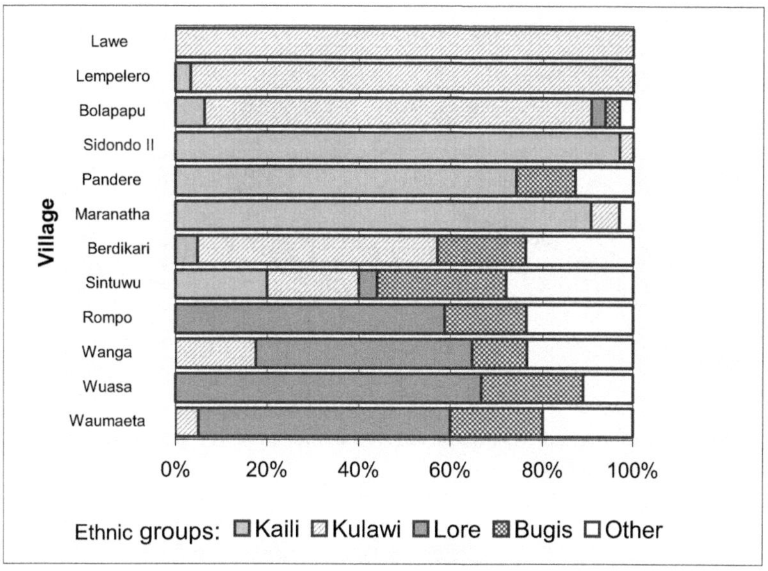

Fig. 3. Percentage of main ethnic groups in the sample villages (Data source: Household survey of sub-project A4, STORMA, Schwarze et al. 2000-2001; see also Zeller et al. 2002)

The investigated region is not homogenous in terms of ethnic composition. The main groups are Kulawi (32 %), Kaili (31 %), Lore (16 %), Bugis (10 %), Toraja (2 %), Minahasa (2 %), Javanese (1 %) and 6 % Other. Whilst the first three ethnic groups originate from within the research area, Bugis and Toraja have their origins in South Sulawesi and Minahasa in North Sulawesi. Fig. 3 shows the regional differentiation for the sample villages.

The different sub districts have different dominant ethnic groups. One can further see that there is a more heterogeneous composition in Palolo and Lore Utara than in Sigi-Biromaru and Kulawi. These population patterns have developed especially due to migration processes. During the past 20 years, in total 3.323 households migrated to this region from other provinces, 76 % of them during the 1990's. 62 % of these households are migrants from other provinces in Sulawesi; 19 % are spontaneous migrants from other islands in Indonesia; and 18 % are transmigrants from Java and Bali (see fig. 4.).

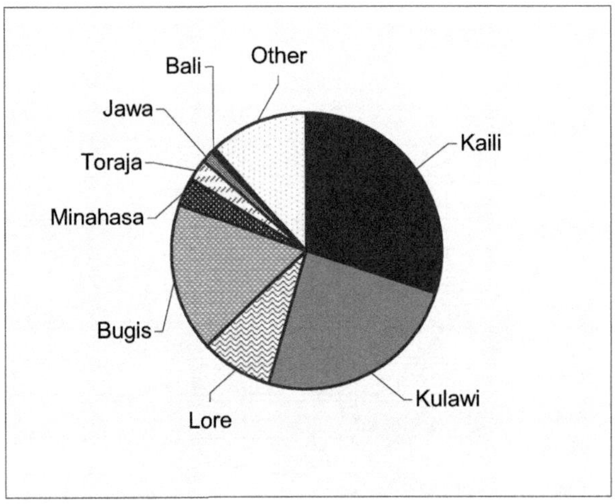

Fig. 4. Ethnicity of migrants (Data source: Household survey of sub-project A4, STORMA, Schwarze et al. 2000-2001)

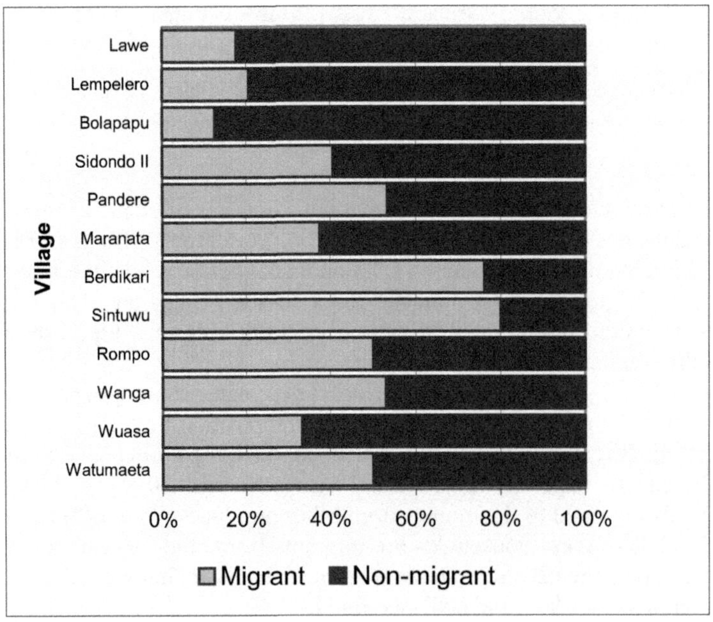

Fig. 5. Percentage of migrants in the sample villages (Data source: Household survey of sub-project A4, STORMA, Schwarze et al. 2000-2001)

The districts Sigi-Biromaru and Lore Utara each received about 1,000 migrant households, from within Sulawesi as well as from other islands. Six hundred households were resettled in Lore Utara as part of the national transmigration program. The districts Palolo and Kulawi received 807 respectively 450 migrant households, almost all of them from within Sulawesi. There was hardly any in-migration in Lore Selatan (Faust et al.). Thus, more than half of the population in the study area are migrants and except of Rompo, all sample villages on the so called "Napu site", in the east of the Lore Lindu National Park, are actually dominated by migrants (see fig. 5.). On the other hand, except for Pandere, on the Sigi-Biromaru and Kulawi site, in the west of the park, local population dominates all sample villages. Three quarters of all migrants came to the study area since the 1970s. The road construction from Palu to Poso in the beginning of the 80s brought an increased influx to the region. In Watumaeta, Wanga and Wuasa more than 50 % of the migrants came within the last ten years. The main motives of immigration are the following: 27 % because of availability of land, 21 % marriage, 18 % job opportunities, 11 % transmigration project. Out of the remaining 23 % the main reasons are following the spouse or natural disaster, neither of which cannot be regarded as "free decision".

The phenomenon of out-migration has occurred not very often. Among all out-migrants 90 % who left their villages settled again inside Central Sulawesi, one third at Palu. Thus, they are all to be seen as local migrants. As motives for out-migration one third of the migrants stated "self-employment in agriculture" and one fourth "education". The core of the migrants (ca. 75 %) originate from Central Sulawesi, only 18,1 % are originated from South Sulawesi, dominated by the ethnic group of Bugis, who migrated primarily because of land scarcity in South Sulawesi and because of the cocoa price boom 1995–97.

As explained in the conceptual framework, the differentiation of settlement and population structure is based on the development of the cultural landscape over time. The influences and changes differ from sub-district to sub-district and even from village to village, which shows the complexity of the regional analysis of this region. To get a more realistic picture of the development and the current status quo it is not enough to regard the region around Lore Lindu National Park as a homogenous unit. As it will be shown below, smaller units of different development of cultural landscape down to village level can be defined. To understand the actual spatial patterns of the region it is necessary to reflect the political and cultural background detailed on local level. This will be done based on selected case studies in the following.

5 The Core Phases of Historical Impacts

5.1 Pre-Colonial Period

Before the Dutch conquest, population density in the investigated area was very low. The people settled in kin group communities relatively isolated from each other, mainly in the surrounding mountains of the respective valleys. The kingdoms that ruled the kin groups represented more or less the territory of the districts that were established after independence. The main kingdoms were those of Sigi, Kulawi, Sedoa and Pekurehua (now Napu). Smaller kingdoms like the one of Pakuli, south of Palu, cannot be recognised anymore in the current administrative structure.

Only subsistence crops like cassava, corn and dry land rice were cultivated by doing shifting cultivation with rotation periods of up to ten years. The kin groups moved their huts with their fields, so that no permanent settlements existed at that time. In the Palu and Kulawi valley wet rice cultivation was already known as well. So far, an explanation why sawah cultivation was not known in the neighbouring Poso region could not be found.

Economic exchange within and out of the respective kingdoms was low but it existed. Most intensive exchange relations existed between upland Kulawi and lowland Sigi. The pre-colonial period can be regarded as quite stagnant in terms of migration. As the people of the different valleys lived relatively isolated, apart from marriage related migration of nobles within the research area, spatial mobility was negligible.

5.2 Dutch Rule

The Dutch colonialists, who influenced the Indonesian archipelago from 1602 to 1942, largely ignored this region for centuries. When they did station colonial officials in Sulawesi at the beginning of the 20th century, the short interval of their presence (1905 to 1942) however left a legacy of enormous structural modifications to the cultural landscape (Kreisel et al. 2002). Dutch colonial power extensively influenced the traditional society in the study area, as it did elsewhere during its expansion to the so-called Outer Islands, which commented on by Doel (1994: 96) as follows:

"The expansion in Indonesian foreign trade was remarkable for both its strong momentum and its exceptional duration: about one-quarter of a century of uninterrupted expansion. It goes without saying that this expansion had a profound impact on economic structure in the regions immediately concerned, in particular among the Buitenbezittingen. Production became oriented towards foreign exports to an unprecedented degree and the Buitenbezittingen acquired their characteristic position as prime earners of foreign exchange in the Indonesian archipelago, a position which has since then grown permanently."

The colonial conquest in Central Sulawesi took place at a time of change in colonial politics. By the turn of the 20th century, the so-called Ethical Policy influenced Dutch administration in the Malay archipelago. Its core element was a change from a policy of total exploitation of the colony without any regard to the local population towards a "Volksverheffing", which means that the local people should participate in the socio-economic status of a "developed" European country like the Netherlands.

Whilst the implementation of their Ethical Policy did not achieve much more than conserving the old structure of power, the influences on land use and agricultural production and settlement patterns are perceptible. New transportation facilities were closely linked to economic needs, like roads to export crop cultivation areas. However, the need of sufficient subsistence production was recognised by the Dutch administrators, which indicates one of the effects of the Ethical Policy. Thus, the agrarian exploitation in this area was limited by moral concerns about the people's nutrition.

Concerning migration, the Dutch started local and inter-island resettlements that were continued after Indonesia's independence by the post-colonial governments. The driving forces during the colonial period were an easier governing and the exploitation of the people (local resettlements) or the improvement of land use (inter-island resettlements). While the local resettlements from the mountains to the valleys took place in large scale and led to a row of new permanent settlements, there was only one case of inter-island transmigration from Java to the research area (see fig. 6).

The colonial rulers did not act as a stabilising force in the study area but confused the local population through a range of new influences. But as far as the sources reveal, there were no attempts of the colonialists to change the cultural landscape of the investigation area completely and without any respect towards the local population. There might be three reasons for this. First of all, one might assume that the impacts on the region would have been much worse, seen from the perspective of the local population, without this new colonial "ethical policy" of the Netherlands that accompanied the start of the 20th century. Secondly, the colonial governments needed the benevolence of the local principals to rule this part of the colony as easy as possible. For a third reason, the mission, in this case, mainly the Salvation Army, which emphasised the necessity to respect the local population, played a major role by preparing the ground for the Dutch administration, and not at last for the facilitation of the resettlements from the hardly accessible mountainous areas.

Under the rule of the Dutch, people were forced to build permanent settlements with a compact village character. Concentrated in these new founded settlements, it was much easier to collect taxes or to recruit forced labour. With regard to agriculture, the Dutch re-vitalised old wet rice fields and expanded, especially in the Palu valley, the total area of wet rice production by building a huge irrigation system. The coffee plants were first distributed to the noble families. Later, also common people started to cultivate coffee. Besides, the introduction of coffee as a new (cash) crop, the introduction of money, the collection of taxes and the introduction of luxury goods like sugar by the missionaries created a more and more capitalist economic system. In addition to the above-mentioned planned migration, people now also started to move spontaneously. However, it were usually Bugis, Chinese and Arab traders coming via Palu or Poso who used the new infrastructure of the Dutch to reach the more remote valleys like Kulawi or Napu.

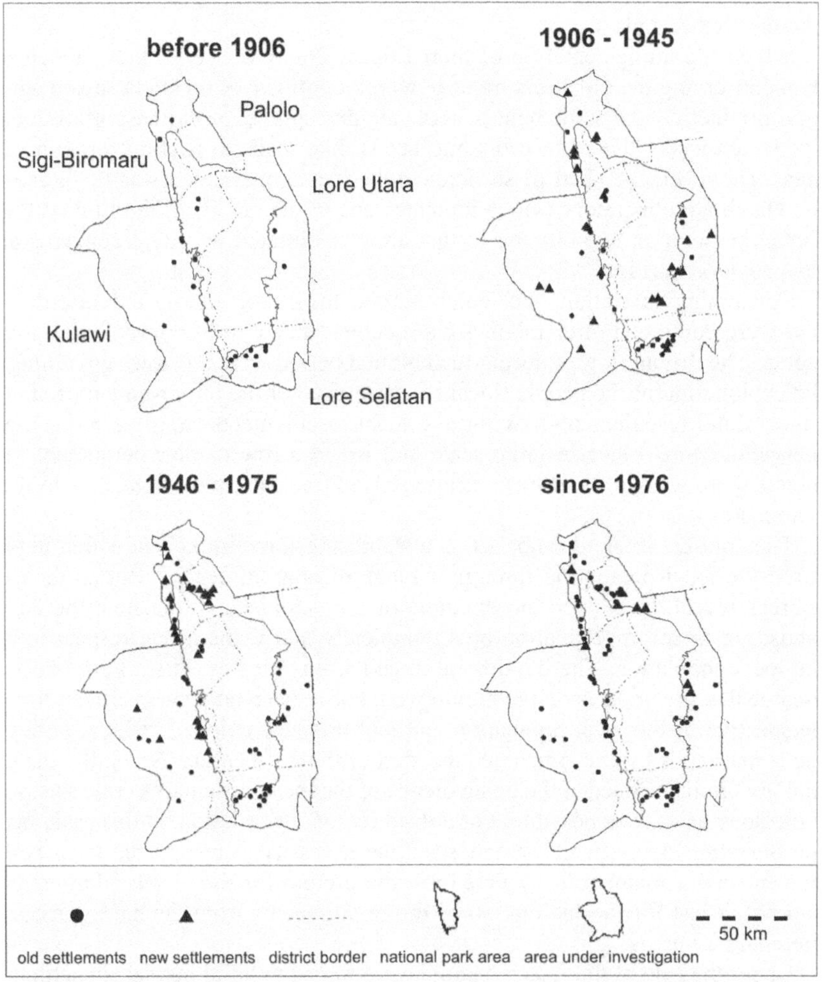

Fig. 6. Process of village foundations around Lore Lindu National Park (Draft: H. Faust, R. Weber; Data source: Village survey of sub-project A3, STORMA, Maertens et al. 2001)

5.3 Japanese Rule

The Japanese colonial rule in Central Sulawesi has to be regarded under the circumstances of World War II. The conquest of the region should open new resources for the needs of war. The most important impacts of the three years of Japanese rule in the investigated region were the compulsory production of cotton and an enormous emphasis on discipline. Besides, the implementations of the Dutch colonial government, e.g. compulsory labour, were continued during the Japanese era. In contrast to the Dutch, the Japanese did not continue to establish new villages.

With regard to the short presence of this colonial power, the introduction of cotton and moreover the order that every household had to plant a special amount of this crop were considerable. A part of the people cultivated cotton trees on newly established fields, while others turned fields of subsistence crops into cotton fields. Even if in the first case the field size for subsistence production was not reduced, the compulsory cultivation of cotton led to a dangerous neglect of food crops. In this context, contemporary witnesses tell about serious shortage of food. Partly, only cassava and forest products served as nutriment. Besides a shortage of food, the people also faced a lack of supply with clothes. Trade to the hinterland was interrupted, so that after the introduction of fabric clothes by the Dutch the villagers were forced to revert to their traditional bark clothes, not only for ceremonial occasions but also for their daily life.

The second most important impact mentioned above, the new discipline introduced with a system of physical punishments that was not known to this extent during the Dutch rule. Not bowing in front of the Japanese commander's house, not cultivating cotton, not joining the marching exercises, not working long and hard enough and many other occasions were regarded as sufficient reason to physically punish the people. Thus, an atmosphere of extreme fear affected daily life in the region.

It can be assumed that the local population does not regard the Japanese rule as more painful than the Dutch one because it was later in time and can be remembered better. Rather, the Japanese just acted more brutally than the Dutch, which may be ascribed to the influence of the Ethical Policy of the Dutch.

5.4 Independence

During the first decades of independence the population growth in Central Sulawesi and in the research region was relatively low, but rose immensely starting from the 1970s. Inter-island resettlements and, on a much larger scale, spontaneous migration during the 1990s are the major cause of this demographic change, as today more than half of the population in the study area are migrants. Main push factors for spontaneous migration are land scarcity and lack of job opportunities at the place of origin, whilst land availability and patron-client or family relations are the most important pull factors. In contrary to local migration, spontaneous migration from outside of our research area is a phenomenon that took place to its largest extent during the youngest history of the last twenty years.

So, we can identify for this type of spatial mobility a phase of low (colonial era), a phase of moderate (ca. 1950s till 1970s), and a phase of high influx (1980s until today). The last phase is mainly dominated by immigrants from South Sulawesi who search for free land which can be cultivated with cocoa, a crop that was cultivated in South Sulawesi much earlier than in Central Sulawesi. The second phase also includes migration of refugees from the rebellion of Kahar Muzakar in South Sulawesi and the Permesta rebellion in North Sulawesi. Currently, land scarcity also becomes increasingly a problem in the study area, which leads to an expansion of agricultural land use into protected forest areas of Lore Lindu National Park.

Adaptation from migrant groups takes place mainly in the case of newcomers from South Sulawesi in terms of agriculture. In general, this adjustment happens for the introduction of new crops, e.g. cocoa, and cultivation methods (e.g. use of fertilizer, planting in a row). Besides this, there are also many cases where the autochthonous know about differences of land use compared with the allochthonous, but do not adapt these innovations. Concerning local resettlements and (semi-) spontaneous local migration, a dynamic phase from the 1960s until the 1980s can be extracted from the study. The resettlements led to foundations of new villages as well as increase of the population of already existing villages, which produced a growth of cultivated land by approaching the limits of arable land.

The historical overview displayed above shows that there are numerous impacts on land use, settlement and population structure that changed the cultural landscape of an area bordering the rainforest in Central Sulawesi. These results already contain a wide range of important information that is necessary to understand the current status quo of use and management of natural resources in the study area. Going a step further, it also allows then trying a projection of how this region could develop in the future.

However, the influences and changes differ from sub-district to sub-district and even from village to village, which shows the complexity of the regional analysis of this region. Thus, if we want to get a more realistic picture of the development, the current status quo and thus a more precise base for projections in the investigated area, it does not suffice to regard the region around Lore Lindu National Park as a whole, as a homogenous unit. As it will be shown below, smaller units of different development of cultural landscape down to village level can be defined.

6 The Regional Differentiation of Historical Impacts on Sub District and Village Level – Two Examples

6.1 Characteristics of Kulawi

Sub-district Kulawi can be divided into three regions: the northern Kulawi (usually called Kulawi) with an enclave around Lake Lindu inside the national park, the southern Kulawi (Gimpu valley) and the remote Pipikoro area in the Southwest. Since 2002, Pipikoro forms an own sub-district. Three main languages, Moma (northern Kulawi, except Lindu area), Uma (Gimpu and Pipikoro area) and Tado (Lindu area) are spoken. Most of the villages (except Lindu and Pipikoro area) are located along the main road from Palu.

Before the Dutch entered Kulawi in 1906, the people of Kulawi stood under the control of the king (*raja*) of Sigi who was, at that time, the most powerful principal in the whole of Central Sulawesi. After the Dutch defeated the Kulawi soldiers in the war at Mount Momi, the *raja* of Kulawi surrendered and asked his people to follow the orders of the Dutch. Those first settlers stayed in Namo, north of the capital village Bolapapu. Later, they established Bolapapu as the centre of the northern part of Kulawi and Gimpu as centre of the southern part. As in other regions of the research area, pre-colonial settlements were not yet permanent. The people lived in kinship communities and planted dry land rice, corn, cassava and sago in slash-and-burn shifting cultivation. However, sawah was already cultivated as well.

The Dutch resettled the people from the mountains to the valley and introduced coffee trees, at first only to the royal families, later on to the common people. While wet rice cultivation is seldom in villages of the Pipikoro area due to the morphological conditions, this crop mainly covers the plains in the North and South of Kulawi. In villages whose borders reach from the western until the eastern mountain range, where the national park starts, the people used to cultivate two fields, one on each side of the valley. However, after the establishment of Lore Lindu National Park people chose different solutions to keep their economic status ranging from gathering rattan to opening of more distant fields in the west or to following the government's local resettlement program to the enclave of Lake Lindu. Besides cocoa, coffee is an important crop in whole Kulawi up until today.

Before the Dutch arrival, trade exchange systems to Palu valley existed already in a small extent. Usually buffalos or gold were exchanged with salt and other goods. Information varies on whether money was already known before the arrival of the Dutch. It may be possible that only in the northern part of Kulawi money was already known before the Dutch conquest because it is located closer to the Palu valley. By the start of World War II in Southeast Asia it became difficult to get fabric clothes because the trade with Kulawi was interrupted. People started to wear bark clothes as they did before the arrival of the Dutch.

The road that the Dutch built in the 1920s connected the Northern part of Kulawi with Palu. Today, most of the villages of the two plains, Kulawi and Gimpu, can be reached on asphalt roads. Only Pipikoro area is still lacking sufficient transport facilities. This slows down the economic development of this part of Kulawi by rising transport costs, and leads to out-migration, which has been going on for decades. During the 1980s and 1990s, most of these migrants searched a new living in less populated areas like Palolo or Napu valley.

With the arrival of the Dutch, first Arab and Chinese traders settled down in Bolapapu, the capital of the sub-district. After independence a main phase of immigration of Bugis started with the DI/TII rebellion in South Sulawesi. However, their places of destination are located in the southern part of Kulawi as the refugees entered the sub-district straight from South via the mountains and not via Palu. Besides this phase and the general influx of migrants from South Sulawesi in the 1990s caused by the improvement of the former gravel main road into an asphalt road, immigration to Kulawi was constantly moderate over time. This constant migration led to an earlier start of land scarcity than in other regions of the Lore Lindu area, and thus only a few of the Bugis who moved to the research area during the 1990s selected Kulawi as their destination. This is one reason why in Kulawi the percentage of locals is still quite high.

A subsidiary reason is the stronghold of the Christian religion in most parts of Kulawi. The Salvation Army opened its first school for common people in Kulawi in 1916 and started to introduce Christianity, and even as recently as the late 1950s everybody converted to Christianity. While the adat law as a whole was untouched by the Dutch, adat rituals/ceremonies that interfered with the beliefs of Christian religion, like praying to stones and trees, slicing of prisoners or slaves, or the initiation rites of hitting off the front teeth of girls, was abolished. Today, Christians form the majority in most of the villages in Kulawi, except a Muslim centre of Bugis people in Gimpu valley.

The implementation of the border of Lore Lindu National Park caused uncertainty among the people like in other parts of the investigated area. However, the protected forest is less affected with clear cutting than in other sub-districts.

Case study – village Bolapapu

The name Bolapapu has been used since 1905, but people settled in this area already before, not down in the valley but up the surrounding hills. The former name of the settlement is Bola, which means "houses". A folk tale says that people from Lake Lindu burned down (*papu*) the huts of Bola and as a result the inhabitants created the today's village name. When the Dutch came in 1906, they started to establish Bolapapu at its present location. Bolapapu has remained the local centre of northern Kulawi.

The village consists of five hamlets, which are spread over more than eight kilometres along the main road, from the centre of Bolapapu (hamlet 1) to the North. Besides hamlet 1 and 2, which grew together and now form one compact village complex, the housing area of one hamlet is clearly divided from the one of the next hamlet by fields and forest. While the plain area is relatively large in the first two hamlets, the valley gets smaller on its way to the North. Until 1985, Salua was a hamlet 6 of Bolapapu. It became an own village due to population growth and the far distance to the centre of Bolapapu. The majority of the people in hamlet 1 until 3 are Christian, while hamlet 4 and 5 are dominated by Muslims. An Arab trader introduced Islam there but most of the today's Muslims are converted Kulawi people.

Terracing of wet rice fields in hamlet 1 started after population growth and expansion of the area of governmental buildings led to a conversion of rice fields on the plain land. Hamlet 1, and more and more hamlet 2 as well, is the obvious centre of the village and thus developed in totally different way from the more distant hamlets 3 to 5. Most of the village's trade infrastructure is concentrated here and dominated by Bugis and Chinese people. Kulawi people, in contrast, dominate the governmental institutions or are farmers.

Hamlet 3 consists of three settlements, called Boma, Boya 1, and Boya 2. All three locations are the same age. Before the Dutch came, the people lived in the valley and cultivated sawah. The Dutch forced all of them to move up to the road. While coffee – besides wet rice – was a main crop after the Dutch had introduced it to the people, cocoa started to become important in 1982. An officer from the Agriculture Dept. promoted this crop, and as the people were promised that cocoa would have a good future, they changed corn-cassava-vegetable fields into cocoa fields although the price for coffee was still higher than the price for cocoa at that time. Today, cocoa and coffee are the main crops, followed by wet rice.

Before the people of hamlet 4, called Laone, moved along the main road (after the colonial period), the Dutch resettled them from Namo, their place of origin, to Tolumanu, about 1 km away from the main road. The Dutch built a horse track to Lindu and gave coffee and cocoa seeds to the royal family. The common people cultivated wet rice or dry land rice fields and lived there for about 20 years. In the 1950s, they moved closer to the main road. Today, no one lives in Tolumanu anymore. Cocoa and coffee fields now cover that area which is called Kintahai (old settlement) since the people moved to Laone. There are still visible remains of the former houses and of the track to Lindu. Behind Kintahai to the East, wet rice fields cultivated by the villagers of Laone cover a valley that has no road access. When the people moved from Kintahai to Laone, they first settled around the school that was built in 1958. Later, the hamlet developed from the school up towards the main road. During the Dutch period, Arab Muslims moved to the village to teach Islam. Today, Laone is the settlement with the highest Muslim density in whole Kulawi (95%). Most of the population of hamlet 3 and 4 was born there or in the neighbouring area or moved from hamlet 1 or 2 to there.

Two settlements, Salua and Sadaunta, form hamlet 5, which is situated along the main road. About seven years ago, Sapoo was hamlet 6, while Tangkulowi, situated in the western mountains, was called hamlet 5. In order to decrease population density and to stop shifting cultivation in the area, the people of Tangkulowi were resettled to Palolo valley and to Luro, a location near hamlet 1 of Bolapapu. The inhabitants of Sadaunta, from where a horse track leads to Lake Lindu, engage mainly in dry land cultivation. A mix of rice, corn, cassava and vegetables was planted in former times, coffee became more and more important and is still the main crop of the hamlet. When the government forbade planting new coffee trees inside the national park and only allowed farmers to collect the trees inside the national park and only allowed farmers to collect the harvest of their already existing fields, the people gave up most of those fields because the harvest they could obtain was very small. Instead of this, they opened new fields at the western mountains although the distance to the fields there is about four times bigger. Sapoo and Sadaunta have a village structure since 1970. Before houses existed in this area, large parts of the land were already planted with coffee. Only one or two years later, the first Bugis moved to the area and started logging and worked as carpenters. In order to prevent young people from entering the protected forest, the villagers started a program to re-process former fields.

Major out-migration from Bolapapu took place to Lake Lindu (local resettlement program) in the 1980s when arable land became scarce because of the implementation of the border of the later Lore Lindu National Park. Most of the people who left Bolapapu spontaneously moved because of marriage or because they got a government job in Palu.

Preliminary Summary

Kulawi can be divided in at least three parts, which differ in terms of urban character, transport infrastructure, land use and population structure: Kulawi valley, Gimpu valley and Pipikoro. The village of Bolapapu, however, cannot be taken as representative example of all villages of the sub-area Kulawi. It is the very centre of this northern part of the sub-district. Such a dense administrative infrastructure can be found in no other village of this part. However, taking into account the factor trade, Bolapapu as a centre has its equivalent in Gimpu valley. The example of village policies on the encroachment of the national park also shows the importance of a study of land use in the region on a smaller scale. By stepping one spatial level down, Bolapapu shows a differentiation in terms of centrality. Only the hamlets 1 and 2 form the village's centre while the remaining hamlets illustrate not only a spatial distance to the centre but also a functional one. The village is not homogeneous in confession either. While Christianity outweighs in the hamlets 1 to 3, hamlet 4 and 5 builds a Muslim stronghold whereas hamlet 4 even represents the community with the highest percentage of Muslims in the whole sub-district.

6.2 Characteristics of Palolo

As in other parts of the study area, the pre-colonial population of Palolo did not live in the valley but in the surrounding mountains. Slash-and-burn shifting cultivations of rice, cassava, maize and vegetables as well as the use of forest products by hunting and gathering formed the subsistence economy. The people lived in small kinship communities spread over the forest area that was territory of the kingdom of Sigi. Between Palolo and Napu valley were almost no relations at that time due to the physical barrier and the rumours about the violent To Pekurehua (today known as To Napu, people of Napu) – although Napu valley was actually subordinated to the raja of Sigi.

Trade exchanges took place between the inhabitants of the upland with their lowland neighbours in the Palu valley. Before first traders (mostly Chinese, Arab or Bugis) came to Palolo, the upland people went to Palu valley to trade their buffalos etc. in order to get salt or fish etc. in exchange.

Roughly, three phases of village foundations form the recent landscape of settlements in sub-district Palolo. First permanent settlements were founded after the Dutch gave an order to the people to move down from the mountains. The first settlements like Kapiroe, Bunga or Menusi are all located close to Palu valley at the western end of Palolo valley. The rest of the valley until the physical border to Napu valley in the East was still covered by forest and grassland and partly served as hunting area. During the Dutch period, the inhabitants of these new settlements were forced to start cultivating wet rice fields and to give up their dry land rice fields.

A second phase of village foundations started at the end of the 1950s when parts of some of the former villages got the status of an independent village, e.g. Berdikari that belonged to the territory of Menusi before, as well as Makmur (capital of sub-district Palolo) or Ampera. At the end of the 1960s and during the 1970s, a third phase can be extracted. Some of these new villages are again the result of separations from already existing villages of the second phase (e.g. Bahagia, a transmigration settlement that belonged to the territory of Berdikari before) or they are founded on land that was still not yet territory of one of the already existing villages (e.g. Sintuwu).

Besides spontaneous immigration, the short period of the second and third phase is based on out-migration of the densely populated oldest villages of the sub-district and resettlements from other regions. Thus, in Palolo valley, in different villages, different ethnic groups formed the first village communities, ranging from local Kaili people to Kulawi or Javanese people. A very important factor concerning the development of settlements in Palolo was the Indonesian government of the New Order era that gave an order to turn the whole valley into cultivated land. However, especially the Kulawi people in Berdikari are proud of the name of their village, which means "Berdiri di atas kaki sendiri" ("to stand on one's own feet").[4] This name says that the migrants from Kulawi got no material support from the government unlike local transmigration projects (e.g. the village Kamarora).

During the 1960s and 1970s, the government pushed wet rice production, but today productivity is slow because of insufficient irrigation. Whereas broad parts of the mountains were covered by clove trees during the 1980s, a decrease in the price for clove led to conversion of this plant into cocoa trees starting in the 1990s. Today, cocoa is *the* cash crop of the sub-district.

[4] The first name of the territory was Karawa Maluo (wide grass land). The district government changed it into Berdikari after the village got its official administrative status.

Case study – village Berdikari

In 1954, the founders of Berdikari, seven Kulawi families from the remote Pipikoro region, who stayed in Menusi before, opened forest in today's hamlet 1 and turned it into dry land rice fields. The grassland was turned into wet rice fields. However, at that time they spent most time on exploiting rattan that was sold to Palu.

In 1964, they started to build a village road and to clear land for a school and football field. In 1968, the population of Karawa Maluo (later: Berdikari) had mounted to 40 families. On the wish of the inhabitants Karawa Maluo became an independent village in the same year and from then on was called Berdikari. During the 1960s, people from South Sulawesi (Bugis, Toraja) had already moved to the area but they left again because of problems with the Kulawi adat (Toraja) or because a padi mill that the Bugis wanted to build existed already. These migrants did not stay at the centre of the village but in today's hamlet 2.

Until 1969, the people still lived in little huts on their fields. That this area developed quite fast was due to the first village chief who got the nickname *tangan besi* (iron hand). Described as quite cruel, he did not hesitate to beat people who did not hurry up opening sawah fields and who did not attend *kerja bakti* (joint work for village issues without payment). Some people even left the village again and moved somewhere else because of this village chief. He also gave an order to settle permanently along the new built road. He promoted coffee and sawah as the villagers' future. Around 1990, the cultivation of dry land rice stopped because the people changed their crops more and more to clove, coffee and cocoa. Clove was introduced and promoted by the Dept. of Agriculture. First cocoa trees were planted about 15 years ago by Bugis migrants. After clove lost its price, the people cut the trees and replaced them with cocoa trees. Since 1995, the inhabitants of Berdikari are prospering from their cocoa harvest. The monetary crisis in Southeast Asia in 1998 let the price for cocoa increase to 20.000 Rupiah per kilogram and thus more and more people were able to invest in status symbols like motorcycles.

With the increase of population growth (mostly due to immigration), a second leader organized the life of the community in Air panas in 1978. One year later, Air panas became hamlet 2 of Berdikari. The vast majority of the inhabitants of that hamlet originates from Kulawi. They had already fields in Air panas but still stayed in hamlet 1. Later, they moved to hamlet 2 to reduce the distance to their fields. Before a big flood in 1986, perennial crops were only planted around the houses. After the flood destroyed a big part of the wet rice fields, the inhabitants of hamlet 2 started to engage in planting esp. cocoa trees on different fields.

At that time, today's hamlet 3 was included in hamlet 2. It became an independent hamlet in 1992 because of the ongoing population growth and the big distance to Air panas. Today, hamlet 3 is mostly populated by people from South Sulawesi (Toraja, Palopo, Bugis). The area of hamlet 3 was first used as hunting area for Kulawi people in the 1970s. In the same decade, first Bugis stayed here and gathered rattan in the forest. The migrants could ask the village head of Berdikari to settle down and open land. He decided on the location. At the moment, there is a government program running to move all houses along the main road and to build side roads if there is need. Around sixty families still do not live along the road but on their fields, up to 7 km away from the road.

Since 1965, hamlet 2 and 3 were divided by a new settlement of transmigrants from East Java. Actually, they should be resettled to Rachmat (near Sintuwu), but as this land was very swampy and infrastructure was poor, they moved to Berdikari and formed Berdikari 2. In 1977, the transmigrants still attended the village elections of Berdikari. One year later, the name Berdikari 2 changed into Bahagia and it became an independent village.

There are almost no reports about problems between the different groups of migrants. However, Berdikari is also mostly divided along ethnic lines. While Kulawi people form the vast majority of hamlet 1 and 2, people from South Sulawesi can almost only be found in hamlet 3. The Javanese people are also assembled only in Bahagia that is an independent village anyway. Despite of that, a good administrative relation can be detected between the hamlets, which might be due to the commitment of the present village chief who takes his job seriously.

Preliminary summary

The cultural landscape of Palolo developed relatively differently compared with, for example, Kulawi. While in Kulawi permanent settlements were already established in the whole sub-district during the Dutch rule, major parts of Palolo remained untouched by the Dutch at that time. Palolo faced a quite short but very dynamic phase of village foundations after independence. The sub-district also evolved into a centre of immigration in Lore Lindu region and thus established a wide range of villages with different ethnic majority. So, Palolo also forms a patchwork of different cultural impacts. This fact itself reveals already the diversity of the cultural landscape in Palolo. However, Berdikari forms no homogeneous unit as well. Families of the founders of Berdikari originating from Kulawi dominate the centre of the village, while especially hamlet three became a stronghold of migrants from South Sulawesi. Today, a new village consisting of transmigrants from Java also spatially divides the village.

7 Discussion and Conclusion

The evidence of the empirical studies presented exemplarily for the sub-districts of Kulawi and Palolo emphasises the theoretical approach of this paper. Both, local settings and external impacts form the current cultural landscape of the investigated area. In addition, internal and external impacts have to be regarded in historical dimension by a distinction of different layers of space and time. Each subregion (Kulawi, Palolo etc.) consists of elements of common changes of cultural landscape. But the today's status quo is based on processes of permanence and change, stability and destabilisation that vary from one sub-region to the other. Natural and cultural settings change between sub-districts, and within each subdistrict they also change between sub-areas (like, for example, Kulawi valley, Gimpu valley and Pipikoro in sub-district Kulawi). Furthermore, even different villages or hamlets, although situated close to each other, develop their own specific cultural landscape as exemplified in the case of Bolapapu.

Starting with the example of the development of village foundations (see fig. 6), a spatially very different process can be recognised over time. Most precolonial settlements existed in Kulawi and Lore Selatan whilst in Palolo valley no settlements existed at all. In contrast, a short but intensive phase of village foundation happened in Palolo during the 1960s to 1980s, an exceptional process that cannot be observed in any of the other sub-districts. Such a facet of cultural landscape is closely linked to a diversity of migration from, to, and within the region. In general, the area can be roughly divided in a region of quite low immigration over time and since independence out-migration mainly to Palolo (Kulawi), a region in transition, where currently immigration decreases from its high level and out-migration to the Eastern part of the research area gets higher (Sigi-Biromaru), a region of high immigration rate since the 1960s and low out-migration (Palolo), and a region of low immigration over time rising exponentially during the 1990s and low out-migration (Lore Utara).

Although the Lore Lindu region, compared with other parts of Sulawesi, represents an immigration area, migrant influxes and out-fluxes diverge among its different sub-regions as well as between the villages. If we take the example of sub-district Kulawi, differences of the general development in terms of immigration of people from South Sulawesi on the one hand and the case of the village Bolapapu on the other hand illuminate Lore Lindu as a heterogeneous region. The southern part of Kulawi (Gimpu valley) experienced a high rate of immigration from South Sulawesi during the period of Kahar Muzakar rebellion, but this view changes if we regard the northern and the southern part of the sub-district separately. The centres of Bugis migration are villages of the southern part. Bolapapu however, counts a relatively small number of those migrants (see Fig. 2), although the village as the main centre of Kulawi offers an attractive base for trade that Bugis people are famous for. In this case of Bugis migration local politics play a crucial role for distribution of the newcomers. Thus, even in the southern part of Kulawi Bugis population is not allocated equally.

In terms of land use, differences can be observed as well. Due to the foundations of irrigation for wet rice cultivation established by the Dutch, Sigi-Biromaru represents the granary of Lore Lindu region. In contrast to other sub-districts, Coffee is still a very important cash crop and became a part of the sub-districts cultural identity. Cocoa is *the* dominant crop in sub-district Palolo because of the big amount of migrants from South Sulawesi who converted mainly wet rice fields into cocoa fields. In Lore Utara, coffee followed by cocoa are the most important crops although their productivity is relatively low. Special village politics based on more or less strong custom's councils influence the natural resource management. In Bolapapu, presenting other possibilities, e.g. cultivation of land in other areas, prevented an encroachment of the national park. In Berdikari, a strong village leadership lead to an increase of wet rice production instead of gathering forest produces.

The importance of the factor "education" concerning land use, prosperity and foresight is exemplified in the village Pandere that was not described in detail in the previous chapter. The discrepancy of the educational status between the hamlets is revealing. While migrant families from North Sulawesi invest in their children's future, locals prefer a basic education as their children will become farmers anyway and thus do not need to "waste" their time at a higher school while they can help on the family's fields. In addition, cultural traditions and their conservation or reformation by the village leaders influence land use individually from village to village. In Pandere, the local's council of customary law (*lembaga adat*) banned the use of the sickle that was introduced by migrants and thus slowed down the productivity of rice cultivation. The *lembaga adat* of Wanga decided in cooperation with a local NGO that *tanah adat* that the ancestors' land shall not be cultivated but conserved as cultural relict.

Not at last, interaction of the villagers with the national park also differs between sub-districts and between villages. Generally, the western border (Sigi-Biromaru, Kulawi) of Lore Lindu National Park is less encroached than the eastern one (Palolo, Lore Utara). If village territories contain additional forest not bordering the park, like in Northern Kulawi, or if climate conditions are less suitable for extensive cultivation of the cash crops cocoa or coffee, like in Sigi-Biromaru, violations of the protected forest area occur less. In Wanga however, this general distinction get more complex if one raises the question of who converts protected forest land into fields. It is not the villagers of Wanga who enter the park but those of the neighbouring local and inter-island resettlement villages that were built on parts of Wanga territory. However, as the example of a program to reprocess fallow land in Bolapapu shows, if easy accessible land outside the national park becomes scarce, village policy for example decides in which direction such a problem gets solved.

The case study villages reveal the danger of a simplistic view of the real situation of the cultural landscape, as the development of settlement structures cannot be explained without focussing migration patterns and cultural settings. An appreciation of processes in land use needs the consideration of political decisions, customs and external influences. An answer on the question of the stability of the rainforest margins is depending on all of those variables, which, at the end, are all based on the acting of individuals and groups. This study shows that society itself is the major driving force of the genesis of the cultural landscape "Lore Lindu region", and a society is structured that is as diverse as the cultural landscape, across space and time.

References

Acciaioli G (1989) Searching for good fortune: the making of a Bugis Shore community at Lake Lindu, Central Sulawesi. Canberra

Adriani N (1915) Maatschappelijke, speciaal economische verandering der bevolking van Midden-Celebes, sedert de invoering van het nederlandsch gezag aldaar. Tijdschrift van het Koninklijk Nederlandsch Aardrijkskundig Genootschap, tweede serie deel XXXII: 457-475

ANZDEC (1997) Central Sulawesi Integrated Area Development and Conservation Project. TA NO 2518 – INO. Draft Final Report – Annex (Working Papers), Indonesia

Aragon LV (1992) Divine justice cosmology, ritual and protestant mission in Central Sulawesi, Indonesia. Urabana

Aragon LV (1996a) "Japanese Time" and the Mica Mine Occupation Experiences in the Central Sulawesi Highlands. Journal of Southeast Asian Studies, 27, 1: 49-63

Aragon LV (1996b) Twisting the gift translating precolonial into colonial exchanges in Central Sulawesi, Indonesia. American Ethnologist 23, 1: 43-60

Bobek H, Schmithüsen J (1949) Die Landschaft im logischen System der Geographie. In: Erdkunde, 3: 112-120

Bobek H (1948) Stellung und Bedeutung der Sozialgeographie. Erdkunde, 2: 118-125

Boomgard P (1993) Upliftment down the drain? Effects of welfare measures in late colonial Indonesia. In: Dirkse JP (ed) Development and social welfare: Indonesia's Experiences Under the New Order. Leiden, pp 247-253

Boomgard P et al. (eds) (1997) Paper landscapes: explorations in the environmental history of Indonesia. Leiden (= Verhandelingen van het Koninklijk Instituut voor Taal-, Land- en Volkenkunde)

Coté J (1979) The colonization and schooling of the To Pamona of Central Sulawesi, 1894 to 1924. Melbourne

Coté J (1996) Colonising Central Sulawesi the ‚Ethical Policy' and imperialist expansion, 1890-1910. Itinerario 20 (3): 87-108

Crang M (1998) Cultural Geography. London, New York

Cribb R (1994) Introduction: the late colonial state in Indonesia. In: Cribb R (1994) The late colonial state in Indonesia political and environmental foundations of the Netherlands Indies, 1880-1942. Leiden, pp 1-9

Doel H W van den (1994) Military rule in the Netherlands Indies. In: Cribb R (1994) The late colonial state in Indonesia political and environmental foundations of the Netherlands Indies, 1880-1942. Leiden, pp 57-78

Faust H, Maertens M, Weber R, Nuryartono N, van Rheenen T, Birner R (forthcoming) Migration and Land Use in the Lore Lindu National Park Area, Central Sulawesi - "Effects for Stabilisation or Destabilisation?" Bogor (= STORMA Discussion Paper Series Sub-program A)

Frank G (1911) Niederländisch-Indien: eine Finanzquelle für das Mutterland. Würzburg

Geertz C (1973) The interpretation of cultures. New York Basic Books

Geertz C (1980) Negara the theatre state in nineteenth-century Bali. Princeton, New Jersey

Geertz C (1983) Local knowledge. New York Basic Books

Gerke S (1993) Ethnic Relations and Cultural Dynamics in East Kalimantan. Journal of Indonesian and Malay Studies 2, London

Grubauer A (1913) Unter Kopfjägern in Central-Celebes: ethnologische Streifzüge in Süd-ost- und Central-Celebes. Leipzig

Haggett P (1972) Locational Analysis in Human Geography. London

Haggett P (1991) Geographie eine moderne Synthese. 2. edn, Stuttgart

Henley DEF (1989) The idea of Celebes in history. Clayton, Victoria (=Centre of Southeast Asian Studies, Working Paper no 59)

Hissink (1909) Nota van toelichting, betreffende de zelfbesturende landschappen Paloe, Dolo, Sigi en Beromaroe. Tijdschrift voor Indische Taal-, Land- en Volkenkunde (1912): 58-129

Hobart A (1995) Banaspati Raja "Lord of the Forest" and Community Well-being in Bali. In: Werlen B, Wälty S (eds) (1995) Kulturen und Raum - Theoretische Ansätze und empirische Kulturforschung in Indonesien. Zürich, pp 333-356

Hutterer KL, Rambo T, Lovelace G (eds) (1985) Cultural Values and Human Ecology in Southeast Asia. University of Michigan

Jäckel W (1990) Tauschhandel und Gesellschaftsform: Entwicklung des Markthandels bei den Toraja (Indonesien). Kassel (=Urbs et Regio 55)

Kaudern W (1925) Structures and settlements in Central Celebes: results of the author's expedition to Celebes 1917-1920. Göteborg (= Ethnographical studies in Celebes 1)

Kolff GH van der (1929) European influence on native agriculture. In: Schierke B (ed) The effect of western influence on native civilisations in the Malay Archipelago. Batavia, pp 101-203

Koninklijk Instituut Voor De Taal-, Land En Volkenkunde Van Nederlandsch Indie (KTLVK) (1911) Regeeringsrapport over de onderafdeeling Paloe (1910). Adatrechtsbundel 1, Serie O: 130-144

Kotilainen EM (1992) "When the bones are left": a study of Material Culture of Central Sulawesi. Helsinki (= Transactions of The Finnish Anthropological Society 31)

Kreisel W, Weber R, Faust H (forthcoming) Colonial interventions on cultural landscape of Central Sulawesi by „Ethical Policy" – Impacts of the Dutch rule in Palu and Kulawi valley 1905-1942. Asian Journal of Social Sciences 31, 2

Kruyt AC (1903) Gegevens voor het bevolkingsvraagstuk van een gedeelte van Midden-Celebes. Tijdschrift van het Koninklijk Nederlandsch Aardrijkskundig Genootschap, tweede serie deel XX: 190-205

Kruyt AC (1926) Pakawa, een landstreek in de onderafdeeling Paloe. In: Tijdschrift Koninklijk Nederlandsch Aardrijkskundig Genootschap 43, 4: 526-544

Kruyt AC (1929) The influence of Western civilisation on the inhabitants of Poso (Central Celebes). In: Schierke B (ed) The effect of western influence on native civilisations in the Malay Archipelago. Batavia pp 1-9

Kruyt AC (1935) De Bewoners van Midden-Celebes, en de culturen, die invloed op hen hebben uitgeoefend.

Kruyt AC (1938) De West-Toradjas op Midden-Celebes. 4 vols, Amsterdam (= Verhandelingen der Koninklijke Nederlandsche Akademie van Wetenschapen te Amsterdam, Afdeeling Letterkunde, Nieuwe Reeks, Deel XL)

Kuitenbrouwer M (1991) The Netherlands and the rise of modern imperialism colonies and foreign policy, 1870-1902. New York

Lindblad JT (1995) Colonial rule and economic development a review of the recent historiography on Indonesia. Jahrbuch für Wirtschaftsgeschichte 1995, 1: 9-22

Locher-Scholten E (1981) Ethiek in fragmenten. Vijf studies over koloniaal denken en doen van Nederlanders in de Indonesische Archipel 1877-1942. Zeist

Maddison A, Prince G (eds) (1989) Economic growth in Indonesia 1820-1940. Leiden (= Verhandelingen van het Koninklijk Instituut voor Taal-, Land- en Volkenkunde 137)

Metzner J (1981) Palu (Sulawesi) Problematik der Landnutzung in einem klimatischen Trockental am Äquator. Erdkunde 35: 42-54

Mitchell D (2000) Cultural Geography – A Critical Introduction. Oxford, Malden, Massachusetts

Pamei J (1997) Perang Bulu Momi Perlawangan Towoalangi Raja Kulawi Terhadap Belanda. Paper presented at a seminar on the war at Mount Momi at the 8th of Novemer 1997, Kulawi

Peet R (1998) Modern Geographical Thought. Oxford, Blackwell Publishers

Poeze HA, Schoorl P (eds) (1991) Excursies in Celebes een bundel bij het afscheid van dr. J. Noorduyn als directeur-secretaris van het Koninklijk Instituut voor Taal-, Land- en Volkenkunde. Leiden (=Verhandelingen van het Koninklijk Instituut voor Taal-, Land- en Volkenkunde 147)

Prince GHA (1995) Dutch Policy in Colonial Indonesia 1900-1942. Jahrbuch für Wirtschaftsgeschichte 1995, issue 1: 23-44

Prince GHA (1996) Monetary policy in colonial Indonesia and the position of the Java bank. In: Lindblad T (ed) Historical foundations of a national economy in Indonesia, 1880s-1990s. Amsterdam et al., pp 55-70

Ramseyer U (1995) Wie Ki Sudamala nach Sidemen kam. Geschichte und endogener Wandel in einem balinesischen Dorf. In: Werlen B, Wälty S (eds) (1995) Kulturen und Raum - Theoretische Ansätze und empirische Kulturforschung in Indonesien. Zürich pp 267-284

Ricklefs MC (1981) A history of modern Indonesia c. 1300 to the present. London

Sauer CO (1925) „The Morphology of Landscape“. In: University of California Publications in Geography 2, pp 19-54, reprinted in Leighly (1963), pp 351-379

Sauer CO (1956) The agency of man on Earth. In: Thomas Jr WL (ed) Man´s role in changing the face of the earth. Chicago, pp 49-69

Schlüter O (1919) Stellung der Geographie des Menschen in der erdkundlichen Wissenschaft. Geogr. Abende 5, Berlin

Schlüter O (1920) Die Erdkunde in ihrem Verhältnis zu den Natur- und Geisteswissenschaften. Geogr. Anzeig., 21: 145-152, 213-218

Schlüter O (1928) Die analytische Geographie der Kulturlandschaft. Z.G.E. Berlin, Sonderband zur Hundertjahrfeier, pp 388-411

Schrauwers A (1999) „It's not Economical“: the Market Roots of a Moral Economy in Highland Sulawesi. In: Li TM (ed) Transforming the Indonesian Uplands Marginality, Power and Production. Amsterdam, pp 105-129

Vroklage BAG (1936) Borneo, Celebes und Molukken. Münster/Westfalen (=Die sozialen Verhältnisse Indonesiens. Eine kulturgeschichtliche Untersuchung 1)

Weber R. (2000) Die Kolonialzeit im heutigen Indonesien und ihre Auswirkungen auf Bevölkerung und Kulturlandschaft in Zentral-Sulawesi. Göttingen [=unpublished diploma thesis]

Wedema S (1998) "Ethiek" und Macht die niederländisch-indische Kolonialverwaltung und Indonesiens Emanzipationsbestrebungen 1901-1927.

Werlen B (1993) Society, Action and Space. An Alternative Human Geography. London

Werlen B, Wälty S (eds) (1995) Kulturen und Raum - Theoretische Ansätze und empirische Kulturforschung in Indonesien. Chur/Zürich (= Konkrete Fremde - Studien zur Erforschung und Vermittlung anderer Kulturen 10; 400 pp)

Williams R (1958) Culture and society, 1780-1950. London

Zeller M, Schwarze S, van Rheenen T (2002) Statistical sampling frame and methods used for the selection of villages and households in the scope of the research program on stability of rainforest margins in Indonesia. (= STORMA Discussion Paper Series Subprogram A, no 1)

STORMA Sources

Maertens M, Chairil M, Birner R (2001) Village Survey Sub-Project A3

Schwarze S, Nuryartono N, Zeller M, van Rheenen T (2000-2001) Household Survey Sub-Project A4

Robo and the Water Buffalo: The Lost Souls of the Pekurehua of the Napu Valley

Satyawan Sunito

1 Introduction

Currently the push toward the frontier areas, which until twenty years ago were still largely untouched by commercial agriculture, has taken a massive form. In most cases the process has resulted in an agrarian differentiation that is not in favor of the local or indigenous communities. Even worse, in many cases the local communities, it seems, do not have the institutions and social organizations to defend their own interest, and their environment on which they partly depend. Cornered by the influx of land hungry migrants and by the shortage of land, many local communities start to demand rights on their ancestors' land, large parts of which are under the category of state forestland, protection forest or even as National Park. In the district North Lore, Central Sulawesi, on the borders of the Lore Lindu National Park, local discontent toward the one sided decision regarding borders of the national park has resulted in activities in mapping the ancestors' land and the establishment of forest zones where local people have use-rights. However, there are still no institutions and organizations able to manage this land. Meanwhile, unregulated extraction of timber and the expansion of agriculture land into the forest and the National Park keep going on.

This paper tries to reconstruct a picture of the local social institutions and organizations in the field of resource management of the direct past. The term "direct -past" refers to the period with a distinctive although extinct complex of institutions and organizations directly pre-dating the present period. The direct past is still part of the local oral history, and in this particular case is still part of the life experience of the older generation of the population. It is the conviction of the author that this historical journey to the past is essential to understand better the current social phenomena and processes. The data used in this paper are part of a larger on-going research project of the author on local social organization in natural resource management conducted in a village bordering the Lore-Lindu National Park.

2 Methodological Considerations

Debate on the character of the peasantry and its position in the larger political and economical context have a long history. Breman (1997) nicely put the whole process of conceptualization of the peasantry in Southeast Asia in five/ typologies: the village colonized, nationalized, developmentalized, anthropologized and globalized. Coined first by colonial researchers but still persistent in the academic as well as in the policy-making circles is the idea of the peasantry as an inward looking, homogeneous rural community, partly subsistence, guided by a communalistic ethos. This concept has slowly changed to an idea of a dynamic rural community subject to continued change, even pre-dating colonial interventions by internal as well as external factors. This change came largely as a product of long term detailed studies on local communities (Huskens & White, 1989; Tjondronegoro & WiradSi, 1984; Breman, 1997; White, 2002). From these studies the rural community comes out as much less homogeneous, already socially stratified from early times, with pronounced differences in land ownership and strong ties with markets and political centers.

The debate that originally focused on lowland rural communities has been extended to the uplands. Albeit further distanced from the centers of economy and politics, the upland rural communities show no less dynamism than their lowland neighbors (Hefner, 1990). The dichotomy of Java and outer-Java in the study of rural communities has been based on the differences of population structure, ecological setting and historical development. In relation to market stimulus, the dynamic nature of the farmers in the so-called outer islands has been put forward in contrast with the passive farmers in Java. The farmers of the outer island were described as progressive, quick in taking advantage of the economic stimulus brought by the colonial and market expansion (Geertz, 1963; Kahn, 1980).

More recently, village forest communities have been included in the debate. These studies refute the conventional wisdom, sponsored by the government, of village forest communities as governed by customary-rights, entirely subsistence, isolated communities and static as communities without history. The government uses the term *suku-suku terasing* or isolated tribes. Instead of that, new studies reveal that change is more the rule than exception. On the Kantu of Borneo, Dove explained land tenure, organization of production and agricultural changes in the face of political stability introduced by the expansion of colonial power in the beginning of the 20th century (Dove, 1988). From her studies of the uplands of Sulawesi, Li concludes that the process of change is not a recent phenomenon. As factors that promote changes in local institutions, Li suggested that factors such as famine, war, slavery triggered population mobility. Further, she points to market influences and the expansion of colonial authority (Li, 2002).

The author of this paper follows the line of this more critical approach of peasant studies, which have been characterized by the combination of synchronic and diachronic approach, where systems of adaptation, social and political institutions and cultural expressions are regarded as inseparable. The objective of this paper is to try an historical reconstruction on the institutions and social organizations related to people and environment interactions. This is to avoid what Brosius called an "ahistorization" of existing institutions and organizations (sited by Eghenter, 2000:354, see also Smith & Wishnie, 2000). This paper draws its information from a larger study conducted by the author on continuity and change in institutions and social organizations in natural resource management, in a village on the border of the Lore Lindu National Park

The data of the historical aspect were gathered in the form of oral history from key informants and partly from locally written material (Tokare, 1990). Land use maps were used to get more understanding of the land use, to stimulate the informant's memory and as a check and re-check instrument through the process of interview. Questionnaires were used to gather information and data on social organization, pattern of cooperation between farmers and the history of their land tenure. As far as possible, the researcher participated in meetings and discussion forums on land and forest management that involved the village community. For this paper, which is more or less a pure historical account, only the historical data are used.

3 Research Site

The village Wuasa is located in the sub-district (*kecamatan*) Lore-Utara that is part of the district (*Kabupaten*) Poso, Central-Sulawesi. In fact the current village Wuasa is the center of sub-district Lore-Utara, where the sub-district head (*Camat*) holds his office. It is situated in the Napu valley, 1100 meters above sea level, which forms the northeastern border of the Lore-Lindu National Park. Located 105 km. south of Palu, Wuasa is connected to the provincial capital by a good tar road. As will be dealt later, the road connection built just 17 years ago in 1982 has exerted a large influence on the population and economic live of the people in the valley. Another road - also of very resent date - connects Wuasa with the district capital city Poso, some 80 km. to the east (Sunito et.al.1999: 3) Part of Wuasa belongs to an old settlement that predates the Dutch period. The Wuasa as it is was built by the Dutch to harbor resettled people from indigenous villages such as Lamba and Lengaro. Before they were resettled, the villages Lamba and Lengaro were located on the grasslands to the south east of Wuasa. Remnants of those villages are still recognized by the local people.

According to the village statistical data of the year 2000, the village Wuasa covers a land area of 2839 hectares. The settlement proper comprises 54 hectares. Total inhabitant are 2318 persons in 518 households. The majority of the population call themselves the Pekurehua, a term used for village communities in the Napu valley. The adjoining valleys in the south are the Besoa or Behoa and the Bada with the Besoa and Bada people. In the north there is the Sedoa valley with the Sedoa or Tawailia people. The Pekurehua, Besoa, Bada and Sedoa / Tawailia are considered closely connected culturally and used to be called the Lore people (Tokare, 1990). The Pekurehua are Christians. According to the village data, there are 414 Muslims in Wuasa, which reflect a migrant population predominantly originating from South Sulawesi. The majority of these migrants came in the 1980s and 1990s with the improvement of the road connections. Recently there has been a stream of refugees from the Poso area, due to the ethno-religious conflict in that region.

On the north, east and south the village Wuasa is surrounded by rice fields. On the west side there is a relatively narrow strip of gardens, which separates the settlement from the hilly border of the Lore Lindu National Park. Further to the south there is the rolling grassland that extends to the hills that separate the Napu and Besoa valleys.

4 Adaptation System of the Direct Past

The system of adaptation of the direct past was mainly land-based, forest-based and animal husbandry. This adaptation system serves the subsistent as well as social and ritual needs and supports the monetary needs of the community. The land-based system can be divided in three systems: the shifting -cultivation; the permanent gardens; and the irrigated rice fields. The forest-based system was mainly the extraction of resin from Damar (*Agathis)* trees, the harvest of rattan and hunting. The most important animal husbandry was the water buffalo and the horse. Cattle, according to informants, came later in the 1920s, imported by the locals from the Poso region.

The **shifting cultivation** was probably the most important land-based system, an assumption based on the fact that the most elaborate rituals and celebrations were related to the cycle of shifting cultivation. It is also around shifting cultivation that an important organization of production had been developed. Shifting cultivation was done in land complexes with different vegetation coverage in the surrounding area of the village. People recognize at least 3 categories of vegetation cover: Pendulu (old forest), Lopo (secondary forest) and Halua (field or garden). People recognize land complexes by names that were given according to geographical characteristics or the brooks or rivers that run through them. Large parts of these land complexes are now not accessible for exploitation because they came under the category of protection forest and the Lore Lindu National Park. However, the names of the land complexes are still alive in people's memory. Under the shifting cultivation, land was used only for 1 harvest and than left fallow for 3-4 years. The main crop was rice, with maize, different tuber crops and some vegetables. Perennials or tree crops were not planted in the shifting cultivation fields, which had to do with the land tenure system, as will be explained at the end of this section. Heavy fencing of the land, which was done cooperatively, was important against wild pigs. When the Dutch introduced coffee, it was first planted in the forest, without clearing the trees. Rows of openings in the under growth were made and coffee was planted. These old coffee gardens can still be observed today.

The **permanent gardens** were located closer to the village. They can be recognized by the earthen walls or ditches that surround them to prevent the obstruction of animals. A permanent garden could be an end process of shifting cultivation. Permanent gardens were planted with seasonal crops mixed with tree crops such as bananas and later coffee. People in Napu have already planted vegetables such as potato and red beans for a long time, probably from the Dutch time. Information is still lacking on the origin of this cultivation and its role in the household as well in the local economy.

The **irrigated rice fields** - *sawah* or in the native language *bonde-keuwai* - already existed before the Dutch time. The old rice field complex of Wuasa is located at the N'talosi, at the southeastern part of the village. The other rice fields that can be observed now were developed as a government program starting in 1970.

Tapping *Agathis* resin was one of the main sources of cash for many in Wuasa and other villages in Napu. The trees were not planted and can be found abundantly in the now national park. In the Dutch colonial time the trees were taxed and for that reason were registered. After more than a decade of not being tapped because lack of demand, there is now a new beginning of interest for this product. **Rattan** was another forest product that was harvested from the forest, although more for domestic use than the market, as is the case today.

The most important domesticated animals, each with its own function and value, were the **water buffalo, the horse and the pig**. The water buffalo played and still plays to a lesser degree an important role in almost every aspect of life. In relation to agriculture, it is used for land preparation in the rice fields. Currently people use mainly hand tractors. As far as the memory of the present people goes, the horse was said to be used only to transport humans, and then only on social occasions. It was more an object of prestige than an economic asset. Only after the introduction of the idea and technique of the packhorse by the immigrants from South Sulawesi late in the first half of the 20th century, the horse acquired an economic value. In Wuasa today, one will only occasionally see horses.

The adaptation system harbors a multitude of **tenure systems** that can be regarded as a continuum with individual based ownership at one extreme and communal tenure at the other end. The land where houses were built within the settlement – the *kintal* – always has been and still is regarded as the private property of individual ownership. The permanent gardens were regarded as in private ownership too, with the condition that it is still used by the owner. A permanent garden that had been neglected for a long time was free to be obtained by others. For the shifting cultivation complexes, in contrast, there was the communal tenure system. The traditional norm did not attached special rights to persons that pioneered the clearing of a patch of forest. Shifting cultivation land abandoned for fallow would turn automatically into communal land, free to every one to made use of. Every member of the community was free to clear and use forestland for cultivation or to use abandoned agricultural land, within the boundary of the village. The coffee gardens in the forest have got a form of private ownership status, in contrast to the shifting cultivation land. Even then, forest coffee gardens would be turned again into the public sphere when they were neglected for a long time. Damar (*Agathis)*trees have also got a form of private ownership status, fostered by the Dutch administrations that levied tax from the product. It is even said that the

Dutch nailed license badges on exploited Damar trees. The direct past of the Pekurehua of the Napu valley had seen a multitude of land tenure systems, a combined product of local system, market influences and state intervention.

5 The Robo Social Organization

5.1 Definition of Robo

The term *Robo* has different meanings and connotations according to the context of the communication. With the term *Robo*, people will first refer to the dry field and the irrigated rice field complexes of their village, where people in the past had their shifting cultivation fields. People will refer to field complexes such as Pembala, Tohaka and Salunamangge on the west side of the village Wuasa and the Popahudua and Pewuloa on the north side, where currently the bulk of the irrigated rice fields of Wuasa are located. On the east part of the village there are the land complexes Tambua and Dodonga, bordering the river Tambua (which in the down stream became the Lariang) and a small stream called Dodonga. On the southern part of the village there are among others the Pupulangka, Kabureana and Malewuko.

The second meaning refers to the *Robo* as the complex of land cleared and cultivated in a cooperative way by a large number of households. It has to be a large size field complex to be called a *Robo.* A field complex cultivated by 10 households is still not a *Robo*. It requires 30 to 40 households, with an average of 0,5 ha. each, if a sizable field complex is to be called a *Robo*. The name of the land complexes or the *Robos* are borrowed from the brooks and rivers on the site. Or as with Kabureana, from its natural characteristics: land that is always flooded (by the Pembala river).

A third meaning, not often used, is the hierarchical connotation of *Robo*. *Robo,* at its most overarching level, refers to the territory of an ethnic group or a tribal group. Next, there is the Robo as territory of a village or in the local language is called *winua*. At its lowest level, *Robo* refers to the complexes of fields cultivated by a large group of households. In the first three meanings, *Robo* is given a spatial connotation. That is, *Robo* as locations with clear boundaries, as complexes of agriculture land and *Robo* as territory of a community.

In its fourth meaning, *Robo* is understood as a social organization of a large number of households that decided cooperatively to clear and cultivate a complex of land. There were always a number of *Robos* at any time. And the *Robo* was not a permanent social organization. After the harvest was over, many *Robo* organizations would be dismantled. New *Robos* would be called into life when the need exists at the start of the new planting season.

Somewhere in the 1950s and 1960s the term Robo was replaced with the term *Partei,* from the Dutch word *partij* (party) as part of the influence from the Minahasa (North Sulawesi) brought by the many teachers and government functionaries coming from Minahasa. People thereafter talk of the *Partei* for the social organization and *Robo* for the complex of land or field.

5.2 Robo Membership and Leadership

A *Robo* organization was formed on the initiatives of a person who – in most cases - would become the leader of the *Robo*. The person announced his plan to established new fields in a particular location. Households would join the effort and with that a *Robo* was formed. Every planting season there would be a number of *Robos* active in a village such as Wuasa. Each *Robo* cleared and cultivated land in one of the land complexes in the surroundings of Wuasa. According to all informants the *Robo* was an open organization. It was not based on kinship, or other social category. Households were free to join whichever *Robo* initiative taker they like. This was also in congruence with the temporary character of the organization. However, looking closer at this social organization, some pattern in the membership can be found.

The oldest key informants in Wuasa are the last ones that experienced membership of *Robo* social organizations as full adults. One of them was even one of the last existing *Robo* ex-leaders. According these key informants it seems that in many cases the nucleus or even the majority members of a *Robo* were close kin members of the *Robo* leader. Those were the people on whom somebody relied when help was needed. However, people tend to laugh when confronted with the question of whether kinship was a requirement for membership of a *Robo*. In a village community like Wuasa almost all people were and still are somehow related, through blood or marriage. The Pekurehua does not have special regulations on marriage, except the incest taboo and the prohibition of marriage to the category of first cousins. This assertion can be regarded as supporting the notion of *Robo* as an organization with an open membership. The information across all the informants is consistent with the notion that it was free for every person or household to join whichever *Robo* there was. However the fact that close kin members form the nucleus of the *Robo* organization informs us that there was a degree of permanence in the membership of a *Robo*. This statement will be supported by the characteristic of the second category of members of the *Robo*.

The second category members of a *Robo* were individuals and households in a dependent relationship with the *Robo* leader. Dependent relationship between members of the village communities was created through indebtedness. The most important sources of indebtedness were, and still are, the festivities and rituals that every household has to perform at least as part of the life-cycle of its members. In the Pekurehua tradition, every important event – marriages, funerals, important phases in agriculture - was celebrated with the sacrifice and the slaughter of water buffalo and pigs. Bride price as well as fines are still calculated in numbers of water buffalo. Households that do not posses enough water buffalo or pigs have to borrow from elsewhere to meet their social obligations, and with that a relation of indebtedness is created. Individuals who possess large herds of water buffalo were in many cases also leaders of *Robos*. People and households indebted to a water-buffalo rich person, who was also by chance leader of a *Robo*, were under pressure of obligation to joint the *Robo*.

The above pattern of membership relativate the notion of the open character of organization for the *Robo*. There were no specific norms that regulate the recruitment of members, nor barricades to the entrance of a *Robo*. However, an important part of the membership of the *Robo* came from both the category of kin group and dependent farmers.

By their social-economic characteristics, the leaders of *Robos* were always individuals with important positions in their community. They were the ones in possession of large numbers of water buffalo, and with that also the ones that had dependent households as a source of labor and as a circle of loyal supporters. Let us look at the roles that a *Robo* leader was supposed to fulfill.

First, the *Robo* leader took the initiative in announcing its plan to clear new agricultural land and the location of it. The decision was necessarily a product of deliberations between other *Robo* leaders in the same village, so as not to hinder one and the other. In principle all the land complexes, which by shifting cultivation activities gradually become established as agricultural land, were available for everyone. In reality, leaders of *Robos* and individual households had their preferences. More of this will be dealt in the next section. Suffice to say that there was a tendency for households as well as *Robos* leaders to concentrate their agricultural activities in limited numbers of land complexes. In one case, a land complex became identical with a *Robo* leader, because this particular person operated for a long period of time, although not exclusively, inside that one particular land complex.

Second, the *Robo* leader represented its *Robo* in dealings with the village administration. First of all in coordinating the activities of the different *Robos* in the village that operates in the same season. Each *Robo* needed the assistance of other *Robos* in clearing and preparing the fields. The village head, in cooperation with the *Robo* leaders, coordinated the mutual cooperation between *Robos* such that no one *Robo* would disadvantaged and thus lag behind in planting activity. In addition, the village administration used the *Robo* leaders as bridges to channel information to the population on village matters and government regulation and programs. **Third**, the *Robo* leader took care that proper attention was given to the rituals prescribed by the tradition to guard against evil spirits and guarantee the success of the harvest. **Fourth**, the *Robo* leader mobilized and coordinated his members in activities such as clearing land, preparing land, planting and at the end harvest the main products of the land. **Fifth**, one of the most important roles of the *Robo* leader was to fulfill the conditions to enable the mutual cooperation between *Robos*. To be able to invite other *Robos* to give a hand in the heavy work of land clearing, the receiving *Robo* had to prepare water buffalo, pigs and other requirements for consumption and gifts. In fulfilling those conditions the *Robo* leader was expected to carry the largest burden. This was also true in the case of preparing animal offerings and consumption for the rituals related to the clearing of land and for guaranteeing good harvest. This was the reason why the *Robo* leaders were chosen or came from the category people that possessed large numbers of water buffalo.

There were no procedures for the selection of the *Robo* leader. People used to say that a *Robo* leader was not chosen by the members, but established himself due to personal qualities. A person grew into the leadership because he fulfilled the conditions for that position. A *Robo* leader had to fulfill certain characteristics. As already mentioned, the possession of large numbers of water buffalo was a precondition. That was already a sign of a strong character, and a powerful means to establish influence in the economic and social spheres. Beside this economic precondition, for the leadership of a *Robo* there were still other qualities needed. One was the ability to gain trust from others. This was an important quality, for the person represented its members toward the outside world. He would also be a focal point in conflict resolution within the *Robo* as well as between *Robos*. Another related quality was the ability to bind the loyalty of the *Robo* members and to be easily accepted by others outside the *Robo.* A quality that relates directly with the objective of the Robo was the ability to coordinate people for the whole process of agricultural work.

There are not many ex-*Robo* leaders left in Wuasa. Umana Wellem, with his white hair and short beard, is probably one of the last *Robo* leaders still alive and still active as a farmer in Wuasa. One can still see him and his wife, both on horseback, going to their field in the morning and back to their big house in the village late in the afternoon. Only very few people in Wuasa - and Lore-Utara in general – still use the horse as personal transportation. Imposing horns of water buffalo adorn the walls of the old house of Umana Wellem, old symbols of wealth and influence. Tambua, one of the large land complexes of Wuasa, was also known as *Robo* Umana Wellem, indicating the length of the period Umana Wellem and his *Robo* practiced shifting cultivation in that particular land complex.

5.3 Robo as a Mutual-Help Organization

As a rule, shifting cultivation was conducted in the context of the *Robo*. Once the location was chosen, each member of the *Robo* decided for himself the size of land to be cultivated. Thereafter, the members of the *Robo* as a group cleared the land, put borders between individual plots of land, building together a system of fences to guard against wild pigs and as a group doing the planting. Weeding and other activities related to the tending of crops were the work of the individual households. It was in relation to this that individual households cultivated land in accordance with the available household labor. That was not the case with cattle rich household, as will be dealt later in this section. Harvesting was again a collective activity. As in the case with land clearing and planting, other *Robos* were invited in the harvest to lighten and - as will be dealt later - to put more joy in the work.

In other words, the *Robo* provided the needed labor for its member households. Through internally mobilizing the members of the *Robo*, and externally by inviting other *Robos* to share their labor. However, it is difficult to be content with this explanation. The need for labor cannot be the sole answer to the existence of the *Robo.* Other shifting cultivation communities in the archipelago survive perfectly without. Nor is the argument of security satisfying, for the same reason. Part of the answer will reveal itself in the course of analysis. Whatever the argument behind the *Robo* is, for the individual households the *Robo* is there to provide solutions for its multifaceted needs.

5.4 Robo as Center of Social and Ritual Life

Shifting cultivation was by far the most important land management system until a combination of factors - land certification, the opening of the region, the increase of the market value of land and government program for irrigated rice fields - ended its practice around the 1970s. The *Robo* was the social organization par excellence where work- and social-relations - between members and between *Robos* - took place through the whole shifting cultivation cycle. The *Robo* was also the context where the rituals accompanying the agricultural cycle took place. For a large part of the time people lived in their field huts, working and watching over their fields, with *Robo* members as their neighbors. Even government messages were channeled to the people through the *Robo* leader. People went back to the village only on Sundays, to attend the church services. The many aspects of life and rituals that the *Robo* represents will be discussed.

In the past people live much of their life in their field huts, alternating in one of the agriculture land complexes of Wuasa. The middle-aged individual in Wuasa can still recall the time where they went to school from their field huts. As already mentioned, all the important phases of the shifting cultivation cycle were accomplished through cooperation between members of the *Robo* and other *Robos*. Agricultural work – whether clearing land, planting, building fences or harvesting – was always a large group activity. Especially the able bodied members, men and women, were constantly on the move from one *Robo* to the other as part of the exchange of labor. As matter of course other activities flourish too on these occasions. As one informant recalls it: those times of exchange of labor were very exciting, going from one *Robo* to the other as if from one festivity to the other, there was always lavish food wherever people went. Especially for the young people, it was a time to show off their strength and abilities, but also a time to indulge in festivities and love affairs. This period of work and festivities could last for two to three weeks. A quieter time with largely routine activities followed where each household took care of their own crops. Until the harvest came, which marked the end of the agriculture season, with its own festivities. There were special events connected to certain phases of the shifting cultivation cycle:

- Before a *Robo* started its planting activity, a dance festivity was organized taking place on the land complex that had just been cleared and prepared for cultivation. Members of other *Robos* were invited, as they will assist in the planting activity. The young men and women dance the *Dondi – Modondi* – that is still very popular nowadays under the name *Dero*. The older people dance the slower and elegant *Dengki*.
- After a *Robo* had completed planting its land complex and the last seed had been put in the ground, people expressed their joy by making noise with their

digging sticks or whatever tools they had in their hands. This expression of joy would be followed by a special event called *Mombe' buri,* where people smeared each other with charcoal, which was easy to find on the ground from the burning phase. Men, women, young and old were all involved in this joyful event. The young used this event to go after the ones they had special feelings for, express their affection and try to figure out whether their approach would be answered positively by the other side. Another event related to the end of the planting was the *Baku Binti,* a kind of sport where two contenders kicked each other's calf until one gives up to the winner. There was neither anger nor hatred involved; both parties congratulated each other after the bout.

- A ritual called *Mobahi* would be conducted after weeding activities were done. It was a ritual conducted as an expression of gratitude to the spirits that inhibit the tools that people had used for weeding. Small packages of glutinous rice would be placed with the tools. The same ritual would be conducted too after the harvest, this time involving all the tools that people had used in the forego-ing season.
- Each *Robo* celebrates the harvest in a ritual annex festivity called *Mangore* where water buffalo were slaughtered and for the first time the new harvested rice consumed. Other *Robos* were invited in this ritual annex festivity.
- The whole village holds the grand harvest festivity, the *Mande Parewou*, after all *Robos* have finished their harvest.

In comparison, the other two land use systems - the permanent gardens and the irrigated rice cultivation - do not involve the same elaborate rituals and social events. There are a number of reasons behind these differences. First, the perma-nent gardens and the irrigated rice cultivation in the past did not fulfill the same importance as the shifting cultivation as producer of consumption products. Sec-ond, although it is still hypothetical, is the argument that, as shifting cultivation is a much older land-use system, it is already more integrated in the social, political and religious system of the Pekurehua. Third, of the three-land use systems, only the shifting cultivation is thoroughly managed by collective action.

5.5 The Role of the Robo in Strengthening the Social Structure

The *Robo* shapes the condition where the *Robo* leader could use the labor of the members for their own interest. As already mentioned, there were two dominant categories of *Robo* membership. The category kin group, and the households in debt relation with the *Robo* leader. Within the *Robo* context, however, all the members become indebted to the leader, because he has provided the animals for the rituals and has invited other *Robos* to share their labor. In this condition all the members alike were socially obliged to give in return their labor for the benefit of the *Robo* leader. Based on this indebtedness relation, the Robo leader could made use of the members' labor to clear, cultivate and harvest more land than his own household labor permits. According to informants average households cultivated between 50 and 75 are (0.5 – 0.75 hectares), whereas the *Robo* leader had the possibility to cultivate around 1.5 hectares. This condition permitted the *Robo* leader to produce agriculture surpluses. The *Robo* had become one of the mechanisms where dependent relationships enabled the leader to produce more agricultural products, which eventually strengthens the material base of the social status already enjoyed by the *Robo* leader.

Looking back to the *Robo*, the more politically conscious key informants perceived the *Robo* system as an unjust labor exchange system or an asymmetrical exchange system where the Robo leader received more labor from the members than the other way round. These key informants see the change from the *Robo* to the *Mapalus* system as positive. The *Mapalus* is a loose and open social organization of 8-20 members for mutual help, made up of close friends or kin group. It is especially popular among young people and new households. Besides mutual help in farm activity, the group hires out its labor to whoever needs labor for agricultural activities. The *Mapalus* social organization came up in the 70s and was introduced by government officials and teachers originating from Minahasa in North Sulawesi, where this form of social organization already has a long history.

6 The Water Buffalo

6.1 Tenure and System of Husbandry

For people entering the Napu valley at the close of the 20th century, it is difficult to imagine that until approximately 40 years ago the currently almost empty grasslands were littered with free grazing deer and water buffalo. Until then, it was not without danger walking to other villages in the evening because of the free wandering water buffalo. Inside the settlement, cows, horses and pigs were also not kept within enclosures. Accordingly, gardens of houses were not used for cultivation. The plants would not survive the pillage of the animals. In the mid 1950s the government urged people to put the animals within enclosures in the grassland, releasing the settlement from free wandering animals, giving the settlement the right condition to develop as a settlement as perceived by the government. Cattle were added to the composition of animals traditionally domesticated by the people, only after the 1920s. In contrast to the belief that it was part of the Dutch program, people held that cattle were imported from the coastal area of Poso by local people.

The traditional husbandry system of water buffalo was an extensive one. The animals were not put into corrals, but instead they were set free in certain locations in the grassland and in the forest margins. In more or less permanent herds the buffalo would wander in the grassland; they took shelter in the forest margins during the hottest part of the day and grassed on the riversides. Each village had their traditional grazing land for their animals. The grazing land of Wuasa was the location called Banga and later in Petandua, the village Kadua in Hawane, the village Alitupu in Ga, the villages Maholo and Winowanga in Habingka. Villages in the Napu valley rank differently in term of the numbers of water buffalo they possess. Sedoa and Alitupu were famous in the past for their large numbers of buffalo. It was not uncommon that herds of these animals upon request were sent to other villages to be used for land preparation. This system of husbandry had the effect that individual animals could astray from its flock and grow wild. For the owners this means a loss, but for the young adults and adventurous it was a chance to generate money and to prove their strength and bravery. They look for these half wild buffalo, drive them into a temporary constructed corral, tame them and make arrangements with the owner to divide the capture between them. It is an enterprise not without danger, which produces income in the form of buffalo and recognition of bravery among the peer group. For the young adults it is a means to start their own herd of water buffalo.

The ownership of water buffalo in the past was more or less common. But there were vast differences in the number of possessions. The nobles by far possessed the greater part of the buffalo. Young men, members of dependent households of the nobles, got the responsibility to observe the movement of the herds. They got compensations for their job, in the form of one calf a year. The common people put their small numbers of buffalo in to the large herds of the nobles. People recognized their animals from the ear incisions that were made while the animals were young. The Pekurehua developed a range of incision types, and by combining them in one ear or both ears there would be no misunderstanding of ownership. On top of that people recognized their animals also from the form of their horns and the shape of their bodies. Moreover, buffalo from one owner tended to flock into one herd, the calves and the young animals following the adults. Which made it easy to round them up, in case there was the need to use them.

According to the present cattle keeper of Wuasa, it is now more difficult to catch the animals because they are now much wilder than in the past when they were wandering freely. In the past the animals would come to lick salt stones brought by the keepers. Now it is not advisable to approach the animals on foot. The keeper and his helpers have to use horses to catch them. According to the cattle keeper this phenomenon is the consequence of indiscriminate hunting in the 1970s when better roads began to attract people from outside and raise the price of dried buffalo meat. For both the nobles and the common people, the selling of dried buffalo meat has become a source of easy money. Some people of the village elite organize and sell their knowledge of the surrounding grasslands to guide deer and buffalo hunting parties for people from the big city in four-wheel-drive vehicles. The hunting decimates the deer and buffalo from the environment in a short time. The rest of the animals become shy of people, difficult and dangerous to approach.

People acquire water buffalo in different ways. In general the nobles get their buffalo through inheritance. At the time when this study was conducted, in Wuasa there are only around 50 buffalo left, owned by 4 families. Two old noble families are the biggest owners. One of them is a known *Robo* leader. The common people get their buffalo through buying, bride price, penalties from others, etc. In the case of the present cattle keeper of Wuasa, he has the right to receive one new born animal a year. The person will receive money when there is no new born animal in a particular year.

In 1985 the government set up a cattle-breeding program for Wuasa. A 1000 hectares grassland was set-aside in Petandua for this purpose. The plan was to distribute Balinese cows among the household through a credit scheme. The plan was never executed. But the grassland is there and all the cattle of Wuasa are there too. The remaining water buffalo are still wandering inside and outside the Petandua grassland.

6.2 The Water Buffalo in the Production System

The water buffalo was used to prepare the irrigated rice fields for planting. Land preparation was done through a technique called *murambai* (an imported term often used is *paruja*). In this technique circa twenty water buffalo were let into a rice field in a compact herd by herdsman's, the animals, then driven in circles while trampling the soil with their heavy hooves. It is difficult to assume that the Pekurehua as late as the 20th century are not aware of the existence of the plough. A plausible explanation is the adaptation to local conditions. The original rice fields of Wuasa are of the deep type. People will sink in the muddy soil to their waist, which makes the use of the plough inefficient. Another informant stated that in the case of a newly constructed rice field, people resort to the *murambai* technique for land preparation, because there are still too many tree roots and tree stumps in the soil that can be damaging for a plough. The *murambai* technique predominated until the beginning of 1980s, when new irrigated rice fields were constructed under a government program. The introduction of the hand-tractors for ploughing the land was part of the same program. In a short time, due to the diminishing of the buffalo population, hand-tractors have replaced the *murambai*. However, the original rice field of Wuasa – the N'talosi – is to small to explain the existence of such a large number of water buffalo. Obviously the water buffalo fulfills also other roles outside agriculture.

6.3 The Water Buffalo in the Social and Ritual Life

Water buffalo play an important role in most aspects of live of the Pekurehua. Besides its role as draught animal, the water buffalo is also indispensable in the social and ritual aspects of agriculture. Water buffalo fulfill also an important role in the sphere of the life cycle. Furthermore, traditional sanctions and punishments are considered in terms of water buffalo. In other words, water buffalo enjoy an all-important place in the culture and society of the Pekurehua, in the past and in lesser degree at present. The following description of the role of the water buffalo in the life and culture of the Pekurehua, has to be perceived as a phenomenon from the direct past, unless stated differently.

Bride price was counted in numbers of water buffalo. The number of water buffalo involved suggests the social status of the parties involved in the marriage. The Dutch colonial government already decreased the allowed numbers of animals for bride price. After independence it was further decreased. The number of water buffalo for bride price in the circles of the nobles (*Tuana*) of the Napu valley is between 8 to 12 animals; while in the circles of the commoners (*Kabul/Galara*) it is between 4 to 6 animals; and in the circles of the under class or "slave" (*Hawi*) it is two animals. In the former days, a bride price of 12 water buffalo was expected to be accompanied by a slave (Tokare, 1990:24), which is an interesting detail pointing to the fact that the Pekurehua in the period pre-dating the Dutch already was a highly hierarchical society. The same source added that at present (in the 1990s) such a high bride price is not realistic due to the absence of water buffalo. Moreover the government has already lowered the maximum to 7 to 8 animals and the lowest between 5 to 6 animals (ibid.24). The community reacts to the absent of water buffalo by inventing new norms. According to these new norms, the bride price is still calculated in numbers of water buffalo; however, in reality, other goods and money of much less value will represent most of the animals. An example from Bada, a valley south from Napu, shows how a symbolic bride price of 6 buffalo is calculated in reality: (1) a pair of water buffalo consist of a male and female, (2) an amount of money that symbolizes the value of three water buffalo, (3) one water buffalo, the delivery of which can be postponed, (4) one pig, (5) a local traditional sword, and (6) seven large machetes (Ibid.:31).

Water buffalo also played and still play an important role in the funeral ritual, especially when it concerns the noble social category. During the funeral process that can take many days, many buffalo, but also pigs, were slaughtered for the consumption of the guests, not only from the village of the deceased, but also from the surrounding villages.

Tradition prescribes purification rituals through the sacrifice of water buffalo as punishment against the offending of important taboos, such as taboos around marriage and adultery. It is believed that the breaking of those taboos will endanger the whole community. In the case of breaking the marriage taboos – such as marriage between too closely connected kin members - the culprits had to undergone certain rituals where the sacrifice of water buffalo are the center of it. In one ritual, the offender had to sacrifice one white buffalo or a pig of the same color. While the animal is slaughtered in the river, the offender and his family took a ritual bath down stream. The sin would be purified by the combined power of water and blood of the sacrificed animal.

Local customary law (*hukum adat*) calculates penalties as punishment for transgressing norms in terms of numbers of water buffalo. In a recent case of suspected adultery in the research village, the suspected man had to pay three cows to the girl's family within three months, with the condition that he will pay another three water buffalo, in case the girl gets pregnant.

Offending the *Ombo* is still considered as endangering the whole community. The *Ombo* is a fixed period of mourning connected with the death of a noble or important person. In which case a wide circle around the house of the deceased will be decided, within which boundary it is strictly forbidden for the members of the community to perform many of the daily activities. Such as placing fish traps, working in the field, making noise, etc. The penalty for offending against the *Ombo* is to pay one water buffalo, which will be consumed by the public at the close of the *Ombo* period.

The involvement of the water buffalo in the production system and in ritual life represents a whole process of its own, which integrates elements of cooperation, sports, competition and ritual. This complex phenomenon is partly a direct consequence of the Pekurehua water buffalo husbandry system described earlier.

The wet rice-planting season started with the rounding up of the water buffalo that would be used for the *murambai/paruja*, the land preparation. It was an activity where the whole village participates. The main teams that did the rounding up of the animals at the grassland consisted of the rice field owners. After rounding up the buffalo, the animals were driven to the corrals of the owners. The whole village participated in the process of driving the water buffalo from the grassland to the location where the corrals were constructed near the village. Young and old hand in hand and shouting made a human fence to drove the buffalo into the corrals. Early the next day the buffalo were set free in the vicinity of the corrals to forage for food. Then the young and the old animals would be set apart, and the rest would be driven to the rice fields for the *murambai*. The group of herdsman and the household members that own the rice field, children and adults alike, join together holding the circa 20 buffalo in a compact group and go around the rice field, at the same time stamping with their powerful hooves the deep muddy soil. Holding sticks and tree branches, the crowd wave and shout to spur the buffalo to move faster. In the afternoon, the buffalo are driven back again into their corral. If all the rice fields are already covered by the *murambai*, then the buffalo will be driven back to the grassland.

The relation between the owner of the water buffalo and the members of the community that make use of the animals was much more than a single stranded contractual relation between owner and user.

In the case of *murambai*, the owner of the water buffalo practically stayed passive as a bystander. The water buffalo for the period of *murambai* became more or less public goods, subject to village-large activity and rituals. At the end, the owner of the water buffalo would receive compensation, one tenth of the harvest. People called the part of the harvest for the buffalo owner *biti beula* or the legs of the water buffalo. An interesting aspect of the water buffalo in its roles in the production, social and ritual aspects, was that it always provoked a community-large activity. There are a couple of factors responsible for that. First, the physical character of the water buffalo and its extensive system of husbandry make it necessary for people to act together in order to make use of the animals. Second, the nature of the technology people use in preparing land and the role of the water buffalo. Third, the water buffalo's role as sacrifice animal and/or for consumption purposes. The end result is that water buffalo might be an agent of community integration.

6.4 The Role in Maintaining the Social Structure

The *Pekurehua* was a more hierarchically structured society in the past than at present. There were hierarchical social categories like the *Tuana* or the nobles, *Galara* the free citizen and the *Hawi* the slave, legitimized by local mythology and genealogy. Despite the existence of these hierarchical social categories, there were no privileges associated with the high-ranking status. On the other hand, there were no discriminatory norms against the lower social class. Even more striking, the ownership of important resources, such as land and domesticated animals, was not restricted to particular social classes. Moreover, the greater part of the land was managed as a communal land; even the trees and plants were not subject to private ownership after the user abandoned the land. That is why before commercial tree products like coffee and cocoa were introduced and private land ownership was implemented, the *Pekurehua* never planted useful- or commercial-trees on their shifting cultivation land.

To the elite of the *Pekurehua* the role of the water buffalo was essential for the mobilization of labor to clear and cultivate more land than their own household labor permitted, a mechanism through which the elite could accumulate economic surpluses to support their social status. The overall importance of the water buffalo in the lives of the *Pekurehua* made it easy for the nobles -who were the largest owners of water buffalo - to create debt relations with other members of the community. Everybody, the rich as well the poor, will some day need water buffalo for different reasons. People have to come into debt relations with the nobles, if they cannot generate the needed water buffalo from own resources.

The question is where the nobles got their water buffalo. Why were there large differences in the possession of water buffalo in a system characterized by open access to of all of its resources? One of the answers is inheritance. The large herds of water buffalo were inherited from the former generations. And the herds reproduce themselves continuously. This is not a strong argument, although the only one for the moment. Key informants report on animal diseases that have swept through the valley at regular periods and reduced the size of buffalo herds substantially. It would take years before the former situation is restored, a situation that can ruin the noble's monopoly of water buffalo. In this case, one can imagine that a situation is created where everybody, nobles, commoners and slaves have to start from the same line. Will that create a changing of the guard in the community? An internal social re-structuring? Or were the social status of the nobles and the old debt relations strong enough to give the nobles the edge over the other social categories? On the other hand, one has to consider also the practice of plundering villages of neighboring regions, of which the Pekurehua were famous in the past. Plundering villages in the past could be seen as a solution for difficult times, where the nobles could take the leadership role and hence took the most benefit of it, including restocking their water buffalo.

Is it possible for the commoners to accumulate water buffalo? Theoretically, there were many roads open for people from any social category to accumulate water buffalo. We have already mentioned going after half wild water buffalo and sharing the result with the original owners. Youngsters can start their own flock by working as herdsman, and received calves as compensation for their work. In reality, however, there were obstacles confronting a commoner with ambitions to become large owner of water buffalo. There were epidemics, mentioned earlier, that were surely more damaging for the small beginners that the large owner. More interesting is the information that the nobles in the past took conscious steps to prevent competition from the lower classes. A common practice was to sabotage the growth process of the herds from the lower class by different means. For example, by borrowing the most productive water buffalo(s) and severely delaying the return of it/them. Another common practice was through magic that prevented the reproduction of female offspring. Apparently, through these strategies the nobles were successful in holding their monopoly on the large-scale possession of water buffalo, which is one of the pillars upon which the nobles maintain their dominant position within the community.

7 The End of the Robo and Water Buffalo

Starting in the 1960s the Indonesian government introduced fundamental changes in the land tenure system. Land certification based on individual ownership was launched. The gradually improvement of road connections with centers such as Palu and Poso triggered the inflow of land hungry migrants and the increase of market economy. Tree cash crops such as coffee and later cocoa were introduced and planted in the formerly shifting cultivation land. Shifting cultivation came to an end, and with it the *Robo* social organization. The improvement of the roads increased also the market value of water buffalo, in live form as well as dry meat. This became a source of easy money for the buffalo owners. In no time the water buffalo almost disappeared from the Napu valley. In the 1970s much of the low land bordering the settlement was transformed into irrigated rice fields. In many cases people gave their water buffalo in exchange for hand tractors, another factor that contributed to the decrease of the population of water buffalo. The hand tractors became the dominant tool for land preparation in the irrigated rice fields.

New social organizations were introduced in the field of resource management. The *Mapalus* is one of these social organization, a purely mutual help organization that consists of a small number of people related by blood, marriage, friendship or by proximity of farmland. Young people form *Mapalus* groups as a way to earn money, by offering their labor to any one that needs a work force for their field. During planting season many women *Mapalus* groups come up for mutual help and offer their labor in turn, for money to people outside the group. Extension officers initiated the farmers' groups/ *kelompok tani.* It was introduced as part of the expansion and intensification of irrigated rice. It has a formal status, with a management structure and permanent members. Their activities include the coordination of their members and the technical aspects of the production process of the irrigated rice fields. However, of the five farmers groups that were established mid 1980s only one is still active.

8 Discussion

The above picture of the mode of subsistence, tenure system and social organization is a picture of the institutions in the village Wuasa during the Dutch time, which endured the Republican time until it died out in the decade of 1960-1970. Wuasa itself was partly a creation of the Dutch colonial administration at the start of the 20th century. There is no clear description of the agriculture system, land tenure, and other social and cultural institutions of the Pekurehua from the time predating the Dutch intervention, a time when people still lived in their original fortified villages, located on the open grassland of the valley. The Dutch subsequently resettled part of these people into mixed villages like Wuasa. The consequence of this is that the above-reconstructed Pekurehua institutions cannot be regarded as the original or pristine ethnographic situation (Marvin Harris, 1979:536-539). It is quite possible that different tenure systems, and other institutions existed at the time before the Dutch entered the Napu valley. As reported by Dove on the Kantu (ibid.) for example, the open access or the communal property right system cannot be held as an earlier form of tenure system in an assumed unilineal development to a private property rights system.

The above reconstruction of the direct past gives a picture of an integrated whole, with mechanisms for production, as well as mechanisms for the maintenance of the social structure, which also provide the avenues to establish social relationships and enable the individual to prove himself in the many aspects of life. Listening to the elaborations from the key informants on the *Robo* and the many aspects of the water buffalo, one cannot deny the impression of the existence of a colorful life, in a regular cycle according to the agriculture seasons. From a cultural-ecological perspective, the *robo* and the buffalo can be regarded as what Steward called the "culture core" of a society.

The latest gulf of changes, starting in 1960s, signaled the end of shifting cultivation and, with it, the *Robo*. The environment became fragmented in separate fields under private ownership, each more or less with its own land-use agenda. New organizations that have been introduced since then, are of the specialized types, such as the *mapalus* and the formal farmers' groups for irrigated rice farmers, which focus their activities on the process of production in their members fields alone. These organizations are composed of farmers that have interest only in their particular piece of land.

Other organizations are supposed to oversee the community as a whole, such as the *adat* council, the village administration and the new Village Representative Body also known as *Badan Perwakilan Desa*. However, since the 1970s the Adat Council has already been reduced to a body that has to preserve the traditional custom on marriage and resolve marriage and divorce related disputes. Only recently motivated by the autonomy issue there are maneuvers on the side of the Adat Council to get a bigger say in public decisions. The village administration has been always in the hands of what had been the nobles in the past. The powerful position of the village administration has been used – in different degrees – to serve the self-interest of the position holders. The Village Representative Body is still in its infancy, having being installed just over one year ago. The composition of the body has been engineered to represent the different social political categories of the community. There are representative of the dominant political parties, from the Adat Council, from the Christians and Moslems, from young people and women. Some members of this body complain of the difficulty of monitoring the activity of the village administration. The different churches represented in the village focus their activity exclusively on religious matters. All the groups organized by the church - young people, the women, the elders - on the neighborhood level never venture outside the pure religious activity. Only near Christmas and New Year some of these groups organize *mapalus* groups to generate money for the yearly festivities.

What can one learn from this historical perspective, for the understanding of the current conditions, especially in regard to local natural environment management. Looking from this historical perspective, it is advisable to be cautious in attaching categories as traditional or indigenous to institutions and social organizations seemingly inherited from the past, such as the *Mapalus* mutual help groups. The introduction of this new social organization came more or less at the same time with the ending both of the shifting cultivation and the *Robo* social organization of the production. With it most of the cultural and religious tradition that surrounded it also came to an end.

The *Robo* can be regarded as the platform where common perceptions and interests of the community towards its natural resources were formed and where social and cultural expressions took place. The *Robo* and other institutions of the direct past, such as the communal land tenure system, the water buffalo and the animal husbandry system provided the framework through which each fully fledged member of the community oversaw and experienced the whole environment, that traditionally belonged to the community, as an integral living space and resource. From handed down stories and their own experience each member of the community had the knowledge of every part of that environment, and saw the fingerprints of the ancestors and their own labor all over the environment. From the point of view of the household, the whole living space that traditionally belongs to the community is the subject of the household economic decision-making.

The *Robo* however was a non-permanent and loosely organized entity, without developed systems of authority, decision-making and division of labor. Apparently a condition of relative welfare and abundance of resources, did not require the Pekurehua of the direct past to develop strong institutions and organizations for the management of what the community considered as their resources.

The changes since the end of the 20th century that engulfed the *Pekurehua* in all aspects of its life have left the community segmented through all its elements. The households have been tied down and focused their energy and interest on their small private landholdings. Modern institutions and social organizations, such as individual land rights, modern government administration and specialized organizations, replaced the ones from the direct past. The religious life has been removed from its agriculture and environmental context and organized in many different churches, all with its own networks of neighborhood bible groups. Above that there is the diligent migrant community that has a large impact on land entitlement and land use. Within this new framework, individual perception on the resources and the natural environment logically have become fragmented too. In the eyes of the individual, the whole new environment will look more like a checker board of small plots of land, each under the authority of individual households, or public-space and state-land, of which the authority rests on institutions far removed from the sphere of influence of the individual member of the community. Nobody has an overall picture anymore of the environment, let alone decision power over it.

It seems that the changes that brings fragmentation in almost all aspects of life do not produce, as a compensation, institutions and social organizations that allow members of the community to perceive the resources and environment of the village as an integrated whole, which eventually allows the community members to participate in the management of these resources and environment. Much of the degradation of the natural environment can be blamed on this condition. However, the solution cannot be found by appealing to the "tradition". As already explained, those institutions and social organization from the direct past were part of a totally different perception and system of management of what can be perceived as community resources. With this perspective in mind, it will be difficult if not impossible to start a village wide sustainable management of resources and environment by appealing to the "tradition", or by employing "traditional" institutions and organizations.

Currently, social organizations are starting that seem to have the ambition to fulfill the role of integrating the fragmented community. These small local organizations are primarily active in the field of environment protection. These local organizations employ primordial sentiments, such as ethnic identity and its glorious past to appeal to the people's solidarity. It tries to revive local knowledge, especially on the boundaries of their ancestors' land, and on sustainable land use. Two objectives motivate the activists of these embryo organizations. First, the advocacy of sustainable land-use and environment management. Second, the empowerment of the indigenous community, by appealing among other things to their ancestors' land rights. These objectives do not always go hand in hand. There are activists that question the appeal of the ancestors land, because of their commitment to the integrity of the Lore Lindu National Park. They like to place the ancestors land rights into a larger context in which the integrity of the National Park has to be taken into consideration.

With the consideration that the community at present is undergoing a quite rapid process of pluralization in ethnic as well as religious aspects, there is the question whether in the long run the approach of appealing to the "tradition" of these embryo organizations will turn self-defeating because of its potential to alienate non-indigenous groups in the communities, and, thereby, to build wedges between ethnic groups, which will simply defeat the integrational objective – primarily in environment management - of these embryo organizations. It seems much more logical to start with the development of a totally new common perception of the current conditions of man – nature relation and, based on that, the establishment of a community wide new participatory sustainable natural resource management mechanism.

References

Breman J (1997) The Village in Focus. In: Breman JC, Kloos P, Ashwani Saith (eds) The Village in Asia Revisited. Oxford University Press, Delhi, pp 15-75

Dove MR (1988) The Kantu system of Land Tenure: the Evolution of Tribal Land Rights in Borneo. In: Fortman L, Bruce JW (eds) Whose Trees? Proprietary Dimensions of Forestry Westview Press, pp 86-95

Eghenter C (2000) What is Tana Ulen Good for? Consideration on Indigenous Forest management, Conservation, and Research in the Interior of Indonesian Borneo. Human Ecology 28,3, pp 331-357

Geertz C (1963) Agricultural Involution: the Processes of Ecological Change in Indonesia. University of California Press, Berkeley & Los Angeles

Harris M (1979) The Rise of Anthropological Theory: a History of Theories of Culture, 3th edn. Routledge & Kegan Paul, London

Hefner RW (1990) The Political Economy of Mountain Java: an Interpretive History. University of California Press, Berkeley

Huskens F, White B (1989) Ekonomi Politik Pembangunan Pedesaan dan Struktur Agraria di Jawa. Prisma 4,XVIII: 15-37

Kahn JS (1980) Minangkabau Social Formations: Indonesian Peasant and the World Economy. Cambridge University Press, Cambridge

Li TM (2002) Local Histories, Global Markets: Cocoa and Class in Upland Sulawesi. Development and Change 33,3: 415-437

Li TM (1999) Transforming the Indonesian up lands: marginality, power and production. Hardwood Academic Publishers, Amsterdam

Smith E Alden MW (2000) Conservation and Subsistence in Small-Scale Societies. Annu. Rev. Anthropol. 29: 493-524

Sunito S, Mappatoba M, Rosmawaty AS, Hasan DH (1999) Wuasa: case study of a village in Lore-Lindu Region. Institut Pertanian Bogor, Universitas Tadulako, George-August Universitat Gottingen, University of Kassel

Tjondronegoro SMP, Gunawan W (eds) (1984) Dua Abat Penguasaan Tanah: Pola Penguasaan Tanah Pertanian di Jawa dari Masa ke Masa. PT. Gramedia untuk Yayasan Obor Indonesia.

Tokare IS DS (1990) Sejarah Lembah Lore Di Kabupaten Dati II Poso, Propinsi DATI I Sulawesi Tengah. (Unpublished document)

White B (2002) Agrarian Debates and Agrarian Research in Java, Past and Present. In: Suhendar E et al. (eds) Menuju Keadilan Agraria: 70 Tahun Gunawan Wiradi . Akatiga, Bandung

Orang Kampung and *Pendatang*: Analysis of Demographic Structure and Migration in Two Forest-Margin Villages, Central Sulawesi

Melani Abdulkadir-Sunito

1 Introduction

This Chapter focuses on change of ethnic composition, which is one aspect of demographic structure, and its effects on the unity of communities. This Chapter is a component of ongoing research on the impact of spontaneous migration on the cultural landscape and agrarian structure in two forest-margin villages in Central Sulawesi. The research is part of a multi-disciplinary research program on the "Stability of Rainforest Margin in Indonesia" and focuses particularly on the margins of the Lore Lindu National Park (LLNP) in Central Sulawesi. The Institut Pertanian Bogor, Universitas Tadulako (Indonesia), University of Göttingen, and University of Kassel (Germany) execute the program.

Concerns over tension and possibility of violence outbreaks between local hosts and migrants have been expressed by many (e.g., Acciaioli 1999; Li 2002). Particular attention is due to the research area's proximity to lowland Poso, where in the past three years there have been numerous ethno-religious conflicts. An ongoing land-use conflict in Dongi-Dongi, which is an area of LLNP between Kamarora and Napu valley, is observed since mid 2001 (Merrill 2002), which appears to encourage encroachment into the other areas of LLNP margin. Issues to be discussed in this chapter are (i) How has migration changed the proportion of *orang kampung* and *pendatang*[1]? (ii) What effect would the rapid change in proportion to assimilation processes and community's social resilience? (iii) What are the divisive factors that initiate the situation of us vs. them?

This Chapter is organized as follows. Firstly, a setting of research gives a brief summary of the existing literature on the identity, conflict and resilience (Section 2). The next section describes the methods used for the analysis. An overview of the research area is given; the demographic structure is analyzed as well as changes based on population data of different sources, and their impact on settlements and inter-ethnic relation. Cases are then presented, in which the term *orang kampung* and *pendatang* reflects the "us vs. them" feeling that creates distrust and prejudice among villagers, as well as power abuse by village elite.

---

[1] *Orang kampung* (=local people) is a term used to describe the autochthon, *pendatang* (=newcomers) are the migrants (allochthon). Further description shall be presented in the paper.

2 Setting of Research

Migration is as much a spatial movement as a cultural one. It entails a change of social setting or a cultural milieu (Davis 1976), and brings together people with quite different attitudes, and behavior patterns (Weeks 1978). The process of interaction among different groups in the area of destination, is one concern. Parrillo (1983) describes the interaction process between the majority and the minority. The majority defines the situation of interaction with acceptance, avoidance, exploitation, or paternalism, whereas the minority's response may either be acceptance, avoidance, or deviance. The reality, however, may not be a mechanistic causal relation while acceptance may belie avoidance.

Mennell (1994) discussed it in terms of the we-identity and suggests that identification, being related to group membership, is multi layered. Citing the Norbert Elias hypothesis, Mennell stated that people generally feel the greatest emotional identification in relation to their survival units -- the tribe, the lineage, or the local community. In a similar way, Gordon (in Fitzpatrick 1983) noted that beyond the range of the family, the basis for identity is the ethnic group, which is based on a number of common bonds such as values, religion, historical experiences and race. Historically, identification progresses in the direction of more inclusive layers of identity. Thus, identification with kinship groups and local communities preceded that with state societies. Yet, the less inclusive layers of identity can be socially constructed and become effectively more charged than was formerly the case identification with a nation-state (Mennell 1994).

The contest between retaining own cultural identity and assimilation is another concern. Gordon (in Fitzpatrick 1983) noted that assimilation occurs in seven stages or processes: (i) cultural assimilation (or acculturation), (ii) A structural assimilation, (iii) marital assimilation, (iv) identificational assimilation, (v) attitude receptional assimilation, (vi) behavior receptional assimilation, and (vii) civic assimilation. Cultural assimilation, in which "...newcomers accommodate the basic values and behavior patterns of their new society..." is usually the first type to occur and could occur indefinitely without any of the other types occurring. In other words, newcomers can live in a society with a distinct culture and function effectively in it while their day-to-day lives remain quite isolated from primary relationships with members of the host society.

What eventually happens to minority groups in the host society varies. Parrillo (1983) stated that characteristics such as a strong cohesive bond or an ideology might help retain the group's cultural identity. Charras (1993) mentioned that where migrants represent only a small minority, interaction is often more integrated than in communities with two sizeable ethnic groups.

In view of the community as a whole, Locke *et al* (2000) holds a similar view concerning resilience and social resilience. Resilience is the ability of a population to absorb and respond positively to external changes, such as those associated with population changes. Social resilience is defined as a community's or an individual's ability to withstand shocks and stress without significant upheaval. The two definitions are used in the framework of community cohesion, whereby multi-ethnic groups exists interdependently. The breakdown of social resilience occurs when a complexity of identity, and ethnicity reflects inequality. As Esman (1994) as quoted in Tirtosudarmo (2001) wrote:

"Ethnicity as politics is meaningful only in a relational framework. As a collective identity is shaped not only by self-definition but also by constraints imposed by outsiders. Thus, the internal 'we' must be distinguished from the external 'they'. Where no relevant others exist, the need for solidarity disappears and society fragments on the basis of internal differences. The need for ethnic solidarity arises only when strange, threatening, competitive outsiders must be confronted. Only then are distinctions between 'us' and 'them' noted, explained, and evaluated to reinforce internal coherence and to signal and explain differences with outsiders and competitors. Only then does the needs for internal solidarity and discipline become evident".

Human Rights Watch (1998) in reporting communal conflict in West Kalimantan put forth three explanations concerning the background of conflict, namely cultural, economic, and political. The cultural explanation focuses on the cultural difference between the conflicting ethnic, especially attitudes towards settling disputes using violence. The economic explanation looks at the gradual dispossession and marginalization of the local people due to various development policies of the government. Increasing competition for resources causes opposition towards outsiders. Migrants, in this relation, are regarded as representation of intruders to be ousted. The political explanation holds that while cultural and socioeconomic factors are important, the scale of the violence can only be explained by the intervention of a third party. In addition to the above explanation, Tirtosudarmo (2001) pointed out the elimination of pluralism and ethnic heterogeneity discourse in Indonesia.

Without excluding the cultural and political explanations, this paper deliberately leans on the socioeconomic marginalization argument when considering the seeds of conflict situation in two forest margin villages in Central Sulawesi. Specifically, a better infrastructure that permits greater penetration to the interior by spontaneous migration that has altered the population balance of an area and increased competition for resources.

3 Methods

The research combined qualitative- and quantitative methods of collecting data. The qualitative research methods comprise of observations, transect walks, and interviews. Focusing on village history and changes, key individuals (administrators, elders, *adat* leaders, and pioneer migrants) were interviewed. Most were interviewed more than once, as new questions in light of other people's accounts need to be verified. This circulatory way of interviewing is a means to "construct a general narrative" (Hefner 1990). A semi-structured questionnaire was used to interview migrants who were chosen by cluster random sampling, based on year of arrival, ethnicity, and area of origin.

The quantitative research method is to do a village-level population census. The census was conducted at each RT (= neighborhood unit) with the help of trained village administrators in May 2001. With the other population data, Village Census is then used to construct time-line changes. This other population data comprises of population censuses (Central Bureau of Statistics/BPS), population record of subdistrict level (2000, 2001), population data in Village Monograph (1985, 1995, and 2000), and population registration of Watumaeta (1985, 1987, and 1997). For the analysis of census data, the SPSS program was used.The fieldwork took five months in three separate visits between March 2001 and August 2002.

4 Results

The Research Area. Watumaeta is about 102 km Southeast of Palu on the intersecting road to Poso, and about 3 km from Wuasa, the capital of subdistrict Lore Utara. Further South, on the road connecting Wuasa to Bada valley is Rompo, a distance about 130 km from Palu. The 30-km distance from Wuasa to Rompo takes 2 hours by vehicle due to the poor condition of the road. Rompo was part of subdistrict Lore Utara until September 2001, when with seven other villages, which formerly were part of sub-district Lore Utara, it separated to become subdistrict Lore Tengah with the capital Doda.

Both Watumaeta and Rompo are directly bordering LLNP. In 1978, when the border of LLNP was established, part of Watumaeta (the *powanuangaa*[2]) was integrated into LLNP. Although similar processes occurred in Rompo, their effect is different due to different types of land use and different levels of population density. At that time *powanuangaa* has already coffee, whereas Rompo was mainly covered with primary forest.

Watumaeta was formally established as *desa* in 1950 after being a hamlet (*dusun*) of Sedoa for about 20 years (Togea 2002). Settlements, however, dated back to early 1930s when families from Sedoa and Wuasa cleared forests to make dry-land plots (*ladang).* Even earlier, Watumaeta used to be hunting ground of deer and waterbuffaloes for people of Wuasa, Sedoa, and Kulawi.

The autochthon of Watumaeta are the Napu (= Pekurehua) ethnic group. In the past, Watumaetans established trade relation to three areas: (1) to Tambarana (subdistrict Poso Pesisir) through Lake Patawu (Sedoa) in the Northeast, (2) to Lake Lindu and other areas of Kulawi to the West, and (3) to Poso in the Southeast Lindu (and Kulawi in general) was the mission area of Salvation Army missionaries from Watumaeta. Poso area was also a destination of education, with Tentena specifically for Christian education.

With the establishment of road connecting Napu and Palu in 1982, Poso's pivotal role changes. Nowadays, the main destination for education and the place for

[2] The *powanuangaa* (*hutan waris,* endowment forest) is believed to be the old-day settlement site. Yet, no one can explain when or why the settlement was abandoned.

agriculture trade are Palu. Meanwhile, the construction of road connecting Napu to Poso (1985-1990) enables a faster influx of migrants, especially from South Sulawesi, to Central Sulawesi uplands.

The waves of migration to Watumaeta mirror this change. The few migrants of the 1970s mostly came from Lindu, their parents being Watumaetans migrated to Lindu in 1950s. From 1980 onwards, migrants arrived in a considerable greater numbers. The Bugis migrants who arrived in the early 1980s were relatives of a Sinjai-Bugis who earlier worked in Palolo, and Bugis families from Palu. One of them established the first, and now the largest, Palu to Napu transportation company. Migrants who arrived from 1995 to 1998 are mostly Bugis who came directly from South Sulawesi. The latest wave of migrants came between 2000 and 2002 and was a combination of the Javanese/Sundanese ex-transmigrants from Kebun Kopi (Parigi, Central Sulawesi), the educated migrants came from Palu and the refugees from Poso areas.

A diversity of employment is observed. Though mostly rice and dry-land farmers, people work in small shops (*kios*), food-stalls (*warung)*, as intermediate traders, and in the transportation service. Some young locals also gather rattan, an arduous work with little income. Highland vegetables and oranges, besides coffee and cocoa adjoin the most common crops of maize and kidney bean. This change reflects a more intensive type of cultivation, with a higher use of chemical-inputs.

Rompo was formally established as *desa* in 1917. The autochthon of Rompo belongs to the Besoa ethnic group. The first settlers of Rompo came from Bariri an enclave village inside LLNP. Its traditional trade relation is Lawua (Kulawi) through Torire and Hanggira, and (2) Poso through Watutau and Sangginora. The existence of a missionary school in 1927 made Rompo a center-point of activities for the neighborhood villages of Talabosa, Katu, and Torire until the 1950s.

Interestingly, over half of Rompo residents in 1940s were the Bugis-Masamba Moslems. They were, initially, part of inter-village traders crisscrossing Besoa valley to trade, who then resided and married the locals. The current village Head has Moslem half-siblings, as his mother was once married to a Masamba. In mid 1950s to early 1960s, they gradually moved to lowlands Tabalu and Saatu (subdistrict Poso Pesisir) due to hardship and the Permesta rebellion[3]. Many Rompoans also migrated during this period; one recalls that only seven families were left Rompo by the mid 1960s.

---

[3] Permesta was one of the rebellions in the Outer Islands (Sumatra, Sulawesi) in mid 1950s, caused by the growing resentment against the central government, which was regarded as dominated by Javanese intent on "exploiting" the natural wealth of the export-producing areas (Crouch 1978). Started in South Sulawesi in mid 1957, it was largely confined to Northern Sulawesi by 1958. Several regional and local military commanders took control of local governments and succeeded in rallying considerable popular support behind their defiance of the central government. In 1958 the central government forces quickly occupied rebel strongholds in Sulawesi -- moved eastward from Gorontalo to capture Manado in June 1958 (Cribb 2000). Though most of the rebel forces were soundly defeated within a few months, the guerilla activities continued until 1961 (Crouch 1978).

For over two decades roads to Rompo were hardly passable, making Rompo less interesting as a destination to migrate to. The road connecting Wuasa to Rompo was established in the early 1990s, but even today it is in relatively poor condition.

In 1997/8, there was a plan to resettle people of Katu village who were living in an enclave inside the LLNP, to the Parawali area of Rompo. When this plan failed, Parawali was further destined as a transmigration site. Within this scheme, in 1999 came the Luwuk (Palopo district, South Sulawesi) migrants. These Javanese, Balinese and Toraja ex-transmigrants decided to leave Luwuk due to communal disturbances in the area. They came as a group, headed by the current Village Secretary. Over 100 families registered and approximately 60 families finally settled in Rompo in September 1999. This, however, did not last long as in Poso conflicts started in April 2000 and many migrants, loosing again their sense of security either returned to South Sulawesi or to villages with better road accessibility around Napu. The failure to deliver promises of housing, financial aids and agricultural support aggravated the situation. By August 2002, there were less than 10 migrant families. Most of the returned migrants are those still having resources in area of origin. Most of those who stayed are the so-called *"migran putus"*, namely, those who have sold all their belongings before moving to Rompo.

Besides farming in wet ricefield and/or dry land, about 60% of households rely on income from rattan gathering (compare to 90% in 1990s). Coffee and maize are the two most planted crops, with cocoa gaining favor in the past two years. Few farmers regularly use chemical fertilizer for their crops. Even among migrants who are familiar with fertilizer and pesticide in their area of origin, the practice was abandoned.

Demographic Evidences from Village-Census. The phases of population growth in Watumaeta is described as a "stable-increase". The wave of migrants of mid 1980s is reflected in the annual rates of population growth, which was 9.6% from 1982 to 1983 and 14.2% between 1983 and 1987. Meanwhile, the 17.4% annual rates of growth between 1997 and 2000 reflected the wave of migrants who came in the mid 1990s. In between the waves, from 1988 to 1996, the population grew steadily at a much lower rate of 2.1%.

The percentage of allochthon households in Watumaeta increased from 40% in 1985 to 63% in 2001. In terms of ethnicity, the decade of 1990 saw a significant flow of migrants from South Sulawesi. In the year 1985, the Bugis and Makassarese migrants consisted 18.5% of total population. In 2001, it was over 60%. Hence, within a short time the Bugis became a dominant ethnic group in Watumaeta.

In **Rompo** the phases are defined as a "contraction", in which a high in-migration in 1999 was followed by out-migration. Between 1987 to 1995 the annual rates of population growth in Rompo was 4.7%, whilst between 1995-2001 was a low 1.5%. The population of Rompo in the past 2 years (2000-2002) is actually decreasing due to out-migration.

The term *pendatang* is not a simple differentiation based on birthplace (= life-time migrant), but also of migrant's year of arrival, own and spouse's ethnicity, and religion. A calculation based on the 2001 Village Census is then made by differentiating birthplace into born in the village (= autochthon) and outside the village (= lifetime migrant, allochthon). The lifetime migrants are then differentiated into born in (1) same subdistrict, (2) neighborhood subdistricts, (3) same district, (4) other districts in the same province and (5) other provinces. Same ethnicity is indicated by those born in the same subdistrict, and/or neighborhood subdistrict. Migrants from these areas are largely regarded as *orang kampung*. Same district (= Poso) indicated past relation and closeness, i.e. with Tentena and Poso Pesisir. In both cases, same religion is also the case. It is no wonder that Christian refugees who have fled to Lore Utara are called "lost brothers" (Togea 2000). Yet, relation to lowland Poso also reflects past conflicts and invasions, and in this sense Poso is neither a complete *orang kampung*, nor *pendatang*. Other district in Central Sulawesi or other province represents not only difference of ethnicity, but also of religion -- with exception of the Toraja Christians.

In Table 1 the result is presented. Of the 48% lifetime migrants of Rompo, 26% is reclassified as *pendatang*. In Watumaeta of the 75% lifetime migrants 61% is reclassified as *pendatang*. This comparison explains why *orang kampung* in Rompo hardly feel the existence of *pendatang* among them, whereas *orang kampung* in Watumaeta feel that *pendatang* have, "...formed an ever-smaller ring around the locals settlement", which can be considered as a strong expression of anxiety.

Table 1. Percentage of Population by Birthplace, Watumaeta and Rompo, 2001

Village	0	1	2	3	4	5	total
Watumaeta	24.98	6.02	2.48	5.49	14.26	46.77	100.00
Rompo	52.19	9.77	6,94	4.88	2.31	23.91	100.00

Notes: 0 = same village; 1 = same subdistrict; 2 = village at subdistricts Lore Tengah and Lindu (Watumaeta) or Lore Utara and Lore Selatan (Rompo); 3 = same district (Poso), other than subdistricts in no.2; 4 = other districts in Central Sulawesi (same province); 5 = other province

Inconsistency of Population Data. In Table 2, the 1996 CSIAD[4] data noted a much lower percentage of migrants compared to the Village Registration just one year later. Interviews confirm that large influx of migrants was not known to occur in that year. In comparing Village Registration of 1987 to 1997, there appears to be a higher out-migration than in-migration in between 1987-97. Indeed, it is unusual that the CSIAD data noted a similar proportion of allocthon in 1996 between the remote Rompo and the highly accessible Watumaeta. The facts lead to the conviction that CSIAD data is somewhat inaccurate.

[4] Abbreviation of Central Sulawesi Integrated Area Development and Conservation Project.

Table 2. Population by Allochton-Autochton and Different Sources, Watumaeta and Rompo

Village (Year - Source of Data)	Pop.	HH	% Aut.HH	% All.HH
Watumaeta				
1985 - Village Registration	365	65	60	40
1987 - Village Registration	466	93	47	53
1996 - CSIAD	570	80	79	21
1997 - Village Registration	564	108	55	45
2001 - own village Census	1,125	272	28	72
Rompo				
1996 - CSIAD	361	37	86	14
2001 - own village Census	363	111	58	42

Notes: (1) Allochton household (in Village Registration and own village Census) is defined as both husband and wife, or more than 50% of household members born outside the village.
Sources: Sunito et al (1999) for CSIAD data ; Togea (1985) for 1985 Village Registration data ; Togea (1987) for 1987 Village Registration data ; Dirjen PMD-Depdaneg (1997) for 1997 Village Registration data.

In Table 3, it appears that Village Monographs of both Watumaeta and Rompo tend to overenumerate[5] their total population. Monographs also appear to be the most unreliable source of population data. Meanwhile, data of Subdistrict Statistics tend to overlook additional population due to migration, except in the case of refugees. In December 2001, 230 new refugee households (but only of 33 persons!) was noted[6], of which only 230 were noted.

The importance of data keeping is often largely ignored, but the 1985 and 1987 Population Registration was maintained by one former Village Secretary. The 1997 population registration kept by Village Head was never updated. In both villages, none keep the result of 2000 Census.

Table 3. Population Data by Different Sources, Watumaeta and Rompo, 2000-2002

Village (Year - Source of Data)	HH	Male	Female	Total
Watumaeta				
March 2000 - Village Monographs	464	598	717	1,315
2000 Census (Oct.2000)	223	571	501	1,072
Dec.2000 - Subdistrict Statistics (BPS)	223	572	502	1,074
May 2001 - own village Census	272	607	518	1,125
June 2001 - Subdistrict Statistics (BPS)	223	574	504	1,078
Dec.2001 - Subdistrict Statistics (BPS)	453	592	519	1,111
May 2002 - Village Monographs	414	689	627	1,316

[5] Incorrect reporting where the number of people is counted as higher than actual.

[6] This is quite a bizarre situation, because when confronted with these numbers the officer of Subdistrict Statistics could not give a clear explanation but insisted that the number was right. As a comparison, the Village Secretary of Watumaeta said that of the 912 refugees (289 families) still resided in February 2002, he estimated about 60 families decided to move permanently to Watumaeta.

Table 3. (cont.)

Rompo				
Feb.2000 - Village Monographs	144	361	274	625
2000 Census (Oct.2000)	87	192	177	369
Dec.2000 - Subdistrict Statistics (BPS)	87	193	178	371
May 2001 - own village Census	111	205	158	363
June 2001 - Subdistrict Statistics (BPS)	87	194	178	372
Dec.2001 - Subdistrict Statistics (BPS)	87	194	179	373

Source: Dirjen PMD-Depdaneg (2001a) for March 2000 Watumaeta Village Monographs data; Dirjen PMD-Depdaneg (2001b) for Feb. 2000 Rompo Village Monographs data; CBS (2000a) for 2000 Census data; CBS (2000b) for Dec. 2000 Subdistrict Statistics data; CBS (2001a) for June 2001 Watumaeta & Rompo Subdistrict Statistics data; CBS (2001b) for Dec. 2001 Watumaeta & Rompo Subdistrict Statistics data

Settlement. In Rompo, migrants assimilated well with locals. A small concentration of migrants' houses is built across the river, north of the village center. Yet this neighborhood is formed around several pre-existing local families, as this area is the most plausible extension of settlement near the village center and on the Wuasa - Rompo roadside.

In Watumaeta, migrants settled in the periphery of village. The 1980s migrants settled mainly in the northside (bordering Sedoa) and in east side (= *Dusun* 2). The *Dusun* 2 is an interesting face-off area where the closer to village center the more houses of *orang kampung*, whereas the farther the more of *pendatang*. The 1990s migrants moved further east (along Napu-Poso road, bordering Alitupu). They formed a separate settlement, which then became the third hamlet (*Dusun* 3).

Dusun 2 Watumaeta is also bordering Tambua, which is part of *Dusun* 4 Wuasa, where most Bugis of Wuasa resides. In fact, the road called *Jalan Ekonomi* in to reach *Dusun* 4 than any other road in Wuasa itself. The cocoa and coffee harvest of Wuasa's Bugis are brought through this road and sold to intermediate traders in Watumaeta. What it implies is that Bugis of the two villages are connected closer to each other than to their respective village.

This spatial segregation between locals and outsiders, as Charras (1993) observed in Southern Sumatra corresponds to the social distance that the two communities retain. In West Kalimantan, the tendency of one ethnic group to live together, separate from the local communities, is interpreted as proof of their inability to adapt and unwillingness to integrate into or respect the locals (Human Rights Watch 1998).

The new refugee site of Mangkapa is another separate settlement in the northside periphery of Watumaeta. This is not only an interesting policy on handling refugees, but also on how refugees obtained the land (Box 1). Meanwhile, the new RT 9, where most of the year 2000 migrants reside is on the South-side periphery (bordering Wuasa). The fact is that some land of RT 9 is questioned as part of LLNP, which has merely heightened the resentment toward migrants.

Further, the migrants' settlement in all but to the west side of village center gives little room for *orang kampung* Watumaeta to expand, because this side is directly bordering LLNP. It is along this side that the flatland *powanuangaa* is currently being claimed. Though no agreement is met with TNLL, observation noted open patches of maize or cocoa gardens inside LLNP, indicating that encroachment has occurred.

Assimilation and Tensions. Two things indicated the occurrence of assimilation, inter-ethnic marriage and the appointment of *pendatang* in political positions, such as village administrators or as members of Council of Representatives (=BPD). In this aspect there is a stark difference between Rompo and Watumaeta. Inter-ethnic marriage is common in Rompo, but not in Watumaeta. This is because most *pendatang* in Rompo are of the same religion with the locals, whereas three-quarters of *pendatang* in Watumaeta are not. Inter-ethnic marriage in Watumaeta occurs only among Watumaetans and Poso refugees, which is of same religion, although they do have different *adat* (tradition, customary law). To resolve this problem, in 2001 the Council of Adat (*Lembaga Adat)* set a rule that the bride's origin determine the choice of *adat* (Napu or Poso) used in the *adat* wedding ceremony.

Box 1:How The Refugee Got Their Land

A new settlement for the refugees developed since early 2002 in Mangkapa, Watumaeta. International organizations like Care, USAID, and Oxfam gave aid to erect houses. The land on the East side of Mangkapa was bought from a landowner in Wuasa, and the West side was *tanah jemaat (*=land belongs to the church congregation).

The *tanah jemaat* was a swapping deal with the Village Head (VH)'s *tanah bengkok (*=land assigned to a village official as a kind of salary; the official only has usufruct rights of it*)*, which he donated. The reason for land swapping was that *tanah bengkok* is not on the roadside. Some informants also pointed to the fact that *tanah bengkok*, located in Dusun 3, is in the middle of a Moslem neighborhood, whereas most refugees are Christian.

The land swapping itself is a debatable issue. Several autochthons voiced disagreement because it was made without the consent of all church members. To secure the deal, refugees who inhabit Mangkapa West paid an 'administrative fee' to VH to issue a letter of ownership. Five months later, the letter has still not been issued, and the money has never proceeded to the church treasurer.

In the East side of Mangkapa, plots (1 to 2 ares) sold land to the refugees by VH under authorization of its owner. To guarantee that all plots are sold -- and a 10 percent commission is obtained, some non-refugee migrants were persuaded to buy plots. Meanwhile, VH also own a house in Mangkapa East, and as rumor goes, another 3 or 4 houses of his children.

In Rompo inter-ethnic marriages became a way of gaining status as *orang kampung*. For example, two prominent cases are a Bugis kiosk owner who marries daughter of a pioneer migrant and a Toraja migrant who marries daughter of a prominent village elite. The latter become member of BPD. Whether intermarriage is a strategy of migrants for access to land, as observed by Charras (1993) in Southern Sumatra, is a presumption, but certainly need a separate study to verify. However, local custom, which dictates that the groom live in his wife's house and work in his parents' fields, supports this notion.

On the political sphere, in Rompo two important administrative positions -- the Village Secretary and Head of Welfare Affair (*Kaur Kesra*) are occupied by migrants, as well as the former LKMD[7] head. In Watumaeta no administrative position is held by migrant, but the position of Hamlet Head (*Kepala Dusun*) of Dusun 3 -- a position which is always in the hand of a Bugis. This is for practical reason, as all inhabitants of Dusun 3 are migrants, mostly of Bugis ethnic and many do not speak fluent *Bahasa* Indonesian. Yet, *Kepala Dusun*'s influence extended his hamlet. He was often called upon to other hamlets to settle disputes between Bugis Moslems and local Christians, was known to send the troublemaker migrants away to South Sulawesi. He is also a member of Inter-Religious Forum, a newly established organization set up at subdistrict level to settle conflicts between religious group.

In both villages, one in five legislative position in the Council of Village Representative (BPD) is a *pendatang*. Migrant leaders in Watumaeta express resentment over this proportion, thinking they are not well represented in BPD. They pointed out the election system as the cause, for only village administrators, hamlet heads and church leaders cast their votes.

Resentment is expressed in several ways. In conversation, people differentiate the old-timer *pendatang*, with the more recent ones. The former is said to have a better attitude, and make efforts to interact with locals. They also request permission to stay, on arrival, which many of the more recent migrants do not bother to do. Among *pendatang*, the old-timer Bugis resents the newcomers who are able to take advantage of cocoa boom. Some old-timers recall that in late 1980s, the Agriculture Agency advised them not to plant cocoa because it is unsuitable for the upland area. Those who had planted cocoa replaced it with coffee. Had they ignored this advice, some lament, by now they would be as rich as the new migrants.

Among youngsters, Poso conflict seems to bolster their courage to express resentment towards *pendatang*, especially when alcohol induced (Box 2). What Acciaioli (1999) observed in Lindu, i.e. stones hurled anonymously into Bugis migrants' houses[8], also occurs in Watumaeta. Of this, one old-timer migrant reacted lightly,

---

[7] Abbreviation of *Lembaga Ketahanan Masyarakat Desa*, or Village Security Council, made up of heads of key village-level sectors. The group administered all village-level development. In 1999, the new law on regional government (UU no.22/1999) replace LKMD (whose members are appointed) with BPD (whose members are elected).

[8] as an expression of resentment over Bugis domination of intermediate marketing (Acciaioli 1999). In the case if Watumaeta of no particular reason.

"... (throwing stones at houses) is nothing. As long as not one's head. A few bricks and a kilo of cement will fix a house. But (a wound in the) head is another matter".

Yet, most *pendatang* did not react as lightly. They remember such accident, and regarded it as proof of insufficient effort by village government (*pemerintah*) to protect the *pendatang*, as well as elders' powerlessness to discipline their youths.

Box 2: Cocoa Tree Pulled Off and An Early 'Harvest'

After clearing his allotted plots, Mon, a Toraja migrant in Rompo planted his cocoa seedlings in between maize. He visited the plot every week to do weeding. On one of those visit about four months later, Mon found some cocoa seedlings wilted. His first thought was of insects or termites, but at closer look he discovered that somebody had deliberately pulled out the tree and replanted them to look intact. He does not know who did it, but suspects the local farmers who often passes his plot on the way to their maize plots. When he mentioned this accident to other migrants he was informed that they had also the same complaint. Another migrant reports a case of neighborhood plots set on fire to clear weeds, which accidentally, or intentionally, burned some cocoa trees in the next plot.

In Watumaeta, one morning Dono found his harvest-ready *daun bawang* (= onion green) scattered on the roadside. A neighbor told him that the night before, some drunken youngsters had come there after a party. He did not report the accident to Hamlet Head because he could not name any suspects. A few weeks later, Ajan found out that rows of ready-to-harvest vegetables in his plot simply disappeared. He mentioned this case to his neighbor, a long-time migrant and an extension officer, who advised him to not to make a fuss over it and simply regard it as a donation to whoever needed the vegetables. The extension officer himself has experienced some losses in his orange harvest every now and then.

Poso conflicts create a sense of insecurity among *orang kampung* and *pendatang* alike. At one time people set up barricades and the night watch was intensified. Though they prided themselves on the fact that even at the height of conflict Napu was peaceful, suspicion of strangers abounds. Cases in hand are what occurs in Watumaeta and Wuasa (ca. March-May 2002) with the arrival of some Moslem preachers who dress like *Laskar Jihad*[9]. The quick reaction of members of Inter-religious Forum and security officers managed to quell the agitated youths already gathering in front of the mosque. Yet, the anxiety spread. In this situation,

[9] Laskar Jihad, is a Java-based radical militia group focused on assisting oppressed Muslims wherever they may need help. In November 2001 Laskar Jihad fighters entered the area around Poso, and have since allegedly been involved in the attack and burnings of Christian villages around Poso (HPCR 2001)

using the Poso issue to expel *pendatang*, as the case in Rompo hereunder, is warranted.

The notion of *pergi tanpa ijin* is not applied rigidly to other migrants who leave. Many migrants simply left the area when rumors of attacks circulated in May 2000 and returned a few months later. Some have moved their families to Wuasa and occasionally come to work on their land.

Box 3:A Peculiar Eviction

Rahim, a Bugis, own the largest kiosk in the village, has been in the area for over 5 years. He was also a 'rattan boss', one who bought rattan from the pickers and provides supply of sugar, rice, and coffee. As often happens in this business, many rattan gatherers are in debt.

In 2001, due to Poso riots, the family left for Palu. He was regarded as *pergi tanpa ijin* (leaving the village without informing village administrators) and denied return. Two village elite, rumored to be the largest debtors, say that he not only *pergi tanpa ijin* but also was found out making some explosives in his kiosk, with the help of his Bugis friends. Many other villagers denied knowing of such situation, even **doubting** it ever happened. None of them openly contested the information.

In March 2002 Rahim sold his kiosk, as a capital to set up a new kiosk in Baliura (Besoa area), and **asked** Village Head to help him sell his garden-plot. He had asked the village administration(*pemerintah desa)* to help him collect the debt, although he has practically given up the idea of getting back even part of it.

In the case of Watumaeta, the land dealings by Village Head are creating conflicts in three ways: (1) *orang kampung - pendatang*, (2) *orang kampung* - administration, and (3) *pendatang* to administration. *Orang kampung* is stereotyped as easily selling land when cash is needed. Village Head is seen as favoring *pendatang* who gets all helps in obtaining locals' land. Migrants often face problem regarding the land they just purchased. *Head of Dusun* 3 said that the most frequent problems between *orang kampung* and *pendatang* that he encounters is due to land purchase -- sold land turns out to be smaller than agreed upon, or same plot is sold several times each with a 'legal' letter issued by Village Head.

The case described in Box 4 is one of the most talked about cases. In other case, Village Head sold plots belong to his in-laws' family and justified it with Agrarian Law that specify village administrators has the rights to reallocate fallow land at will. All this is made possible by several things. One, that he gives positions to family members, to retain his power. Two, that he skillfully manipulates the familial relationship among *orang kampung* to silence the opposition[10].

Box 4: Underhanded Land Dealings

Udin's brother, Sam, bought land from Village Head (VH) who said that the land is his wife's inheritance from her grandparents. Sam paid VH 1.5 million IDR, the money from selling land in the South. When Udin and some other Bugis men came to help Sam clearing the land, Geso, a local, came and claimed that the land belongs to his father. "Do not incite troubles on other's homeland," Geso warned, as he demanded them to leave the area.

Sam reported this to VH, who simply gave him another plot. This time VH said the land was his own. And this time Sam believed him because soybean had been planted in part of the plot, and large trees were already cut out.

One day a few weeks later, a PPA (Forestry) officer came, informed Sam that his plot is inside LLNP area, and asks who sold it to him. A trial was then held in *baruga* (village hall). Udin was the prime witness. As a brother and head of RT, Udin witnessed the deal and his signature was in the receipt of purchase. Yet, the VH had told him to "take care of his brother". He said, "If I confess that the land was bought from VH, I can be expelled from this village". So he hid the receipt, said it was not sold but borrowed. Udin's confession actually saved VH from harsher penalty, and alters the charge into VH's mishandling of LLNP plot. No one remembers clearly the outcome. Either it was stalled by the onset of Poso conflicts, or VH was given a light sentence. Sam was allowed to continue cultivating the land.

5 Conclusion

Using demographic data, this paper shows evidences how migration rapidly changes the balance of *orang kampung* and *pendatang* in Watumaeta. Since the dominant 'minority' in Watumaeta, the Bugis of South Sulawesi, is of different religion, ethnicity and religion became divisive factors. Combined with spatial segregation in settlement, resentment, and illegal land dealings, the situation of "us vs. them" emerged. It is difficult to accurately judge how critical the situation is.

[10] Inheritance of land in Napu is such that one cannot sell the plot without consent of all sibling but daughters has a decisive opinion. In selling his in-laws' plots, his wife gave consent, and the rest of the family is reluctant to protest because it would sever family ties. As most families of *orang kampung* are related through intermarriage, a silence is assured.

There have been attempts to dispel tensions, such as formation of Inter-Religion Forum, migrants' attempt to slow the flow of in-migration, and encouragement of locals to improve their economic situation by planting tree-crops and vegetables following the example of their fellow migrants. Yet, it is clear that *orang kampung* have lost a large part of their land resources.

In Rompo, although a mixed settlement is formed, cultural as well as structural assimilation is achieved, and out-migration has become a homogenizing process, it is not extended to all *pendatang*. External influence, i.e., Poso conflict has increased religion as a divisive factor.

The effect is a weakening social resilience of the village as a community. Further, there is an escalation of conflict in Watumaeta, because of its proximity to the Poso area, and encroachment of protected rainforest due to perceived resource (land) scarcity among *orang kampung*.

References

Acciaioli G (1999) Principles and Strategies of Bugis Migration: some Contextual Factors Relating to Ethnic Conflict. Masyarakat Indonesia 2: 239-268

CBS (Central Bureau of Statistics/BPS) (2000a) Laporan Pendataan Sensus Penduduk 2000 - Kecamatan Lore Utara (*Subdistrict Report on the 2000 Population Census*). (Unpublished report)

CBS (Central Bureau of Statistics/BPS) (2000b) Lembar Isian Perkembangan Penduduk - Kecamatan Lore Utara (1 Juli - 31 Desember 2000) *(Subdistrict Bi-annual Report of Population*). (Unpublished report)

CBS (Central Bureau of Statistics/BPS) (2001a) Lembar Isian Data Tengah Tahunan Penduduk - Kecamatan Lore Utara (1 Januari - 30 Juni 2001) (*Subdistrict Bi-annual Report of Population).* (Unpublished report)

CBS (Central Bureau of Statistics/BPS) (2001b) Lembar Isian Data Tengah Tahunan Penduduk - Kecamatan Lore Utara (1 Juli - 31 Desember 2001) (*Subdistrict Bi-annual Report of Population*). (Unpublished report)

Charras M, Pain M (eds) (1993) Spontaneous Settlements in Indonesia: agricultural Pioneers in Southern Sumatra. Orstrom-CNRS/Departemen Transmigrasi

Cribb R (2000) Historical Atlas of Indonesia. Curzon Press, London and New Asian Library, Singapore

Crouch H (1978) The Army and Politics in Indonesia. Cornell University Press, Ithaca and London

Davis G (1976) Parigi: a Social History of the Balinese Movement to Central Sulawesi, 1907-1974. Ph.D. Dissertation, Stanford University

Dirjen PMD, Depdaneg (1997) Buku Induk Penduduk. Watumaeta 1997 (*Population Registration*). (Unpublished report)

Dirjen PMD, Depdaneg (2001a) Daftar Isian Data Dasar Profil Desa/Kelurahan - Watumaeta 2000 *(Village Monographs)*. Unpublished report

Dirjen PMD, Depdaneg (2001b) Daftar Isian Data Dasar Profil Desa/Kelurahan - Rompo 2000 *(Village Monographs)*. (Unpublished report)

Esman MJ (1994) Ethnic Politics. Cornell University Press, Ithaca and London

Fitzpatrick JP (1983) Identity: The Abiding Problem. In: Stimson J, Stimson A (eds) Sociology: Contemporary Readings. FE Peacock Publishers, Itasca

Hefner RW (1990) The Political Economy of Mountain Java: an Interpretive History. University of California Press, Berkeley LosAngeles London

Human Rights Watch (1998) Indonesia: Communal Violence in West Kalimantan. Report 9(10), Dec.1997 (http://www.hrw.org, 9/19/02)

HPCR - Program on Humanitarian Policy and Conflict Research (2001) Sulawesi: The Conflict in Central Sulawesi. (http://www.hsph.harvard.ude/hpcr/cpi, 9/19/02)

Li TM (2002) Local Histories, Global Markets: Cocoa and Class in Upland Sulawesi. Development and Change, 33(3): 415-437

Locke C, Adger WN, Kelly PM (2000) Changing Places: environmental Impact of Migration. Environment (Sept. 2000) 42i7:24

Mennell S (1994) The Formation of We-Images: a Process Theory. In: Calhoun C (ed) Social Theory and the Politics of Identity. Blackwell, Oxford (UK) Cambridge (USA)

Merrill R (2002) Land-Use Conflict in Lore Lindu National Park and Dongi-dongi: creating Ecological Mess. NRM Headline News, Sept 4, 2002

Parrillo N (1983) Minorities. In: Stimson J, Stimson A (eds) Sociology: Cotemporary Readings. FE Peacock Publishers, Itasca

Sunito S, Mamar S, Mappatoba M, Abdulkadir M (1999) Socio-Economic Aspects of Village Communities in and Around Lore-Lindu National Park: Base-line Study. Institut Pertanian Bogor - Universitas Tadulako - Georg-August Universität Göttingen - Universität Kassel

Tirtosudarmo R (2001) Pengantar: Heterogenitas Etnik dan Konflik Komunal di Kalimantan Barat (*Introduction: Ethnic Heterogeneity and Communal Conflict in West Kalimantan*). In: Sudagung HS Mengurai Pertikaian Etnis: Migrasi Swakarsa Etnis Madura ke Kalimantan Barat. Institut Studi Arus Informasi Yayasan Adikarya Ikapi, The Ford Foundation

Togea A (1985) Penduduk Watumaeta Tahun 1985 (*1985 Population of Watumaeta*). (Unpublished)

Togea A (1987) Penduduk Watumaeta Tahun 1987 (*1987 Population of Watumaeta*) (Unpublished)

Togea A (2002) Sejarah Singkat Jemaat/Desa Watumaeta *(A Brief History of Watumaeta).* (Unpublished report)

Weeks JR (1978) Population: an Introduction to Concepts and Issues. Wadsworth Publishing Company, Inc., Belmont (USA)

"*Revolusi cokelat*": Social Formation, Agrarian Structure, and Forest Margins in Upland Sulawesi, Indonesia[1]

MT Felix Sitorus

1 Introduction

Since agrarian structure refers both the composition of agrarian resources and the social relationship regarding distribution of access to such resources, it clearly assumes the articulation of a social formation, which is "a multiplex entity in which several modes of production co-exist, one of which, however, is dominant" (Worsley 1988). Referring to the elements of the mode of production, namely forces and relations of production (Worsley 1988; Taylor 1989), agrarian resources are connected with the forces of production, particularly the means of production, and social relationship is connected with the relations of production. Hypothetically, then, it can be said that any change in the structure of social formation implies a change in agrarian structure.

The hypotheses mentioned will be examined here through a case study of a forest margin community living in a village named Sintuwu. In a context of a local social formation, the community of Sintuwu is one of many which have experienced a shift of the dominant mode of production from the subsistence to the petty commodity one through the expansion of cocoa plantation over the food crop cultivation in the upland Sulawesi during the 1980s. Some have viewed the expansion phenomenon as the "cocoa boom" (Li 2002), which is an economic term indicating the dramatic effect of a high market price. But I myself interpret it as a revolution labeled "*Revolusi Cokelat*" (Cocoa Revolution), since it implies a revolutionary change in the mode of production as well as the agrarian structure.

Rural Indonesia has learned a lesson from "Green Revolution" regarding the implication of a revolutionary change of social formation in agrarian structure. Many researchers clearly show that modernization of wet-rice cultivation in Java in the 1970s, which implied the shift of domination from subsistence to petty commodity production or petty capitalist, brought about a social polarization phenomenon or at least a sharp social stratification among the peasantry, indicated by the increase of marginal and of landless peasants on the one hand and the concur-

---

[1] This article is one of the reports under "Project A2: Social Organization and Processes of Ecological Stabilization and Destabilization" of Stability of Tropical Rainforest Margin (STORMA) Research Program, a collaboration between Georg-August University of Goettingen (Germany), University of Kassel (Germany), Bogor Agricultural University (Indonesia) and University of Tadulako (Indonesia).

rent accumulation of land resources in the hands of a small number of big farmers on the other (Hayami and Kikuchi 1981; Husken 1998; Wiradi and Makali 1984).

The phenomenon of social polarization among the wet-rice peasantry of Java who experienced the "Green Revolution" probably happened among the upland peasantry of Sulawesi who experienced "Revolusi Cokelat". Li concludes that the "cocoa boom" has led the upland community of Sulawesi toward agrarian differentiation or, specifically, class formation based on land ownership (Li 2002). Consequently, it can be suggested that a minority of "landed" people and a majority of "landless" people have emerged in upland Sulawesi as the result of "Revolusi Cokelat".

Since the social polarization is accompanied by the declining of traditional village welfare institutions and, consequently, of economic opportunities for the lowest social stratum, namely marginal and landless peasants (Collier et al. 1978), a way out for coping with the lack of a livelihood would be appropriate. In rural Java, where the villagers experience the lack of land resources, the peasants have been practicing multiple occupation by expanding their economic activities into the non-farm sector (White 1991). Such a way out probably cannot be replicated for the context of upland Sulawesi, where the non-farm sector is not well developed yet. For the marginalized peasants of upland areas, one possibility for expanding economic activities is probably the illegal encroachment on the forest margins for a new cultivation area.

The case study of Sintuwu, one of forest margin villages situated next to Lore Lindu National Park (LLNP) in upland Central Sulawesi, presented here, aims to describe the above-mentioned suggestions. Based on a qualitative work, with the support of a small survey[2], it analyzes respectively the change of social formation or, specifically, the shift of the dominant mode of production named as "Revolusi Cokelat", the change in local agrarian structure as one of its major implications, and its subsequent impact on the utilization of agrarian resources, particularly land resource in the forest margin.

2 Sintuwu: A Multi-Ethnic Village

Situated in Palolo, Sintuwu has been a part of one populous district in Donggala Regency, at least since the beginning of the 1990s. Palolo, one out of 18 districts in Donggala Regency, Central Sulawesi, is a frontier area situated along the eastern side of LLNP. It is a valley which gradually has been occupied by both spontaneous and programmed immigration since the 1960s. Immigration has made the Palolo valley, including Sintuwu and other villages, a high density population area, which is a deviant situation, considering the domination of dry land cultiva-

[2] A small survey was held for the purpose of data collection on land holding and ownership and modes of land obtainment. Survey samples were 30 households elected randomly in village Sintuwu. Table 2, 3 and 4 of this article are based on the results of the survey.

tion in its ecological setting.[3] Elsewhere in rural Indonesia, as mentioned by some researchers (Koentjaraningrat 1987; Geertz 1963), a high density of population is usually characteristic of a wet-rice field ecology.

As was indicated, Sintuwu was established during the last 30 years through a process of multi-ethnic immigration, which makes the village a multi-ethnic one (Table 1). Although the village community consists of more than four ethnic groups, it is clear that Kaili and Bugis appear to be the major ethnic groups. Since the Kaili peoples are the pioneer immigrants, they claim themselves as the "indigenous people" and, at the same time, give the Bugis people the label of "newcomer people".

Table 1. Household distribution by ethnic groups and hamlets in Sintuwu, 2001.

Ethnic groups	Number of households			
	Hamlet-1	Hamlet-2	Hamlet-3	Total
Kaili[a]	31	70	88	189 (57.62)
Kulawi	30	0	0	30 (9.15)
Bugis	42	51	2	95 (28.96)
Toraja	3	0	0	3 (0.91)
Others	6	5	0	11 (3.35)
Total	112	126	90	328 (100.00)

Notes: [a]Including Kaili Taa (98 households), Kaili Tara (60 households), Kaili Ledo (25 households), and other Kaili (Ija, Unde, and Rai: 6 households).

Situated on the Gumbasa river banks, the first settlement in Sintuwu was pioneered in 1961 by five Kaili households. These households were forced by local government to move out of village Bakubakulu of Sigi-Biromaru District, after breaking the local law by practicing shifting cultivation in the protected forest area. Following the pioneers, the Kaili immigration then continually flowed mainly from Sigi-Biromaru to Sintuwu in the 1960s and had its peak in the 1970s. In the 1980s the immigration flow was still going on, but in a decreasing trend.

In 1968 the first Bugis immigrant entered Sintuwu to be a laborer in a small saw-mill company. Following the first one, in the 1970s a number of Bugis took their families to settle in Sintuwu. They were labor forces for a logging company, which logged the forest area now included in LLNP area. With immigration at its peak in the 1980s, most of Bugis immigrants came from Bone, Soppeng, and Barru districts of South Celebes throughout Palu city. The relative economic success demonstrated by the former immigrants was a pull factor which drew more Bugis to Sintuwu.

[3] In 1999 population density in Palolo district (339.1 sq km) was about 79 persons per sq km Compared with the 1996 situation (69 persons per sq km), the 1999 situation represented an increase of about 13 percent. Meanwhile in 1996 the population density in Sintuwu (19.23 sq km) was 54 persons per sq km and three years later (1999), the number had increased to 65 persons per sq km. It means the growth of population density during the last three years in Sintuwu is 20 percent.

Following the Bugis were the Kulawi immigrants. The first Kulawi immigrants were a group of rattan gatherers came from a village of Kulawi District throughout the forest area now included in LLNP. With immigration at its peak at the last of the 1980s, most of Kulawi immigrants came from some villages of Kulawi District, a district situated along the western side of LLNP.

As the immigration flow into Sintuwu was going on, each of the ethnic groups had established its own exclusive settlement. Administratively, the villagers are distributed into three hamlets, namely Hamlet-1, Hamlet-2, and Hamlet-3. As indicated in Table 1, Hamlet-3 is exclusively a Kaili settlement, while the others are mixed settlement. But sociologically Hamlet-1 and Hamlet-2 are not really mixed settlements, since each of the ethnic groups has established its own settlement cluster exclusively. Actually Hamlet-1 consists of three main settlement clusters by ethnic groups (Bugis, Kaili, Kulawi) and Hamlet-2 consists of two main settlement clusters (Kaili, Bugis).

Considering the settlement pattern, it can be said that Sintuwu is not really an integrated community, since each of the ethnic groups tends to develop its own social life exclusively. Referring to the concept of "sodality" (Tjondronegoro 1984), one can conclude that the hamlets of Sintuwu are not the real sodality units, since social solidarity, particularly the mechanistic one, is not working, regardless of ethnicity. Social solidarity is likely to work exclusively among the members of each of the ethnic groups, especially in the unit of settlement cluster.

3 "*Revolusi Cokelat*": The Change of Social Formation

"Revolusi Cokelat", on the whole, is really a fundamentally rapid change in rural ecology due to a rapid expansion of cocoa cultivation. Consequently, it does not include only the change in sociological aspect of a community but also the cultural, economical, and political aspects. Moreover, it also includes the fundamental and rapid change in natural aspects, such as biology, climate, chemistry, and physics.

As the sociological perspectives are various, I choose to see "Revolusi Cokelat" as the change of social formation or, more specifically, the shift of the dominant mode of production which implies a fundamental change in the local agrarian structure. Assuming that the local social formation consists of three distinguishable modes of production, namely subsistence, petty commodity, and capitalist productions (Kahn 1974), it seems clear that "Revolusi Cokelat" can be regarded as the domination shifting from the subsistence production to the petty commodity or the petty capitalist one.

During the 1960s and the 1970s, as articulated in the local economic activities, the social formation of Sintuwu was characterized by domination of the subsistence production over the petty commodity one. The main subsistence production activities are the wet-rice cultivation in wet land areas and the corn cultivation in dry land areas. Rice and corn were produced, on the whole, for household consumption. Meanwhile, in the petty commodity sector, the villagers cultivate the perennial cash crops, mainly coffee. These kinds of cash crops were cultivated extensively, with minimum care, so that there was only low productivity. Nevertheless, coffee was a significant source of cash for the villagers, at least in the first half of the 1980s.

Cocoa was introduced about the mid-1980s and thereafter was cultivated extensively as a new "golden crop". Unlike the coffee plants, cocoa plants need more intensive care, so that its cultivation needs more intensive capital. Briefly, as is the case in Sintuwu, cocoa is cultivated in a more capitalistic way than coffee is. Consequently, as the cocoa cultivation was rapidly extended all around the village, the domination of subsistence production fell. Thus, at least since the early of the 1990s, the petty commodity or petty capitalist production has risen to become the dominant mode.

Since the exact data about the area of cocoa plantation are not available, one can take the data of land utilization composition as an indicator for the purpose of approximation. At least in the 1990s, Sintuwu experienced a drastic change in the composition of land utilization. In 1998 the official data of Palolo District indicate that dry land cultivation (647 ha, 34 percent), together with plantation area (491 ha, 25 percent), have dominated the village ecology, while wet-rice field contributes just a very little portion (75 ha, 4 percent).[4] Since dry land cultivation also includes cocoa cultivation in a multiple cropping system, and the plantation area is dominated by cocoa plants, it can be said that nearly 60 percent of the village area was taken up with cocoa plantation. Moreover, land plots known as house gardens are also cultivated with cocoa plants so that the area of cocoa cultivation in Sintuwu is now larger than what official data indicates.

Nevertheless, the existing composition of land utilization patterns in Sintuwu is a result of the shift of the dominant mode of production from subsistence production, namely wet-rice cultivation (and other food crops cultivation), to the petty commodity (or the petty capitalist) one, namely cocoa cultivation. During 1992-1998 nearly 72 percent of wet-rice fields in Sintuwu was converted to cocoa plantation, so that the wet-rice fields area drastically declined from 270 hectares in 1992 to as little as 75 hectares in 1998.[5] The main reason for such land conversion is the good price of cocoa, at least during the first half of 1990s. Sometimes the decrease of wet-rice fields productivity, due to the lack of water for irrigation, is proposed as a reason, but it is only a minor one.[6]

---

[4] Land distribution in Sintuwu by land utilisation patterns (1998): wet-rice fields 75 ha (3.90%); dry land cultivation 647 ha (33.60%); plantations 491 ha (25.50%), housegardens 30 ha (6.20%), iddle land 120 ha (6.20%); others including housing and roads 560 ha (29.10%) (Kecamatanan Palolo Dalam Angka/*Palolo District in Figures,* 1999).

[5] The 1992 data of wet-rice fields area in Sintuwu is cited from "Data Statistik Kecamatan Sigi-Biromaru" (*Statistical Data of Sigi-Biromaru District*) 1992. Sigi-Biromaru was the former district to which Sintuwu was included.

[6] Referring to an official letter issued by the regent of Donggala Regency, in June 1998 the head of Palolo District had issued an official letter (No. 525/0406/PMD/1998) concerning the prevention of land conversion from irrigated rice fields to plantation areas.

Without doubt, cocoa cultivation was a profit-making farm enterprise during the 1990s in Sintuwu. It was in the decade mentioned that the enlargement of cocoa plantations in Sintuwu drastically expanded in various ways. One of the ways was the conversion of subsistence agriculture area, namely wet-rice fields, corn fields (dry land), and house gardens into cocoa plantation. Although statistical data about corn fields and house garden conversion are not available, based on field observation, it can be said that such land conversion was done totally or partially. Partial conversion was done by introducing a multiple cropping system, in which cocoa plants and food crops, particularly corn, were put together in the same land parcel. Land conversion into cocoa plantations was mainly done by the Bugis, who bought rice-fields and dry land from mainly the Kaili.

The other way of enlargement, which is labeled as illegal, is the encroachment on forest area. This is mainly done by the Kaili people by reclaiming and grabbing forest area now included as LLNP. There are no exact data about the encroachment scale, but informants estimate that by mid-2001 the villagers of Sintuwu had cultivated nearly 100 hectares of cocoa inside the LLNP.

It is clear that "Revolusi Cokelat" has changed the economic structure of Sintuwu from the domination of subsistence to the domination of petty commodity or petty capitalist production. More specifically, on the whole, the village economy has shifted from the peasant economy, which is domestically oriented, to the small-holder economy, which is market (business) oriented. Concurrently, as the "cocoa economy" dominates the economic activities, the natures of capitalist production had gradually came to be significant among the village community. Two of the most notable features are the phenomenon of land privatization and the phenomenon of surplus accumulation as well as capital formation.

As will be described in more detail in the next section, it can be said briefly here that land privatization relates with the ownership of land resources, which is the main means of production. In the subsistence production, which is mostly articulated by the Kaili people, land resources are under collective ownership. Each household holds one or more parcels of land under the approval of village head, who has been given the authority for distributing the land resources impartially to his/her peoples. As the cocoa cultivation dominates the economic activities, and concurrently most of the land resources inside the village have been transferred to the Bugis people through purchase, the phenomenon of private ownership of land resources has become clear through land certification.

Transfer of land from Kaili to Bugis, as mentioned above, clearly indicates the second phenomenon, namely surplus accumulation particularly among the Bugis. Normally, as the Bugis peoples become successful in accumulating surplus out of the cocoa plantation, they subsequently use the surplus to accumulate capital by purchasing more land from the "indigenous people". What then comes to be clear about the shift of mode of production, as is indicated by all of the mentioned phenomenon, is that it really refers to the shift of economic power from the hand of Kaili people ("indigenous pepole") to the hand of Bugis people ("newcomer people").

4 Structural Change: "New Landed" and "New Landless"

From a sociological point of view, change in local agrarian structure is one of the most significant effects of "Revolusi Cokelat" in Sintuwu. As the "Green Revolution" brought about a sharp social stratification or even a social polarization among the wet-paddy community in rural Java, it is the case that "Revolusi Cokelat" has brought a phenomenon of new social stratification to the upland community of Sintuwu. Assuming that land ownership is the base of social stratification, it can be said that this revolution has both downgraded the Kaili households from the status of "landed" to the status of "landless" and upgraded the Bugis households from the status of "landless" to the status of "landed". Through the "Revolusi Cokelat", the Bugis peoples have experienced a social rise to be the "new landed", while the Kaili peoples have experienced a social decline to be the "new landless". Although it is not supported by quantitative data on land ownership, qualitatively I can say that in the 1960s and even the 1970s, when the Kaili immigration had its peak, the Kaili peoples luxuriated in land resources abundance. Under the collective regime of land ownership, which perceives land as the collective resource belongs to the community, in the beginning everyone could clear the forest area as much as he or she was able. One just needed to ask the chief of the pioneer settlement, or then the village head, about a piece of forest area to clear and to cultivate. After considering the availability of land resources the chief, who had the authority over land distribution and the responsibility for ensuring the land-based livelihood for his/her people, would immediately approve such an application. This pattern of land acquirement, however, led the Kaili immigrants to be the "landed" in Sintuwu.

But in the 1980s, the distributive authority of the village head, generally, was not valid anymore as a way of obtaining land (Sitorus 2002). It was the case that a number of Kaili immigrants still obtained land from the village head in the 1980s, but the land pieces were as small as the house-gardens. Survey data presented in Table 2 indicates that grant (from the village head and/or relatives: 29 percent of parcels) is no longer the general way to obtain land among the Kaili of Sintuwu, but purchasing is (33 percent of parcels) the significant one. Generally, they purchased the pieces of land mainly from their own relatives. Sharecropping and forest grabbing are alternative ways of land obtainment for the Kaili people, especially for the members of the landless stratum.

For the Bugis people, land obtainment through the distributive authority of the village head is not valid. The exception is the first Bugis immigrant, who obtained a piece of land through such a mechanism. In the first years of arrival in Sintuwu, the Bugis households generally obtained the land for cultivation by practicing the land tenancy institutions such as sharecropping and leasing. However, for the Bugis peoples, sharecropping or leasing is not merely a survival but an accumulation strategy through which they get money or financial capital. Once they get enough money, as has happened since the early part of the 1980s, they purchase the land parcels from the Kaili households. As shown in Table 2, land purchasing then is practiced by the Bugis households as the main mode of land obtainment

(59 percent of parcels). Such a mechanism has led the Bugis peoples to be the "new landed", with more and more land parcels in their hands. Concurrently, the regime of land ownership has also experienced a revolutionary shift. It has shifted from the collective type, in which land distribution is subjected to the political power, namely distributive authority of the village head, to the private type, in which land distribution is subjected to the power of the market.

Table 2. Modes of land obtainment by ethnic group in Sintuwu, 2001

Ethnic groups	Number of land parcels by modes of obtainment				Total parcels
	Grant	Purchasing	Share-cropping	Grabbing	
Kaili (n_1=16)	12 (28.57)	14 (33.33)	8 (19.05)	8 (19.05)	42 (100.00)
Bugis (n_2=8)	7 (25.93)	16 (59.26)	4 (14.81)	0 (0.00)	27 (100.00)
Others (n_3=6)	4 (26.67)	8 (53.33)	1 (6.67)	2 (13.33)	15 (100.00)
All (n=30)	23 (27.38)	38 (45.24)	13 (15.48)	10 (11.90)	84 (100.00)

The emergence of Bugis people as the "new landed" has its main explanation in the "Revolusi Cokelat". As Sintuwu has experienced a rapid expansion of cocoa cultivation since the end of the 1980s, the social group that derives the largest benefit from this phenomenon is the Bugis. They have started to cultivate cocoa since the early part of 1980s and also have started to accumulate land resources in their hand. Consequently, when the market opportunity came for cocoa in the end of the 1980s, the Bugis peoples were posited in a condition ready for a revolutionary expanding of cocoa cultivation. Moreover, as was shown by Acciaioli, the Bugis people is one of the most active ethnic groups in seeking economic opportunities (Acciaioli 1998).

However, comparing with the Kaili peoples, the Bugis peoples had more socio-economic advantages to participate in the "revolution", such as technological skill, farming managerial skill, and economic capital particularly land and money. Once the Bugis peoples took the initiative and led the "revolution", they consequently grew to be the leaders of the cocoa economy as well as the leaders of village economy. Through a process of surplus accumulation, in a relatively short period they have grown to be "petty capitalists" or "rural capitalists". Continually they accumulate capital both in term of surplus and of land resources so that most of the cocoa plantations inside the village are under their ownership.

Particularly it seems that land accumulation among the Bugis is facilitated by the Kailis' habit of selling their land in order to get a large amount of cash money for covering the budget of *adat* (customary) rituals, such as marriage and funeral events. Normatively, a norm of intra-ethnical land transaction must be a framework, but actually it is not working well, since there are no significant numbers of wealthy Kaili households to rely on as the land purchasers. Meanwhile, there are a significant number of wealthy Bugis who already have cash to purchase land from the Kaili peoples at any time. Since a marriage or funeral ceremony needs a large sum of money for budgeting the *adat* prerequisites, and most of the Kaili households have not enough cash to spend, the fastest way to get money is to sell a piece of land to the wealthy Bugis.

However, in the context of "Revolusi Cokelat", land market in Sintuwu has socially and economically led the Kaili and the Bugis in a different direction from each other. Survey data presented in Table 3 give indications of how far the Kaili households are sinking towards the status of "landless" and how far the Bugis households are climbing towards the status of "landed". Bugis households hold a total of 2.32 hectares of land on average, including ownership of a sub-total of 0.95 hectares of dry land. Meanwhile the Kaili households hold only a total of 1.86 hectares of land on average, including ownership of a sub-total of 0.81 hectares of dry land (0.69 ha) and wet-rice field (0.12 ha). The Bugis households of Sintuwu in general have no wet-rice fields anymore since they were converted to cocoa plantation.

Table 3. Average of land holding by ethnic groups and land utilization patterns in Sintuwu, 2001 (ha)

Ethnic groups	Dry land		Wet rice fields		Grabbed forest	Total
	Owned	Share-cropped	Owned	Share-cropped		
Kaili ($n_1 = 16$)	0.69	0.62	0.12	0.06	0.37	1.86
Bugis ($n_2 = 8$)	0.95	0.81	0.00	0.00	0.56	2.32
Others ($n_3 = 6$)	1.37	0.00	0.10	0.17	0.42	2.06
All (n = 30)	0.90	0.55	0.08	0.06	0.43	2.03

The Bugis households have been growing as the economic elite in Sintuwu today. They are establishing hegemony over the village economy through a process of land accumulation as well as accumulation of surplus produced by cocoa plantation. A Bugis household has never sold a piece of land to other ethnic groups members. In case a Bugis household needs much money, usually for the purpose of funding a *naik-haji* (pilgrimage) to Makkah, the household would not sell the land, but pawn it exclusively to another Bugis household. That way gives the household an opportunity to get back the pawned land at the end of pawning period. Moreover the economic hegemony of Bugis ethnic group is being established through a domination in the cocoa economy. In spite of holding the largest cocoa plantation on average, the Bugis people also control the local market of cocoa. Some large Bugis farmers also take the role of "middleman trader", who collect cocoa from the smaller plantation operators and then sell it in a relatively large volume to the larger traders, such as exporters and inter-island traders in Palu city.

5 Forest Margins: Alternative Bases of Socio-Economic Security

Having sold out their land to the Bugis people, where do the Kaili people go? In the case of the "Green Revolution" in rural Java, non-farm economic activities have come to be the solution for the marginalized peasants. In the case of Sintuwu or Palolo valley as a whole, non-farm economic activities are less developed, so there is no economic field for them except agriculture. But in the agriculture sector now the Kaili peoples are facing the problem of land scarcity, since most of their land parcels inside the village have been transferred to the Bugis peoples. In other words, the Kaili people have lost the bases of their socio-economic security on a substantial scale. Consequently, the economic opportunities for them have decreased.

Actually, the Kaili peoples have not experienced the declining of the bases of socio-economic security only in terms of economic capital, namely land resources, but also in terms of social capital, namely access to land tenancy institution. The root of such problems, on the one hand, is the fact that most of the Kaili peoples practice the *bapetak* production relationship exclusively over the wet-rice cultivation[7]. On the other hand, the total area of wet-rice cultivation under the Kaili's control is severely reduced in Sintuwu, since a large part of rice fields have been transferred to the Bugis and subsequently converted to cocoa plantation. It is very clear that the space for *bapetak* has been narrowed, thus reducing the opportunity to work it. Such a situation is worsened by the Bugis' exclusiveness regarding access to land resources. Once the Bugis people have the land on their hand, there is no chance for other ethnic group members to get access to it. In addition the Kaili peoples are not familiar enough with land tenancy institutions practiced over dry land cultivation, such as *bapajak* (land leasing), *bagi-hasil* (sharecropping), and *pinjam-pakai* (land borrowing), in the way that the Bugis are.

It is clear that under the "cocoa economy" the Kaili peoples are experiencing the lower socio-economic security, while the Bugis peoples are experiencing the higher one. However, cocoa plantation is now an indicator of socio-economic welfare among the people of Sintuwu. The Kaili peoples, of course, are eager to cultivate cocoa extensively, as has been done by the Bugis peoples. But once again they are faced with the problem of land scarcity inside the village area, since most of the land parcels are now controlled by the Bugis peoples. Such a situation causes a sort of jealousy on the Kaili side and, to some extent, also causes a sort of social tension between the two major ethnic groups.

---

[7] *Bapetak* is commonly practiced among the Kaili households exclusively over the wet-rice fields. It is a control over a piece of wet-rice field based on a prerogative right for planting and then harvesting. All of work sessions between planting and harvesting, including financial cost, are responsibilities of the land owner. Norm of yield sharing in Sintuwu is that one quintals of each *petak* (land parcel) yields (rice) goes to the *bapetak* right holder, whatever the productivity. Since it is exclusively demarcated by either neighborhood or kinship, *bapetak* relationship is likely an equalization mechanism among the Kaili people regarding the fulfillment of subsistence.

Coping with the "declining bases of socio-economic security", the Kaili peoples are seeking a new basis outside the village and are realizing that forest area is the best alternative. Taking such alternative bases, the Kaili peoples then encroach onto the forest area now included in LLNP area and cultivate cocoa extensively there. Such encroachment however has brought the Kaili people into land dispute with LLNP Office. On the one hand, the LLNP Office proclaims that the Kaili people have transgressed against the national park borderlines to cultivate the forest area and regards the Kaili as the land grabber. On the other hand the Kaili people do not agree with the latest park borderlines, which were established in 1982 along the logging road, separating the village and the forest area. The villagers argue that the borderline has proceeded about two kilometers into the village area, measured from the old borderline -- that was established by the colonial administration -- just on the top of Manjapi Hill inside the forest area. In fact, the Kaili had cleared and cultivated the forest area before the last official park borderline was fixed in 1982.

Although cultivating forest area is an illegal activity, locally it is formalized by the village head through issuing *Surat Keterangan Pengolahan Lahan* (Letter of Land Utilization) for each of the Kaili households who apply to cut the forest down for cocoa plantation.[8] Since the Office of LLNP is a public body, so that the LLNP is consequently under the public regime of ownership, it can be said that the Kaili peoples are now transgressing the borders of land rights in order to fight for private ownership in the state domain.

Anyhow, it seems that expansion into the forest area has worked effectively to cope with the scarcity of land as the main bases of socio-economic security inside the village. As shown in Table 4, the Kaili household samples that illegally cultivate the forest area (0.92 hectares in average) legally hold only a total of 0.62 hectares of land inside the village. This number is much smaller than a total of 2.01 hectares of land legally held by the Kaili household samples who are not illegal cultivators.

Compared with Kaili, a Bugis household has never grabbed a piece of land inside the forest area. The Bugis people always put the law first concerning with land holding. Moreover, the Bugis people calculate that land clearing activity inside the forest area needs much time, labor force, and money. Purchasing is more efficient than grabbing. This is the way taken by two Bugis household samples who had "illegal cultivation field" inside the LLNP area (Table 4). Actually they had purchased the field from the Kaili people before 1982, when the newest park borderlines had not yet been established.

---

[8] Each of applicants will be charged about Rp 100 000 for each hectare of forest area asked for cultivation. However the letter is not officially accredited by supra-village government institutions, especially by National Agency for Land Affairs (Ind., *Badan Pertanahan Nasional*) at regency level. Meanwhile the village head argues that such a letter is not a land ownership document but a land use document which prevents the cultivators from borderline disputes each others.

Table 4. Average of land holding among cultivator and non-cultivator of forest area by ethnic groups in Sintuwu, 2001 (ha)

Status, ethnic groups	Forest illegally cultivated	Land owned		Land sharecropped		Total legally land holding
		Dry land	Wet rice field	Dry land	Wet rice field	
Cultivator:						
1. Kaili (n_{11}=6)	0.96	0.54	0.00	0.08	0.00	0.62
2. Bugis(n_{12}=2)	2.25	0.50	0.00	0.00	0.00	0.50
3. Others n_{13}=2)	1.25	0.00	0.00	0.00	0.00	0.00
All (n_1=10)	1.29	0.42	0.00	0.05	0.00	0.47
Non-cultivator:						
1. Kaili (n_{21}=10)	0.00	0.78	0.19	0.95	0.09	2.01
2. Bugis (n_{22}=6)	0.00	1.10	0.00	1.08	0.00	2.18
3. Others (n_{23}=4)	0.00	2.06	0.15	0.00	0.25	2.46
All (n_2=20)	0.00	1.13	0.13	0.80	0.10	2.16

It seems that land grabbing inside the forest area has been considered as a way of land obtainment especially for the Kaili people. When the Bugis people come to dominate land holding inside the village, so that the Kaili have not enough land to cultivate anymore, there is no other way to cope with the land scarcity but to grab the forest area around the village margin. Although the agriculture density in Sintuwu is not as dense as is the case in rural Java, the access of landless Kaili to land resources is relatively limited. Such an explanation of this phenomenon is that the agrarian institutions, both the land tenancies and agricultural labor relationships which give the landless Kaili access to cocoa plantation held by the Bugis, are not well developed there. Hence, grabbing forest area is a way of accumulating a socio-economic security basis among the Kaili peoples who have been continually marginalized by the "Revolusi Cokelat".

6 Concluding Remarks

"Revolusi Cokelat", as exemplified by the case of Sintuwu, implies the change of local social formation. It was specifically indicated by the shift of the dominant mode of production from the subsistence production articulated as the wet-rice and corn cultivation to the petty commodity or petty capitalist production articulated as the cocoa plantation. Since a mode of production in agricultural society assumes the land resource as the main means (force) of production and the social structure, namely relation of production, is mainly based on the access to land resource, the "Revolusi Cokelat" consequently brought about a fundamental change in the local agrarian structure. This revolution has significantly facilitated the shift of land ownership regime from the collective type, in which land distribution is subjected to the authority of local political leader (village head), to the private one, in which land distribution is subjected to the power of the market (land market).

As the collective regime exclusively enabled the "indigenous people", namely Kaili, to accumulate land resource, and the private regime mainly enabled the "newcomer people", namely Bugis, to do the same, the shift of regime however implies the transfer of land resource, by purchasing, from the hand of the Kaili to the hand of the Bugis. Consequently, it implies the fundamental change of local agrarian structure, through which the Kaili people were downgraded from "landed" to "landless", while the Bugis were upgraded from "landless" to "landed". In the context of "Revolusi Cokelat", which established cocoa plantation as the main base of socio-economic security, such a change in the agrarian structure implies both the decrease of socio-economic security among the Kaili and the increase of such security among the Bugis.

Looking at the cocoa cultivation as a solution for the structural inequalities inside the village, the Kaili peoples face the lack of ownership of land resources, since most of such resources are controlled by the exclusive Bugis, who utilize it as the cocoa plantations. Moreover, land tenancy institutions, such as land leasing, land borrowing, and sharecropping are not in favor of the Kaili. This situation leads the Kaili peoples to look for a new basis of socio-economic security and, finally, they realize that the forest area in the margins of LLNP is the best alternative. Based on the traditional argument of forest borderline, and also informal as well as formal support from the village head, the marginalized Kaili peoples then encroach on the forest area and convert it to cocoa plantation. Perceiving the LLNP is under the public regime of ownership , it can be said that the Kaili peoples are transgressing the land right borders in order to fight for private ownership in the state domain.

It is clear that from the point of view of the Kaili, as exemplified by the case of Sintuwu, forest encroachment or conversion is not a problem, but rather a solution to the problem of social inequality as well as socio-economic insecurity. By reclaiming the forest area, and covering it with cocoa plantation, the Kaili of Sintuwu to some extent have succeeded in solving their problem of both access to land resource and socio-economic security. This has achieved social stability inside the village, with, however, forest margins destabilization, at least in the short term, as its cost.

However, in the context of human-nature interrelationships, analysis of "Revolusi Cokelat" reveals a warning that every "human-made ecological problem" inside the forest area, such as the "damaging forest encroachment", has its roots in the "sociological problem" outside the forest or inside the village community. Consequently, any solution proposed to solve the forest problem perceived as man-made must be tested for sociological soundness.

References

Acciaioli GL (1998) Bugis enterpreneurialism and resource use: structure and practice. Antropologi Indonesia 57: 81-90

Geertz C (1963) Agriculture involution: the process of ecological change in Indonesia. University of California Press, Berkeley Los Angeles

Hayami Y, Kikuchi M (1981) Asian village economy at the crossroads: an economic approach to institutional change. University of Tokyo Press, Tokyo

Husken F (1998) Masyarakat desa dalam perubahan zaman: sejarah diferensiasi sosial di Jawa 1830-1980 (Village society in the changing era: the history of social differentiation in Java 1830-1980). Grasindo, Jakarta

Kahn JS (1974) Economic integration and the peasant economy: the Minangkabau (Indonesia) blacksmith. Ph.D. thesis, University of London (London School of Economics)

Koentjaraningrat (1987) Manusia dan kebudayaan di Indonesia (Human and culture in Indonesia). Djambatan, Jakarta

Li TM (2002) Local histories, global market: cocoa and class in upland Sulawesi. Development and Change 33: 415-437

Sitorus MTF (2002) Land, ethnicity and the compeeting power: agrarian dynamics in forest margin communities in Central Celebes, Indonesia. STORMA discussion paper series no. 5

Taylor JG (1989) From modernization to modes of production: a critique of the sociologies of development and underdevelopment. Macmillan, London and Basingstoke

Tjondronegoro SMP (1984) Social organization and planned development in rural Java: a study of the organizational phenomenon in Kecamatan Cibadak, West Java and Kecamatan Kendal, Central Java. Oxford University Press, Singapore

White BNF (1991) Economic diversification and agrarian change in rural Java. In: Alexander P, Boomgard P, White B (eds) In the shadow of agriculture: non-farm activities in the Javanese economy, past and present. Royal Tropical Institute, Amsterdam

Wiradi G, Makali (1984) Penguasaan tanah dan kelembagaan (Land tenure and institutions). In: Kasryno F (ed) Prospek pembangunan ekonomi pedesaan (Prospect of rural economic development). Yayasan Obor Indonesia, Jakarta

Worsley P (1988) The three worlds: culture and world development. George Weidenfeld & Nicholson, London

Traditional Land Tenure among the Black Thai and its Implication on the Land Allocation in Yen Chau District, Son La Province, Northwest Vietnam

Thomas Wirth, Dao Chau Thu & Andreas Neef

1 Introduction

Since the mid-1950s Vietnam has experienced a number of changes in its land policy. After the French quit the country, the North Vietnamese government started a collectivisation process following the socialist model. After the reunification, the Government of Vietnam (GOV) extended the collectivisation process to the south. This resulted in serious food shortages in the late 1970s (Pingali and Vo Tong Xuan 1992). As a reaction, the de-collectivisation process started in 1981 with the Directive 100 which gradually shifted responsibility for production from the agricultural cooperatives to farm households. Land allocated to co-operatives could be subcontracted to individual households. By the end of 1987, 30% of the agricultural land in Vietnam was already under private use (Nguyen Van Tiem 1992). The second step began with the Resolution 10 issued in 1988. It restored the farm household as the main unit of agricultural production, which led to a large-scale decollectivisation in most parts of the country (Tran Thi Van Anh & Nguyen Manh Huan 1995). In the third stage, land use rights were allocated to farm households with the Land Law, enacted in 1993, providing long-term tenure security of 20 years for annual crops and aquaculture and 50 years for forest and perennial crops. The concomitantly issued so-called *red book certificates* (RBC) guarantee the rights to exchange, transfer, inherit, mortgage, and lease land use rights. The land allocation process was complemented by additional reforms in the institutional sector ranging from improved supply with and access to high-yielding varieties, fertilisers and pesticides to the development of a rural credit system (Neef et al. 2000).

Apart from the issues of food security and land reform, Vietnam faced the problem of rapid deforestation. The share of forest and woodlands decreased from 1965 to 1990 from 42% to 28% in the whole country and since then has been slightly increasing (FAOSTAT 2002, General Statistic Office 1999). In Son La Province the deforestation was even stronger. In 1994, forest covered only 10% of the total land area and 76% of the land were considered as wasted land (General Statistical Office 1999). Through a number of national reforestation programs, the GOV aims to substantially increase the amount of forested land. At the same time, it tightens the control on land use through the land law of 1993. Therefore, the land allocation process in Yen Chau district went hand in hand with land use planning (SFDP 1997). By combining these two measures, the local authorities aim to clarify the boundaries between agricultural land and forest and to protect remaining forest and the reforested land from further destruction.

Neither the latest changes nor any former changes in the land policy replaced the traditional tenure regime, but led to a co-existence and interaction of traditional/customary and modern/official tenure systems which is a common phenomenon in many countries. Several authors have dealt with this kind of 'legal pluralism' in a conceptual and empirical way (e.g. Merry 1988, von Benda-Beckmann 1991, Meinzen-Dick and Pradhan 2002). The concept of legal pluralism in resource tenure and management is based on the view that "in all societies, management and the use of natural resources occur within a complex context of normative regulations" (von Benda-Beckmann 1991, p 76). Several regulations, norms and laws can coexist and interact in a specific locality. These can be state laws, traditional laws, religious laws, project laws, donor laws, customs and social norms. Complex and overlapping legal contexts can impose potential constraints on the activities of various actors involved in natural resource management (von Benda-Beckmann 1991). On the other hand, legal pluralism can allow the actors to choose the law which supports their position, so-called forum shopping (Meinzen-Dick and Pradhan 2002). The extent to which forum shopping is possible depends on the administrative position, the social status and those of opponents, but allow several solutions for each claim.

This situation is also true for the case of Yen Chau where the new land law meets with traditional laws and customs of the local population and local adaptations of former state laws. This paper contributes to the ongoing discourse on legal pluralism by discussing the case of the Black Thai ethnic minority group in Yen Chau. The focus is set on both ways of interaction, the adaptations of the traditional land tenure to the official land policy and the influence of the traditional land tenure on the local implementation of the official land policy.

2 Material and Methods

The paper is based on a one year field work in eight villages in Yen Chau District, of which 4 villages are Thai villages. Despite the ethnic group, the villages vary in altitude, market access and the availability of wetlands. The project focusses on the relationship between tenure of natural resources, investment and resource management. It is part of the Uplands Program (SFB 564), a Collaborative Research Program involving German, Vietnamese and Thai research institutions. A broad range of methods were used: a complete household survey by means of standardised questionnaires including 390 households, selected Rapid Rural Appraisal tools, semi- and unstructured interviews and genealogies in two Thai villages. Semi-structured interviews with local authorities and administration staff, analysis of legal documents and the land register were used to cover the official side of the research objectives.

Yen Chau is a district of Son La province in Northwest Vietnam situated along the national road no. 6 at the border to Laos, 250 km from Hanoi (see figure 1). Its economy is based on agriculture with rice production for subsistence and maize, fruit trees and animal husbandry for cash income generation. The altitude of Yen Chau varies from 200 m to 1420 m above sea level with only a small portion of flat area. The total area of Yen Chau is 84.000 ha of which around 14.000 ha is agricultural land (Statistical Department of Yen Chau 2001, Land Administration Department 2001). The population is estimated at 55,000 in 11,300 households and it is composed of 5 different ethnic groups: 54% of the population are Black Thai, 18% Kinh (Vietnamese), 12% Hmong, 13% Sinh Mun and 0.5% Kho'mu (Land Administration Department 2001, Luibrand 2001). In the communes along the national road no. 6 the share of the Black Thai is higher, ranging from 73% to 96% (Statistic Department Yen Chau 2001)

Table 1. Characteristics of the Thai villages in the research region

Name	Tat Heo	Dong	Sai	Na Pa / Nhuung
Commune	Chieng Pan	Sap Vat	Sap Vat	Sap Vat
Number of Households	16	47	82	63
Altitude [m asl]	700	300	280	200/500
Market Access	7 km dirt road to main road, 3 km to Yen Chau on main road	1 km footpath from Yen Chau, access by motorbike possible	On the main road 3 km from Yen Chau	7 km footpath from main road, 3 km on the road from Yen Chau/+ 4 km footpath (dirt road constructed in 2002)
Availability of wetlands	1 harvest wetlands for all households	1 or 2 harvest wetlands for all households	1 or 2 harvest wetlands for all farm-households	1 or 2 harvest wetlands for some households
Genealogies	No	Yes	No	Yes

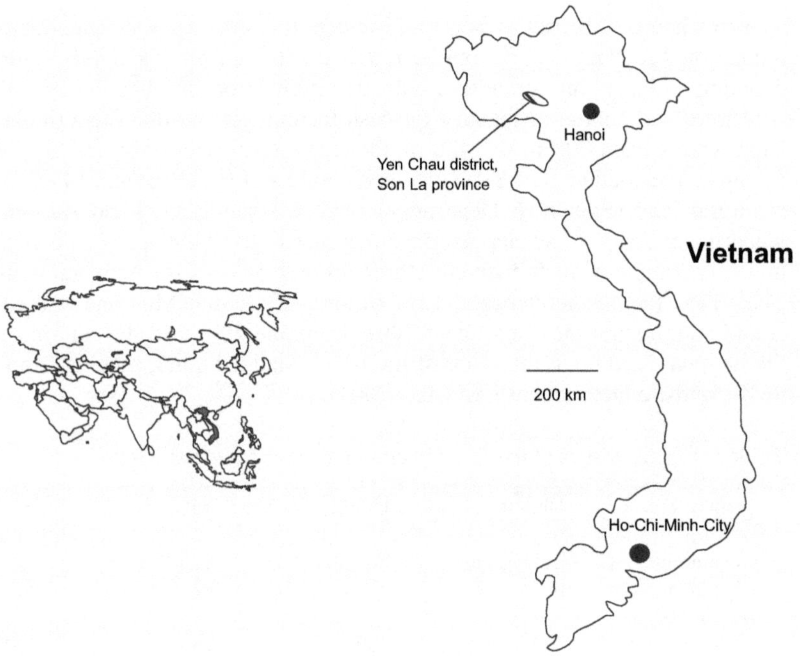

Fig. 1. Location of the study site

The farming system in mountainous Southeast Asia is often divided in two major parts, lowland and upland cultivation (e.g. Towa et al. 2001). In this paper, we will divide the farming system into wetland and rainfed cultivation, as we feel that this fits better to farming practice than the usual geographic distinction. The wetlands are irrigated and allow 1 or 2 harvests per year, depending on the water availability. Generally, the farmers grow paddy rice on these fields, but a few farmers also grow vegetables, maize and potatoes on their wetlands. Major annual and biannual crops on rainfed land are maize, cassava, upland rice and sugarcane. As perennial crops, farmers grow fruit trees (mango, longan, lychee and tamarind) and tea.

3 Development of the Traditional Land Tenure System from 1900 to De-collectivisation

The traditional Thai society before the August Revolution in 1945 was characterised by a feudal structure (Le Sy Giao 1979). The society comprised of two classes: nobles and farmers. The basic administrative unit was the muong, a small municipality where the land belonged to the noble family, which allocated land use rights to the farmers. In some areas with no influence by a noble family a society of free farmers developed, e.g. in Quy Chau (Phan Chi Than 1979). Yen

Chau was situated between the two muong Moc Chau and Mai Son and thus also had no feudal structure. The land belonged to clans of free farmers and the clan council, made up of the oldest male household members, allocated the land to the households according to their needs on a regularly basis, usually every five years. The members of the clan relied on the same customs and rites while the genealogical relations between its members often remained unclear. However, everybody agreed on the fact that they existed.

Different factors have influenced this traditional land tenure system in the last hundred years and led to changes and adaptations. **Increased population pressure** and enhanced **market integration** have led to an expansion of annual rainfed fields for the cultivation of upland rice, cassava and maize. In contrast to the paddy rice land in the valley, these fields were under a private property regime with limited rights. The clan council had to grant the right to the households to clear the forest, which was under the common property regime of the clan. The owner had the right to bequeath the fields to his children, but he needed the allowance of the clan for selling it. The same rights were also possible for wetlands, when the households constructed terraces and an irrigation system on his private fields. However, for paddy land in general, the traditional reallocation mechanism were kept. During the **wars** against the French and the Americans, many newly founded households became uxorilocal[1] as many men lost their brothers and fathers in the war and the ties to their home village became weak. Some of them also were admitted by the new clan and became clan members. Since the reunification with South Vietnam in 1975, they returned to the virilocal[2] tradition, but uxorilocal households are still more frequent than before the war.

The most important state intervention was the **collectivisation** process, which started after the liberalisation of North Vietnam in 1954. In a first step cooperatives were founded at village level. Later, the state tried to establish bigger cooperatives including several villages and further specialisation in production. In Yen Chau, these bigger cooperatives proved difficult to manage and were therefore separated after a few years into the previously existing village cooperatives. The cooperatives aimed to replace the households as production units, which never happened completely in Yen Chau. Table 1 shows the main modes of access to the wetlands (mostly paddy fields) and to annual rainfed. In both villages, the cooperative was much more active in the cultivation of the wetlands, and a substantial part of the annual rainfed remained under "private ownership" within the traditional tenure system. Additionally, it shows, that the collectivisation process was much stronger in Sai, a village at the national road no. 6 and center of the commune than in Na Pa, which is 7 km by foot from Sai. This was against the official policy, but these local adaptations anticipated somehow the subsequent decollectivisation process (see below).

[1] Living within the wife's community

[2] Living within the husband's community

Table 2. Modes of access in two villages for annual rainfed land and paddy fields

	Na Pa				Sai			
	Annual rainfed		Wetland		Annual rainfed		Wetland	
	Plots [n]	Area [ha]	Plots [n]	Area [ha]	Plots [n]	Area [ha]	Plots [n]	Area [ha]
From cooperative	38	11	26	1.9	92	30.3	172	11.5
	(12.2)	(17.3)	(30.6)	(29.7)	(52.6)	(48.5)	(83.0)	(87.1)
Inheritance	63	14.1	8	0.5	68	25.9	30	1.4
	(20.2)	(22.2)	(9.4)	(7.8)	(38.9)	(41.4)	(14.5)	(10.6)
Clearing	195	35.8	48	3.8	9	4.9	0	0
	(62.5)	(56.5)	(56.5)	(59.4)	(5.1)	(7.8)	(0.0)	(0.0)
Others / no answer	16	2.5	3	0.2	6	1.4	5	0.3
	(5.1)	(3.9)	(3.5)	(3.1)	(3.4)	(2.2)	(2.4)	(2.3)
Total	312	63.4	85	6.4	175	62.5	207	13.2
	(100.0)	(100.0)	(100.0)	(100.0)	(100.0)	(100.0)	(100.0)	(100.0)

4 The Land Allocation Process – Equity Issues

The land allocation process is based on the land law from 1993, which on the one hand grants long-term use rights of land and forest to individuals, households, group of households and other organizations, and on the other hand enforces the State's control over land use and imposes taxes on land, replacing taxes on agricultural production. Newly established cadastral agencies together with the Departments of Agriculture and Rural Development and Forest Protection Department were responsible for the process in each locality. The general process was organized in a decentralized way through devolving decision making power to provincial and district People's Committees. In Yen Chau, the land allocation process was first carried out in the districts along the national road No. 6 between May and July 1994. The results of this land allocation were unsatisfactory due to inaccurate estimations of plot size (SFDP 1997). Additionally, paddy land was excluded from the land allocation by a general agreement within the district (Sikor and Dao Minh Truong 2000, SFDP 1997). This practice was not compatible with national policy, but followed the tradition of the Black Thai. As a result, the land allocation process was redone in 1999, covering the whole district. Two crucial steps can be identified in the whole process. First, the development of an appropriate methodology and the formulation of guidelines (preparation) and, second, the application for land and the following field visit (see figure 1). While the District PC (People's Committee) was responsible for the first step, strongly supported by the GTZ-GFA Project SFDP Song Da (Social Forestry Development Project), the latter was mainly in the hand of the local community.

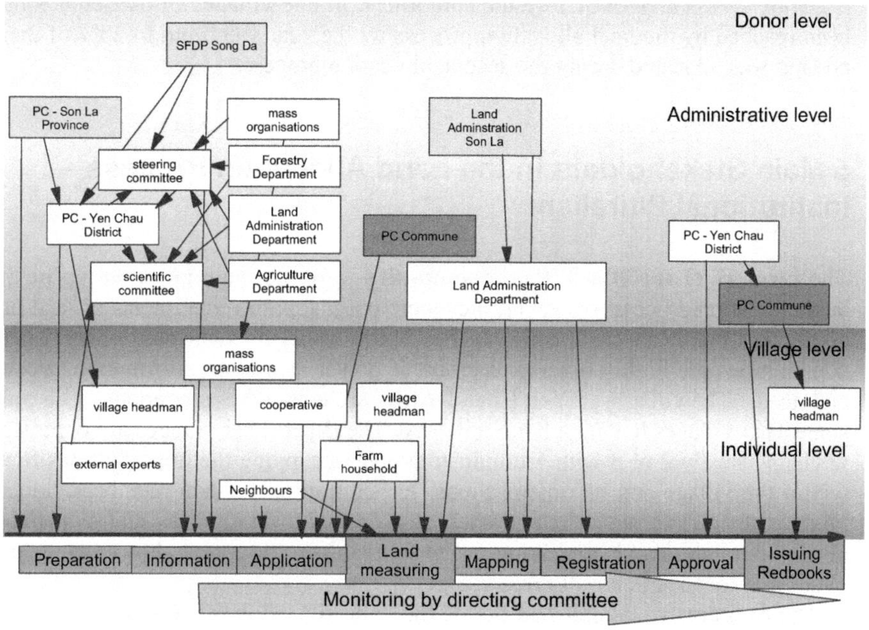

Fig. 2. The land allocation process and its involved stakeholders (simplified)

During the preparation phase, these laws influenced the results of the negotiation process within the administrative and donor system. National and provincial guidelines and decrees set the frame within which the locally involved stakeholders could develop their own guidelines and methodologies for the land allocation process. Interests of the local elite in the villages were supported by the commune staff, which are members of both the administrative system and the local community. The PC of Yen Chau feared the possibility of social unrest when the changes for the local community through the land allocation would be to big (Sikor and Dao Minh Truong 2000). Therefore the negotiation for the first land allocation led to a legalisation of the existing land tenure system of the Black Thai. In the second negotiation process, the donor law, represented by the SFDP Song Da had a big influence on the result, as their proposed methodology was approved by the PC and the influence of the local elite was much smaller than before. The outcome of this process, however, represented only the official part of the negotiations. In the later stage of the land allocation process, additional negotiations took place in each village after the farmers were informed by village assemblies in the "application"-phase (see Figure 1). This process was dominated by the local elites as access to information proved crucial at this step. The study of the SFDP (1997) showed that many farmers forgot a lot of information as the time period between the application and the village assembly was (too) long. In addition, the farmers were confused by the fact that the former land allocation process became invalid and everything needed to be redone. This is reflected by the fact that most households stated that land was allocated to them, even if they did actually not receive *red book certificates* (RBC) by the authorities. In one village, all households had been reached by the land allocation process by the year 1994, but to 25% of them no land was allocated during the second allocation process.

5 Main Stakeholders in the Land Allocation Process – Institutional Pluralism

The **farmers of the Black Thai community** were the main target group of the land allocation process. However, they were not just passive recipients of land use certificates but also a main actor, as they had to apply for their land rights. These communities are highly diverse in terms of power, access to information, wealth and needs. The villages in Yen Chau District are ethnicly homogenous, with only some Thai villages having a Kinh minority. The Black Thai villages are organised in clans, usually 2 or 3 with 1 dominating clan occupying the important positions within the village and its organisations. In addition to the clan members, households with links through their mothers resp. wives to these clans and sometimes allochthone households are living in the same village. Figure 2 shows one example of this social structure for one village.

The **cooperative**, established during the collectivisation process, was by far the most important institution in the Black Thai villages. They were installed during the collectivisation process. The decollectivisation process restored the farm households as the main production unit in agriculture. As a consequence the cooperative lost power and influence on agricultural production and the use of the land. Currently, several types of cooperatives can be found in Vietnam, from cooperatives providing little service for its members to still fully functioning cooperatives (Kirsch 1997). Today, the most frequent type of cooperatives in Yen Chau are service cooperatives, supporting farmers in marketing, providing inputs and organising extension for the village. Officially, the cooperatives were no longer included in the decision making processes at district or communal level (DARD Yen Chau, 2002). However, in most cases the members of the traditional clan council were members in the board of the cooperative. In a few cases, outsiders in key positions could limit their influence. Even the official tasks of the cooperative are reduced, in the perception of the villagers, the cooperative is the successor of the clan council and therefore authorised for many decisions on land allocation and management of the communal resources. This important role is represented by the fact that the cooperatives hold the land reserves for future generations and rent or sell this land to the households. In most villages, the head of the cooperative is also village headman, which further strengthens the position of the cooperative.

The village units of the **mass-organisations** like Farmers' Union, Women Union, Veteran Union and Youth Union and the **Communist Party** are important means to gain information for their members. On district level, representatives of these organisations were involved in the negotiation process described below. A stronger influence on the realisation of the land allocation process in the village was exerted by the **village security council**, which is linked to the District police. Among other tasks, these councils are responsible for the solution of land conflicts. Often sons of important families are members of this council and the village headman or an older member of the council, the leader, takes the decisions in land disputes. According to the village headman, land disputes are settled this way without notification or inclusion of the district police or local authorities.

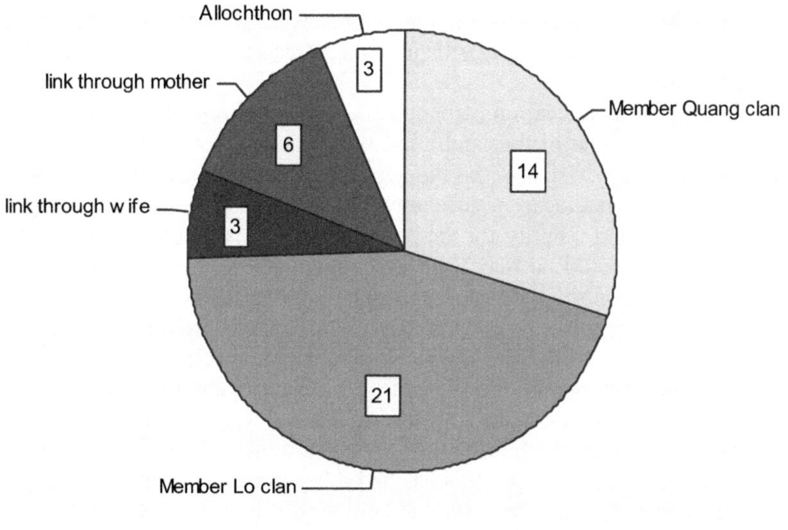

Fig. 3. Social structure of Dong village

At administrative level, the **People's Committee** of the District is the most important stakeholder, as they are responsible for the whole process and, finally, for the approval of the RBC's. For the supervision of the whole process, the PC was supported by the **steering committee**. In this committee representatives of the mass-organisations and the involved departments coordinated the work. This was important as it was intended to allocate land and forest in the same process to save capital and labour. Due to many uncertainties in the forest land allocation, this objective could not be completely achieved. The **Land Administration Department** was responsible for the technical side of the land allocation process, like measuring the fields, mapping and registration of the plots.

6 Outcome of the Land Allocation Process – Coexistence between Traditional and Official Tenure Systems

For the Black Thai people, the result of the first land allocation process was relatively favourable. They received the rainfed fields, which had often been cultivated by their respective families for decades and in the paddy fields they could apply the traditional reallocation mechanism among the clan members to maintain a rather equal situation for all involved households. Through the intervention of the administration in many rainfed fields, land conflicts were settled. In all the villages, key-persons indicated that since the land allocation process, village internal land conflicts decreased substantially. Between the villages, the borders were clarified too, but some conflicts remained unresolved. In particular, the boundaries of Thai villages to neighbouring villages of a different ethnic group are still often contested today.

The second land allocation strengthened the government's control over the land, an important goal of the land allocation process. But it only fixed the land tenure situation that existed at the moment of the land allocation. The traditional system of regular reallocation officially disappeared and with it an important means to ensure food security for the poorer households. In the future all changes in access to land would be registered and supervised by the state administrative system. With this control, the state wants (1) to protect the poor from losing access to land by market regulations and (2) to avoid land concentration. The question of whether the state can fulfil this task better than the local community by their traditional institutions remains open and needs to be examined in the future.

Until now, the land allocation process has not had a significant influence on the endowment of land for the households. Whereas minor changes have occurred in terms of access to land, the process has had a big impact on tenure security. Generally, the households have received the land which they occupied before. For this land tenure security was guaranteed by the traditional institutions in the village. Even a land market existed prior to land reform, which was able to play an important part in promoting long-term-investments (Carter and Yao 1999).

Traditionally, the Black Thai farmers gained access to paddy land through the reallocation system or by building new terraces on their annual rainfed fields or in the forest (Sikor and Dao Minh Truong 2000). The latter is of minor importance in the villages along the national road no. 6, as the most suitable places for paddy cultivation were used under their communal system. In the remoter villages, however, this mode of access is more important (see also table 1). For the annual rainfed fields, which did not remain under individual management, the cooperative generally returned the fields to their former owners or their heirs. Through the forest land allocation, some households gained access to land, which did not belong to their family before. In Yen Chau, forest land with no forest cover was allocated to individual households for reforestation within two national programs, the former 327-program ('Regreening of the barren hills'-program) and Dao No-Project, a part of the 5 mio ha program, an ambitious government project aimed at establishing 2 million ha of protection forest and 3 million ha of production forest by the year 2010 (cf. Neef and Schwarzmeier 2001). Most of the reforested land in the 327-program already belonged traditionally to the involved households. With the Dao No-project some households gained access to formerly unused or communally used land. Many of the reforested areas are far away from the villages. In one village, this area even had previously not belonged to the village. As considerable parts of this land have actually not been reforested, but are used as agricultural land for annual crops, these households gained access to agricultural land by the Dao No-project.

The effects of the land allocation process on tenure security are diverse. Prior to the land allocation, tenure security was granted by the traditional institutions in the village. The members of the local communities had high security on their annual rainfed, which also protected their long-term investments in this land, e.g. building rice terraces, planting fruit trees. For members linked through their wives or mothers, tenure security was reduced, especially in the densely populated main valley, as these farmers were more vulnerable in land disputes. During the cooperative period, allochthon households often came into the village with support by the district administration to work in the management of the cooperative. Usually, these households faced a lot of difficulties during the integration in the local community and often returned to their native village. Those who succeeded in installing themselves in the village had a quite strong position in the village and therefore also their tenure security on the rainfed fields. But this case only appeared in one village in the survey. More often, no allochthon households live in the villages or non-farm households bought some residential land in villages along the national road no. 6. Changes in tenure security were caused by two major factors. First, the traditional access to land was legalized for the households obtained RBCs (see Table 2). The tenure rights are now recognized by the state and can therefore be protected by administration support. Through the five rights of the RBCs (exchange, transfer, inheritance, mortgage, and lease), the legal situation for the households improved. Officially, selling the land is now possible, but in reality many restrictions from the local authorities and the local communities limit this possibility. Second, the land allocation process strengthened the state control on land use. Households can lose their use-rights, when they use the land differently than indicated in their RBCs, but also many households have some agricultural land in areas, which are officially declared as forests. The tenure security on this land has weakened considerably, as the households are now in danger of losing this land by state intervention. Household control rights on land use are now weaker than before.

Table 3. Overview of the results of the land allocation process in Yen Chau (Land Administration Department 2001)

Commune	Total of Households	Households with redbooks	Percentage of households with redbooks	Land allocated with redbooks		
				Agricultural Land [ha]	Forestry Land [ha]	Resident Land [ha]
Chieng Khoi	538	537	99.8	769.1	79.1	19.7
Chieng Dong	1170	1161	99.2	1183.6	6.36	39.8
Chieng Pan	815	627	76.9	832.1	120.9	23.8
Sap Vat	707	590	83.5	757.6	73.6	20.2
Muong Lum	383	383	100.0	428.0	0.0	11.8
Tu Nang	1255	1255	100.0	1903.3	16.8	40.0
Chieng Hac	681	675	99.1	1126.6	202.5	24.3
Border communes	5783	5315	91.9	7010.2	526.86	203.9
Total District	11332	10543	93.0	14010.5	1026.12	383.5

The diverse groups within the local community are differently affected by these changes in tenure security. Households without RBC and much agricultural land on declared forest land are worst off. For these households, tenure security is very low and they can easily lose their land. The households in this situation have rather weak links to the ruling clans and seemed to be excluded from information on the land allocation procedures. Officially, everybody disagrees on this, but no other reasons could be identified during the field research. For newly established households, tenure security was reduced. By law, they can gain access to land through officially recognized land transfers, either by inheritance or through the land market or through rented land. While they lack the capital to buy land, inherited land usually is not registered at the Land Administration Department. Therefore, they have no officially recognized land use rights. As regards tenure security on their land, they depend on the traditional system. For many households related by their wives or mothers to the ruling clan and a RBC, tenure security improved by the official recognition of their land use rights. Plot measurements and clarification of boundaries reduced village internal land disputes. Additionally, they have now better possibilities to get state support to defend their claims, as they became less dependent on the ruling clan. The long-established households of the ruling clan just legalized most of their traditional land rights, therefore their tenure security did not change for them.

The village Dong is dominated by two clans, the clan Lo and the clan Quang, each consists of three groups. Currently, the Lo clan is holding the power, but in the past the power was shared between the strongest groups of the village (2 groups from the Lo and 1 group of Quang). Based on the tradition, each group has its claim on the village territory. During the war, the Lo clan allowed 6 women of their clan to settle in the village with their husbands (uxorilocal). One of these couples changed later to the Lo clan and became a member. Descendants of two couples remained in the village and build their claim, but still depend on the Lo clan. The sons of the other two couples moved out of the village with only one remaining to work the land of his parents. These now form the group of the 7 households linked through their mother. Later, the same happened twice again. In 1994, the village had 44 households and land was allocated to all of them. After the second land allocation only 34 of now 47 households received a RBC. Figure 3 shows the distribution of the RBCs in the village.

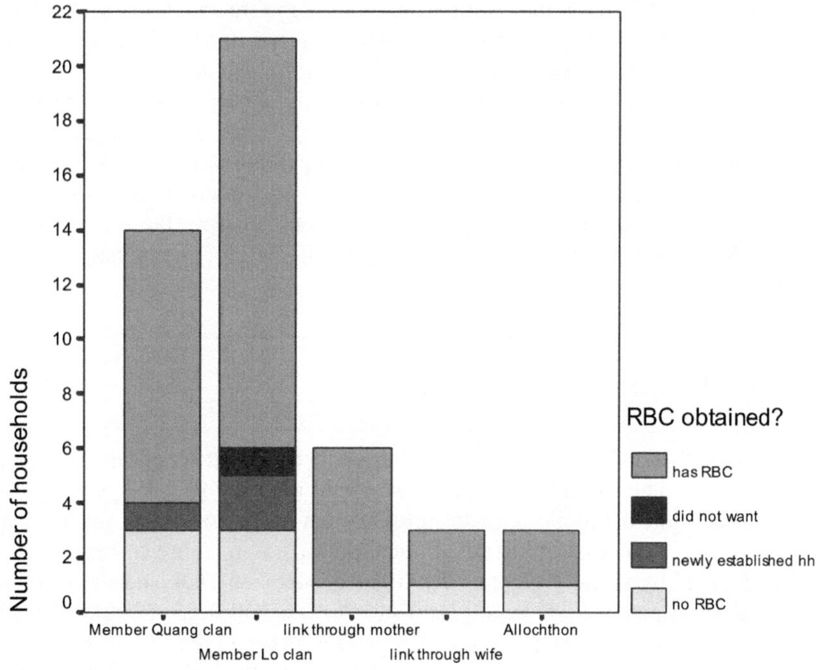

Fig. 4. Distribution of RBCs in the different clans of Dong

Two of three allochthon households have a RBC. One of them, arrived in the village to work in the management board of the cooperative and was thus able to obtain an influential position in the village, which he still holds, even without official position. The other household with an RBC are relatives from his native village who moved to him after they had problems in their former village. Therefore, these two households were able to enhance their tenure security. The remaining allochthon household has weak links to the other allochthon households and no relations to the other families and seemed to be excluded from the information on the land allocation process. The two wives of the households, which are linked through them, are offsprings of important families. One of them is the daughter of the village headman. Therefore they received a RBC, opposite to the third household, which is also rather isolated in the village. The biggest groups of households which did not receive a RBC are members of the Quang clan and Lo clan (3 households each). All of the Quang clan belong to the weakest group in the village. These households did not apply for their own RBC and depend now legally on the land of their father and brother respectively. The same is also true for the three households of the Lo clan. However, this case neither happened in the strongest Quang group nor in the strongest Lo group, holding the power in their respective clans. The households which are linked through their mothers to the ruling clan are well off in tenure security. Only one household has not received a RBC and depends now on his brother, which holds the redbook in the family. The households with RBC in this group have either good information channels outside the village (teacher and his brother) or their fathers gained influence in the cooperative and their mothers are members of the strong groups within the Lo clan (3 households). This group of households is also important in the legitimization of the actual power of the ruling groups within the Lo clan. This example shows the importance of access to information, which depends mainly on two factors, the personal relation to the powerful villagers and the interactions with outsiders, mainly with the administrative system. Households missing these two factors are threatened by the exclusion from information and therefore handicapped in the realization of their rights.

7 Conclusions and Outlook

Obviously, the land allocation process has created a situation of legal pluralism in Yen Chau, where the traditional tenure system and the official system presently co-exist and interact in the same locality. Results have shown that outcomes of the land allocation process and effects of legal pluralism on tenure security have to be differentiated by social groups in the villages. Meinzen-Dick and Pradhan (2002) discussed the potential of legal pluralism to cope with uncertainties. They state that through forum shopping farmers are able to defend their claims in the system which suits best to their current situation. However, the potential for forum shopping is limited to the households which received RBC. The others can only defend their claims in the traditional system, but became vulnerable to claims which are not embedded in the local legal system. Currently, the danger that households without RBC will lose their land rights because of other villagers, using their advantage in forum shopping, is low. The traditional tenure system is still securing their land use rights within the community.

Due to several reasons (no registration of land transfers, illegal land occupation, forgotten land application), an increasing gap between the land register and the current tenure situation in the villages can be observed. This gap indicates the strong influence of the traditional land tenure institutions. At the same time, an individualization process can be observed, mainly caused through the allocation of wetlands to individual households.

At this stage, it is difficult to predict how the coexistence of the two tenure systems will influence the future development, but two alternative scenarios can be assumed: (1) the differences between the land register and the local land tenure situation increase which would create a great potential for land conflicts and (2) the traditional land tenure institutions will gradually adapt to the new institutional environment and a balance between the official tenure policy and the traditional tenure system can be found.

8 Acknowledgements

The authors would like to thank all villagers and local authorities for their hospitality and patience in answering our questions during the field research. The financial support of the Deutsche Forschungsgemeinschaft, Germany, in carrying out this study is gratefully acknowledged.

References

von Benda-Beckmann F (1991) Legal uncertainty and land management. In Savenije H and Huijsman A (eds) Making haste slowly: Strengthening local environmental management in agricultural development. KIT Publishers, Amsterdam, pp 75-88.

Carter, MR and Yao Y (1999) Market versus Administrative Reallocation of agricultural land in a period of rapid industrialization. Policy Research Working Paper 203, The World Bank, Washington

Department of Agriculture and Rural Development DARD Yen Chau (2002) Interview June 2002.

FAOSTAT (2002) Query on Land use at www.fao.org

General Statistic Office (1999) Environment Statistic Compendium of Vietnam. Statistical Publishing House Hanoi

Kirsch, OC (1997) Vietnam: Agricultural Cooperatives in Transitional Economies. Diskussionsschriften der Forschungsstelle für Internationale Wirtschafts- und Agrarentwicklung eV (FIA) Nr. 59. Heidelberg

Land Administration Department of Yen Chau (2001) Interview in March 2001

Le Sy Giao (1979) Vai net quan he xa hoi cua nguoi Thai o muong Ca Gia, Thanh Hoa. (Features on the social relation of the Thai in muong Ca Gia) ER 2(1979):63-69

Luibrand A (2001) Ten Years of Doi Moi How did the Black Thai participate in the rural reform process in Vietnam. Ph.D. thesis, Stuttgart

Meinzen-Dick RS and Pradhan R (2002) Legal Pluralism and Dynamic Property Rights. CAPRi Working Paper No. 22. International Food Policy Research Institute, Washington

Merry SE (1988) Legal pluralism. Law and Society Review 22: 869-896
Neef A and Schwarzmeier R (2001) Land tenure systems and rights in trees and forests: Interdependencies, dynamics and the role of development cooperation – Case studies from mainland Southeast Asia. Deutsche Gesellschaft für technische Zusammenarbeit, Eschborn
Neef A, Sangkapitux C and Kirchmann K (2000) Does land tenure security enhance sustainable land management? Evidence from mountainous regions of Thailand and Vietnam. Discussion Paper 2/2000, Research in Development Economics and Policy. Institute of Agricultural Economics and Social Sciences in the Tropics and Subtropics (University of Hohenheim) Stuttgart
Nguyen Van Tiem (1992) Agrarian Policy in Agriculture of Vietnam Since the August Revolution 1945. Ministry of Agriculture and Food Industry, Hanoi
Phan Chi Thanh (1979) Mot hinh thuc tong toc o nguoi Thai Quy Chau, Nghe Tinh.(Akind of genealogy of the Thai people in Quy Chau, Nghe Tinh) ER 4(1979): 46-52
Pingali, PL and Vo Tong Xuan (1992) Vietnam: Decollectivization and rice productivity growth. Economic Development and Cultural Change 40(4): 697-718
Social Forestry Development Project (SFDP) Song Da (1997) Land allocation in the Song Da Watershed
Sikor T and Dao Minh Truong (2000) Sticky Rice, Collective Fields: Community-based development among the Black Thai. Agricultural Publishing House, Hanoi
Statistical Department of Yen Chau (2001) Interview in March 2001
Towa T, Trung M, Nguyen and Otsuka K (2001) Management of State Land and Privatisation in Vietnam. In: Otsuka Keijiro and Place Frank (eds) Land Tenure and Natural Resource Management. John Hopkins University Press, Baltimore London, pp 234-270
Tran Thi Van Anh and Nguyen Manh Huan (1995) Changing Rural Institutions and Social Relations. In: Kerkvliet BJT and Porter DJ (eds) Vietnam's Rural Transformation. Westview Press, Oxford

Local Ethnic Minority Networks for Sustainable Resource Management: The Pang Ma Pha Hilltribe Network Organization in Northern Thailand

Hans-Dieter Bechstedt & Patcharin Nawichai

Preliminary Remarks

'The most essential ingredient for the promotion of low-external-input and sustainable agriculture is the existence of strong, farmer-based networks in the rural communities.' (Haverkort 1993)

'Networks' and 'networking', as Haverkort notes, is '*nothing new under the sun*', and networking between farmers is '*as old as farming itself*'. However, in recent years new forms of farmer networks are arising partly due to new political space in some countries, such as Thailand, and partly due to government bureaucracies that have largely failed to deliver adequate services. In Northern Thailand, for example, ethnic communities came under increasing pressure through new government legislation banning traditional forms of agricultural production in the name of conservation. Here, traditional forms of practices, information exchange and cooperation are revived and linked with new forms and opportunities. They grew out of the need to fight for the protection of farmers' land, for conflict resolution over resource use among themselves or vis-à-vis other, often powerful and well-connected interest groups, for civil rights (e.g. citizenship), for protecting their community from drug trafficking and abuses, and for getting access to relevant information, new technologies, or development funds.

The following chapter analyses one of the major hilltribe network organizations in northern Thailand, based on an extensive field survey conducted by members of The Uplands Program. The survey was part of the ongoing Thai-Vietnamese-German Research Program for Sustainable Land Use and Rural Development in Mountainous Regions, funded, among others, by the Deutsche Forschungsgemeinschaft. It focused on an analysis of the Pang Ma Pha Hilltribe Network Organization, its weaknesses and strengths in terms of its development potentials for a more sustainable use of natural resources. Mainly ethnographic methods had been applied. Between August 2001 and October 2002, a field survey was carried out in villages of different ethnic minority groups (Shan, Red and Black Lahu, Lisu, Lawa, and Karen) in Pang Ma Pha District, Mae Hong Son Province, Thailand. Mainly ethnographic methods have been applied during the field study using qualitative research methods for conducting socio-economic village profiles, situation analysis, network and institutional analysis through participant observation and semi-structural interviews. Particular emphasis has been given to an examination of the way network objectives and activities have been established (chapter 3.1), on the challenge of domination and centralization (3.2), the issue of participation and legitimacy (3.3), the role of funding (3.4), the importance of non-financial support and linkages (3.5), the necessary development stages (3.6), monitoring and evaluation (3.7), and the political potentials and constraints (3.8).

Building upon these results, in a later phase of the research program the information received were critically reinterpreted along the lines of an actor-oriented framework (Long 2001) that considers networks as a platform, where different people or groups of people interact according to their different interests, and where objectives, procedures and activities are negotiated due to those internal, but also external factors and power relations (cf. Bechstedt and Nawichai 2003).

1. Networks and Networking

1.1 Definition

A network in the following is understood as a group of individuals and/or organizations who, on a voluntary and mutual basis, exchange information, goods or undertake joint activities, and who have organized themselves for that purpose in such a way that their individual autonomy and independence remains intact (Haverkort et al. 1993, Starkey 1998).

Networking involves making contacts, encourages reciprocal information exchange and voluntary collaboration. Networking is more than simply disseminate information by TV, radio or a journal, or putting together mailing lists, for they are based on mutuality and multi-directional information exchange. Other characteristics are that network members take part on a voluntary basis, and that they carry out join activities that cannot easily be performed alone. There is a great diversity in networking, as networks differ in origin, scope, goals, organizational structure, procedures depending on the specific environment, interests of their members and exposure to outside influencing control.

1.2 Types of Networks

There are many types of networks which can be classified by their (1) membership and geographical scope, (2) subject matters and activities, and (3) organizational structure.

(1) Membership and Geographical Scope

Networks may be designed for, and restricted to, particular categories of people, such as farmers, students, or researchers. Some bring together people with the same or a similar occupational background (horizontal orientation), e.g. farmer networks, while others link people from different fields, such as farmers, researchers, students, academics and policy makers (vertical orientation). Some networks are limited to a geographical, administrative area or an agro-ecological zone.

(2) Subject Matters and Activities

In rural areas, networks have been formed for various purposes, such as promoting sustainable natural resource management, getting access to new crop varieties or new technologies, promoting integrated pest management or community-based forest management, even for cultural purposes or control of drug abuses. Most networks have been established to share certain experiences or information. Others have been set up for advocacy, for strengthening their members' bargaining power vis-à-vis other powerful groups, or for being better able to solve certain conflicts among themselves or between the network and outsiders.

(3) Organizational Structure

Some networks are more formal and centralized, while others are more informal and decentralized, emphasizing direct communication between members. In general, it is assumed that informal networks are better tailored for serving the needs of grassroots movements and people organizations (Starkey 1998).

1.3 Benefits of Networks

From a more conventional and functionalist development perspective, the advantages of networks usually are seen in:

Increasing Competence:

Networks can facilitate the exchange of information, skills, knowledge, experience, or materials through meetings, workshops, publications and cooperative programs. Sharing of skills and experiences increases the overall competence of network members, whether individuals or organizations (Reijntjes 1995, Starkey 1998).

Strengthening Self-Confidence through Reviving Indigenous Knowledge:

Networks can give local people self-confidence by strengthening or reviving indigenous knowledge systems. There are examples that networks have helped to better share or revive local knowledge (Haverkort 1993, Rist 1993). However, there are also examples where indigenous knowledge has been appropriated by outsiders through networks, and had led to a sell-out and further control of rural communities by them (O'Brien and Flora 1992).

Cost-Effective Way of Sharing Information, Less Duplication, and More Impact:

Information exchange through networks can be a powerful and cost-effective way of sharing information and can lead to less duplication of work and efforts. Thus, faster progress and a wider overall impact could be possible (Nelson and Farrington 1994, Starkey 1998).

Improving Interaction, Empowering:

Networks can effectively link people of different levels, disciplines, organizations and backgrounds who would not otherwise have an opportunity to interact. It can empower members who belong to isolated or marginalized social groups. Women's networks can function as a mechanism to counter male domination (Jiggins 1993). Networks can help farmers to better control and direct the agricultural development process. They may even help to protect farmers' intellectual property rights (Haverkort 1993).

Raising Awareness and Motivation:

Networks can raise the awareness that many other people have similar concerns or development problems. They can provide members with a source of peer support, encouragement, motivation and professional recognition (Starkey 1998).

Improving Bargaining Power, Addressing Complex and Macro-Level Problems:

Networks can facilitate e.g. the cooperative purchase of farming inputs, selling and marketing of products. They can increase political bargaining power needed for local or national advocacy, defense of customary or legal rights or policy changes (Deepa 2002). They can help to address complex development or policy problems and issues that are beyond of those working at village levels (Starkey 1998).

Improving the Effectiveness of Research and Extension:

Networks have long been recognized by research and development institutions as an important way of improving the effectiveness of agricultural research and development (Plucknett 1990) and of developing farming practices that respond better to local conditions and needs. Networks have proven an effective link between farming communities and the agricultural extension system (Haverkort 1993). In addition to generating and exchanging knowledge based on farmers' experiences, farmer networks can facilitate the contact between funding and technical cooperation agencies and those in need of resources and support (Starkey 1998).

2 The Pang Ma Pha Hilltribe Network Organization (Mae Hong Son Province, Northern Thailand)

2.1 Background

Overall Problem Situation in Northern Thailand

On the surface, the general situation of the northern highlands of Thailand is an example of a situation in which a centralized government with policy priorities aimed at forest conservation, at national security and integration of the ethnic minorities runs into conflict with local people's traditional modes of agricultural production, i.e. shifting cultivation practices. However, a closer look reveals, in socio-political terms, that the northern regions of Thailand are a particular landscape that reflects growing competition over scarce and increasingly degraded resources between various social groups, with some holding in their hands most of the power in terms of resource control, policies and supremacy over the definition of meaning, while others, particular the ethnic minorities, face a process of increased marginalisation (Bechstedt and Nawichai 2003).

Location, Ethnic Composition

'The land of graceful mountains, of spirits, of a hundred caves, the land, where many hill tribes throng.' (Prasong 1998)

This is the way in which hilltribe people refer to their environment in Pang Ma Pha District, Mae Hong Song Province, located about 60 km from the provincial capital and 180 km from Chiang Mai City. Bordering Burma (Myanmar) to the north and west, Pang Ma Pha District in the past has been under the territorial influence of Khun Sa, once considered to be the world's leading drug lord. The district is populated by diverse ethnic groups such as lowland Thai, lowland and upland Shan, Karen, Black and Red Lahu, Lawa and Lisu (all upland). Its physical geography comprises steep sloping mountains of complex limestone. Given the continuous border tensions between Thailand and Burma, the area has a strong military presence. Two roads have been built, connecting the major Chiang Mai – Mae Hong Son Highway with the Thai-Burmese border, mainly to serve military purposes. Border villages have received special support through a military fund, villagers were trained in self-defense, and soldiers got involved in infrastructural improvements by a special army development unit.

History of Village Settlements

Most villages in Pang Ma Pha District, in particular those from the Lahu, Lisu and Karen minority, were officially established and recognized during the last 20 years by the Department of Local Administration (DOLA). Their inhabitants are migrants from neighboring Burma, who crossed into Thailand over the last 50 years due to continuous political turmoil in their country. They built temporary houses, cleared the surrounding forest and moved on after a number of years when soil fertility and yields began to decline. A few villages, however, are more than 100 years old permanent settlements. Their inhabitants are mostly ethnic Shan people who occupy the flat areas on both sides of the Mae Lana stream, where they cultivate paddy rice.

Production System

Most farmers of the Lahu, Lisu and Karen tribe, particularly those in the more distant villages (as measured in terms of distance to/from the Chiang Mai – Mae Hong Son Highway) are still basically subsistence, shifting cultivation farmers (rotation: 3-6 years) with upland rice (basic staple diet), maize (animal feed) and kidney beans as their major crop grown on sometimes steep fields. Until 20 years ago, the major cash crop was opium. However, opium production was less profitable than sometimes assumed, as land preparations must be thorough, its growth depending on weather condition, with villagers thus sometimes unable to obtain a good harvest for several consecutive years (Prasong 1998). Today, some villages grow coffee beside tea and vegetables (sesame, garlic, cabbage). Livestock raising has become the major income source for cash. Another source of income is occasional agricultural and non-agricultural wage labor, mostly within the same or in neighboring villages. As land is scarce and restricted, and soils are poor and prone to erosion, up to 80% of villagers suffer from rice shortage between 3-6 months per year on average. Village rice banks, whose setting up has been encouraged and supported by the Community Development Department, play a major role in compensating for those deficits.

Natural Resources

The eastern and southern parts of Pang Ma Pha District are part of the Pai Wildlife Sanctuary, the western parts belong to the San Pan Dean Wildlife Sanctuary, while virtually all the other areas fall into the category of National Conservation Forests. In the past, forests in the area were rich and vastly undisturbed. During the last 20-30 years, however, due to the influx of new settlers and a rise of the indigenous population, the demand for natural resources began to increase. Forests degraded rapidly, wildlife, for all bigger species, disappeared, and water shortage began to adversely affect paddy production in the lower areas, especially during the dry season. Certain village groups, encouraged by outsiders, made profit from cutting timber for commercial purposes. Others cleared forestland to convert it into agricultural land. As a result, many conflicts arose regarding forest encroachment, land and water use and demarcation of village/community boundaries. It was reported that at that time agreed rules and regulations governing natural resource use did exist within villages but not between villages (Prasong 1998). In this situation, the Royal Forestry Department (RFD) stepped in taking over the responsibility of forest management without, however, any relevant participation or consultation of the local population. Today, all the land falls under the jurisdiction of the RFD, while farmers have no land titles and, following a strict interpretation of the law, are illegal squatters. Yet, given the present political situation in Thailand towards more democratic rule and local participation, farmers are not under the immediate threat of being evicted. Unlike other areas in northern Thailand, most of the people in Pang Ma Pha District (an estimated 80%-90%) owns Thai citizenship, which otherwise would limit people's movement and their access to government services considerably.

2.2 The PMP Network History

Based on an initiative of the former, GTZ-funded Thai-German Highland Development Program (TG-HDP), the Pang Ma Pha Hilltribe Network Organization (in the following referred to as the PMP Network) started in 1996 by grouping together villagers from three Lahu communities who attempted to solve problems related to the management of forest products. Encouraged by traders from Chiang Mai and an increasing demand from passing motorists along the Highway between Chiang Mai and Mae Hong Son, villagers from the three communities had started to collect and sell an every increasing amount of bamboo shoots, mushrooms, orchids and other non-timber forest products. This had lead to increased conflicts among themselves as well as with neighboring villagers who complained about forest encroachments and depletion.

Given this situation, the TG-HDP in 1996 started to support a 'People's Forum' for the group of forest collectors, joined by two resource persons from the RFD, which was followed by other regular meetings as villagers became increasingly aware of the situation. They held negotiations among themselves resulting in a management plan for forest products collection from what they considered as their community forests. Finally, resource utilization rules and regulations were defined, among others a total ban on the collection of wild orchids (cf. Box 1 below). News soon spread to other neighboring villages and their leaders became likewise interested in joining the PMP Network. By 1998, 20 villages, mostly from the Lahu ethnic group, had joined the Network, and its scope became broader to include conflicts resolution over village boundaries, land use, encroachment of watershed areas, trespassing by livestock, among others.

Box 1: Natural Resource Utilization Rules set up by the Pang Ma Pha Hill-tribe Network

1. Villagers are allowed, for own consumption only, to cut down certain trees in the community forest that is managed by one or more villages. However, they need the permission from the Village Committee members beforehand.
2. Villagers who want to collect bamboo shoots have to leave at least 3 shoots untouched at each bamboo clump.
3. Villagers are not allowed to collect and sell any kind of wild orchids.
4. Villagers are allowed to gather other forest product, such as certain leaves, wild flowers and fruits. They are, however, not allowed to cut down whole trees or branches for that purpose.
5. Field owners whose plots are adjacent to forests are obliged to make firebreak lines before burning their fields.

Villagers of the PMP Network expressed their commitment to follow these rules of natural resources utilization strictly. Based on these rules, each village is authorized to adjust the regulations to what they consider as their own requirements, and to set up their own law enforcement (fine) system.

2.3 The PMP Network Today

By the year 2002, 28 villages had joined the PMP Network, including villagers from different ethnic minority groups, such as Red and Black Lahu (still constituting the majority), lowland and upland Shan, Lisu, Karen and Lawa. Meetings are held monthly, composed of two representatives from each village, and hosted by member villages in rotation, apart from the rainy season when villagers meet in the District town due to difficult access to some communities. With its now extended scope, the PMP Network established four different sections related to: (1) natural resources and environment, (2) culture and tradition, (3) drug prevention and control, and (4) education and accommodation (for students who want to further their studies in Pang Ma Pha District town)[1]. Already in 1997, the PMP Network had adopted an administrative structure with a Steering Committee at its top, consisting of a Chairman (a Lahu Village Headman), a Vice-Chairman (Lahu), a Treasurer (Thai) and a Secretary (Shan).

Main issues the PMP Network today is dealing with are problems related to conflicts over land use and management of natural resources, environmental conservation (forest and watershed protection), the fight against illiteracy, the suppression of drug abuses, and the revitalization and strengthening of cultural and traditional forms of livelihood. In 1998, the TG-HDP was terminated, however, together with AusAid, continued to provide some limited financial support for the PMP Network over the next year. Since 1999, a number of PMP Network activi-

[1] In October 2002, a fifth section on agriculture was added.

ties receive support through the Social Investment Fund (SIF) that has been jointly set up by the Government of Thailand and the World Bank nationwide to mitigate the effect of the economic crisis on the rural population.

3 Challenges and Lessons Learned – An Evaluation

3.1 Establishing Objectives and Activities

In 1997, at the time the PMP Network established its administrative structure and agreed on its main areas of operation, it defined its major objectives as follows: (1) strengthening cooperation and self-help capacity, (2) solving natural resource, i.e. land conflicts, (3) increasing knowledge, skills and information exchange of PMP Network members, and (4) fighting drug abuses. On the initiative of supporting staff members from the TG-HDP as many potential members as possible were included in the discussions. However, until now some PMP Network members appear to be not fully clear about the Network's major objectives and activities. Recently, a review of objectives has been carried out informally among the Steering Committee members in order to assess their continued relevance and the progress made towards meeting them. Yet, according to our own assessment, the PMP Network has at least to a certain degree proven its ability to change objectives as circumstances or demands changed. It has succeeded in initiating concrete activities that are seen as interesting and beneficial to its members and that require their active involvement. Through specific activities, such as setting up fire protection lines and organizing festivals for the revival of cultural and tradition, the PMP Network managed to win the participation of a large number of villagers.

Lessons learned: A network needs to establish clear objectives that will determine its general direction and its core activities. Objectives should be formulated and activities agreed with the participation of all actual and potential members. Network objectives should be openly as well as periodically reviewed by its members and not only among the core group. Without laying down clear and specific objectives, networks will find it difficult to develop a consistent program with distinct targets that could be met. Network members might then become confused and lose interest. Without a clear vision, strategy and objectives networks might become vulnerable to manipulations that serve the interest of dominant individuals or groups. Finally, the impact of a network can only be judged in relation to the network's goals which therefore need to be established beforehand (Haverkort et al. 1993, Starkey 1998).

3.2 Membership, Steering Committee and the Challenge of Domination

A closer look at the PMP Network's Steering Committee reveals a number of weaknesses. Its 16 members have maintained their position since 1997. Attempts to replace them have so far failed due to the resistance of the Chairman. Apart from the Chairman, the Vice-Chairman, the Treasurer, and the Secretary, other Steering Committee members are the Chairman, Vice-Chairman or Secretary of each of the four PAP Network sections. In its planning, operational and managerial issues, the PMP Network depends to a high degree on the initiatives of a core group within the Steering Committee, consisting of the Treasurer, the Secretary, and, in addition, an Advisor. All other members, including the Chairman and his Vice-Chairman, are more or less inactive. Outside contacts, including fund raising and the responsibility of fund management and reporting, rest with the Advisor, a university graduate and former staff member of the TG-HDP.

The PMP Network with its wide range of members, who differ in terms of social background, education, and ethnicity, has found it difficult to obtaining a balance of member involvement. Due to their higher education (ability to write and read in Thai language) and standard of living, their closer relationship to government agencies and urban areas, the lowland Shan people, although a minority within the PMP Network, occasionally dominate discussions and decision-making. In most cases, however, different ideas can be and actually are raised, and different interests are expressed that provide members with the opportunity to learn more about other people's perspectives. That obviously has stimulated each other's thinking, and there are indications that so far competing interests could be bridged and differences due to the undeniable existing of factions as a result of the ethnic divisions could be successfully negotiated and settled.

On the other hand, as a result of a lack of transparency and continuous accusations about mismanagement of funds, the PMP Network leadership's reputation has suffered, though not to the point where most members would withdraw their support. There are no doubts about the Steering Committee's commitment and their dedication; however, some of its members lack the skill for a more transparent and open behavior, thus giving rise to speculations and misunderstandings. Their ability to a self-critical assessment of their management work, as an example, appears limited. So far, no evaluation of the Steering Committee's management work has been conducted.

Some PMP Network members appear to be aware of the danger that the Steering Committee might start acting too much along its own interest, such as using its position to gain outside favors. However, suggestions to dissolve the core group and return to an administrative structure where no authority is given to anybody was discussed during one meeting, but was turned down by the majority as impractical.

Obviously, although the PMP Network has expanded considerably over the last 4-5 years, the formation of a Steering Committee to meet the requirements of funding agencies has not resulted in over-centralization or bureaucratization. In an attempt to counter such a tendency, certain responsibilities have been delegated to ordinary members and core groups formed for each of the four PMP Network sections. However, it turns out to be difficult to overcome an attitude of passivity on the side of the members who tend to hold the view that PMP Network activities are the duty and responsibility of the Steering Committee only. Clearly, more efforts are needed to involve ordinary members in PMP Network activities, such as in contacting local authorities and funding agencies, and disseminating information on PMP Network activities among own communities and beyond.

Lessons learned: Domination and centralization occurs when a network coordinator or a network steering committee starts to control and run the network for its own sake rather than coordinating and facilitating the activities of its members. Many networks, which have started small and informally, tend towards centralization and bureaucracy when growing bigger. Bureaucracy here means that a network committee becomes increasingly concerned with maintaining an office, handling correspondence and contacts. Certain network responsibilities should be delegated to a variety of individual members. This will help to avoid the dangers of excessive centralization, domination, isolation and passive membership. Network members should not expect all network activities to be arranged by the core group. Members should feel themselves responsible for arranging or assisting in appropriate network activities.

Better-off, better-educated and self-confident staff from certain ethnic or interest groups with more contacts with the outside world tends to dominate decision-making and activities. Knowledge of the national language has proven to be an additional barrier to equal participation. Thus, networks can easily be dominated by particular individuals or (ethnic) groups. Networks activities then become limited to those favored by the dominant group. For that reason most development practitioners consider regular rotation of leadership to be important to avoid domination. Moreover, the internal process of management should be evaluated periodically (Haverkort et al. 1993).

Despite the threat of domination, networks appear to need a committed core group of active members who will take initiatives and be responsible for coordination, management and logistics. The individuals of such a core group must have a vision and be prepared to look beyond the immediate objectives of their own group, and be able to link day-to-day activities with the wider perspectives of the network. They must have communication skills and the ability to encourage and motivate others, thus creating an atmosphere of mutual confidence and commitment, and to avoid misunderstandings through open consultations, transparent decision-making and disclosing financial status. Many experiences, not only in Thailand, have demonstrated that efficiency and impact of a network largely depends on such a small group of committed and skilled volunteers (cf. Starkey 1998).

It is important that network members actually share the feeling of being represented by the core group. The core group must interact regularly with the ordinary members to keep up with their ideas and changing needs. Without such interaction, network core groups tend towards centralization, favoritism and bias towards their own specific interests. The core group should regularly, in a self-critical manner, review its progress as well as its own relevance to other network members. It needs to establish an atmosphere of openness, self-criticism and willingness to admit mistakes.

3.3 Participation and Legitimacy

The PMP Network has given itself an open structure. Two representatives of each village, some appointed by a village headman, some others in one way or another elected by their community, from any of the villages in Pang Ma Pha District are welcome as members. So far, representatives of 28 out of 38 villages have joined. They are considered to be full members as soon as they have participated at the regular meetings five times continuously. Outsiders, like university-based researchers, NGO staff members or Government officials are free to join the meetings, although their cooperation or support (except financial) is not actively sought. Through improved ties established with local authorities in recent months, the PMP Network hopes to get more recognition (and funds) from state agencies.

All major issues are prepared by the Steering Committee, then presented to, openly discussed with and decided by the members during monthly meetings. This is not to deny that there are conflicts of interests among PMP Network members, which emerge, in particular, when village members submit their development project proposals[2]. So far, however, different interests could be balanced, as the Steering Committee includes members from both of the two major ethnic groups (Shan and Lahu). Moreover, at several occasions the Steering Committee has proven its ability to mediate successfully disputes related to land and forest use among its Network members.

While PMP Network meetings obviously are conducted in a fairly transparent way, and decision-making is based on consensus, the relationship between member villages and their representatives appears to lack efficient communication. Most of the representatives have not been elected, and reporting is poor. Typically, it is the village head who first enrolls his village for membership and may participate in the beginning, but later assigns someone else to take over. Ideally, suggestions for PMP Network activities and development projects should emerge from village meetings, and villagers should likewise be informed about the final decision-making process. In reality, however, suggestions are made by the village representative(s) without prior consultation of their village, and planning or feedback meetings on village level are the exception. Thus, it must be concluded that the PMP Network, in general terms, rather expresses the needs and aspirations of the local, mostly better-off (male) elite rather than of the highlanders' poorer and more disadvantaged social strata. Network members represent the better-off, local

---

[2] Within the framework of the SIF-Fund Network village members are requested to submit proposals for funding referring to one or more of the four Network sections. Villages then enter into a kind of contact with the Network, enlisting in detail for what purpose the money will be used. Regulations recently have become stiffer after complains were raised about misuse of funds.

rural (male[3]) elite, as villagers tend to vote for the wealthiest and thus most powerful and influential person in a village to become village headman. They consider, for some good reasons, a strong and powerful village leader to be in a better position to represent the village vis-à-vis the outside world. A Network representative, who is not a village headman or his deputy, will not have the authority to call for a village meeting, to speak on behalf of his community, and to contact local government authorities. However, interests of those persons are sometimes less related to tackling the community's most pressing needs, but rather concerned with trade, shop keeping, transportation, construction, or with serving their clientele.

Lessons learned: *'Participation is the heart of any network'* (Haverkort et al. 1993). A sense of cooperation, sharing, belonging and openness is vital to any network's success. Planning, decision-making, the choice of network activities and means of their implementation should be done in a democratic way, and should give all members the feeling that they can influence events. Ultimately, participation in network activities will be determined by the perceived benefits that individuals or certain groups see in it. Network members should be democratically legitimized. They should have been elected by their community or village, and should be able to represent the different interests within their community rather than only those of the elite. Moreover, networks should welcome participation from a broad range of individuals and agencies concerned with the same or similar objectives. If working well, networks can provide a relatively informal, non-threatening framework for exchange, discussion, debate and cooperation between various institutions and organizations.

It is not unusual that network members might share the overall goals and objectives of the network, although they still might have different interests or different agendas. There might be competition for limited network resources and funds, and different views on how to approach and solve certain problems. To maintain network harmony and the spirit of cooperation and to prevent exclusion of some and the domination by others, the network's leading group should be able to tackle and resolve conflicting ideas and views in a way that is fair and transparent to everyone. As a precondition, the network's core group needs the support, trust and recognition of its members.

Success of some other networks in Thailand has demonstrated that much can be achieved by a group of enthusiastic and committed volunteers without much professional training in group management. In the case of the PMP Network there is evidence that occasional misunderstandings and personal biases could be prevented if the Steering Committee would be more experienced in management issues and communication skills, and more aware of the benefits of transparency and democratic legitimization.

---

[3] With only 2 PMP Network representatives (out of 56) who are women, gender-related issues are totally underrepresented.

3.4 Funding

It seems to be unlikely that the PMP Network would have been able to grow and expand to its present size and scope, or to carry out such a diverse number of activities without outside assistance. From 1998 onwards, the PMP Network received Baht 70,000 (US$ 1,667) through the Office of Narcotic Control Board (ONCB, Thai Ministry of Interior) and Baht 100,000 (US$ 2,381) through the TG-HDP. In 1999, AusAid stepped in with another transfer of Baht 450,000 (US$ 10,740). For the period 1999 to 2002, with the assistance of the PMP Network Advisor and other former TG-HDP staff members, the PMP Network applied successfully for a fund through SIF with the total amount of Baht 1.9 Mio. (US$ 45,230). Some limited funds have been provided by various local Government agencies and NGOs.

Apart from covering administration[4] costs, the SIF fund was used for activities in line with the four sections of activities:

1. **Natural resource and environmental conservation:** 'Tree ordination', aquatic animal conservation, tree planting, setting up of fire break lines, establishing rules for land and forest utilization.
2. **Culture and tradition:** Thai language training, training in traditional Shan dancing, use of musical instruments (Lahu), and training in preparation and execution of religious ceremonies (Lahu).
3. **Drug prevention and control**: Information campaign, medical treatment of addicts, agricultural incentives for former drug addicts, youth camp, sport activities.
4. **Education**: training in chicken raising, mushroom cultivation and broom making.

Each village receives the same amount of money and is free to decide by itself which kind of activities to carry out under each of the four sections.

According to our own assessment, the PMP Network appeared not well prepared to absorb an amount of money in the range of the SIF fund provided. Complains over misuses were heard frequently and dissatisfaction was running high among PMP Network members due to alleged irregularities. It was reported that some village representatives, including members of the Steering Committee, had pocket in certain amounts they received without initiating activities in their villages. Since the end of 2001, however, financial management appeared to have improved. Today, there is more control and transparency, and village representatives have to produce evidence about activities carried out. Likewise, the Steering Committee has now been compelled that borrowing out Network money for private clients has to follow strict conditions and requires the approval of all mem-

[4] E.g. salary for an administrator, stationary, per diems and travel expenses provided for participation in meetings, field trips, and M&E, maintaining the student dormitory, including a monthly salary for a guard.

bers[5]. Finally, no evidence was found that the SIF Coordinator, who occasionally participates in Network meetings, uses his position and influence to rule into Network affairs.

Lessons learnt: Networks, particularly those among resource-poor and marginalized populations, do need resources to be effective. Although operating costs should be kept at a minimum, networks often suffer from a near total lack of funds for their activities and for coordination. While core group members are generally committed to sacrifice their time on behalf of the network, they seldom have sufficient funds for management and administration, The SIF fund, for example, which allowed the PMP Network to carry out its various activities, did not allow for covering operating costs. Its coordinator, however, tolerated the use of surplus money from other activities to cover expenses for administration and management. Applications for funding through governmental as well as non-governmental agencies often appear to be a bureaucratic process that takes up much time and energy, and is often beyond the capacity of many ordinary members.

On the other hand, the case of the SIF fund also demonstrates that, if too large amounts of money are channeled through a network, the network's core groups might not adequately be prepared to administer and account for those finances, and competition for funds can easily lead to favoritism and misuse. Easy access to excessive resources can attract 'freeloaders' and 'fair-weather friends' (Haverkort et al. 1993), i.e. people who are more interested in travel opportunities, per diems and easy project money, but do not feel much commitment to the long-term network's objectives. In other cases it was found that 'fund dumping', a favored method by the present Thai Government to raise its popularity, can undermine existing traditional forms of saving and communal funds. To avoid such a negative impact, members should contribute at least some of their own resources, for only then they feel ownership of the network's results.

Networks need to establish necessary contacts, and to develop plans for the funds they require. They, moreover, need recognized, established and maintained structures to secure and manage such funding. These tasks are often beyond the ability of ordinary farmers, and outside assistance might be indicated. To maintain network credibility to members and donors, funding should be managed or supervised not only by the network's treasurer or the core group, but a broader section of members should likewise be included. Apart from keeping financial reports, they should be read out during meetings for illiterate members.

Although it is not the case with respect to the above-mentioned SIF fund, attention needs to be given to the possibility that donor agencies might use their financial power to influence network policies. Resource organization coordinators

[5] Meanwhile, this rule was violated by the Chairman of the Steering Committee himself, what resulted in a joint resignation of the other non-Lahu Steering Committee members in October 2002. Their positions have later been filled by other Lahu fellow members closely related to the acting Chairman, thus putting at risk the very existence of the PMP Network as a multi-ethnical project.

might enjoy being in the position of power, and might exploit their position for influencing strategies, objectives or activities of networks.

Eventually, funding is not everything. This is clearly demonstrated by the low impact of some resource-rich networks established by international institutions. By contrast, some networks without any central funding have achieved a remarkable impact due to the commitment of their members to dedicate time, effort and personal resources to network activities (Starkey 1998).

3.5 Non-financial Support and Linkages

Donor Agencies, NGOs

Apart from financial and other kinds of support through the Thai Government's Office of Narcotics Control Board (ONCB), AusAid, TG-HDP and SIF, staff members of the 'Project for the Recovery of Life and Culture' (PRLC), a NGO based in Mae Hong Son, provided advise in terms of outside contacts and information about, for example, new laws or relevant political events. During its initial phase, the PMP Network depended much on the advisory assistance of staff members from the TG-HDP and, later on, on staff members of the PRLC. On a non-permanent basis, an academic free-lance researcher, running a program on local history and cultural change that is funded by TRF (Thai Research Fund) likewise acts occasionally as an advisor to the PMP Network. Apart from these contacts, however, the PMP Network's Steering Committee did not actively seek contacts to other networks in neighboring districts, such as those of the Karen minority, nor to NGOs or university institutions. PMP Network members tend to view contacts and outside relations first of all in terms of funding opportunities. Existing inter-network linkages are facilitated by staff members of the PRLC. The relationship between the PMP Network and the PRLC, however, appears to be not without frictions. PMP Network members complain that the NGO is using the PMP Network's name to attract money that only partially benefits the PMP Network.

Local Authorities

While the relationship between the PMP Network and local government authorities until recently has been impeded by mutual suspicion, with the arrival of a new, relatively young and open-minded District Chief in June 2001 the situation has improved considerably. He and his Deputy, on several occasions, participated in PMP Network activities and meetings, and spoke out in favor of more support from and cooperation with local state authorities and TAOs (see below). The yearly PMP Network's festival on minority culture and tradition has been invited by him to perform in front of his District Office.

Tambon (Sub-district) Administrative Organizations (TAOs)

Likewise important, there are a number of recently established linkages between the PMP Network and the Tambon Administration Organizations (TAOs). These are elected, sub-district-level administrative bodies, equipped increasingly with decision-making power and resources, and instituted by the Thai Government as a major tool in its process of decentralization and devolution of power (cf. Nelson 1998, Asian Development Bank and the Kingdom of Thailand 2001). The PMP Network's Treasurer is a TAO Executive Member. The PMP Network's drug addiction program is closely cooperating with similar activities carried out by the TAOs. There are attempts underway to ensure the PMP Network's participation in the next TAOs' 5-years Development Plan for the period 2003-2007, which will contain plans for improving irrigation systems, measures to mitigating water shortage, declining soil fertility, preventing forest destruction, and solving the problem of lack of land titles that affects most villages in the District. It is also expected that the TAOs in future will adopt the PMP Network's Natural Resource Utilization Rules (cf. Box 1 above). However, according to our own observations, the various expressions of good intentions voiced by both sides have yet to be translated into active cooperation. Irregularities in the management of funds within and between both organizations need to be overcome as well as bureaucratic obstacles resulting from the fact that PMP Network member villages cover the whole of Pang Ma Pha District with four distinct TAOs within its borders.

Lessons learnt: Ideally, Networks should seek 'horizontal' as well as 'vertical' complementarities and linkages. Horizontal networking is seeking contacts with networks with similar goals and objectives from other geographical areas. Vertical linkages are based on different types of organization, such as NGOs, research stations, and local government agencies of the same geographical or administrative area. Both can help to bring together the views and experiences of local people from different areas as well as those from extension services, researchers, policy makers or people from the private sector of the same area.

Networks can and clearly should benefit from being associated with sympathetic local, regional, national/international governmental and non-governmental agencies or individuals that can provide access to resources, advocacy, legal support, or other useful contacts and training. However, safeguards may be required to prevent such institutional arrangements from leading to undesired outside interventions or domination. Other cases from Thailand have shown that NGO's involvement in farmers' network, for example, can result in new forms of dependency and domination which only serve as justification for the NGO's continuous involvement with the network and thus be dictated by the NGO's own pursuit of organizational survival. NGO staff members, virtually all of urban, middle class background, who work with farmers, may not always be aware that their own biography, their living and working reality separates them considerably from those people who they are supporting. Certain romantic and idealistic ideologies, such as the 'cultural community' approach, still popular among rural Thai NGOs, tend to mask the class differences between urban middle-class supporters and rural folk.

3.6 Development Stages

The PMP Network has evolved out of the desire of some neighboring villages to solve land disputes among themselves and to coordinate the extraction of non-timber forest products in a more sustainable way. From its start, it has received outside support through international and national development agencies in terms of advice and resources. Its membership has grown steadily over the year from 3 villages in 1996 to 28 in 2002. Some observers, however, are of the opinion that the network expanded too fast just for the sake to increase membership and thus to enhance its chances to attract funds. Others argue that expansion was at least partly motivated by the attempt to better control and prevent drug abuses that require broader inter-village cooperation.

With its increase in membership and availability of more resources, new issues were taken up, such as drug control and rehabilitation, revival of culture and tradition, and leadership training for the youth. The PMP Network has proven its ability to adapt to changes in the overall environment and due to internal dynamics. While it expanded in scope and membership and constituted itself as an implementing body for development projects, redesigning of its structure and management became necessary – a challenge that in general was carried out successfully.

At present, the PMP Network's administrative set up does not appear to be a burden. On the other side, the PMP Network might consider to keep its office in a somehow better shape, make more efficient use of the administrator employed, and may improve its outward appearance through publications (e.g. a regular newsletter).

Lessons learnt: Given the great diversity of networks in Thailand, there is no blueprint for their development. While the PMP Network evolved out of the need to settle land conflicts between some villages, other ethnic minority networks were a result of political pressure and are supported by Thai NGOs, e.g. the Hmong Environmental Network, or were created and maintained by state agencies like the Royal Forestry Department, e.g. the Mae Klang Watershed Network (both Chiang Mai Province) to make hilltribe people to comply with official conservation policies.

Based on the experiences of the PMP Networks as well as others, some stages in a network's development can be distinguished. During a preparatory stage it is important that topics are identified that are perceived by members as those of common concern. Likewise, during that initial stage it is essential that as many members as possible participate in the planning. Members should determine the mechanisms and structures chosen for information exchange and collaboration. Networks usually at some stage draw up formal rules and regulations, however, quite often following the requirements of outside donor agencies, as it was the case with the PMP Network. Experiences have shown that it is important to keep overhead costs low and network structures simple, as formalized networks often tend to transform into a bureaucratic superstructure that inhibits rather than promotes the exchange of information and cooperation.

After its setting-up, the network will gradually become operational. While the network might expand in scope and membership, some redesign of its structure and management may be necessary. As a general rule, networks should extend slowly and follow a development path dictated by their own requirements and internal logic rather than by outside pressure or political demands.

3.7 Monitoring and Evaluation

So far, the need for regular and effective monitoring and evaluation of development activities has yet to take roots among the members of the PMP Network. Last year, the Steering Committee assigned three groups with three to four members each to monitor the implementation of activities, such as building firebreak lines, in selected villages were they suspect some reluctance in carrying out these tasks. These groups were also requested to inform villagers on PMP Network goals and activities. They were said to have had some talk with village leaders, but no report has surfaced so far, and many villagers appear unaware. Monitoring guidelines drafted and handed out by the Steering Committee have been mostly ignored.

Lessons learnt: According to the mainstream development discourse, networks, like all organizations, are seen to require regular and thorough monitoring and evaluation of work plans as well as of progress and achievements. Regular M&E is essential for making learning experiences explicit, for improving efficiency and effectiveness, and it is crucial as an early warning system for the identification of shortcomings and emerging conflicts. Critical questions should include:

- Do members perceive that they derive benefits from participating in the network?
- How do structure and activities of the network relate to its objectives?
- How is management related to its members?
- How participatory is planning and implementation of activities?
- Do networking activities improve the performance of individual members or member groups (Haverkort et al. 1993)?

It appears that networks in northern Thailand so far have little experience on this subject. This is partly because, by their very nature, networks tend to be informal and loosely structured. In other cases, networks consider themselves too busy to devote their limited time and resources for introspection. Moreover, self-evaluation procedures require the ability for self-examination and criticism, which, from the perspective of a network's steering committees, in particular, might be considered as undesirable, embarrassing or too challenging. And finally, it often appears difficult to define reliable and measurable indicators of network success (Starkey 1998).

3.8 Political Potentials and Constraints

Although some PMP Network leaders are reported to have contacts with national or regional politicians, unlike other networks in rural Thailand, no direct interference so far has occurred. Members appear to be aware of the danger of becoming a tool for politicians campaigning for supporters, votes and influence. Most PMP Network representatives are village headmen from among their entourage, who usually are a target for canvassing and vote-buying that is still widespread in many rural areas of Thailand[6]. During a PMP Network meeting, members agreed that political alliances should be kept outside the Network and be considered as a member's private matter. Indeed, no evidence was found that the PMP Network was used as a political platform.

On the other hand, the need for having land titles and receiving recognition for community forests in the eyes of most PMP Network members is a continuous and pressing topic that requires involvement in policy decision-making. With the support of PRLC, around 35 village representatives joined protest activities in March 2002 in front of the Chiang Mai City Hall to underline their support for a proposed Community Forestry Bill. Some Steering Committee members, in the same matter, tried to approach Thai Senators in order to win over their support.

Lessons learnt: As examples from other parts of Thailand show, political interference remains a permanent threat to local organizations and networks. In their attempt to seek local support, national as well as local politicians try to let 'money speak', not without success. Some farmer movements have turned into a political tool for party interests. Others, however, have resisted and defended successfully their independence.

Networks have to operate within the political realities of a country. In some countries, policies might tolerate or even encourage the establishment of grassroots networks, while in other countries such attempts are viewed with suspicion, require official clearance or are impeded by strict control through local authorities. In this respect it appears obvious that the spread of farmers' networks in Thailand, unlike its neighboring countries, has been facilitated by the increased political opening, pluralism and freedom of organization during the last 10 years. As a result, over the last decade rural protests have slowly worked their way from villages and forests into the cities, the centers of power and rule. Spearheaded by the north-eastern farmers in 1992-95, in 1996 the rural grassroots movement linked up with slum and worker groups in the cities. Since the end of the 90s, the ethnic minority highland population of northern Thailand has become increasingly better organized, and carried their demands for land rights and citizenship, joined by

[6] The Chairman of the PMP Network, for example, is said to be an ally of a well-known local politician, Member of Parliament and former Minister.

thousands, at least twice into the heart of Chiang Mai to capture media spotlight and public attention.

4 Conclusions

To sum up, since its establishment in 1996 the PMP Network has been, on the one hand, relatively successful in a number of issues, such as in:

- establishing Network objectives, procedures and activities based on a broad consensus among its members, as well as in adjusting and extending the scope of activities due to new situations and demands;
- creating an organizational environment, in which (at least some) Steering Committee members have been able to encourage and motivate others, are well-trusted and respected by a majority of ordinary members, and are viewed as committed and competent;
- initiating concrete activities that are seen as interesting and beneficial to its members and that require the communities' active involvement;
- creating an atmosphere where decision-making is transparent and emerging conflicts due to different interests and ethnic affiliation are openly discussed, negotiated and settled, e.g. land disputes, including agreements on rules and regulations for natural resource utilization;
- avoiding domination, over-centralization and bureaucratization, and
- fending off attempts at political interference.

The PMP Network, on the other hand, has been less successful in a number of other areas (shortcomings of which some members are well aware), such as:

- insufficient attention given to regular reviewing of the relevance of objectives and of an assessment of progress;
- lack of ability for self-critical assessment;
- dependence on outside financial support;
- dependence on non-network members (advisors) in issues such as advocacy, establishing and maintaining contacts and relationships to governmental and non-governmental groups, agencies and organization;
- lack of awareness of the benefits of non-financial support;
- passivity of Network members as well as some of some members of the Network Steering Committee;
- insufficient transparency in financial matters partly due to lack of experience, partly due to fraud;
- language problems;
- domination of the Network through village elites in general and through a certain Lahu faction in particular, and
- lack of legitimacy of Network representatives vis-à-vis their village communities.

References

Asian Development Bank and the Kingdom of Thailand (2001) Restructuring of the Ministry of Agriculture and Cooperatives (MOAC). ADB Technical Assistance Project No. 3261 – Thailand. Consultant Reports for the GTZ and MIDAS, Bangkok

Bechstedt HD and Nawichai, P (2003) Local ethnic minority networks for sustainable resource management: The Pang Ma Pah Hilltribe Network Organization in Northern Thailand - A reinterpretation from an actor-oriented perspective. In: Allmendinger J. (eds): Entstaatlichung und soziale Sicherheit. Verhandlungen des 31. Kongresses der Deutschen Gesellschaft für Soziologie in Leipzig 2002. Leske, Opladen (forthcoming)

Deepa N (2002) Empowerment and poverty reduction: A Sourcebook. World Bank, Washington D.C.

Haverkort B (1993) Farmers' networks: Key to sustainable agriculture. In: Alders C, Haverkort B, Veldhuizen L (eds) Linking with farmers: Networking for low-external input and sustainable agriculture. Intermediate Technology Publications, London

Haverkort B, van Veldhuizen L, Alders C (1993) Networking for low-external input and sustainable agriculture. In: Alders C, Haverkort B, and van Veldhuizen L (eds) Linking with farmers: Networking for low-external input and sustainable agriculture. Intermediate Technology Publications, London

Jantakad P (1998) Natural resource management by network organization. A case study of Pang Ma Pha Hilltribe Network Organization, Tambon Pang Ma Pha, Pang Ma Pha District, Mea Hong Son Province. In: Thai-German Highland Development Programme (TG-HDP), Volume 2, Case studies of experiences in implementing community-based land use planning and local watershed management and sustainable farming systems (1984-1998), Chiang Mai

Jiggins J (1993) Networking with women farmers. In: Alders C, Haverkort B, Veldhuizen L (eds) Linking with farmers: Networking for low-external input and sustainable agriculture. Intermediate Technology Publications, London

Laungaramsri P (1999) The ambiquity of 'watershed': The politics of people and conservation in northern Thailand: A case study of Chom Thong conflict. International Institute for Asian Studies, Amsterdam

Long N (2001) Development sociology. Actor perspectives. Routledge, London and New York

Nelson J and Farrington J (1994) Information exchange networking for agricultural development: A review of concepts and practices. Sayce Publishing, Exeter

Nelson MH (1998) Central authority and local administration in Thailand. Studies in Contemporary Thailand No.6. White Lotus Press, Bangkok

O'Brien WE, Flora CB (1992) Selling appropriate development vs. selling out rural communities: Empowerment and control in indigenous knowledge discourse. Agriculture and Human Values 9 (2). Gainesville

Plucknett DL, Smith NJH, and Ozgediz S (1990) Networking in international agricultural research. In: Food Systems and Agrarian Change. Cornell University, Ithaca

Reijntjes C (1995) Networking for sustainable agriculture. In: ILEIA Newsletter Vol. 11, No.1

Rist S (1993) Supporting indigenous knowledge for sustainable rural development in Bolovia: The case of AGRUCO. In: Alders C, Haverkort B, Veldhuizen L (eds) Linking with farmers: Networking for low-external input and sustainable agriculture. Intermediate Technology Publications, London

Starkey P (1996) Networking for sustainable agriculture: Lessons from animal traction development. Gatekeeper Series 58. IIED, London
Starkey P (1998) Networking for development. IFRTD, London

Creating Political Capital to Promote Devolution in the Forestry Sector - A Case Study of the Forest Communities in Banyumas District, Central Java, Indonesia

Slamet Rosyadi, Regina Birner & Manfred Zeller

1 Introduction

In recent years, devolution of forest management to local communities has become a major policy trend in developing countries. The term devolution is used here to refer to the transfer of responsibility and authority over natural resources from the state to non-governmental bodies at the local level (Meinzen-Dick and Knox 2001, p. 42). Devolution policies aim to address institutional problems that have been identified as major reasons behind the degradation and misuse of forest resources in developing countries such as state property and centralized management of forest resources, corruption in the forestry administration, lack of effective monitoring, and enforcement and deficient incentives for the local communities (McCarthy 2000a; Ligon and Narain 1999; Wibowo and Byron 1999). Different mechanisms have been identified in the relevant literature on the subject by which devolution can lead to a more sustainable forest management in terms of equity, efficiency, and environmental sustainability: the creation of incentives by a fair and democratic distribution of benefits; the creation of accountability; the reduction of transaction costs; the mobilization of local knowledge; the strengthening of local institutions for sustainable resource management; and – in view of a low state capacity – the limitation of the role of the state to the provision of enabling frame conditions and the protection of public interests (compare Ribot 2002, Meinzen-Dick et al. 2001, Birner and Wittmer 2000; World Bank 1997). However, the empirical evidence on the effects of devolution in the forestry sector has been mixed (Banerjee 1997, Meinzen-Dick et al. 2001, Ribot 2002), which implies a need for more theoretical and empirical research on devolution.

Focusing on the optimal levels and forms of devolution, normative analyses of devolution can draw on the economic theory of federalism (Braden and Prost 1997), transaction costs economics (Ostrom et al. 1993, Birner and Wittmer 2000), property rights and collective action theory (Meinzen-Dick et al. 2001) and theories of democracy (compare Ribot 2002). Considering the fact that devolution is associated with a considerable shift in political and administrative power, there is also a need to study why and under which conditions local communities will eventually receive more rights to forest resources. As such studies on the political feasibility of devolution are still scarce (Agrarwal and Ostrom 2001, pp. 75-76), our paper focuses on these questions, using a case study from Indonesia as an empirical example. As theoretical basis of our analysis, we combine the concept of social capital, which has been widely applied in studies on natural resource management, with the concept of political capital, which is derived from political resource theory (Birner and Wittmer 2003).

Under Suharto's "New Order" regime from 1966 to 1998, Indonesia's forest management was characterized by centralized management and a strong focus on exploiting the country's forest resources in the pursuit of macro-economic objectives such as obtaining foreign exchange and promoting growth. This policy fitted well into the political strategy of Suharto's regime, which justified its rather authoritarian and centralized way of government by the economic growth it was able to promote. With the onset of the economic crisis, this legitimization eroded and the reform movement increasingly criticized the prevalence of corruption, collusion, and nepotism in the "New Order" regime (McCarty 2000a). After the fall of Suharto, a far-reaching policy of decentralization and devolution was implemented (Law 22/1999 on Regional Governance and Law 25/1999 on Fiscal Balancing). The forestry legislation was reformed with the aim to stipulate a larger participation of local communities in forest management (Forestry Law 41/1999, Article 4). The transition from an authoritarian to a democratic regime, however, was also accompanied by reduced law enforcement which contributed to rapidly increasing deforestation rates. At the national level, the annual deforestation rate during the reform period reached 2.5 million hectares per year as compared to an annual average of 1.6 million hectares prior to the reform period (Kompas 2002).

In response to this crisis in the forestry sector, the State Forest Corporation (SFC), which is in charge of managing all production forest and most protection forest on Java, issued the policy of community-based forest management (CBFM) (SFC 2000). According to this policy, forest villagers are entitled to plan CBFM activities together with SFC and other stakeholders and receive a share of the benefits (SFC 2001). The forest farmer communities in the Banyumas district of Central Java played a pioneering role in implementing this devolution policy. Therefore, we analyzed the political strategies of these communities and their opponents in order to identify factors and conditions that influence the political feasibility of devolution.

2 Theoretical Framework

2.1 The Concept of Social Capital

In the burgeoning literature on social capital, one can distinguish two major perspectives (compare Wall et al. 1998, Birner and Wittmer 2003): (1) the "actor's perspective" formulated by Bourdieu (1992) in which social capital consists of the resources that individual actors can mobilize due to their belonging to exclusive networks, and (2) the "society's perspective" characterizing the work of Putnam (1993) in which social capital is a public good constituted by the organizations and horizontal networks existing within the society. Coleman (1990) formulated both perspectives, but considered a wider range of forms of social capital, including expectations, norms and sanctions. This paper applies the actor's perspective on social capital, because its purpose is to study the role that social capital plays in the strategies of different corporate actors such as local communities and state agencies. As a study on community forestry in Thailand (Birner and Wittmer 2003) showed, the following types of social capital are particularly valuable for such actors, because they can be used to create political capital:

- community-based organizations,
- non-governmental organizations (NGOs),
- the coalition of such organizations, and
- belonging to exclusive networks (elites).

This paper also adopts Narayan's (1999) division of social capital into bonding and bridging capital. Social cohesion within the group structure is referred to as bonding capital, whereas the social relations that create links across different communities/groups are defined as bridging capital. In addition, we make use of Rubio's (1997) concept of "perverse social capital" which refers to organizations that reward rent-seeking and create incentives for criminal behavior. The Mafia is an example of such "perverse social capital."

2.2 The Concept of Political Capital

As Booth and Richard (1997) showed, the concept of social capital is not refined enough to explain the relation between social organization and political outcomes. Therefore, these authors introduced the concept of political capital and defined it as the attitudes and activities that are able to influence political regimes. In a study on the role of political trust in democratic systems, Newton (2002, p. 213) also concluded that it is useful to keep the concepts of social and political capital apart for analytical purposes. Fuchs et al. (1999) used the concept of political capital in order to explain political participation. Analyzing how local communities can gain influence in national policy processes, Birner and Wittmer (2003) defined political capital as the resources that are an actor can use to influence policy formation processes and achieve outcomes that serve the actor's perceived interests. Their study showed that the following forms of political capital are relevant in explaining political outcomes:

- electoral leverage (e.g., mobilizing voters),
- direct participation in the legislative process (e.g., law petition),
- public protest (e.g., public rallies),
- lobbying of political and administrative decision makers,
- ideological resources and scientific knowledge used in the public discourse, and
- international influence (e.g., donor funds).

2.3 Using Social Capital to create Political Capital

Booth and Richard (1998) and Newton (2001) showed that not all forms of social capital are useful for the creation of political capital. To explain the outcome of controversial political processes, Birner and Wittmer (2003) suggested identifying the types of social capital held by proponents and opponents and analyzing how effectively both parties use their social capital for the creation of political capital. The parties may combine their social capital with other resources – such as economic capital – in order to create political capital. The political outcome may have 'feed-back-effects' on the political capital of both the proponents and the opponents. The opportunities to use social capital for the creation of political capital are influenced by a number of frame conditions such as the type of political regime, the political party system and the electoral system, the scope for participatory elements in political decision-making, the possibility of achieving goals by lobbying, the role of the bureaucracy, freedom of press, the political culture, and the scope for international influence (Birner and Wittmer 2003).

3 Research Area and Methods

3.1 Profile of the Research Area

The Banyumas district is located in the northern part of Central Java Province and characterized by small-scale agriculture. The average farm size is 0.5 ha, and land distribution is unequal. The major crops are paddy, maize, and cassava. Most of the rural population live adjacent to the state forests and relies on farming for their livelihood because alternative income opportunities are limited. The population density is 1,415 people per square kilometer, and the population is growing at a rate of 1.06 percent (Banyumas Statistics 2000).

The representative of the central-level SFC in this district is the East Banyumas SFC which is responsible for managing most of the forest areas in Banyumas district and the forest areas in three other districts. The forests in the operational area of the East Banyumas SFC are dominated by pine and damar trees. Compared to other forest districts with teak forests, the profitability of pine and damar forest is low, and the central SFC has to subsidize the East Banyumas SFC. The forestry sector contributes only 3 % per year to the total gross domestic product (GDP) of Banyumas district (Banyumas Regional Planning and Development Agency 2001). There are about 50 forest village communities (in short, forest communities) located in the district. According to SFC's definition, a forest community is "a group of people living in a forest village and undertaking activities that rely on forest resources in order to sustain their livelihood" (SFC 2001). Forest villages are defined as those that border on forest area.

A major strategy for addressing the socio-economic problems of forest communities which has a considerable tradition in Indonesia is the **tumpangsari** program, which allows farmers to cultivate crops on forest land during the first years after establishing a forest plantation. In the period between 1987 and 2001, there were more than 32,000 forest farmers involved in the **tumpangsari** program of the East Banyumas SFC (East Banyumas SFC 2001a), and they were organized in approximately 1,000 forest farmer groups with an average group size of 30 members.

3.2 Methods

The data collection techniques applied in this study included participant observation through direct involvement in meetings and workshops; in-depth interviews with key informants such as NGO representatives, leaders of forest farmer groups, and SFC staff members; and the analysis of documents issued by the NGO involved and by the East Banyumas SFC. The interviews were conducted from February to June 2001, using interview guidelines that were cross-checked with key informants. They covered general problems of illegal logging, the current organisation status of the forest farmers, underlying motives for promoting or opposing devolution, and the strategies used by the proponents and opponents of devolution to pursue their interests. Interviewees were selected by using snow-ball sampling, a non-probability sampling technique that relies on referrals from former respondents to identify additional interviewees. Data were analyzed using the technique of explanation building (Burns 2000, p. 473) with regard to the strategies the actors applied in the devolution process by stipulating a set of causal links regarding the case.

4 Forest Management Problems and the Struggle for Devolution

4.1 Problems of Forest Management in Banyumas District

The general problems of increasing deforestation mentioned in the introduction can also be observed in Banyumas District. The number of forest trees lost due to illegal logging increased nearly five fold between 1998 and 2000 (see Figure 1), and the resulting financial loss to the East Banyumas SFC amounted to US $ 14 million in the year 2000 (East Banyumas SFC 2001b).

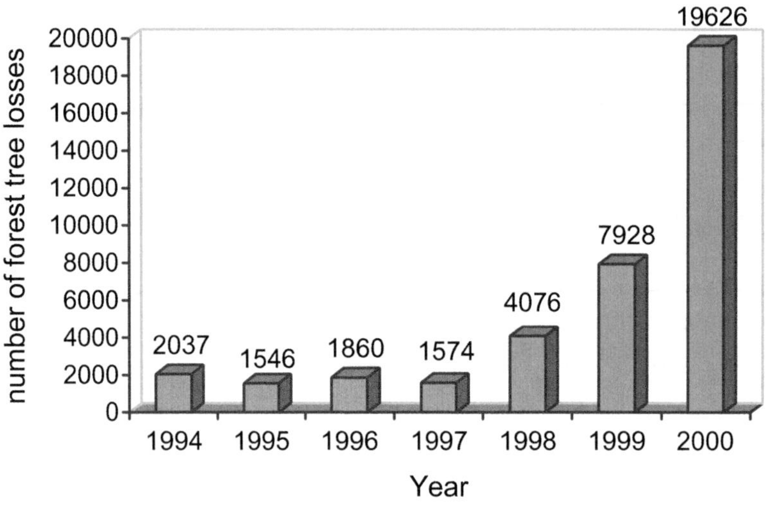

Fig. 1 Forest Tree Losses in the East Banyumas SFC Between 1994 – 2000
Source: East Banyumas SFC (2001b)

As the research showed, it was the combination of the economic crisis and problems of law enforcement arising during the reformation period that contributed to the deforestation problem. One strategy of the local people who were negatively affected by the economic crisis was to engage in illegal logging or to clear additional forestland for farming. According to interviews with forestry officers, 134 ha of the more than 13,000 ha of forestland were converted to farm land in 2000. This suggests that illegal logging was the more important problem. Illegal logging relies on an extensive network of different groups of actors, including forest rangers, financial backers and the military. Enterprises that do not have access to the legal logging trade cooperate with community members who provide labor for large-scale illegal logging. However, community members also practice illegal logging on their own on a small scale, both for meeting their subsistence needs which became more difficult due to the economic crisis and as a means of protest against the behavior of the SFC. The villagers protested above all against the handling of cases in which illegal logging was detected and prosecuted. While the forestry officers in charge had to bear comparatively minor consequences such as relocation to another position, the involved villagers had to go to jail.

The problem of illegal logging also led to conflicts within the villages. As the interviews in the communities showed, villagers who did not participate in illegal logging were concerned about its negative ecological and economic impact such as landslides and floods during the rainy season and decreased water availability for household purposes and cultivation. Villagers who tried to report illegal logging to the authorities often received threats from illegal loggers. Moreover, there was no local institution that was able to resolve this social conflict within the forest communities. The dissatisfaction of the villagers who were not able to resolve the illegal logging problem but experienced its negative ecological and economic consequences was an important motive for them to engage in the struggle for the implementation of the CBFM policy, as will be discussed in the following section.

4.2 The Struggle for Devolution

The struggle of the forest communities for more rights to forest resources started as a response to the failure of the government's forestry programs such as **tumpangsari** and social forestry. In 2000, the Forest Farmer Association of Banyumas was formed as an organizational umbrella for the forest farmer groups existing in the district. The Forest Farmer Association received support from a regional non-governmental organization (NGO) referred to as LPPSLH as well as from two national NGOs. In its first congress report, the Association criticized the government forest programs as follows:

1. The forest farmers are treated as low wage laborers in the whole wood production cycle. The rights of program participants such as support for land preparation and reforestation and fertilizer subsidies were neglected and subject to corruption.
2. The right of forest farmers to receive benefits from the forest output is not recognized; they are instead accused of being responsible for deforestation (FFA 2000).

After the first congress, the Forest Farmer Association worked for about one year to ensure the support of the forest communities in the district before organizing a large "forest meeting" (**rembuk alas**) from May 24-27, 2001, in Sokawera village. The Association invited representatives of government institutions such as the heads of the government districts in the region, the leaders of the regional House of Representatives, the governor of Central Java, the head of the police, representatives of the mass media, students of the regional university, and two national NGOs.

Prior to the meeting, the leaders of the Forest Farmer Association, with the support of the regional NGO, had designed a concept for sustainable forest management which they considered to be a response to the CBFM program of the central level SFC and the decentralization Laws 22 and 23. They emphasized the role of the forests for water quantity and landslide prevention and demanded a change from the existing clear-felling system to a selective-felling system. To improve the socio-economic conditions of the forest farmers, they suggested developing coffee plantations on the existing **tumpangsari** land (on the average 0.25 ha of land for each forest farmer) and proposed that the farmers should receive three-fifths of the revenues from coffee. The remaining revenues should be shared by SFC, the district and the village governments, and by a village forest surveillance institution. The forest farmer group in Sokawera Village had established such an institution prior to the meeting. It comprised members of the village leadership such as traditional community leaders, religious leaders and the teacher. The institution was envisaged to resolve problems and conflicts in forest management.

In the forest meeting, the head of the planning division of the central-level SFC accepted the concept of sustainable forest management proposed by the forest farmers, even though the representatives of the SFC at the district level strongly objected to this plan. In a following meeting attended by all of the leading officers of the East Banyumas SFC, it was revealed that they clearly disagreed with the concept presented by the Forest Farmer Association. At the time of submitting this paper, an agreement respecting the concept had not yet been implemented. However, the approval of the CBFM plan by the central-level SFC can already be considered as an important step towards achieving devolution.

5 Analysis of the Policy Process

Applying the theoretical framework, this section analyzes how the proponents and opponents of the devolution process used their social capital to create political capital in order to pursue their goals.

5.1 Social Capital of the Proponents

Non-Governmental Organizations

Three NGOs were involved in the devolution process in Banyumas district: two national NGOs and the regional NGO LPPSLH. Representatives of the two national NGOs were invited by LPPSLH as key speakers and facilitators because of their experience in dealing with community empowerment. Still, the most important NGO in the devolution process was LPPSLH. The major objective of this NGO is to alleviate poverty, foster social justice and promote democratization. The LPPSLH's target groups include small enterprises, households headed by women, farmers, forest farmers, and groups marginalized within the society. Therefore, LPSSLH claims that it has a strong basis at the grassroots level. LPSSLH also has social capital that helps bridging in the form of good relations to politicians and administrative officers at the regional level because this NGO facilitates the implementation of district government programs.

Community-Based Organizations

During the Suharto regime, the government promoted the establishment of community-based organizations related to natural resource management – such as the above mentioned forest farmer groups – under its corporatism policy. The main aim of establishing such groups was to control the political dynamics in rural forest areas and to prevent state forests from being used illegally by villagers. The forest farmer groups represented a form of social capital upon which reform efforts could build. After the fall of the Suharto regime, these groups developed into organizations that were able to strengthen the farmers' position against the government. They argued that they did not benefit from government policies in the forestry sector. As their access to the forest resources has been limited to date and they lack alternative income sources, they now emphasize land scarcity issues and advocate land or forest tenure rights. The first umbrella organization of the community-based farmers' organizations that emerged after the fall of the Suharto regime in Banyumas District was the Forest Farmer Association referred to above. It was established by 28 forest farmer groups without intervention from the government. The support of the NGOs was found to be dominant both in terms of financial and technical assistance.

In addition to the Forest Farmer Association, the forest farmer group in Sokawera village, which hosted the forest meeting, played an important role. One member of the Sokawera's forest farmer group is a staff member of LPPSLH so that he had a good opportunity to increase his human capital by participating in training courses and workshops as well as to create social capital to form a bridge between his village community and the NGO network. Some members of the forest farmer group at Sokawera were also involved in public demonstrations organized by NGOs and farmers' networks at the national level. They also joined the general Banyumas Farmers' Association, the Communication Forum on Community Forestry and the Association of Community Empowerment. These activities resulted in good contacts with other NGOs and provided additional resources in the form of technical and financial assistance. This economic, human and bridging social capital was essential in enabling the Sokawera Community to host the forest meeting where these resources could be converted into political capital.

5.2 Social Capital of the Opponents

Non-Governmental Organization

The East Banyumas SFC, the major opponent against implementing the CBFM plan, also possessed social capital in form of a relation to a NGO. The East Banyumas SFC relied on the NGO Gempita to assist them in implementing its forestry programs in rural areas. This NGO was established in 1999 by academic staff members of the regional university with the aim of assisting people in forest areas to develop rural enterprises. The East Banyumas SFC preferred to contract this NGO since the other NGOs tended to provoke the forest communities against the SFC. By relying on Gempita, SFC still retained control over the forest villagers' activities. The cooperation by the East Banyumas SFC with Gempita led to conflicts between the leaders of the forest farmer groups assisted by Gempita and those assisted by LPPSLH. Moreover, the East Banyumas SFC used the cooperation with Gempita to develop their own version of CBFM which, for example, does not allow farmers to plant coffee trees on **tumpangsari** land.

Illegal Logging Network

A second category of social capital held by the opponents of the CBFM consists in the illegal logging network depicted in Figure 2. Financial backers such as wood traders and wood enterprises play a critical role in providing finances, transportation and machinery to the illegal loggers. According to a newspaper report, some forest rangers and the leading officers of East Banyumas SFC are involved in the illegal logging network (Suara Merdeka 2001). Financial backers, the police, and the military apparatus provide financial and security services for the operations. The wood industry is able to receive timber from illegal operations at a comparatively low price, which increases their profit margin.

As outlined in the section on problems of forest management, the villagers are involved in the illegal logging network by providing labor for large-scale logging operations and by practicing small-scale logging on their own account. The illegal logging network, as summarized in Figure 2, can be considered as a form of "perverse" social capital because it rewards rent-seeking and creates incentives for criminal behavior.

The forestry officers involved appeared to anticipate that the CBFM program would undermine their authority in forest management as well as reduce their income derived from their involvement in the illegal logging network. They had reason to assume that the CBFM policy would reduce the villagers' preparedness to collaborate in illegal logging, especially by providing alternative income and employment sources. Some forest rangers reported to the head of East Banyumas SFC that the forest communities lacked the capacity for appropriate forest management and placed more emphasis on short-term economic gains, while neglecting ecological goals.

Social Relations within the SFC

Some of East Banyumas SFC's senior officers also had social capital in the form of good relations with the top officers of the central SFC. Since they often visited Central SFC in Jakarta for meetings and coordination, they were able to build up this type of bridging capital.

6 The Creation of Political Capital

Figure 3 illustrates the strategies of the proponents and opponents to convert their resources into political capital.

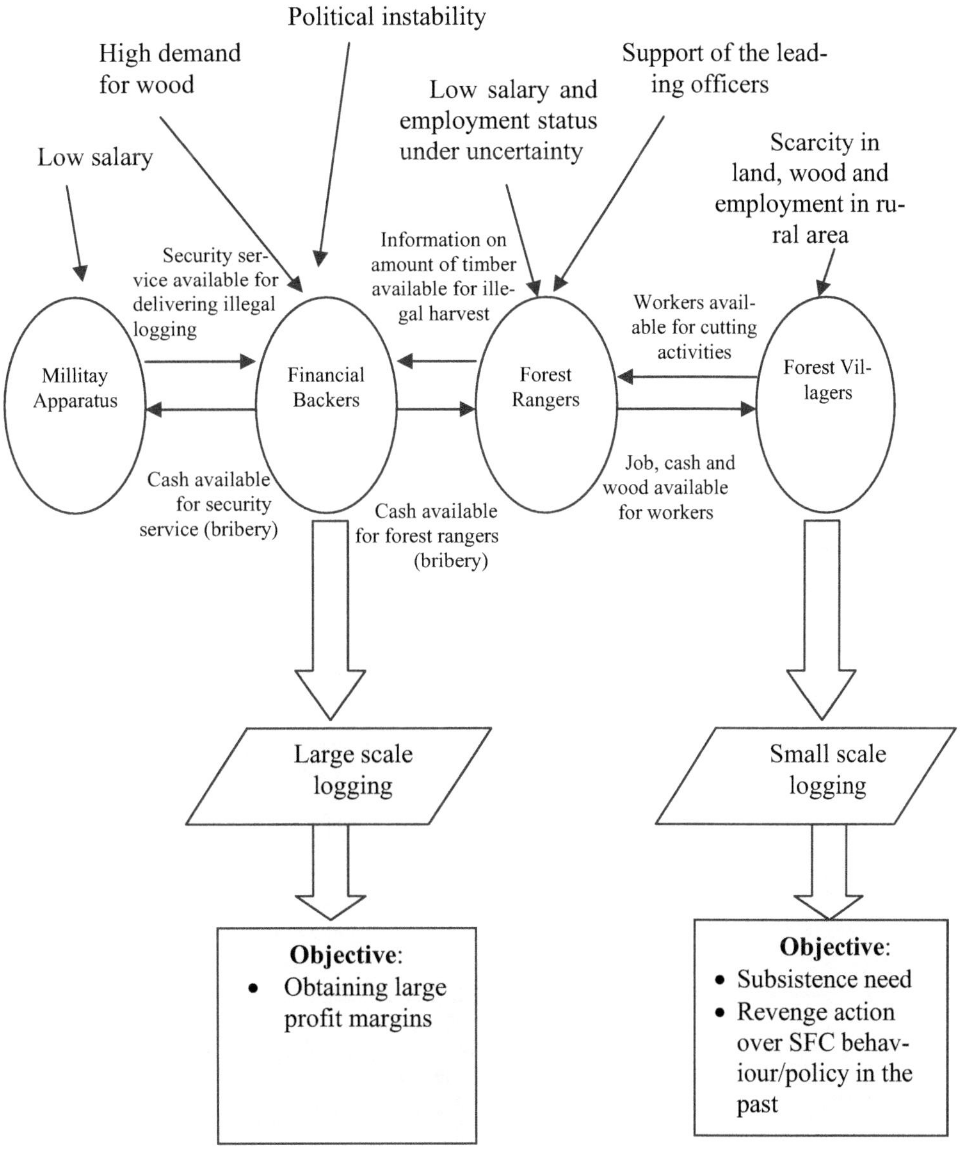

Fig. 2. Illegal Logging Network
Source: adapted from Ford Foundation Team (2001)

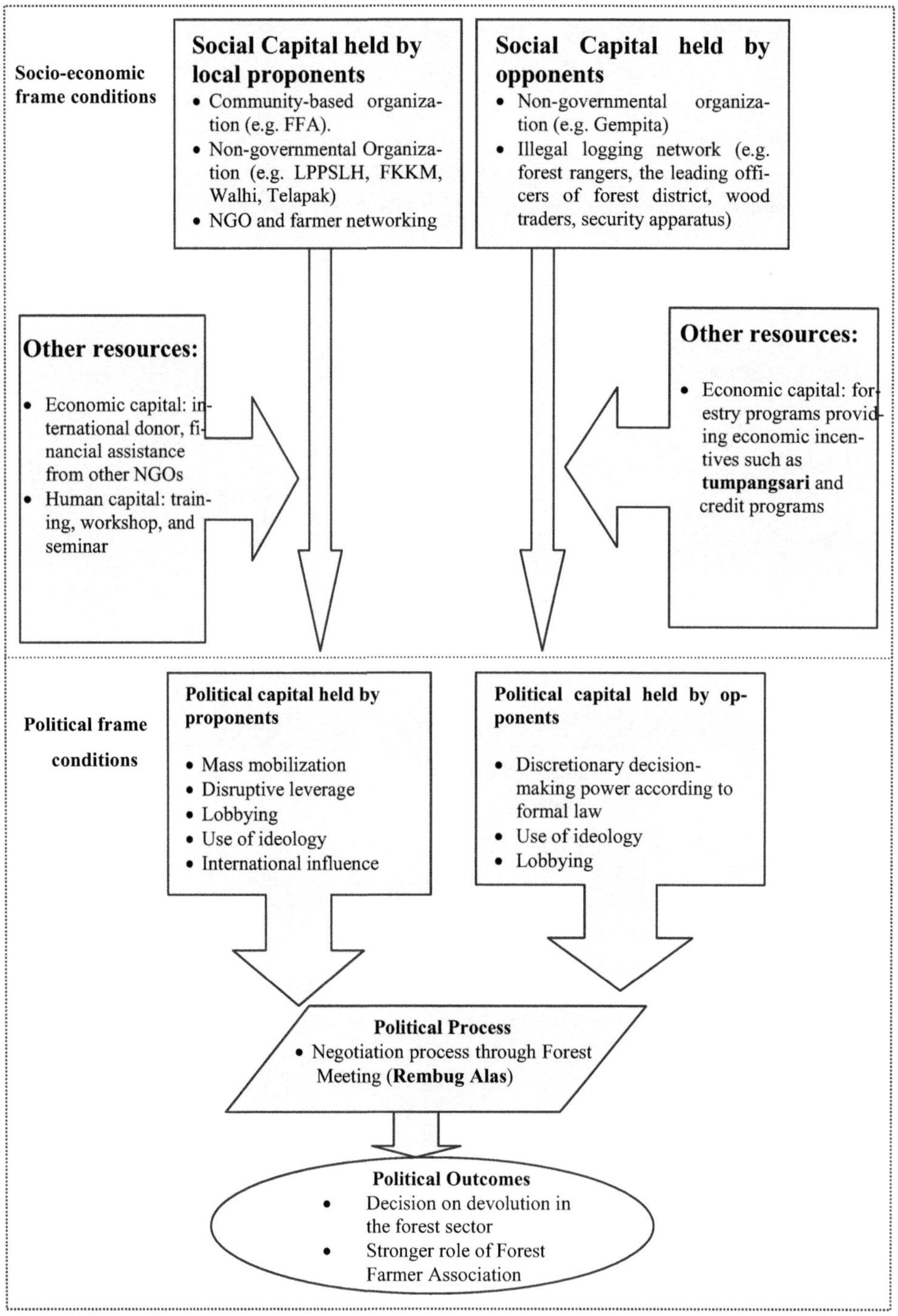

Fig. 3. Use of Social Capital for the Creation of Political Capital
Source: based on Birner and Wittmer (2003)

6.1 Political Capital of the Proponents

Mobilization of the Communities

The mobilization of the forest villagers turned out to be an important political resource. The forest meeting proved to be an effective strategy used by the Forest Farmer Association and the NGOs to mobilize the villagers organized in the forest farmer groups against the SFC in order to implement a devolution policy. In addition, two critical workshops were held by the Forest Farmer Association and LPPSLH. These workshops involved representatives of the forest villages such as members of the official village government institutions. The workshops informed the local population about the new scope for action that was opened by new legislation. The workshops appeared to be an important prerequisite for the mobilization of the villagers.

Disruptive Leverage

In informal meetings with leading members of the forest farmer groups, the regional NGO as well as one national NGO encouraged the forest villagers to take actions against the East Banyumas SFC. They gave examples of actions carried out by forest villagers in other districts, such as the occupation of state forest land and the claiming of forestland that had belonged to their parents. Though the Banyumas forest communities did not undertake such actions themselves, they were able to create a credible threat due to the fact that they were capable of such disruptive activities.

Lobbying

Leading members of the Forest Farmer Association and LPPSLH were able to lobby important state officials such as the head of the forest district, the head of the district government, the head of the regional House of Representatives and the head of the Planning Division of the central level SFC. For these lobbying efforts, LPPSLH were able to use the bridging social capital they had acquired by their facilitation of district government projects. The lobbying efforts were most successful with politicians at the regional level, who may have considered supporting the villagers to be a strategy to receive political capital in the form of votes. During the forest meeting, the head of the Banyumas House of Representatives provided strong support to the forest communities to promote their own CBFM concept. He also urged East Banyumas SFC to pay more attention to the improvement of the socio-economic situation of the forest communities.

Use of Ideology

In their struggle against the power of the SFC, the Forest Farmer Association and LPPSLH framed their argumentation in a discourse focusing on exploitation, underdevelopment, inequity, and impoverishment This discourse strategy was obviously successful both with regard to achieving the support of the local politicians as well as mobilizing the villagers.

International Influence

During the reform era, international donors increasingly provided funds to NGOs that aimed at assisting local people in building democracy at the grassroots level. LPPSLH, as a major actor in the devolution process, essentially relied upon this foreign funding for its activities. Otherwise the NGOs would not have been able to provide financial support to the Forest Farmer Association, the workshops, and the crucial forest meeting.

6.2 Political Capital of the Opponents

Discretionary Decision-Making Power

The head of the forest district, as the top decision-maker, has the legal authority to determine the forest management policy at the district level. This implies considerable discretionary power which represents an important type of political capital. The head of the East Banyumas SFC used this power to declare the Forest Farmer Association an illegal umbrella organization for the forest farmer groups. He also decided that groups who joined the Association would have difficulty participating in forestry programs such as **tumpangsari** and credit programs. This decision put tremendous pressure on the villagers to withdraw from the Association because their livelihoods depend to a large extent on their participation in these forestry programs.

Use of Ideology

To weaken the Forest Farmer Association, the head of the East Banyumas SFC claimed that the leader of the Association was a communist cadre since he tried to mobilize forest farmer groups against SFC. He also connected the Indonesian acronym SETAN, which was used by the Association, with the term "satan". Traditionally, this term was used to defame communist groups. Communist ideologies were prohibited in Indonesia after the defeat of the communist uprising in 1965. The former political regime often accused their opponents of being communists. By using such accusations, the East Banyumas SFC was successful in convincing some forest farmer groups not to join the Association.

Lobbying

The opponents of the CBFM policy were able to use the social capital they had due to their good relations with SFC officers at the central SFC level for lobbying. However, in contrast to the proponents of the devolution program, they started these lobbying efforts only after the forestry meeting was held. Otherwise they might have prevented the approval of the representative of the central SFC during the forestry meeting.

7 Discussion

The case study shows that devolution cannot be achieved without a strategic effort by the proponents of this policy against those who fear losing power and income as a result of devolution. By highlighting the strategies proponents and opponents of devolution can use to create political capital, the study makes it possible to derive strategic insights for other communities and NGOs that are involved in comparable devolution processes. However, it is important to take the macro-political frame conditions into account, as they play an important role with respect to the possibilities the proponents and opponents of devolution have for creating political capital.

In the Indonesian case, the new legislation on decentralization and the new forestry policy created the legal frame conditions for devolution and the people's participation in the management of natural resources. At the same time, the increased freedom of the press made an open discourse concerning state failure in forest management possible (McCarthy 2000a). The role of the military in the political and administrative arena decreased (especially with the amendment of the 1945 Constitution; Article 10), which reduced the opportunities of the illegal logging network to use the military apparatus in order to pursue their political and economic goals. These factors imply that the change from an authoritarian to a democratic political regime increased the opportunities for the forest villagers to create political capital, while at the same time depreciating the political capital held by the opponents. Similar effects may be expected in other countries that undergo a change from an authoritarian to a democratic regime.

The analysis showed that within these changing macro-political frame conditions, two types of social capital were especially relevant for the proponents of devolution: (1) the organization of the forest farmer groups into an umbrella organization (bonding type of social capital), and (2) informal links with powerful groups such as NGOs and the head of the regional House of Representatives (bridging type of social capital). The role of NGOs was essential in mobilizing the village communities and providing resources in the form of economic, human, and social capital that were required to create political capital. They also played an essential role in making the information on the new legislation and policies concerning decentralization and community-based forestry available at the community level. The fact that these NGOs by and large rely on international funds highlights the role of international donor organizations in promoting devolution. International actors also influence the public discourse by highlighting questions of accountability, clean government, and good governance. The open criticism of corruption, collusion, and nepotism formulated both by the NGO movement and the international organizations increased the pressure on the bureaucracy. This was certainly an essential factor influencing the decision of the central level SFC representative to approve the villagers' CBFM plan. These observations show that NGOs and international organizations can play a very important role in "making devolution work." However, this prominent role implies the risk that NGOs and international actors may use local communities as instruments for their own political goals. Moreover, it raises questions with regard to their legitimacy in the political process which can be considered to be an important area for further research (compare Hudson 2001).

With regard to the applicability of the Banyumas experience to other areas, one also has to take into consideration the fact that the local government did not receive substantial financial resources from the forestry operations. By supporting the forest villagers' claim to resource rights, the local government was able to gain political support without forgoing financial resources for their district budget. This option for creating political capital is more limited in regions where forestry is more profitable. Casson and Obidzinski (2002) showed in a recent study in Kalimantan that local authorities progressively institutionalized illegal logging after the implementation of regional autonomy because this practice provided substantial contributions to local government budgets. McCarthy (2000b) reports that the NGOs' strategies to reduce illegal logging in South Aceh eventually failed for similar reasons. This shows that in areas in which forest extraction provides considerable revenues for local governments, the prospects of reducing illegal logging by means of the promotion of devolution are more limited than in the Banyumas case.

The case study also throws light on the strategies applied by the opponents of devolution processes. Typical strategies, which can also be observed in other devolution processes (compare Birner and Wittmer 2003), include: creating or fostering NGOs that do not criticize government agencies; discretionary use of program resources in favor of communities that do not collaborate with advocacy NGOs; discriminating the critics by associating them with a disrespected ideology (communism in this case); using intra-agency communication networks for lobbying with decision-makers at higher levels; maintaining linkages with private sector interest groups behind the scenes; and using without further proof the argument that local communities are incapable of protecting the forest resources.

An interesting message delivered by the case study is that in order to counteract such strategies even an individual village can play an important role as a "change agent" for the initiation of devolution processes. The experience with community forestry in Thailand showed that in that case it was the struggle of one particular village for recognition of traditional rights to forest resources which played an important role in initiating a devolution legislation (compare Brenner et al. 1998). This shows that individual cases of success, which may receive symbolic value in the public discussion, provide an important feed-back mechanism in the creation of political capital in favor of devolution.

8 Concluding Remarks

In conclusion, the study shows that individual cases can provide useful insights in the dynamics of devolution. Even though the reform movement in Indonesia has created favorable frame conditions for devolution at the national level, the implementation of theses policies cannot be taken for granted, especially if regional level agencies are the entities that will loose control over resources. In such a case, initiatives at the local level, supported by actors with powerful relations to the regional and central levels, are crucial for promoting the devolution of rights over forest resources to local users. The case study gives reason to hope that devolution policies can reduce deforestation by reducing the incentives for villagers to participate in illegal logging networks and by empowering those villagers who are concerned about the environmental damage caused by illegal logging. In view of the rapidly increasing deforestation rates in Indonesia, further research on this issue is warranted. Finally, our case study indicates that the concepts of social and political capital are useful analytical tools to understand devolution in the forestry sector better as they capture the political dimension of this process.

References

Agrawal A, Ostrom E (2001) Collective Action, Property Rights and Devolution of Forest and Protected Area Management. In: Meinzen-Dick R, Knox A, Di Gregorio M (eds) Collective Action, Property Rights and Devolution of Natural Resource Management – Exchange of Knowledge and Implications for Policy. DSE, Eurasburg, pp 75-109

Banerjee AK (1997) Decentralization and Devolution of Forest Management in Asia and the Pacific, Asia-Pacific Forestry Sector. Outlook Study, Working Paper APFSOS/WP/21, FAO, Rome and Bangkok

Banyumas Statistics (2000) Banyumas dalam Angka Tahun 2000. Banyumas

Banyumas Regional Planning and Development Agency (2001) Laporan Tahunan Kabupaten Banyumas. Banyumas

Birner R, Wittmer H (2003) Using Social Capital to Create Political Capital: How Do Local Communities Gain Political Influence? A Theoretical Approach and Empirical Evidence from Thailand. In: Dolšak, N, Ostrom E (eds) The Commons in the New Millenium, Challenges and Adaptation. MIT Press, Cambridge and London, pp 291-334

Birner R, Wittmer H (2000) Co-management of Natural Resources: A Transaction Costs Economics Approach to Determine the "Efficient Boundaries of the State". Paper presented at the Annual International Conference of the International Society of the New Institutional Economics (ISNIE) in Tübingen, September 2000

Booth JA, Richard PB (1998) Civil Society, Political Capital and Democratization in Central Africa. The Journal of Politics 70:780-800

Bourdieu P (1992) Ökonomisches Kapital – Kulturelles Kapital – Soziales Kapital, in: Die verborgenen Mechanismen der Macht – Schriften zu Politik und Kultur 1. VSA-Verlag, Hamburg, pp 49-79

Braden JB, Proost S (eds) (1997) The Economic Theory of Environmental Policy in a Federal System. Edward Elgar, Cheltenham, Lyme

Brenner V, Buergen R, Kessler C, Pye O, Schwarzmeier R, Sprung RD (1998) Thailand's Community Forestry Bill: U-Turn or Roudabout in Forest Policy. SEFUT Working Paper No. 3, DFG Graduate College Socio-Economics of Forest Use in the Tropics and Subtropics, Freiburg

Burns R (2000) Introduction to Research Methods. Sage Publication, London

Casson A, Obidzinski K (2002) From New Order to Regional Autonomy - Shifting Dynamics of "Illegal Logging" in Kalimantan, Indonesia. World Development 30 (12): 2133-2151

Coleman J (1988) Social Capital in the Creation of Human Capital. American Journal of Sociology 94:95-120

East Banyumas SFC (2001a) Perkembangan Program Perhutanan Sosial (in Indonesian). East Banyumas SFC, Purwokerto

East Banyumas SFC (2001b) Data Perkembangan Pencurian Kayu (in Indonesian). East Banyumas SFC, Purwokerto

FFA (2000) Laporan Konggres Pertama *Setan Balong*. Purwokerto

Ford Foundation Team (2001) Resume Diskusi Illegal Logging. In: Awang, S.A., Kurniawan, I., Nuh, I.M., (eds), Otonomi Sumberdaya Hutan. Proceeding of the Fifth Regular Meeting of FKKM in Bandar Lampung, Indonesia

Fuchs ER, Minnite LC, Shapiro RY (1999) Political Capital and Political Participation. Columbia University Working Paper No. 9, Columbia University.

Hudson A (2001) NGOs' Transnational Advocacy Networks: From 'Legitimacy' to 'Political Responsibility'? Global Networks 1 (4):331-352.

Kompas (2002) Otonomi Ancam Pengelolaan Hutan, June 1, 2002 http://www.kompas.com/kompas-cetak/206/01/daerah/oton19.htm

Ligon E, Narain U (1999) Government Management of Village Commons. Journal of Environmental Economics and Management 37:272-289

McCarthy JF (2000a) The Changing Regime: Forest Property and Reformasi in Indonesia. Development and Change 31:91-129

McCarthy JF (2000b) "Wild Logging": The Rise and Fall of Logging Networks and Biodiversity Projects on Sumatra's Rainforest Frontier. CIFOR Occasional Paper 31, Bogor

Meinzen-Dick R, Knox A (2001) Collective Action, Property Rights and Devolution of Natural Resource Management: A Conceptual Framework. In: Meinzen-Dick R, Knox A., and Di Gregorio M (eds) Collective Action, Property Rights and Devolution of Natural Resource Management, Exchange of Knowledge and Implications for Policy. DSE, Eurasburg, Germany, pp 41-73

Meinzen-Dick R, Knox A, Di Gregorio M (eds) (2001) Collective Action, Property Rights and Devolution of Natural Resource Management, Exchange of Knowledge and Implications for Policy. DSE, Eurasburg, Germany

Narayan D (1999) Bonds and Bridge: Social Capital and Poverty. World Bank, Poverty Reduction and Economic Management Network, Washington, DC

Newton K (2001) Trust, Social Capital, Civil Society and Democracy. International Political Science Review 22 (2): 201-214

Ostrom E, Schroeder L, Wynne S (1993) Institutional Incentives and Sustainable Development - Infrastructure Policies in Perspective. Westview Press, Boulder

Putnam R (1993) Making Democracy Work: Civic Tradition in Modern Italy. Princeton University Press, Princeton

Ribot J (2002) Democratic Decentralization of Natural Resources - Institutionalizing Public Participation. World Resources Institute, Washington

Rubio M (1997) Perverse Social Capital: Some Evidence from Colombia. Journal of Economic Issues 31 (3):805-816

SFC (2000) Pengelolaan Hutan Bersama Masyarakat. Jakarta

SFC (2001) Transformasi Menuju Perhutani Masa Depan. Jakarta

Suara Merdeka (2001) Dipertanyakan, Kontribusi Perhutani untuk Pemkab, Suara Merdeka Newspaper, December 28, 2001, www.suaramerdeka.com/harian/0112/28/dar23.htm

Wall E, Ferrazzi G, Schryer F (1998) Getting the Goods on Social Capital. Rural Sociology 63 (2):300-322

Wibowo D, Byron NR (1999) Deforestation Mechanism: A Survey. International Journal of Social Economics 26 (3):455-474

World Bank (1997) World Development Report 1997 - The State in a Changing World. Washington, DC

Does Technical Progress in Agriculture have a Forest Saving or a Forest Clearing Effect? Theory and Evidence from Central Sulawesi

Miet Maertens, Manfred Zeller & Regina Birner

1 Introduction

Enhancing agricultural technology is an important policy tool for improving livelihoods and stimulate economic growth in rural areas of developing countries. It is however not completely understood whether technical progress and agricultural intensification would reduce or aggravate pressure on forests (Angelsen et al, 1999). Does technical improvement diminish forest clearing by reducing the area farmers need to make a living? Or does technical progress lead to increased agricultural expansion and deforestation by rendering agriculture more profitable? The question of a 'forest saving' or a 'forest clearing' effect of technical progress is extremely relevant within the framework of interactions between environmental and socio-economic goals. Policy debates have been dominated by the assumption of a win-win relation between technical progress and forest conservation (Angelsen and Kaimowitz 2001).

Several empirical studies have analyzed the impact of technical change on land use and forest resources. It has been demonstrated that irrigation development leads to significant reductions in forest clearing (e.g. Shively 2000; Shively and Martinez 2001; Deininger and Minten 1999). Some studies indicate that the use of Green Revolution inputs has cut down deforestation (e.g. Mertens et al 2000; Coxhead and Jayasuriya 1994). Others argue that chemical input use stimulates agricultural expansion at forest margins (e.g. Ruf 2001; Angelsen et al 1999). Findings are ambiguous and additional empirical evidence is essential. Yet, empirical analysis of the impact of technical progress on land use is complicated by the fact that both land use and the adoption of new technologies are likely to be influenced by the same factors (Angelsen et al 2001). According to the induced innovation literature, technical improvement comes about to facilitate the substitution of relative abundant factors for relative scarce factors and is induced by changes in relative prices, population growth etc. (Koppel 1995; Ruttan and Hayami 1984; Ruthenberg 1980; Boserup 1965). However, those factors are also often mentioned as driving forces for land use change and deforestation (Contreras-Hermosilla, 2000; Sunderlin and Resusodarmo 2000; Bergeron and Pender 1999; Kaimowitz and Angelsen 1998; Scherr and Hazell 1994). This simultaneity of technical progress and land use change causes difficulties in econometric models and can lead to wrong estimations and misinterpretation of regression coefficients.

Technical progress in Southeast Asian agriculture, such as irrigation development and adoption of Green Revolution inputs, has mostly been oriented towards the lowland sector, associated with irrigated paddy rice production (Coxhead and Warr 199; Hayami and Ruttan 1984). Yet, agricultural expansion into forested areas is mostly taking place in the uplands at forest margins. The impact of technical progress in the lowlands on land use at forest margins is not straightforward. Therefore disaggregating the agricultural sector into frontier (upland) and non-frontier (lowland) agriculture is particularly relevant when discussing technological change (Angelsen 1999).

The aim of the paper is to assess the effect of different types of technical change in lowland paddy rice production on land use at forest margins in the Lore Lindu region, Central Sulawesi. To guide the empirical analysis we develop an analytical framework based on a Chayanov-type household model. The model shows how changes in technology affect households' land allocation decisions in the lowlands and in the uplands. We evaluate the analytical findings with the estimation of an econometric model, using advanced estimation techniques to allow for the simultaneity of technical progress and land use change. We discuss the estimation results and spell out some policy implications of the findings.

2 Agriculture, Technology and Deforestation in the Lore Lindu Region

2.1 Data

The Lore Lindu region consists of the Lore Lindu National Park and 5 surrounding districts. The region is located in the humid tropics (around 1° south of the equator) and the most northern district borders Palu, the provincial capital of Central Sulawesi. Eighty of the 119 villages in this area have been selected for a village survey using a stratified random sampling method (for more details on the selection procedure see Zeller et al 2002). The survey respondents were questioned about the current (March-July 2001) socio-economic situation and were asked to recall information from the past. The survey data were complemented with secondary statistical data and geographic data combined in a GIS.

2.2 Land Use Dynamics

The Lore Lindu region covers an area of 7.2 thousand km^2 of which 2.2 thousand km^2 constitute the Lore Lindu National Park (table 1). It is a topographically diverse region with mountain ranges reaching up to 2500 meters interspersed with narrow and outstretched valleys at different elevations. The district boundaries are set along topographic features which make them have distinct natural characteristics. Using the survey data, the actual population size in the region was estimated to be 136 thousand people (table 1). Population density is rather low: 27.3 people per km², excluding the Park's area. The northern districts, Sigi Biromaru and Palolo, are better accessible and are more densely populated. The southern districts Kulawi, Lore Utara and Lore Selatan, are more mountainous, further away from the city Palu, and less densely populated. In general, land is still fairly abundant in this region. We estimated that about 15% of the total area outside the National Park is used for agricultural cultivation (table 1).

Table 1. Land Use and Demographics per District

	Total	Sigi Biromaru	Palolo	Kulawi	Lore Utara	Lore Selatan
Total area (km²)	7,223	529	670	3,085	2,010	929
Total area, excluding TNLL*	5,008	370	506	2,265	1,106	761
Area with slope < 3° (km²)	736	130	88	101	328	88
% agricultural area	15%	53%	33%	11%	10%	3%
Population 2001	136,707	49,868	27,126	33,574	18,743	7,396
Population density (km²)	27.30	134.83	53.63	14.82	16.94	9.72
Population growth 1980/2002	66%	44%	127%	44%	195%	21%

Source: Calculated from STORMA A3 Village Survey Data and Secondary Data
* TNLL: Lore Lindu National Park (Taman Nasional Lore Lindu)

With 87% of the households being farmers, agriculture is the most important income source in this region. Farmers are cultivating food crops as well as cash crops and have some small livestock. One third of the total agricultural area in the Lore Lindu region is allocated to paddy rice production (figure 1). Perennial crops, mainly coffee and cocoa, take about half of the agricultural area. The remaining area consists of fields with annual crops (12%), mainly upland rice and corn, and home-gardens (7%) where different crops are mixed. Rice is the staple food crop while cocoa and coffee are almost entirely destined for export. As in many parts of Southeast Asia, there is a lowland-upland dichotomy in the agricultural sector in this region. The lowland sector is associated with food crop production, mainly paddy rice cultivation. The upland sector corresponds with the cultivation of perennials, mainly coffee and cocoa that are sometimes intercropped in an agroforestry system with fruit or forest trees. Households cultivating paddy rice typically also have one or several upland plots. Due to the topographic diversity in this region, there is not really a distinction between lowland and upland farmers or between lowland and upland villages. Rather, a lowland and upland production system co-exist in most villages.

Over the past two decades, population in this region has grown with 66% (table 1). In this time span, the area paddy rice increased from 18,000 to 24,000 hectares; the area annual crops (corn; upland rice, vegetables) decreased from 10,000 hectares to 6,000 hectares; and the area perennial crops (cocoa, coffee, coconut) tripled from around 10,000 hectares to slightly more than 30,000 hectares (figure 2).

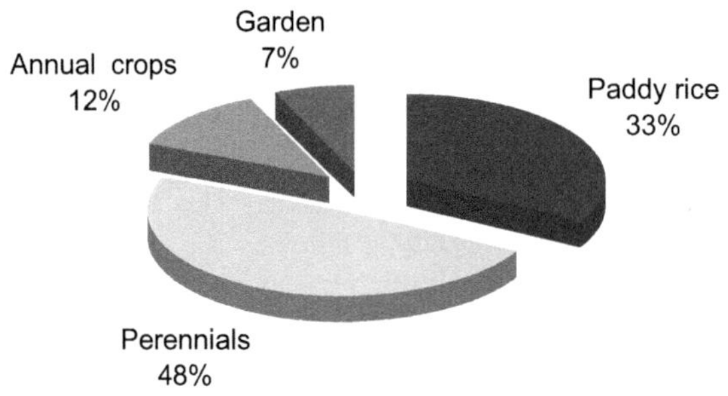

Fig. 1. Allocation of Land to Different Crops

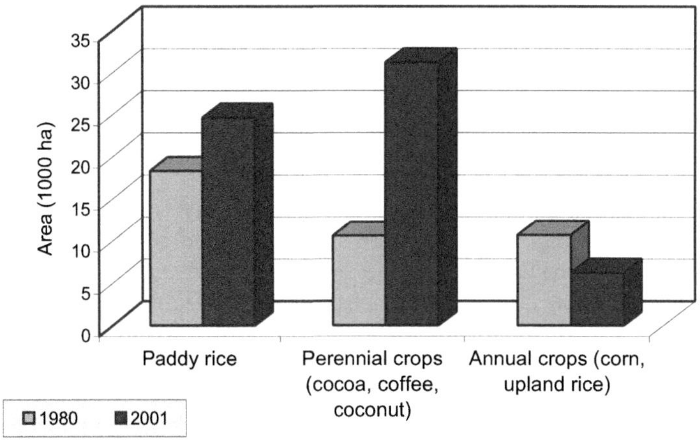

Fig. 2. Land Use Change in the past 2 Decades

Due to technical progress and yield improvements, expansion of lowland agriculture has been relatively small notwithstanding a very high population growth. Overall in Indonesia, paddy rice yields increased from 1.76 ton/ha to 4.39 ton/ha over the past 4 decades (FAO 2002; BPS 2002). Yet, there has been a vast expansion of the agricultural area in the upland sector, which took place at forest margins. During the past 2 decades, a lot of forest in the Lore Lindu region, even inside the National Park, had to gave way to cocoa and coffee plantations. In this paper we want to investigate how different technical changes for paddy rice cultivation have promoted or impeded this agricultural expansion at forest margins.

2.3 Agricultural Technology

Irrigation

There are 3 different systems for paddy rice cultivation: a technical, a semi-technical and a simple system. The technical irrigation system involves a permanent irrigation infrastructure with locks and separate drainpipes and tertiary water-supply channels, which allows complete control of the water level on the fields (Direktorat Jenderal Pengairan 1986). This system allows farmers to harvest 3 times a year: 2 harvests of paddy rice and 1 intermediate harvest. Usually peanuts or green beans are grown as intermediate crop. The semi-technical irrigation infrastructure is also permanent but without tertiary channels such that drainage and water supply cannot be regulated completely. With this system 2 rice harvests or 1 rice harvest and 1 intermediate harvest is possible. The simple irrigation system is a rain-fed system of paddy rice cultivation with temporary ditches and no control of water inflow. The number of harvests, 1 or 2, in this system depends on the natural conditions. Of the 80 villages in the sample, 13 are equipped with a technical irrigation system; 25 with a semi-technical system; and in 29 villages paddy rice cultivation is rain-fed. The remainder (13 villages) have no paddy rice fields at all. In the Lore Lindu region, technical irrigation was established during the 1930s under the Dutch rule and in the 1970s. The semi-technical irrigation infrastructure in the region was mainly built during the 1980s and 1990s by the Department of Public Works.

Green Revolution

The use of Green Revolution inputs is widespread in the Lore Lindu region. Chemical weed and pest control is practiced in 95% of the 80 sampled villages and fertilizers in 74%. HYVs (High Yielding Varieties) for paddy rice are used in 84% of the rice cultivating villages in the sample. These Green Revolution inputs were first introduced in the region in the early 1970s and are made available to the farmers through the government research and extension agency, 'Assessment Institute for Agricultural Technology in Central Sulawesi' (BPTP 2001). Since 1985 this agency has been based in the district Sigi Biromaru and doing research on locally adapted crop varieties and inputs combined with extension towards farmers.

Mechanization

The major device for increased mechanization in paddy rice cultivation has been the introduction of the hand-tractor. As compared to animal traction, the use of the hand-tractor reduces labor requirements to prepare paddy fields by half (Dinas Pertanian 2001). The first hand-tractors were introduced in the Lore Lindu region in 1979. Now, 81% of the paddy rice growing villages in our sample have at least one hand-tractor. The machines are provided by the Agriculture Agency who demonstrates the hand-tractor in the village and teaches the villagers to form a cooperative for purchase, maintenance and use of the equipment (Dinas Pertanian 2001). Also, rice mills and chainsaws are widespread in the this region.

Induced or Directed Innovation?

According to the induced innovation literature, technical progress is induced by changes in relative prices, population growth, land scarcity etc (Koppel 1995; Ruttan and Hayami1984). However, technical progress can also be directed to reach certain macro-economic goals and spread through government extension services (Burmeister 1987). Agricultural research and extension in Indonesia has focused on increasing paddy rice production on the basis of the macro-economic goal of rice self-sufficiency. As described above, government agencies are involved in the establishment of irrigation infrastructure and the introduction of chemical inputs and hand-tractors in the Lore Lindu region.

Table 2. Extension Services and technical progress across population density groups

	Population density groups			Total
	High	Medium	Low	
Population density (per km²)	303	63	10	127
Villages receiving extension services	81%	85%	62%	76%
# extension visits per year	34	25	22	27.28
% of villages using				
chemical fertilizer	96%	81%	42%	74%
HYV for paddy rice*	92%	91%	63%	84%
Share of paddy rice area with*				
technical irrigation*	31%	8%	3%	15%
semi-technical irrigation*	46%	35%	17%	34%
# hand-tractors /100 ha paddy rice*	3.03	2.92	1.74	2.63

Source: Calculated from STORMA A3 Village Survey Data

* Only paddy rice growing villages taken into account for calculation

Source: Calculated from STORMA A3 Village Survey Data
* Only paddy rice growing villages taken into account for calculation

The placement of irrigation and extension programs might not be random but chosen in order to maximize the impact of the program (Pitt et al 1994). If we divide the villages according to population density, we observe that government extension services are biased towards areas where population density is higher (table 3). Also irrigation development and the use of chemical inputs and hand-tractors increases across population density groups. Technical progress in the Lore Lindu area does not really originate within the villages but is rather directed by the government to increase paddy rice production in densely populated areas. However, other factors such as access to capital might also influence the adoption of certain technologies.

3 A Model of Agricultural Land Allocation with Lowland-Upland Dichotomy

3.1 The Analytical Framework

We assume a rural economy with a lowland and an upland agricultural production system, both characterized by Leontief-type production functions with labor (*L*) and capital (*K*) inputs per hectare fixed:

$Y_l = f(L_l, K_l)$ with Y_l: lowland yield per hectare

$Y_u = g(L_u, K_u)$ with Y_u: upland yield per hectare

Lowland and upland production compete for labor (*L*) and capital (*K*), which are assumed to be completely mobile between the two sectors, but not for land (*A*). Expansion of the upland sector is taking place at forest margins while no interaction is assumed between the lowland sector and the forest. When markets are perfect, household behavior can be modeled as a profit maximizing problem in both sub-sectors. Then, as long as no general equilibrium effects are considered, any type of technical progress in the lowlands is likely to cause an expansion of the lowland sector but will have no direct effects on land use at forest margins. However, when markets are not perfect and households cannot separate consumption and production decisions, technical change in the lowland sector can have a large impact on land use in the uplands. The direction and magnitude of such effects critically depends on the type of technology.

To explain this further we use a Chayanov-type model, which is the standard formulation in the agricultural household literature (Singh et al. 1986; Ellis 1993; Deaton 1992; Udry 1997). The choice of such a model is based on the observation that households in the research area are quite well integrated in output markets; selling their entire cocoa and coffee harvest and part of their rice harvest. A model with utility maximizing households is in this setting better suited than models assuming subsistence behavior. Besides, the assumption of perfect markets, as in open-economy models, would be unrealistic too. As is often the case in rural areas of developing countries, off-farm employment opportunities in the Lore Lindu region are very limited and labor markets are highly imperfect. Therefore, a Chayanov-type model that allows for substitution between income and leisure matches the situation better than open-economy models that analyze household behavior as if they were profit maximizing agents acting in a set of perfect markets.

We assume a representative household to maximize utility, which is a concave function of income (*i*) and leisure (*z*): $U(i, z)$. The household derives income from lowland and upland cultivation for which the produce can be sold in the market at prices p_l and p_u. The area lowland (A_l) and upland (A_u) to cultivate and the amount of leisure to consume (*z*) are the household's decision parameters which cannot be negative. We assume an additional labor cost related to an increased walking time to upland plots as more land is being cultivated. We specify this cost as a convex function of the upland agricultural area: xA_u^t with $x > 0$ and $t > 1$. The labor market is assumed to be missing such that leisure plus working time cannot exceed the total household's available time (*T*). Further, we assume that the availability of lowland is limited. The area lowland cultivated can not exceed the area flat land per household in the village (*F*). The household's maximization problem can be written as:

$MaxU(i,z)$ with $i = A_l(p_lY_l - K_l) + A_u(p_uY_u - K_u)$

Subject to: $A_l \geq 0$; $A_u \geq 0$; $z \geq 0$ non-negativity constraints

$T \geq z + A_lL_l + A_uL_u + xA_u^t$ time constraint

$F \geq A_l$ land constraint

The objective function of the maximization problem is:

$$\psi = U(i,z) + \theta(T - z - A_lL_l - A_uL_u - xA_u^t) + \lambda(F - A_l) + \mu_1 A_l + \mu_2 A_u + \eta_3 z$$

leading to the following Kuhn-Tucker first order conditions:

$$\frac{\partial\psi}{\partial z} = U_z - \theta + \mu_3 \leq 0 \;; \quad z\frac{\partial\psi}{\partial z} = 0 \tag{1}$$

$$\frac{\partial\psi}{\partial A_l} = U_i(p_lY_l - K_l) - \theta L_l - \lambda + \mu_1 \leq 0 \;; \quad A_l\frac{\partial\psi}{\partial A_l} = 0 \tag{2}$$

$$\frac{\partial\psi}{\partial A_u} = U_i(p_uY_u - K_u) - \theta(L_u + xtA_u^{t-1}) + \mu_2 \leq 0 \;; \quad A_u\frac{\partial\psi}{\partial A_u} = 0 \tag{3}$$

Assuming that the non-negativity constraints are not binding ($\mu_1 = \mu_2 = \mu_3 = 0$); that the time constraint ($\theta > 0$) and the land constraint ($\lambda > 0$) are binding and substituting (1) in (3), we come to the following equilibrium expressions for A_l, A_u, *c*, and *z*:

$$A_l = F \tag{4}$$

$$A_u = \left[\frac{U_i}{U_z}\left(\frac{p_uY_u - K_u}{xt}\right) - \frac{L_u}{xt}\right]^{\frac{1}{t-1}} \tag{5}$$

$$i = A_l(p_lY_l - K_l) + A_u(p_uY_u - K_u) \tag{6}$$

$$z = T - A_lL_l - A_uL_u - xA_u^t \tag{7}$$

Using this framework we specify the factors influencing land use in general and discuss in particular the effects of technical changes on land use at forest margins. Since technical progress in the Lore Lindu region has mostly been oriented towards paddy rice cultivation, we consider only the effects of changes in lowland technologies. In a lot of villages in the research region, the lowland land-frontier is not closed yet and expansion of paddy rice cultivation is still possible. Therefore, we broadly explore how influences alter if the land constraint in the model is not binding.

3.2 The Effect of Technical Progress on Land Use at Forest Margins

If yields per hectare in the lowland sector (Y_l) increase as a result of technical progress (e.g. irrigation development; fertilizer use) then, according to equation (6), income (i) will rise. Consequently, the marginal rate of substitution of leisure for income (U_i/U_z) will drop and in line with equation (5), this will cause a decrease in the cultivated upland area (A_u). So, a yield increasing technical change in the lowland sector will lead to less expansion in the uplands and reduce pressure on forests. If the technical change is in addition labor intensive, this 'forest saving' effect will be even stronger. The increase in lowland labor requirements (L_l); and the subsequent reduction in leisure (z) will lower U_i/U_z further such that the upland agricultural area (A_u) decreases to a greater extent.

In contradistiction, labor saving technical changes in the lowlands (e.g. mechanization) will have a land expansionary effect in the uplands. If lowland labor requirements (L_l) decrease; leisure (z) increases and the marginal rate of substitution of leisure for income (U_i/U_z) rises. Consequently, the upland agricultural area (A_u) increases.

3.3 Other Factors Influencing Land Use at Forest Margins

According to the framework, land use in the uplands will also be influenced by changes in agricultural prices and population levels. Firstly, population growth can cause an expansion of the upland sector in two ways. If the number of farmers increases, the available lowland area per household (F), and hence also the cultivated lowland area (A_l) decreases. This reduces income (i); increases leisure (z); increases the marginal rate of leisure for income (U_i/U_z); and results in an expansion of the upland agricultural area (A_u). Or else, population growth could add to the household's total available labor time (T); increase leisure (z); decrease the marginal rate of leisure for income (U_i/U_z); and lead to an increase in the upland agricultural area (A_u).

Secondly, agricultural prices affect land use. Higher prices (p_l) for lowland (food) products will increase household income (i) and reduce the upland agricultural area (A_u). Changes in prices (p_u) for upland products (cash crops) will have an ambiguous effect. According to equation (5); an increase in (p_u) causes an increase in (A_u). However, also income (i) increases; the marginal rate of leisure for income (U_i/U_z) decreases and consequently also the upland agricultural area (A_u) decreases. These two opposite effects denote the 'substitution' and the 'income' effect of household behavior (Ellis 1993; Kaimowitz and Angelsen 1998; Angelsen 1999). If upland agriculture becomes more profitable households will substitute leisure for labor and cultivate a larger upland area. But they could also reach a higher level of utility by increasing both income and leisure and by reducing the upland agricultural area. Which of the two effects will prevail strongly depends on household's preferences.

3.4 Relaxing the Land Constraint

Technical progress, population growth and price changes might also influence households' land allocation decisions in the lowlands. The impact of lowland technical progress, population growth or increasing food prices on the lowland agricultural area is the sum of a positive 'substitution effect' and a negative 'income effect'. If the land constraint is not binding, the 'substitution effect' could result in an increase in the cultivated lowland area (A_l). This will increase income (i) and lower leisure (z), which will cause a decrease of the marginal rate of substitution of leisure for income (U_i/U_z). As a result, when land expansion in the lowlands is possible, the 'forest saving' effect of yield increasing technologies or increasing food prices will be strengthened. The 'forest clearing' effect of labor saving technical changes or population growth will be dampened.

3.5 Summary

According to the analytical framework outlined above, yield increasing technical changes in the lowland agricultural sector lead to less agricultural expansion at forest margins. This 'forest saving' effect is stronger if the new technology is labor intensive such that upland labor will be attracted to the lowlands. Pure yield increasing technical changes in the lowlands can reduce pressure on forests, even if expansion of the lowland sector is unfeasible but the 'forest saving' effect is stronger if lowland agriculture can be expanded. Labor saving technical changes in the lowland sector have a 'forest clearing' effect in the uplands, especially when lowland expansion is not possible. Similar theoretical findings were discussed by Angelsen et al (2001).

4 Empirical Analysis

During the past 2 decades, lowland agriculture in the Lore Lindu region became increasingly intensified and mechanized by the development of irrigation infrastructure and the use of hand-tractors and chemical inputs. Besides, there has been a huge agricultural expansion at forest margins in this region. In accordance with the theoretical insights given in the previous chapter, we hypothesize that irrigation development (a labor intensive and yield increasing technical change) has reduced agricultural expansion at forest margins while the introduction of the hand-tractor (a labor saving technical change) has aggravated pressure on forests[1].

[1] Increased use of Green Revolution inputs has also been a cause of agricultural intensification during the last 2 decades. However, in this village level analysis we are unable to assess this because the application of such inputs is determined at the household or farm level rather than the village level.

To test these hypotheses we employ a system of equations in which land use change in the lowlands and in the uplands, population growth and technical change are estimated simultaneously. The need for a simultaneous equation model arises because of the endogenous character of technical change and population growth in the analysis of land use change. Population growth and technical change, whether induced or directed, are likely to be influenced by the same factors that drive land use changes. An overview of the variables in the model is given in table 3. From the analytical model it follows that agricultural land use in the lowlands and in the uplands is influenced by agricultural prices, the availability of land, population growth, and technical change. The time span of our analysis is the past 2 decades and hence current land use depends on land use at the beginning of the period, in 1980. We consider the natural logarithm of the cultivated lowland area (*ln(A_low)*) and the cultivated upland area (*ln(A_up)*) as dependent variables[2]. These are regressed upon a series of explanatory variables. Firstly, land use in the previous period is measured as the natural logarithm of the cultivated lowland area in 1980 (*ln(A_low)80*) and the cultivated upland area in 1980 (*ln(A_up)80*). Secondly, the availability of land is represented by the variables *ln(SLOPE_3)* and *ln(SLOPE_20)*; the natural logarithm of the village area with a slope lower than 3 degrees and 20 degrees respectively. These variables were calculated from a digital elevation model of the area. Thirdly, differences in agricultural prices are assessed by differences in market access and transaction costs. This is proxied in the model by the variable *TT_PALU*, the hours needed to travel from the village center to the capital city and main market. Fourthly, we include the natural logarithm of current population, *ln(POP)*. This variable is endogenous in the model and is a function of the natural logarithm of the 1980 population (*ln(POP)80*); population density in 1980 (*POPD80*); travel time to the city (*TT_PALU*); and access to infrastructure. The latter is represented by the number of years there has been a junior high school (*SLT_YRS*), a doctor (*DOCT_YRS*), and electricity (*ELEC_YRS*) in the village since 1980. In addition, the number of households that were resettled as part of a government transmigration program (*TRANSMIG*), was included as an explanatory variable in the equation.

[2] The cultivated lowland area (current and in 1980) is defined as the sum of the cultivated area rice, corn, and vegetables. The cultivated upland area (current and in 1980) is defined the sum of the cultivated area coconut, cocoa and coffee. We do not have past land use data for other crops but since these are the main crops we underestimate the cultivated area only marginally. Natural logarithms are taken to make the variables have the desirable normality characteristics.

Table 3. Overview of Variables in the Model

		Mean	Std. Dev.	Min.	Max.
Endogenous Variables					
ln(A_low)	Lowland area (natural logarithm)	4.89	1.22	0.69	7.01
ln(A_up)	Upland area (natural logarithm)	5.04	1.35	0	7.06
ln(POP)	Population (natural logarithm)	6.78	0.68	5.46	8.45
IRRSEMI_YRS	# years semi-technical irrigation	3.68	6.36	0	20
HTRAC	# of hand-tractors	5.07	8.23	0	50
Exogenous Variables					
ln(A_low)80	1980 lowland area (natural logarithm)	4.78	1.16	0	6.92
ln(A_up)80	1980 upland area (natural logarithm)	4.22	1.42	0	6.57
ln(SLOPE_3)	flat area (slope < 3°) (natural logarithm)	5.61	1.41	1.85	8.05
ln(SLOPE_20)	flat area (slope < 20°) (natural logarithm)	7.15	1.11	4.51	9.59
TT_PALU	travel time to Palu	5.80	5.65	0.31	17.91
ln(POP)80	1980 population (natural logarithm)	6.30	0.75	4.45	8.28
POPD80	1980 population density	87	117	1.35	570
SLT_YRS	# years junior high school	3.28	6.48	0	20
ELEC_YRS	# years electricity	7.34	5.77	0	20
DOCT_YRS	# years doctor since	6.25	8.04	0	20
TRANSMIG	# household resettled	9.89	40.87	0	200
IRRTECH	Technical irrigation (dummy variable)	0.17	0.38	0	1
IRRSEMI80	Semi- technical irrigation in 1980 (dummy variable)	0.05	0.22	0	1
HTRAC80	# hand-tractors in 1980	0.08	0.42	0	3
HOUS80	% of houses from stone in 1980	5.43	10.77	0	60
CREDIT	credit program (dummy variable)	0.91	0.29	0	1
EDUC	% non-illiterate persons aged > 45	89	15	4	100

Source: Calculated from STORMA A3 Village Survey Data and Secondary Data

Finally, we consider 3 technology variables: the number of hand-tractors (*HTRAC*); the number of years there has been semi-technical irrigation since 1980 (*IRRSEMI_YRS*); and a dummy variable for technical irrigation (*IRRTECH*). The latter is considered exogenous because technical irrigation was established before the period studied here. The variables *IRRSEMI_YRS* and *HTRAC* are endogenous and appear as dependent variable in two equations of the model. The explanatory variables in these equations are: a dummy variable for semi-technical irrigation in 1980 (*IRRSEMI80*); *IRRTECH*; the number of hand-tractors in 1980 (*HTRAC80*); a dummy variable for technical irrigation (*IRRTECH*); land use in the previous period (*ln(A_low)80* and *ln(A_up)80*); land availability (*ln(SLOPE_3)*); population density in 1980 (*POPD80*); and access to physical capital and human capital. The latter is measured as the percentage of people above age 45 that can read and write (*EDUC*). Access to physical capital is represented by a dummy variable indicating whether there has been a credit program in the village (*CREDIT*) and a wealth indicator measured as the percentage of houses built from stone in 1980 (*HOUS80*). The model is estimated using the three-stage-least-squares estimation technique. Four of the 80 observations in the sample are dropped because those 4 villages were not yet existing in 1980.

5 Results

The results of the 3SLS analysis are given in table 4. The overall performance of the model is good with the R^2 of the individual equations between 0.36 and 0.84. The regression results indicate that current land use strongly depends on land use 20 years ago. The variables *ln(A_low)80* and *ln(A_up)80* have a significant positive effect on the cultivated lowland and upland area respectively. Since the regression equations are specified log-log for the variable *ln(POP),* the regression coefficients can be interpreted as the elasticity of the cultivated area with respect to population. We find that a 1% increase in population enlarges the cultivated lowland area with 0.4% and the cultivated upland area with 0.3%. Only the effect on lowland area is significant. Population itself is positively influenced by population in the previous period; better access to the city; improved infrastructure; and government transmigration programs. This emerges from the significant positive effects of the variables *ln(POP)80*; *SLT_YRS*; *ELEC_YRS*; *DOCT_YRS*; and *TRANSMIG* and the significant negative effect of the variable *TT_PALU* on *ln(POP)*. The negative/positive effect of *POPD80* and *ln(SLOPE_3)* point to a higher population growth when more land is available. However, the effects are not significant which could mean that land availability is not the major issue in the rural migration process in this area.

The variable *TT_PALU* has a significant positive effect on *ln(A_low)* and a significant negative effect on *ln(A_up)*. With the city and main market in easy reach, transaction costs are lower and hence inputs can be obtained at a lower cost while output prices, especially for cash crops such as cocoa and coffee, are higher. The results imply that lower transaction costs and higher cash crops prices increase the cultivated upland or cash crop area and aggravate pressure on forests. Contrarily, higher transaction costs increase the cultivated lowland or food crop area. One possible explanation for this observation is that more inputs are used in food crop cultivation where transaction costs to obtain those inputs are lower and that this more capital intensive method of cultivation reduces the cultivated area. Another interpretation is that households further away from the market adopt a food-first strategy and cultivate a larger area of food crops because of risk considerations.

Table 4. Results of 3SLS regression

Equation	Obs	RMSE	"R-sq"	chi2	P
ln(A_low)	76	0.73	0.64	174.73	0.000
ln(A_up)	76	0.93	0.52	111.04	0.000
ln(POP)	76	0.27	0.84	400.77	0.000
IRRSEMI_YRS	76	4.60	0.47	69.11	0.000
HTRAC	76	6.54	0.36	42.44	0.000

	Coef.	P>|z|		Coef.	P>|z|
ln(A_low)			**HTRAC**		
ln(A_low)80	0.529***	0.000	HTRAC80	2.238	0.179
ln(A_up)	-0.028	0.761	ln(A_low)80	1.940***	0.004
ln(POP)	0.395*	0.051	ln(A_up)80	-0.004	0.994
TT_PALU	0.049***	0.006	ln(SLOPE_3	1.888***	0.001
ln(SLOPE_3)	-0.058	0.424	POPD80	0.006	0.351
IRRTECH	-0.042	0.886	HOUS80	0.156**	0.033
IRRSEMI_YRS	0.016	0.460	CREDIT	4.173*	0.088
HTRAC	0.071**	0.011	EDUC	0.075*	0.103
_cons	-0.542	0.683	_cons	-19.362	0.000
ln(A_up)			**IRRSEMI_YRS**		
ln(A_up)80	0.595***	0.000	IRRSEMI80	15.294***	0.000
ln(A_low)	-0.213	0.285	IRRTECH	-2.397	0.149
ln(POP)	0.280	0.317	ln(A_low)80	0.253	0.607
TT_PALU	-0.048*	0.094	ln(A_up)80	-0.522	0.191
ln(SLOPE_20)	0.137	0.280	ln(SLOPE_3	0.747*	0.066
ln(SLOPE_3)	-0.203*	0.078	POPD80	0.017***	0.002
IRRTECH	-1.269***	0.000	HOUS80	-0.029	0.620
IRRSEMI_YRS	-0.045*	0.106	CREDIT	1.868	0.336
HTRAC	0.073*	0.083	EDUC	0.015	0.677
_cons	2.123	0.227	_cons	-3.081	0.450
ln(POP)					
ln(POP)80	0.570***	0.000			
POPD80	-0.001	0.154			
TT_PALU	-0.02***	0.001			
ln(SLOPE_3)	0.037	0.125			
SLT_YRS	0.015***	0.007			
ELEC_YRS	0.013*	0.079			
DOCT_YRS	0.008*	0.098			
TRANSMIG	0.003***	0.000			
_cons	2.935***	0.000			

Source: Calculated from STORMA A3 Village Survey Data and Secondary Data

Further, the availability of flat land reduces agricultural expansion in the uplands, as pointed out by the significant negative effect of the variable *ln(SLOPE_3)* on *ln(A_up)*. The variables *IRRTECH* and *IRRSEMI_YRS* have a significant negative effect on *ln(A_up)* and *HTRAC* has a significant positive effect on both, *ln(A_low)* and *ln(A_up)*. So, the establishment of irrigation infrastructure has reduced agricultural expansion in the uplands while the introduction of the hand-tractor resulted in more expansion, in the lowlands as well as in the uplands.

The number of years there has been semi-technical irrigation in the village is positively influenced by the amount of flat land (*ln(SLOPE_3)*) and the population density in 1980 (*POPD80*). So, irrigation development has been directed towards villages that have a large capacity for paddy rice cultivation but where land is getting more scarce. The number of hand-tractors is higher in villages with a lot of flat land but where there was already a lot of paddy rice cultivation in 1980, as can be revealed from the significant positive effects of the variables *ln(SLOPE_3)* and *ln(A_low)80* on *HTRAC*. In addition, the variables *HOUS80*, *CREDIT* and *EDUC* have a significant positive effect on the number of hand-tractors. This indicates that better access to capital and a higher level of education increases the rate and speed of adoption of hand-tractors.

6 Discussion

6.1 Driving Forces of Agricultural Land Expansion at Forest Margins

The results are in line with what was expected from theory and confirm the specified hypotheses: yield increasing technical changes, such as irrigation development, reduce pressure on forests while labor saving technical changes, such as hand-tractors, aggravate pressure on forests. The effect of technical progress on deforestation strongly depends on the factor intensity of the technical change. The results prove that if a lowland and an upland agricultural production system coexist and farmers can shift resources between those sub-sectors, technical progress in one sub-sector can have a large impact on land use in the other sub-sector. For example, the use of hand-tractors was found to have an expansionary effect of similar magnitude in the lowlands and in the uplands.

Expansion of lowland food production is strongly driven by population growth, while expansion in the uplands is rather influenced by the profitability of cash cropping. The elasticity of cultivated area with respect to population that we find (0.4 for lowland and 0.3 for upland) is quite high compared to findings from the literature. The fact that land is still quite abundant in the Lore Lindu region could bring about this high elasticity. Population growth leads to agricultural expansion rather than to land scarcity and agricultural intensification. The land-abundance also emerges from the observation that land availability is not the major factor in explaining population growth and rural migration.

6.2 Policy Implications

It has often been argued that there is a trade-off between environmental goals on the one hand and socio-economic goals (economic growth and poverty alleviation) on the other hand (e.g. Vosti and Reardon, 1997). Our results from the Lore Lindu region however indicate that yield increasing, labor intensive technical changes could lead to a win-win situation, while labor saving technical changes are more likely to be associated with a lose-lose situation. We have shown that improved irrigation systems for paddy rice cultivation could benefit the environment by reducing agricultural land expansion at the forest margin. At the same time, irrigation development brings about higher rice yields, which could lead to economic growth, higher rural incomes and improved food security. Such welfare enhancing effects related to irrigation development was shown e.g. by Ahmed and Sampath (1992) for Bangladesh. On the other hand, the use of hand-tractors for preparing paddy rice fields, pushes labor to forest margins and causes forest clearing. In addition, the decreased labor demand could suppress agricultural wages and reduce the chances of poor landless workers.

Although our findings are specific for the Lore Lindu region, some generalization is justified, since many remaining forested areas in Southeast Asia share the same features that underlie our results: a lowland-upland agricultural dichotomy, high population growth and a low absorption of the rural labor force in off-farm employment. In such a setting, policy makers should focus on promoting labor intensive technologies for paddy rice cultivation. Increasing the labor demand in rice farming will increase the welfare of the rural poor and the efficiency of rice production (Otsuka et al, 1992). Profitable labor intensive technologies will limit deforestation and are therefore also beneficial from an environmental point of view (Kaimowitz and Angelsen 1998).

7 Conclusion

The aim of this paper has been to assess the effect of different types of technical changes in lowland paddy rice cultivation on land use at forest margins. Both theory and empirical evidence from the Lore Lindu region are in support of the hypotheses that lowland irrigation development reduces upland agricultural expansion and pressure on forests, while the introduction of hand-tractors aggravates land expansion at forest margins. These results point to the possibility of identifying rural policies associated with a win-win relation between environmental goals and socio-economic goals.

8 Acknowledgements

We would like to thank Marhawaty Mappatoba, Natalina Tampangallo, Nunung Nuryatono, Stefan Schwarze, Teunis van Rheenen, and Christine Elbel for fruitful discussions and a pleasant time in Central Sulawesi.

References

Ahmed AU, Sampath RK (1992) Effects of irrigation-induced technological change in Bangladesh rice production. American Journal of Agricultural Economics 74: 144-57.

Angelsen A (1995) Shifting cultivation and deforestation: A study from Indonesia. World Development 23: 1713-29.

Angelsen A (1999) Agricultural expansion and deforestation: Modelling the impact of population, market forces and property rights. Journal of Development Economics 58: 185-218.

Angelsen A, Kaimowitz D (2001) Agricultural technology and forests: A recapitulation. In: Angelsen A, Kaimowitz D (eds) Agricultural technologies and tropical deforestation. CABI Publishing/CIFOR, New York/Jakarta. pp 383-402

Angelsen A, Shitindi EFK, Aarrestad J (1999) Why do farmers expand their land into forests? Theories and evidence from Tanzania. Environment and Development Economics 4: 313-31.

Angelsen A, van Soest D, Kaimowitz D, Bulte E (2001) Technical change and deforestation: A theoretical overview. In: Angelsen A, Kaimowitz D (eds) Agricultural technologies and tropical deforestation. CABI Publishing/CIFOR, New York/Jakarta. pp 19-34

Angelsen A, Kaimowitz D, (2001) Introduction: The role of agricultural technologies in tropical deforestation.. In: Angelsen A, Kaimowitz D (eds) Agricultural technologies and tropical deforestation. CABI Publishing/CIFOR, New York/Jakarta. pp 1-17

Bakosurtanal (1988) Peta Rupabumi Indonesia skala 1:25,000. Badan Koordinasi Survey dan Pemetaan Nasional, Jakarta.

Bappeda (2000) Analisa Peta Dasar Neraca Lahan Kabupten Donggala. Badan Perencanaan Pembangunan Daerah Kabupaten Donggala. Kantor Pertanahan Kabupaten Donggala, . Palu.

Bergeron G, Pender JL (1999) Determinants of land use change: Evidence from a community study in Honduras. EPTD Discussion Paper No 46. International Food Policy Research Institute, Washington DC.

Boserup E (1965) The conditions of agricultural growth. Aldine, Chicago.

BPS. Statistics Indonesia. http://www.bps.go.id. 2002.

BPTP (2001) Badan Pusat Technologia Pertanian. Personal Communication.

Burmeister LL (1987) The South Korean Green Revolution: Induced or Directed Innovation? Economic Development and Cultural Change 35: 767-90.

Contreras-Hermosilla A (2000) The underlying causes of forest decline. CIFOR Occasional Paper No 30. Center for International Forestry Research, Bogor.

Coxhead IA, Jayasuriya S (1994) Technical change in agriculture and land degradation in developing countries: A general equilibrium analysis. Land Economics 70: 20-37.
Coxhead IA, Warr PG (1991) Technical change, land quality, and income distribution: A general equilibrium analysis. American Journal of Agricultural Economics 73: 345-60.
Coxhead IA, Warr PG (1995) Does technical progress in agriculture alleviate poverty? A Philippine case study. Australian Journal of Agricultural Economics 39: 25-54.
Deaton A (1992) Understanding consumption. Oxford University Press, New York.
Deininger KW, Minten B (1999) Poverty, policies, and deforestation: The case of Mexico. Economic Development and Cultural Change 47: 313-44.
Dinas Pertanian (2001). Personal Communication.
Direktorat Jenderal Pengairan (1986) Standar Perencanaan Irigasi. Republik Inodnesia Departemen Pekerjaan Umum Direktorat Jenderal Pengairan, Jakarta.
Ellis F (1993) Peasant economics: Farm households and agrarian development. Wye Studies in Agricultural and Rural Development. Cambridge University Press, Cambridge.
FAO. Statistical Databases FAOSTAT . http://www.fao.org . 2002.
Hayami Y (1974) Conditions for the diffusion of agricultural technology: An Asian perspective. Journal of Economic History 34: 131-48.
Hayami Y, Ruttan VW (1984) The Green Revolution: Inducement and distribution. Pakistan Development Review 23: 37-63.
Kaimowitz D, Angelsen A (1998) Economic models of tropical deforestation: A review. Center for International Forestry Research, Bogor.
Koppel BM (1995) Induced innovation theory, agricultural research, and Asia's Green Revolution: A reassessment. Johns Hopkins University Press, Baltimore and London.
Maertens M, Zeller M, Birner R (2002). Explaining agricultural land use in villages surrounding the Lore Lindu National Park in Central Sulawesi, Indonesia. STORMA Discussion Paper Series No 4. STORMA, Bogor
Mertens B, Sunderlin WD, Ndoye O, Lambin EF (2000) Impact of macroeconomic change on deforestation in South Cameroon: Integration of household survey and remotely-sensed data. World Development 28: 983-99.
Otsuka K, Cordova V, David CC (1992) Green Revolution, land Reform, and household income distribution in the Philippines. Economic Development and Cultural Change 40: 719-41.
Pingali PL (1997a) Agriculture-environment-poverty interactions in the Southeast Asian humid tropics. In: Vosti SA, Reardon T (eds) Sustanability, growth, and poverty alleviation: A policy and agro-ecological perspective. Johns Hopkins University Press, Baltimore and London. pp 208-28.
Pingali PL (1997b) From subsistence to commercial production systems: The transformation of Asian agriculture. American Journal of Agricultural Economics 79: 628-34.
Pitt MM, Rosenzweig MR, Gibbons DM (1993) The determinants and consequences of the placement of government programs in Indonesia. World Bank Economic Review 7: 319-48.
Ruf F (2001) Tree crops as deforestation and reforestation agents: The case of cocoa in Côte d'Ivoire and Sulawesi. In: Angelsen A, Kaimowitz D (eds) Agricultural technologies and tropical deforestation. CABI Publishing/CIFOR, New York/Jakarta. pp 291-315
Ruthenberg H (1980) Farming systems in the tropics. Clarendon Press, Oxford.

Ruttan VW (1986) Technical change and innovation in agriculture. In: Landau R, Rosenberg N (eds) The positive sum strategy: Harnessing technology for economic growth. National Academy Press, Washington DC. pp 333-56..

Ruttan VW, Hayami Y (1984) Toward a theory of induced institutional innovation. Journal of Development Studies 20: 203-23.

Ruttan VW, Hayami Y (1995) Induced innovation theory and agricultural development. In Koppel BM (ed) Induced innovation theory, agricultural research, and Asia's Green Revolution: A reassessment. Johns Hopkins University Press, Baltimore and London. pp 22-36.

Scherr SJ, Hazell P (1994) Sustainable agricultural development strategies in fragile lands. EPTD Discussion Paper No1. International Food Policy Research Institute, Washington DC.

Shively GE (2001) Agricultural change, rural labor markets, and forest clearing: An illustrative case from the Philippines. Land Economics 77: 268-84.

Shively G, Martinez E (2001) Deforestation, irrigation, employment and cautious optimism in Southern Palawan, the Philippines. In: Angelsen A, Kaimowitz D (eds) Agricultural technologies and tropical deforestation. CABI Publishing/CIFOR, New York/Jakarta. pp 335-46

Singh I, Squire L, Strauss J (1986) Agricultural household models Extensions, applications and policy. John Hopkins University Press, Baltimore.

Sonderforschungsbereich (1999). Stabilität von Randzonen Tropischer Regenwälder in Indonesian - Finanzierungsantrag für die Jahre 2000/2-2003/1 (Stability of rainforest margins in Indonesia - Application for funding 2000-2003). Göttingen.

Sunderlin W, Resosudarmo AP (2002) Rates and causes of deforestation in Indonesia: Towards a resolution of the ambiguities. CIFOR Occasional Paper No 9. Center for International Forestry Research, Bogor.

Udry C (1997) Recent advances in empirical microeconomic research in poor countries: An annotated bibliography. Journal of Economic Education 28: 58-75.

von Thünen (1966) Der Isolierte Staat in Beziehung auf Landwirtschaft und Nationalökonomie.

Vosti SA, Reardon T (1997) Sustainability, growth, and poverty alleviation: A policy and agroecological perspective. Johns Hopkins University Press, Baltimore and London.

Warr PG, Coxhead IA (1993) The distributional impact of technical change in Philippine agriculture: A general equilibrium analysis. Food Research Institute Studies 22: 253-74.

Zeller M, Schwarze S, van Rheenen T (2002). Statistical sampling frame and methods used for the selection of villages and households in the scope of the research program on Stability of Rainforest Margins in Indonesia. STORMA Discussion Paper Series No 1. STORMA, Bogor

Encroachments on Primary Forests: Are They Really Driven by Despair?

Teunis van Rheenen, Christine Elbel, Stefan Schwarze, Nunung Nuryartono, Manfred Zeller & Bunasor Sanim

1 Introduction

The Lore Lindu National Park has received much attention nationally and internationally during the past few decades. Maintaining existing levels of biodiversity has become an increasingly challenging task involving decision-makers and stakeholders at different levels of aggregation. A quarter of all mammal species of the earth are facing extinction within the next 30 years and one in eight bird species (12 %) within the next 100 years (Hilton-Taylor, C., 2000; BirdLife International, 2000). The three factors that are responsible for the decrease in biodiversity are habitat loss, overexploitation and the introduction of alien species (Primack 2000). Sulawesi has an outstandingly unique flora and fauna (Stattersfield *et al.* 1998, Myers *et al.* 2000) and the Lore Lindu National Park is a key protected area for the conservation of Sulawesi's biodiversity since it is one of only two protected areas in the Central Province (Coates *et al.* 1997). But the loss of biodiversity is closely linked to other problems and much more is at stake. Policy makers, scientists and the public are increasingly concerned about tropical deforestation and its negative consequences such as climate change, reduced timber supply, flooding, silting, and soil degradation (Kaimowitz and Angelsen 1998). In the case of Central Sulawesi, a combination of numerous factors contribute towards a trend of decreasing biodiversity. Primary forest lands are rapidly being converted to agricultural lands and practices such as slash and burn have increased this trend. The conversion of primary forest for the cultivation of perennial crops, mainly cocoa has been taking place for a long time. Farmers may very well be aware of the negative long term consequences of encroaching on the national park and logging of primary forest trees, but under the existing socio-economic circumstances, often short term private benefits will prevail over long term social costs. A central question in this connection would be:

> *To what extent is it really necessary for households to use land that is situated in the national park or convert land that is still covered by primary forest to agricultural land?*

In other words, are encroachments taking place due to poverty and despair or to increased wealth beyond levels required for minimum household survival. To evaluate this in further detail this paper will reflect on the following questions: (1) which are the major crop production technologies (input intensities, productivity)? and (2) considering the natural resources managed by households, to what extent are households able to meet their food and other requirements? Do existing crop production technologies and resource constraints lead households to such levels of despair that they have no other choice but to encroach on the park? To gain insights into these questions, we will first take a closer look at the research area, then discuss how we define an indication of despair. Following this, results will be presented and discussed.

2 The Research Area and Data Collection

The research area includes five sub-districts (kecamatan) and 117 villages (desa). For the sampling frame use was made of the ANZDEC report (ANZDEC 1997). Three selection criteria (SC) (Zeller *et al.*, 2002) were defined to delineate the sampling frame:

- SC 1: proximity of the village to the Lore Lindu National Park (2 sub-groups: close and not close to the park); 50 % of all villages were close to the park;
- SC 2: population density (2 sub-groups: above or below median population density of the research area);
- and SC 3: ethnic composition of the village population (3 sub-groups: village with 75% or more of indigenous population, village with 75% or more migrant population, and village with a mixed population).
- In 12 villages a total of 301 households were randomly selected.

Districts in the research area show large differences in population density ranging from 76.5 people/km^2 in Sigi Biromaru to only 4.1 in Lore Utara (see Table 1). Wetland rice, maize, cocoa and coffee are the most common crops in the research area. In Sigi Biromaru and Lore Utara more than half of the households reported growing wetland rice. The crop is often grown for home consumption and the market while cocoa and coffee are primarily grown to be sold. As many as 40 different types of intercrops were reported.

Table 1. Characteristics and land use in the research area

	Sigi Biromaru	Palolo	Kulawi	Lore Utara
District level:				
Population	47786	23907	28568	1188 3
No. Of households	9201	4392	5752	2468
Average household size	5.2	5.4	5.0	4.8
Area (km^2)	625	315	3114	2928
Population density (people/km^2)	76.5	75.8	9.2	4.1

Table 1 (cont.)

Households in the sample:					
Number		95	46	78	82
Area owned (ha)		111	128	175	140
Acquired from (%)					
Transmigration project		14	1	2	6
Cleared primary forest		29	24	51	17
Heritage		26	17	39	36
Purchased		26	47	7	31
Gift		5	11	0.4	9
Marriage		0	0	0.3	1
Main land use (%)					
Wetland rice:	Households	56	24	35	54
	Area	45	9	10	21
Maize (sole crop):	Households	5	10	4	24
	Area	3	3	1	7
Cocoa (sole crop):	Households	13	55	37	10
	Area	14	41	14	4
Cocoa with coffee/ bananas or coconuts:	Households	13	30	45	31
	Area	6	13	12	17
Coffee (sole crop):	Households	1	6	33	26
	Area	0.5	4	19	8

Source: survey data of sub-project A4 (socio-economic analysis of households) of STORMA, 2000/2001.

Households were asked how they had acquired their owned plots. In Kulawi a very large percentage of the land (51 %) was acquired through the clearing of primary forests (see Table 1), and for the households in the sample this particularly took place recently (see Figure 1).

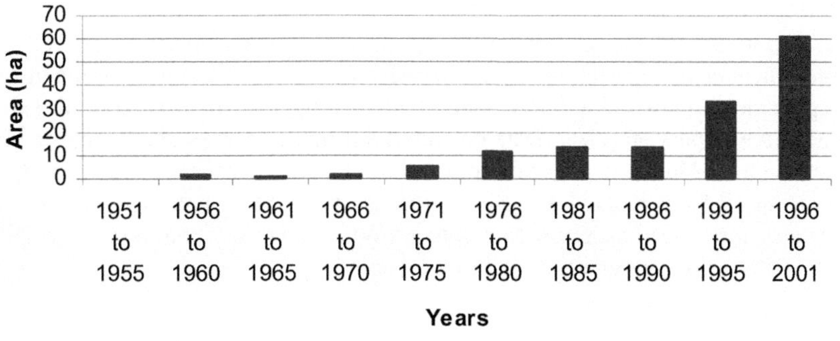

Fig. 1. Area of dry-land acquired by clearing primary forests by farm households in the survey, 1951 to 2001. Source: survey data of sub-project A4 (socio-economic analysis of households) of STORMA, 2000/2001. Note: Information presented does not include land transactions, like purchase and heritage.

Using a methodology developed by Henry *et al.* (2001), households were divided into terciles, poorest, poor and less poor (Abu Shaban 2001). For cocoa and wetland rice we defined different technology sets (see Table 2). For the former crop we included the existence of forest trees on plots, because this is a common practise for Soil and Water Conservation (SWC). All plots were selected on which these technologies could be found and we looked at the plot ownership. Those plots on which external inputs were applied especially belonged to the relatively wealthier households.

Table 2. Technology types according to poverty levels, for cocoa when grown as a mono crop and wetland rice.

Crop	Technology type	% of households belonging to: The poorest	Poor	Less poor
Cocao	FT,F&/P			100 (1)
	NFT,F&/P		31 (4)	69 (9)
	FT,NF,NP	54 (14)	23 (6)	23 (6)
	NFT,NF,NP	47 (24)	29 (15)	24 (12)
Wetland rice	TI,P&F	7 (2)	38 (11)	57 (17)
	STI, F/P	20 (3)	60 (9)	20 (3)
	SI,NF,NP	38 (3)	50 (4)	13 (1)

Source: survey data of sub-project A4 (socio-economic analysis of households) of STORMA, 2000/2001. Notes: Numbers in brackets indicate the number of households. (N)FT = (no) forest trees, (N)F = (no) fertilizer, (N)P = (no) pesticides, TI = technical irrigation, STI = semi-technical irrigation, SI = simple irrigation. For cocoa N=91 and p = 0.010; for wetland rice N=53 and p = 0.030.

A recent survey in the research area conducted by Birner and Mappatoba (2002) showed that more than 90 % of those interviewed were very well aware of the importance of the national park and primary forests for SWC as well as the consequences of logging activities. While biodiversity was not mentioned by the respondents, almost all acknowledged that logging activities could or already had a negative impact on SWC. Data collected during our survey shows that in all districts on a considerable area of cultivated land SWC measures were applied, ranging from 18 % in Kulawi and Sigi Biromaru to 38 % in Palolo. Mixed cropping was also mentioned as a way to conduct SWC and is very common, especially on plots with a slope greater than 15% (see Table 3).

Table 3. Area shares in percent of sole crops and crop-mixtures on all plots and on plots with a slope greater than 15%

	Annual sole crops	Perennial sole crops	2- or 3-crop mixtures with annuals	2- or 3-crop mixture with perennials	2- or 3-crop mixtures with annuals and perennials
all plots	28 (226)	30 (244)	9 (56)	24 (242)	8 (72)
on slopes	6 (22)	40 (119)	16 (45)	29 (92)	9(26)

Source: survey data of sub-project A4 (socio-economic analysis of households) of STORMA, 2000/2001. Number of cases in brackets.

According to main land use purposes (MLUP), three household types were defined: households that used land for (1) food crops, (2) food and tree crops, and (3) only tree crops. If a household cultivated 66 % or more of its area with food crops, then the main land use purpose was defined as being used for food crops; idem for food and tree crops, and for tree crops. Although, when considering all land cultivated, a significant difference could be found between the household types. However, when considering only land from cleared primary forests, the significance was higher. The MLUP for the poorest households was food crops, whereas for the less poor households this was tree crops (see Table 4).

Table 4. Main land use purposes (MLUP) for all land and plots cleared from primary forests (PF) according to poverty levels

	% of farm households according to poverty level:					
	Poorest		Poor		Less poor	
MLUP	All Land	PF	All Land	PF	All Land	PF
Food crop	36 (35)	31 (12)	33 (33)	31 (8)	29 (25)	7 (1)
Food & treecrop	36 (35)	39 (15)	24 (24)	8 (2)	21 (18)	13 (2)
Tree crop	29 (28)	31 (12)	44 (44)	62 (16)	49 (42)	80 (12)

Source: survey data of sub-project A4 (socio-economic analysis of households) of STORMA, 2000/2001. Notes: The number in brackets indicates the number of households. Land is considered for a certain MLUP when more than 66 % of the area is grown to that crop or crop combination. (All land: $N = 284$, $p = 0.038$; Primary forests: $N = 80$, $p = 0.003$)

3 Quantifying Despair with Linear Programming

Land use changes are often studied either with the aid of econometric or programming techniques. With the former trends from the past are analyzed and with extrapolations short to medium term *"predictions"* can be done. The latter is more suitable to explore different options for development and to evaluate the role of existing and alternative production technologies or technologies that are still in the research and development pipeline. A commonly used technique of analysis for these studies is Multiple Goal Linear Programming (MGLP). It has been used in many different parts of the world to evaluate the role of existing and alternative technologies. This technique is based on the common Linear Programming technique (Hazell and Norton, 1986) and has been used in land use studies at different levels of aggregation ranging from the farm (e.g. Stroosnijder and Van Rheenen 1993; Sissoko 1998), village (e.g. Stroosnijder and Van Rheenen, 2001), regional (e.g. Van Keulen and Veeneklaas 1993; Alfaro *et al.* 1994), national (e.g. Veeneklaas 1990) to the supra-national level (Rabbinge and Van Latesteijn 1992). LP has proven to be a powerful tool to integrate different scientific disciplines. However, these studies have in common a very high level of detail, which makes them very data and time demanding.

A core component of any LP problem is the Activity Matrix (AM). The AM specifies the inputs that are required to obtain certain levels of outputs. For example, how much land, labor, fertilizers and pesticides are required to produce a certain amount of wetland rice? These Input-Output Coefficients (IOCs) will depend on management styles of farm households as well as their resource endowments, for example the quality of their land. No two households are identical, so ideally speaking, an AM should be developed for every household in the research area, but due to time, budget and logistic constraints this of course is not feasible. We decided to concentrate in the initial stages of the analysis on the development of a so-called Base Activity Matrix (BAM). While developing the BAM we primarily focused on the main crops and crop combinations that can be found in the research area. A distinction was made for irrigated and non-irrigated land units. The technologies that were defined and quantified for the BAM were derived from household survey data (see Section 2). The BAM was adjusted in the following for different poverty groups, location and whether or not households grow wetland rice (Table 5). Coefficients for average household endowments, namely labor, wetland and dryland, as well as cash constraints, were calculated for each of the eight household types, including the aggregate of all households represented by matrix A in Table 5. Moreover, we conducted t-tests to check for differences in prices for major crops (coffee, cocoa, rice, maize) across poverty groups, location or access to wetland rice. Whenever a price for a specific household type significantly differed between groups, we included the average price for that household type into the matrix. If there was no significant difference, we used the sample average.

Table 5. Distinction between different matrices

Matrix	Household description
A	Average farm household
Poverty level:	
P1	Poorest
P2	Less poor
P3	Poor
Location and access to wetland rice	
L1	Far from the park with wetland rice
L2	Far from the park without wetland rice
L3	Close to the park with wetland rice
L4	Close to the park without wetland rice

Note: Input-output coefficients for cocoa and wetland rice are specified according to household poverty level and according to household location and access to wetland rice. For all other crops input-output coefficients are for the average household.

The Despair Index (DI). Why do people living in the vicinity of the LLNP encroach on the park or areas still covered by primary forests? Is their socio-economic situation so desperate that they are left with no other alternative? That is one of the central questions for all those concerned with the integrity of the national park. We operationalize this question by asking whether the household income is sufficient to meet basic needs. Are households – considering their natural resource endowments and knowledge of and access to crop and livestock production technologies - able to meet their minimum household requirements? As an estimate of the minimum expenditures needed to meet basic needs, we used the poverty line of IDR 66,580 $capita^{-1}.month^{-1}$ as quantified by Widyanti et al. (2001) for Southeast Sulawesi. However, as their estimate is for 1999, we adjusted this by the Consumer Price Index to arrive at a poverty line of IDR 76,755 $capita^{-1}.month^{-1}$ for mid-2001 (mid-term of the survey). Since household size varies between the different types, we adjusted the minimum expenditures. As long as household are not able to meet their minimum expenditures considering their resource endowments, they might be forced to encroach on the forests out of despair, which under such circumstances would then hardly be surprising. The DI gives a quantitative snapshot indication of the need for farm households to use resources of the park. It is calculated using the following formula:

$$\textit{Despair Index (DI)} = \frac{\textit{Household income that can be generated with the resources managed by the household}}{\textit{Minimum expenditures to satisfy basic needs}}$$

4 Construction of Policy Scenarios

Construction of the BAM gave insight concerning existing crop production technologies. The most important crops found in the research area in terms of area cultivated and income generated are rice, maize, cocoa and coffee. These crops are either grown as sole crops or as intercrops (Table 6). While acknowledging that there are variations in the research area in the BAM, we will include only the IOCs for the most important crops and crop combinations.

Table 6. Percentage of plots by crop / crop combination

Crop / crop combination	%	Crop / crop combination	%
Wetland rice**	19.0	Bananas / other perennial crops	1.5
Cocoa**	13.0	Other 2 crop mix of annuals & perennials	1.3
Coffee**	7.7	Bananas**	1.1
Other tree crops	7.4	Cocoa / Maize**	1.1
Cocoa / coffee**	6.6	Cocoa / Coffee / Maize**	1.1
Cocoa / coffee / other crops	4.7	Bananas / other crops	1.0
Other perennial crops	3.6	Coffee / other crops	0.9
Maize**	3.6	Upland rice**	0.6
Cocoa / bananas**	3.0	Other 2 annual crop mixtures	0.6
Other annual crops	3.0	Other 2 perennial crop mixtures	0.5
Cocoa / other perennial crops	3.0	Cocoa / other annual crops	0.5
Cocoa / Bananas / Coconut	2.2	Maize / other perennial crops	0.4
Cocoa / Coffee / Bananas	2.1	Cassava**	0.4
Pasture	1.7	Cocoa / maize with other crops	0.3
Cocoa / Coconuts**	1.7	Coffee / maize**	0.2
Coconuts**	1.7		

Source: survey data of sub-project A4 (socio-economic analysis of households) of STORMA, 2000/2001. Note: ** Crop / crop combination included in the LP model.

The IOCs included in the BAM reflect a situation, which we estimate could exist for a number of decades. The IOCs were quantified with survey data and where information was missing use was made of literature (e.g. Rehm and Espig, 1991), expert knowledge and educated guestimates. Generally speaking, the use of external inputs such as urea and pesticides is still very low except for wetland rice, where 77 % of all households use fertilizer. For the other major crops in the research area, less than 20 % of farmers reported that they used fertilizers. The use of external inputs will depend on a whole range of factors such as availability of the external inputs, cash constraints, and of course whether or not it really is necessary. The soil in the margins of the LLNP is in general fertile and where possible farmers will often fallow land when they consider it no longer suitable for crop cultivation.

To obtain an idea of the potential role of existing technologies for the generation of household income, the gross margins of the main cropping activities were calculated. For the off-farm activities we used average wages according to household types and for the base situation we specified that the number of days per month spent on off-farm activities could not exceed the actual situation. For livestock the average gross margins generated by households was included. Taking into consideration land and labour constraints the gross margins objective function was optimised for households with different specifications (see Table 5).

The value of the DI will depend on factors such as off-farm employment opportunities, the access that household members have to input and output markets, the price of input and output products, and the resource constraints. Households on a daily basis face numerous uncertainties, such as the fluctuations in prices they obtain for their products. Prices vary greatly in the research area, not only over time, but different households also received different prices. To cope with such uncertainties Figure 2 shows that households have different sources of income. Income of crop activities is the most important income followed by self-employment (income from trading and small scale industries).

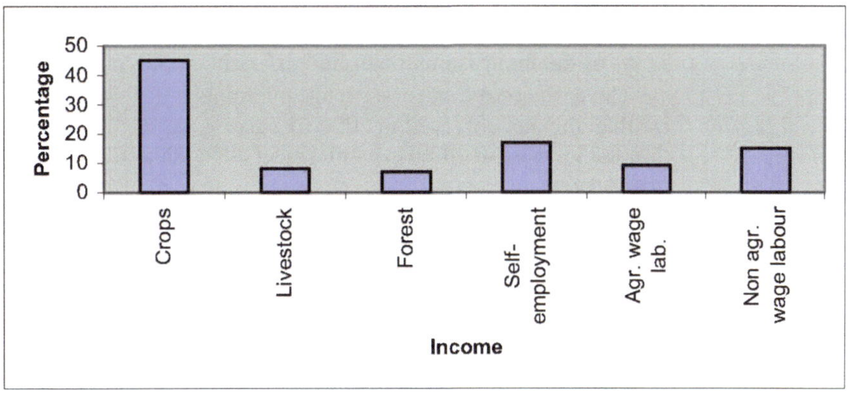

Fig. 2. Sources from where household income is derived in the forest margins of the Lore Lindu National Park, Sulawesi, Indonesia (2000/01). Source: Schwarze et al. (2002).

To explore the impact that a number of different factors could have on gross margins and the DI four different policy scenarios (PS) were constructed (Table 7).

Table 7. Description of Policy Scenarios (PS)

PS	Description
PS A	**Actual situation (as observed in the survey for the 1 + 7 household types)** *Assumptions:* • Prices, cash constraints and access to off-farm employment are included as they were found in the household survey of 2000/01. • Areas that can be cultivated to cocoa, maize and rice cannot exceed the levels as they were found in the household survey of 2000/01.
PS 1	- Strong improvement in off-farm employment and cash availability *Assumptions:* • Opportunities for off-farm employment improve. Up to 50 % of the family labour can find off-farm employment • The cash constraint for agricultural inputs is 1.5 times higher than the one observed in 2000/01 survey
PS 2	**Modest increase in off-farm employment and cash availability** *Assumptions:* • Off-farm employment opportunities: where 20 % of the family labour can find off-farm employment • The cash constraint for agricultural inputs is assumed to be at the current 2000/01 cash constraint level.
PS 3	**Reductions in farm size and decline in cash availability** *Assumptions:* • Due to population growth farm size decreases • Land available for upland and wetland agriculture is 50 % of what it was in 2000/01 • Cash available for agricultural inputs decreases by 50 % of what it was in 2000/01 • Income from livestock activities is assumed to be at 50 % of what it was in 2000/01 • No change is assumed in off-farm employment opportunities
PS4	**Lax enforcement of forest protection leading to deforestation** *Assumptions:* • Farm household are able to expand their arable area by up to 1 hectare. • No change in off-farm employment and cash availability compared to 2000/01.

In Table 8 the basic characteristics of the household types are shown. These were represented in the activity matrix of the programming model.

Table 8. Characteristics of household types that were included in the LP model

	Base:	Poverty group[1]:			Location[2]:			
Household Type	A	P1	P2	P3	L1	L2	L3	L4
	286	98	101	86	77	79	61	68
% of households below the poverty line	63	77	71	39	72	74	46	58
Labour units	3.42	3.14	3.40	3.78	3.24	3.43	3.72	3.32
Land (ha)	1.54	1.32	1.27	2.12	1.46	1.44	1.88	1.47
Main crops grown (ha):								
Wetland rice	0.34	0.14	0.33	0.59	0.70	0.00	0.70	0.00
Maize as sole crop	0.05	0.04	0.07	0.04	0.01	0.03	0.10	0.09
Cocoa as sole crop	0.29	0.18	0.30	0.41	0.45	0.30	0.11	0.27
Coffee as sole crop	0.13	0.22	0.07	0.11	0.03	0.24	0.16	0.09
Coffee with Cocoa	0.14	0.08	0.10	0.25	0.02	0.11	0.18	0.27

Source: survey data of sub-project A4 (socio-economic analysis of households) of STORMA, 2000/2001. Notes: (1) Poverty groups: P1 = Poorest; P2 = Poor; P3 = Less poor. (2) Location: L1 = Far from the park with wetland rice; L2 = Far from the park without wetland rice; L3 = Close to the park with wetland rice; L4 = Close to the park without wetland rice

Table 9 shows the results of the different model runs for the optimized farm gross margin and the DI, differentiated by farm household types and policy scenarios. The base scenario (PS A) for the average household (A) yields a gross margin of IDR 7,410,000, which is very close to the average income observed in the sample households.

Table 9. Gross Margins (GM in 10^7 IDR) and Despair Index (DI) for different Policy Scenarios (PS)

		Matrix[1]							
PS[2]		A	P1	P2	P3	L1	L2	L3	L4
PS A	GM	0.741	0.611	0.627	0.968	0.623	0.870	0.677	0.961
	DI	1.519	1.286	1.339	1.821	1.358	1.775	1.285	1.984
PS 1	GM	1.377	1.112	1.175	1.740	1.184	1.580	1.373	1.657
	DI	2.821	2.340	2.512	3.274	2.582	3.225	2.605	3.420
PS 2	GM	1.085	0.870	0.897	1.512	0.900	1.303	1.080	1.423
	DI	2.224	1.831	1.917	2.845	1.961	2.659	2.050	2.937
PS 3	GM	0.536	0.426	0.448	0.747	0.438	0.636	0.532	0.713
	DI	1.098	0.897	0.959	1.406	0.955	1.298	1.009	1.472
PS 4	GM	1.447	1.192	1.300	1.898	1.342	1.559	1.496	1.988
	DI	3.025	2.507	2.779	3.572	2.926	3.181	2.839	4.103

Notes: (1) see Table 5 (2) see Table 7.

5 Discussion and Conclusions

The results presented in Table 9 show that the DI fluctuates between 0.9 and 4.1. The DI, as it was calculated here is a snapshot indication of the ability of households to meet minimum expenditure requirements under different policy scenarios. These different situations have been called policy scenarios because they indicate the "what if?" situation. For example, PS 4 shows the attractiveness for farm households to encroach on another hectare of primary forests. In the case of PS 3, where farm size is halved and the type of crops that can be grown are not specified, farm households are still able to achieve a DI of approximately 1. This is because the selection of crops will be the optimal one considering existing crop technologies and actual prices. But this will generate an average income of only IDR 5,360,000. As the DI measures an average, it does not consider distributional issues. Already in the base situation A, 63 percent of households were below the poverty line (Table 8).

A low level of DI indicates, however, that there is a problem, and some caution is required when drawing conclusions. For example, a household may very well be able to just meet its requirements today, but its members might feel very vulnerable to increasing poverty and on the basis of that perception might consider it necessary to encroach on the park. As Pritchet *et al.* (1999) explain, many households, while not currently in poverty, recognize that they are vulnerable and that events could easily push them into poverty – a bad harvest, a lost job, an unexpected expense, an illness, a lull in business. While Pritchet *et al.* (1999) develop a quantitative methodology to explore the notion of vulnerability, they also admit that like the notion of poverty, the concept of vulnerability to poverty is complex and multifaceted and will never be adequately summarized in a single measure. So the DI as calculated in this paper is very much a "snapshot measurement". It could be seen as a volume control on an alarm. The lower the DI, the louder the alarm should ring. A very low DI tells us that household members possibly are left with no other choice but to encroach on the park or areas still covered by primary forest. It gives also an indication of the level of necessity to search for alternative opportunities for income generation. Quantification of the minimum level of income required for household survival is a topic that easily lends itself for debate. If one were to include aspects such as vulnerability when considering the DI, it would have been more appropriate to work with a so-called "subjective poverty line".

When considering the DI there are other considerations. A very low DI could indicate the need to look for alternative higher yielding production technologies. But, new technologies, which could be more productive, need not necessarily lead to fewer encroachments on the park. For example, if the new technology were greatly to reduce the labor requirements, this would create or increase the surplus level. Now, if there are no other employment opportunities, the extra labor could still be used for encroachments on the park. Kaimowitz and Angelsen (1998) point out that the direct effect of technological change on individual farmer behavior depends on the characteristics both of the technology and the labor and output markets. In their review of economic models of tropical deforestation they find that agricultural research and extension policies designed to limit deforestation should focus on promoting profitable technologies more easily applicable to areas that have already been de-forested and intensive in labor and capital requirements. For example, our model results showed that when cash became the limiting constraint, perennial crops were selected rather than annual crops. This is because cocoa once planted requires fewer external inputs, cash as well as labor, than crops such as peanuts or maize.

References

Abu Shaban A (2001) Rural poverty and poverty outreach of social safety net programs in Central Sulawesi - Indonesia. Unpublished M.Sc. Thesis, Institute of Rural Development, University of Göttingen, September 2001.

ANZDEC (1997) Report on the Central Sulawesi integrated area development and conservation project. Palu, Sulawesi.

Alfaro R, Bouma J, Fresco LO, Jansen DM, Kroonenberg SB, Van Leeuwen ACJ, Schipper RA, Sevenhuysen RJ, Stoorvogel JJ, Watson V (1994) Sustainable land use planning in Costa Rica: a methodological case study on farm and regional level. In: Fresco LO, Stroosnijder L, Bouma J and Van Keulen H (eds) The future of the land: mobilizing and integrating knowledge for land use options. John Wiley & Sons Ltd, Chicester

BirdLife International (2000) Threatened Birds of the World. Lynx Edicions and BirdLife International, Barcelona and Cambridge (UK)

Birner R, Mappatoba M (2002) Community Agreements on Conservation in Central Sulawesi: A Coase Solution to Externalities or a Case of Empowered Deliberative Democracy? STORMA Discussion Paper Series, No. 3 (2002) (http://www.storma.de/DPS/index.htm)

Coates B, Bishop KD, Gardner D (1997) A guide to the birds of Wallacea: Sulawesi, the Moluccas and Lesser Sunda Islands, Indonesia. Dove Publications, Alderley, Queensland

Hazell, PBR, Norton RD (1986) Mathematical Programming for Economic Analysis in Agriculture. Macmillan Publishing Company, New York and Collier Macmillan Publishers, London

Henry C, Sharma M, Lapenu C, Zeller, M (2001) Assessing the Relative Poverty of Microfinance Clients. A CGAP Operation Tool. Consultative Group to Assist the Poorest (CGAP), The World Bank, Washington D.C.

Hilton-Taylor C (compiler) (2000) IUCN Red List of Threatened Species. IUCN, Gland (Switzerland) Cambridge (UK). Available at http://www.redlist.org

Kaimowitz D, Angelsen A (1998) Economic Models of Tropical Deforestation. A Review. Centre for International Forestry Research

Myers N, Mittermeier RA, Mittermeier CG, da Foncesca GAB and Kent J (2000) Biodiversity hotspots for conservation priority. Nature 403: 853-858

Primack R (2000) A Primer of Conservation Biology. Sinauer Associates, Sunderland (MA)

Pritchet L, Suryahadi A, Sumarto S (1999) Quantifying Vulnerability to Poverty: A Proposed Measure, Applied to Indonesia. The World Bank and the Social Monitoring and Early Response Unit, Jakarta. (http://econ.worldbank.org/docs/1192.pdf)

Rabbinge R, Van Latesteijn HC (1992) Long-term options for land use in the European Community. Agricultural Systems 40: 195-210

Rehm S, Espig G (1991) The Cultivated Plants of the Tropics and Subtropics. Verlag Josef Margraf, Weikersheim (Germany)

Stattersfield AJ, Crosby NJ, Long AG, Wage DC (1998) Endemic Bird Areas of the World: Priorities for Bird Conservation. Birdlife International, Cambridge (UK)

Schwarze S, Heß S, Van Rheenen T, Zeller M (2002) Determinants of Income Generation Activities of Rural Households in Central Sulawesi, Indonesia. Paper to be presented at: Deutscher Tropentag. October 9 - 11, 2002 in Witzenhausen, Germany

Stroosnijder L, Van Rheenen T (1993) Making farming systems a more objective and quantitative research tool. In: F.W.T. Penning de Vries *et al.* (eds) System Approaches for Agricultural Development. Kluwer Academic Publishers, The Netherlands, pp 341-353

Stroosnijder L, Van Rheenen T (2001) Agro-silvo-pastoral land use in Sahelian Villages. (Advances in GeoEcology 33) Catena Verlag GMBH, Reiskirchen, pp 408.

Sissoko K (1998) Et demain l'Agriculture? Options techniques et mesures politiques pour un développement agricole durable en Afrique subsaharienne. Tropical Resource Management Papers no. 23. Wageningen

Van Keulen H and Veeneklaas FR (1993) Options for agricultural development: a case study for Mali's fifth Region. In: Penning de Vries FWT *et al.* (eds) System Approaches for Agricultural Development. Kluwer Academic Publishers, The Netherlands pp 367-380

Veeneklaas FR (1990) Dovetailing technical and economic analysis. PhD thesis. Erasmus University Rotterdam, The Netherlands

Widyanti W, Sumarto S, Suryahadi A (2001) Short term poverty dynamics: Evidence from rural Indonesia. SMERU Working Paper

Zeller M (1999) Economic Analysis of Land Use Systems of Rural Households. In: Sonderforschungsbereich (SFB) 1999. Stabilität von Randzonen Tropischer Regenwälder in Indonesien – Finanzierungsantrag für die Jahre 2000/2-2003/1 (Stability of rainforest margins in Indonesia – Application for funding 2000-2003). Göttingen

Zeller M, Schwarze S, Van Rheenen T (2002) Statistical Sampling Frame and Methods Used for the Selection of Villages and Households in the Scope of the Research Program on Stability of Rainforest Margins in Indonesia (STORMA). STORMA Discussion Paper Series, No. 3 (2002) (http://www.storma.de/en/newsandevents/index.htm)

Chapter 2:

Biodiversity and Conservation

Rain Forest Margins and their Dynamics in South-East Ethiopia

Klaus Müller-Hohenstein & Asferachew Abate

1 Introduction

Research on tropical forests started many decades ago. According to Whitmore (1991) it began with taxonomy and floristic and structural inventories and was followed by classification of forest communities. Subsequent successional relationships between these communities generated special attention. Then investigation into ecophysiology became a very active field. Fluxes of water and nutrients were measured. So step-by-step basic knowledge about the functioning of tropical forests was compiled. And especially foresters contributed to this knowledge concerning production and regeneration issues. To date many scientists agree upon the necessity of a holistic approach, an ecosystem approach for understanding the multidimensional complexity of tropical forests in time and space, and they also agree upon the many "white spaces" left to investigate, especially functional relationships between plants and animals and the stability of these tropical ecosystems.

Only in the last some two decades has there been a growing consciousness about a worldwide loss of natural forests, not only in the humid tropical lowland and mountain regions, but also in the tropics with marked dry seasons. This loss concerns the most biologically diverse terrestrial ecosystems. Man started his ongoing influence on these ecosystems, the tropical forests, already a long time ago. People have lived in and around forests for centuries and used them in many ways. They are largely dependent, either indirectly because of the water storage-, climate regulation- and soil conservation-functions of these forest ecosystems or directly, using timber and firewood, medicinal plants, food for stockbreeding. The latter have direct financial value and may be the only and therefore decisive resources for these people. Large areas have been totally cleared in order to obtain agriculturally suitable land. Many scientists believe that almost no untouched natural forest is left.

The underlying causes of forest clearing or degradation also differ quite a lot from one region to another, but almost everywhere growing population pressure increases the demand on the natural forest resources and non-sustainable ways of use and production contribute largely to the present situation. Therefore UNEP (2001) believes that at least in the next three decades a global trend will lead to increasing biological diversity losses at an unpredictable but very high rate. Over 5,000 botanists from over 100 countries underlined in a Congress resolution in St.Louis, Missouri 1999 that as many as two-thirds of the world's plant species are in danger of extinction, especially due to habitat fragmentation and destruction. And of course forest degradation is accompanied by soil degradation, erosion and mesoclimatic changes, so that successful rehabilitation measures are difficult to instigate. The growing awareness about this present situation has led to research topics, which are today of the highest importance. The main keywords are "biodiversity", "conservation" and "sustainable use".

In a multidisciplinary approach, soil scientists, foresters, plant physiologists and biogeographers from Ethiopian and German research institutions (Addis Abeba University, Ethiopian Agricultural Research Organisation (EARO) and University of Bayreuth, sponsored by DFG and DAAD) have collected for the last two years data about aboveground and belowground biomass, sap flow, nutrient uptake and photosynthesis of different tree species in natural forest fragments (e.g. *Podocarpus falcatus, Croton macrostachys*) as well as in different plantations with exotic trees (*Eucalyptus saligna, Cupressus lusitanica, Pinus patula*). The results of this "traditional" research should contribute to a better understanding of the ecosystems of these forests and plantations. But hopefully one outcome will be a guideline for Ethiopian foresters, comprising recommendations on how to manage forests and plantations in the future and how to protect remaining parts of the natural forest. To achieve these objectives, it is not merely sufficient to know everything about floristic diversity, structural characteristics and functional relationships within the forests. Even more important is sufficient knowledge and understanding of the different land-use systems and agricultural practices of the population and about the socio-economic and political conditions they live in.

Therefore the focus of this contribution lies in the rather complicated socio-economic situation of the people living in and around the forests. The main objective is to describe and to explain the present situation of the forests and plantations in the project area, especially at the forest margins, to present some facts on the actual dynamics of the growing anthropogenic influences in our Ethiopian study area and to discuss these findings with special respect to biodiversity. Included is also as much as is already known about traditional forest-related knowledge, because this may be vital for all forms of sustainable use and the conservation of natural forests and their resources in the future.

2 The Study Area

The study area (Oromia region, Arusi zone, Munessa administration) comprises the Munessa-Shashamene forest, one of the 58 Priority Forest Areas of Ethiopia and its surroundings, an area of about 2,000 km². This area is located about 250 km south of Addis Abeba and covers the northwest exposed slopes of the Rift Valley, north of Shashamene and east of lake Langano. It can be divided very roughly into three different natural landscapes, the so-called "lowland savannah" close to the lake, a large, only slightly dissected plain at about 1,700 m asl, the Munessa Forest on the partly steep slopes up to 2,700 m asl and finally the adjacent "mountain savannah", a more undulated plain (Fig.1). Apart from a few lacustric sediments in the Rift valley bottom, plains and slopes are built out of tertiary and quaternary volcanic rocks.

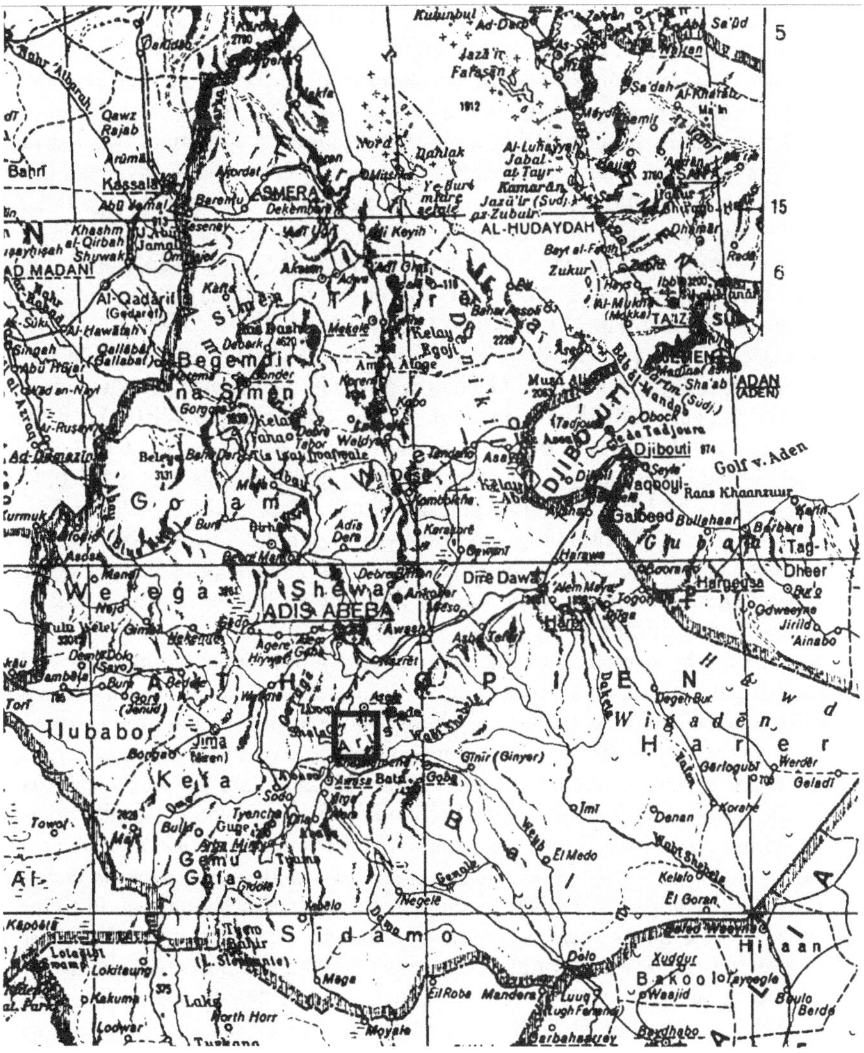

Fig. 1. Location of the study area in Ethiopia (square).

The relief character contributes to a large variety of climatic conditions. According to the observations of our research team in Kuke (main area of permanent research plots at 2,300 asl on a large forest gap in the Munessa forest) the average annual temperature is about 15° C (coldest month July 13° C, warmest April 17° C). This means, that in the lowland savannah the average annual temperature rises not above 18° C (with maximum temperatures of over 40° C), in the mountain savannah the average annual temperature reaches hardly 10° C. In the colder season frost is here quite common during nighttime. Rainfall in a "normal" year reaches 1,500 mm in Kuke. There is certainly much less in the lowland savannah (about 800–1,000 mm) and not very much more in the mountain savannah (1,500–1,800 mm). The distribution of these amounts of rainfall is bimodal. The main rainy season extends from June to October with a maximum occurring in July and August; the dry season extends from November to February. Single rainfall events can be very intense (63.2 mm in 24 h in Kuke, 17th of July 2001).

Relief characteristics and the climatic condition are mainly responsible for the development of different soils. Generally all soils are reddish-brown with very high clay content (> 50%) and almost no stones. According to WRB (World Reference Base) in the lowland savannah mazic vertisols dominate (pH 6.0–7.5, CEC 30 –60) in the lower parts of the natural forests mollic nitisols (pH 5.5–7.0, CEC 30–100), in the higher parts humic umbrisols (pH 4.8–5.5, CEC 30–70) and in the open mountain savannah mollic cambisols (pH 5.0–6.0, CEC 20–100). Traces of soil erosion can be seen in the open savannah areas. It also seems that the contents of organic matter diminish as soon as the forest cover has been cleared, but these trends have to be confirmed by further studies.

Reports on the vegetation of the study area have been included in different forest and vegetation surveys. According to Friis (1986 a) the forests of the whole Shashamene-Munessa area belong to the upland rain forests and are classified as „humid mixed forest". These forests contain a mixture of *Podocarpus falcatus* and broadleaved species in the canopy. Relatively abundant are pioneer species such as *Croton macrostachys* and also *Discopodium penninervium*. Locally *Arundinaria alpina* is common. Among the epiphytes, ferns and orchids dominate as well as some *Peperomia spec.* There are smaller trees and shrubs such as *Bersama abyssinica, Calpurnia aurea* . Different parasites Loranthaceae) are quite common as well as some lianas such as *Urera hypselodendron.* Chaffey (1979) proposed the following names for the altitudinal zones from the lowland savannah up to the mountain savannah: *Acacia-Croton*-savannah up to 1700 m, submontane *Podocarpus-Croton*-forest up to 2100 m, montane forests with dominating *Podocarpus* up to 2400 m, montane forests with *Podocarpus* and *Arundinaria* up to 2500 m, montane forests with *Rapanea melanophloes* and *Arundinaria* up to 2700 m and montane forests with *Hagenia abyssinica* and *Juniperus procera* up to 3000 m. However due to anthropogenic influences it is hardly possible to establish this altitudinal sequence in such a clear zonation as described by Chaffey (1979). As already Friis (1992) underlined, the original extent of these floristically defined forest communities is difficult to identify because of the widespread clearing and degradation of the forests in the Ethiopian highlands.

The population in and around the study area belongs to at least 12 different tribes of the Oromo, the most important ethnical group in the South of Ethiopia. They are mainly sedentary farmers with limited areas for crop-cultivation and pasture-grounds. Although there is no official census, population density in the lowland savannah is much higher (more than 20/km^2) than in the mountain savannah (less than 10/km^2). The rate of population increase is high (close to 3 %). Twice a week people in the lowland savannah have the chance to visit a large market in Goljota, a village situated close to the forests and linked by a good earth road to the asphalted road to Addis Abeba. In the study area's part of the mountain savannah lies only one small village (Lakicha) without any commercial activities, except an almost insignificant little market once a week, mainly for sheep and goats and some handmade tools (Table 1). Transport is almost exclusively possible on horseback. At first sight, nobody seems to live in the forest itself, separating the two populated parts of the study area.

Although the population groups in the lowland and in the mountain savannah have the same ethnic roots, they developed different land-use systems. In the lowland savannah people grow first of all maize and other grain crops. They also sometimes even have big herds (cattle, sheep, goats) grazing in the remaining woodland fragments and on harvested fields. Sometimes there are small home gardens close to the huts and in a few places irrigation is possible. Both contribute to a larger variety of cultivated plants, namely vegetables. In the mountain savannah the dominating land use is cattle breeding on large grassland areas. Wheat and barley are also grown but almost exclusively for their own consumption. The same is true for some vegetables and spices cultivated in small gardens.

3 Methods

It has already been mentioned that a comprehensive research program concerning the ecology of the Munessa forest started two years ago and that the main objective is to find options for the conservation and the sustainable use of these forests. Such a target needs a holistic environmental approach and a multidisciplinary team. Basic research has to be combined with applied research in order to propose an ecologically justified and economically as well as socially acceptable resource management. Nevertheless up to now so-called holistic approaches have only been a (valuable) theoretical framework for ecosystem research (Beck & Müller-Hohenstein 2001). This theoretical framework is kept in mind here and in the following presented case study the social and economic aspects of the population are of special interest.

The necessary data about climate and soils have been collected with standard methods. Most of the results mentioned in this respect come from our research group. Identification of plant specimens was possible with the cited Floras from East Africa. Vernacular names used by the Oromo people have often been very helpful. Plants have been named according to the Honey Bee Flora and the volumes of the Flora of Ethiopia and Eritrea. (Comprehensive data about soil analysis, plant lists and tables, based on a grid system and treated with a GIS, will soon be published in a larger contribution on the study area).

All detailed information concerning land and forest use has been collected directly from the population, living not further away than 5 km from the lower respectively upper forest margin. For practical reasons "forest margin" is understood as a 1 km large transition zone between the closed contiguous forest and the almost treeless pastures and agricultural lands. However, this does not mean that there are no more traces of anthropogenic influences beyond this distance (burning, felling, selective use of trees and so on). More than 10 families (older men and women, sometimes „opinion leaders" of a whole village) on each side of the forest have been asked more or less the same questions in order to get a complete and sound picture of the actual situation which helps to understand why the forest margins have their special physiognomy.

The most prominent questions concerned:

- form and extent of land use
- form and extent of forest use (separately for forest products such as timber, firewood, lianas and non-timber products such as honey, mushrooms, medical plants)
- forest products required for the construction of huts, the production of tools for land use purposes as well as various household items for cooking, for furniture and so on
- the difficulties of currently getting enough of all these resources
- the possibilities of replacing some of the traditional resources with new ones (e.g. modern remedies, plastic ropes)

Furthermore several foresters, forest guards and the head of the governmental Forest Company (Degaga forest district) have been asked the same questions and some more related to

- the stability of the actual forest margins
- the possibilities of controlling illegal forest use and illegal forest clearing
- the development of the forest margins according to their own experiences

Finally some information came from the merchants as well as the customers at the market places in Goljota and Lakicha.

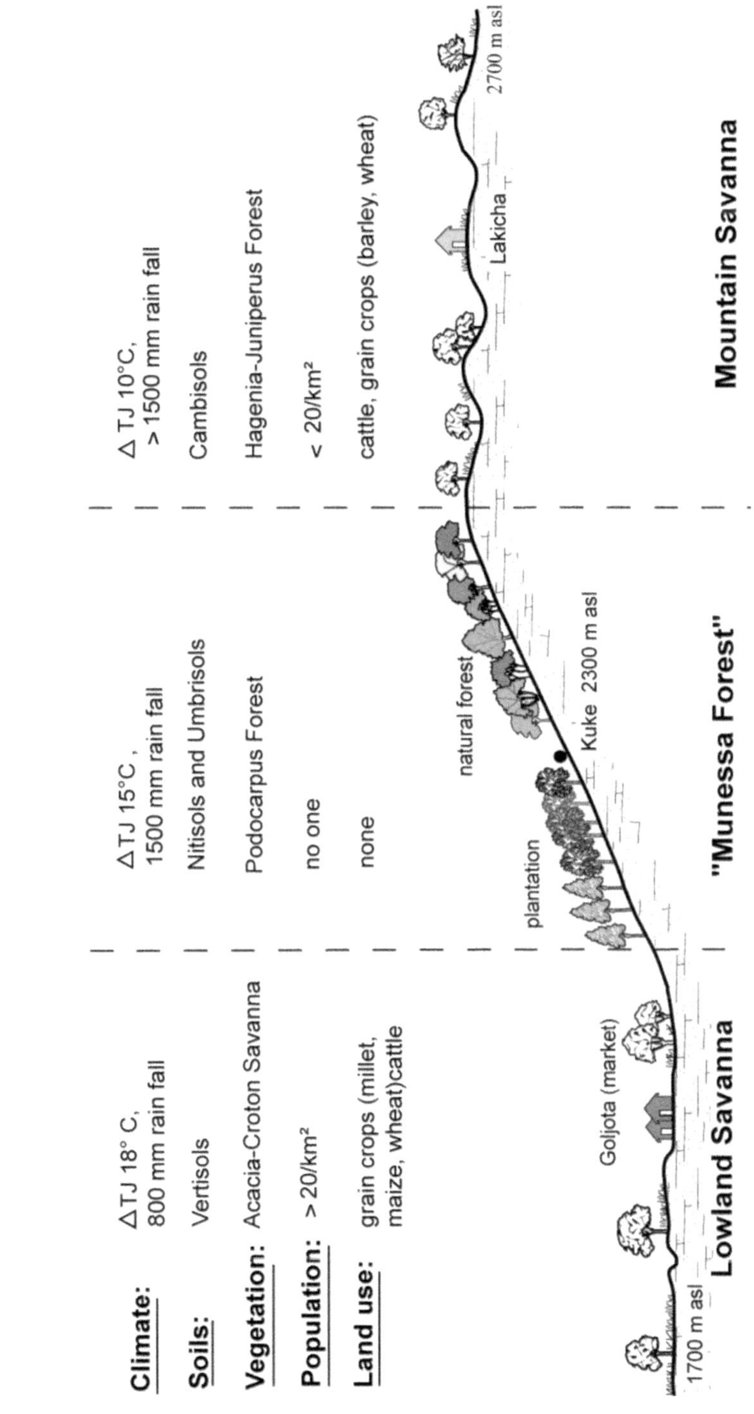

Fig. 2. The study area "Munessa Forest" and adjacent savannahhs

4 Results: The Present Situation at the Lower and the Upper Forest Margin of the Munessa Forest Area and its Dynamics

To understand the present situation of the two forest margins in the study area, two transects (Fig. 3, Fig. 4) and two tables (Table 1, Table 2) have been drawn and composed respectively. Fig. 3 shows both sides of the lower forest margin, starting from the open "lowland savannah" and going up to parts of the degraded natural forest at 2200 m asl.

The lowland savannah is quite intensively cultivated up to close to the forest margin. There are only traces left of the former natural drought-deciduous Acacia woodlands and the evergreen forests. They consist of small remnants composed of *Acacia* species such as *Acacia tortilis*, *A. seyal, A. etbaica* and some other thorny trees and shrubs (*Balanites aegyptiaca, Carissa edulis, Maytenus senegalensis*). On the roadsides, sometimes dense thickets with *Vernonia auriculifera, V. adoensis, Hibiscus trionum, Pterolobium stellatum* and different *Cassia* species protect the adjacent agricultural fields. Mainly evergreen species such as *Ficus thonningii, Rhus glutinosa, Dovyalis abyssinica, Salix subserrata, Syzygium guineense* and *Milletia ferruginea* accompany the only small but often deep incised river valleys. Highly characteristic for almost all parts of the lowland savannah are quite regularly distributed large single trees, mainly *Ficus* species (*F. vasta, F. sycomorus*) but also *Celtis africana, Ekebergia capensis, Prunus africana, Apodytes dimidiata, Olea africana* and – closer to the forest margin – *Podocarpus falcatus. Ficus vasta*, the most often left huge tree on farmland is a multifunctional tree, providing shade for cattle and timber but also serving as a meeting point of social and religious importance.

Cattle, sheep and goats are grazing almost everywhere, especially in the remaining woodlands and on harvested fields. But more important is the agriculture. People grow almost all kinds of grain crops, especially maize and millet. In a few places the water resources allow irrigation. On this basis many vegetables are grown as well as tobacco and sugar cane. The homegardens close to the traditional houses are protected by living fences out of *Agave sisalana*, *Euphorbia tirucalli* and *E. abyssinica* or by dead fences out of branches from *Acacia* species or – only recently – branches from *Cupressus lusitanica*. Also, close to the houses, especially the more modern ones with tin roofs, *Eucalyptus* species have been planted for their personal consumption or as a cash crop.

In Table 1 all plant species and plant products are recorded, which have been offered on the only regular market place near the forest margin in the study area in Goljota (twice a week market day). On the 1st of June 2002 no fewer than 36 plant products came from the lowland savannah, only 11 from the mountain savannah and among these only 2 exclusively from this area (e.g. barley). Finally only 7 other plant products came from outside the study area via merchants, e.g. coffee, ginger and khat. There is only one remarkable permanent shopkeeper in Goljota with a very limited variety of goods coming from the larger towns. In Table 2 all plant species identified as important for different purposes during the interviews mentioned above are listed. Cultivated plants are not included. Both tables – although they might not be 100% complete and according to seasonal changes, the market offer will change, too – prove that the population in the lowland savannah depends to a very high degree on the resources from the cultivated and non-cultivated, spontaneous growing plants. This is even truer for the population in the mountain savannah.

Everywhere in the study area the lower forest margin is very sharp, due to forest plantations with exotic tree species (*Eucalyptus globulus, E. camaldulensis, E. saligna, Cupressus lusitanica, Pinus patula).* They were already established some decades ago in a very regular way (1,600–2,500 trees /ha, thinned to a final stocking of about 500–600 trees/ha) and constitute actually a major proportion of the Munessa forest at altitudes up to 2300 m asl. Today the Shashamene-Munessa Forest Enterprise manages these plantations and tries to stop illegal woodcutting and timber production. There is some natural regeneration in the forest plantations and especially in gaps pioneer trees, such as *Croton macrostachys*, develop very quickly. However, the main objective of the forest company is to produce timber from the above mentioned tree species.

In places where the forest company has not yet planted exotic trees, the remnants of the natural forests are influenced and degraded to some extent.In particular, grazing, which disturbs the regeneration of many native tree species, is indicated by some weeds such as *Kalanchoe densiflora, K. lanceolata, Achyranthes aspera* and *Hypoestes forskaolii.* Nevertheless, the originally humid mixed forest is still indicated by a large number of native tree species such as *Podocarpus falcatus* and *Prunus africana* in the upper canopy and many smaller trees and shrubs such as *Bersama abyssinica, Croton macrostachys, Ekebergia capensis, Allophyllus abyssinicus, Dovyalis abyssinica, Maesa lanceolata, Olinia rochetiana, Teclea nobilis* and *Calpurnea aurea.*

Higher up, as is shown now in Fig. 4, the composition of the natural forest changes. The marked changes in species composition are associated with higher rainfall (> 1,300 mm/a) and have been found in other forests, too (Friis 1991). *Podocarpus falcatus* is still the most impressive and dominating tree, but now *Dombeya torrida, Nuxia congesta, Galineria coffeoides, Schefflera volkensii* and *Rapanea simensis* have become more frequent. Around 2,500 m asl *Arundinaria alpina* is quite common in dense but not very large stands. There are more epiphytes (ferns, mosses, some orchids and *Peperomia spec.*) and lianas. The ground cover again is not very dense, but there are many species indicating an increasing pressure of the grazing cattle (*Cirsium schimperi, C. dender, Bothriocline schimperi, Bidens macroptera, Kalanchoe petitiana* and *Commelina benghalensis*), especially on now quite regularly appearing open spaces with thickets of smaller bushes (*Vernonia leopoldii, V. amygdalina, Rosa abyssinica, Bersama abyssinica*) close to the forest margin. Now and then small areas are covered with young shoots from *Rapanea simensis*, which are used for the construction of traditional houses.

Even more characteristic now are a lot of still standing tall but dead trees and small clearings in the more or less closed forest. Sometimes these clearings are fenced with dead *Arundinaria alpina*; sometimes first houses have been built together with small homegardens (with ensete, cabbage, different spices) and some granaries. In fact, it is now more difficult to identify an exact forest boundary. Step by step some agricultural fields with barley or potatoes replace the forest, until a little bit further away from the last closed forest fragments, pasture land with single trees, dead and alive, dominate the landscape. Among these trees the most prominent are now *Hagenia abyssinica, Prunus africana, Olea africana, Hypericum revolutum* and *Erica arborea.* Around the huts still some homegardens with living fences (*Discopodium penninervium*) are characteristic. Planted *Eucalyptus* or *Cupressus* trees however are very rare and the houses are almost without exception made out of native trees and covered with grasses or barley straw.

The predominant source of income for the population of the mountain savannah is cattle breeding. All the other agricultural products are for their own consumption. Only after a rich harvest is some barley sold in the market place down in the lowland savannah. There is no bigger market in this part of the mountain savannah and no real road leading to the more populated regions. There are no cash crops, irrigation is unknown, fertilizers are not available, nor are there modern tools. Some people go to the forests and get not only the firewood they need (what the forest company allows as well as the grazing in the forest), but also illegally get some timber which they prepare for construction purposes only with traditional axes. This is almost the only chance to get some money, but it is also a chance to end up in a prison for some time.

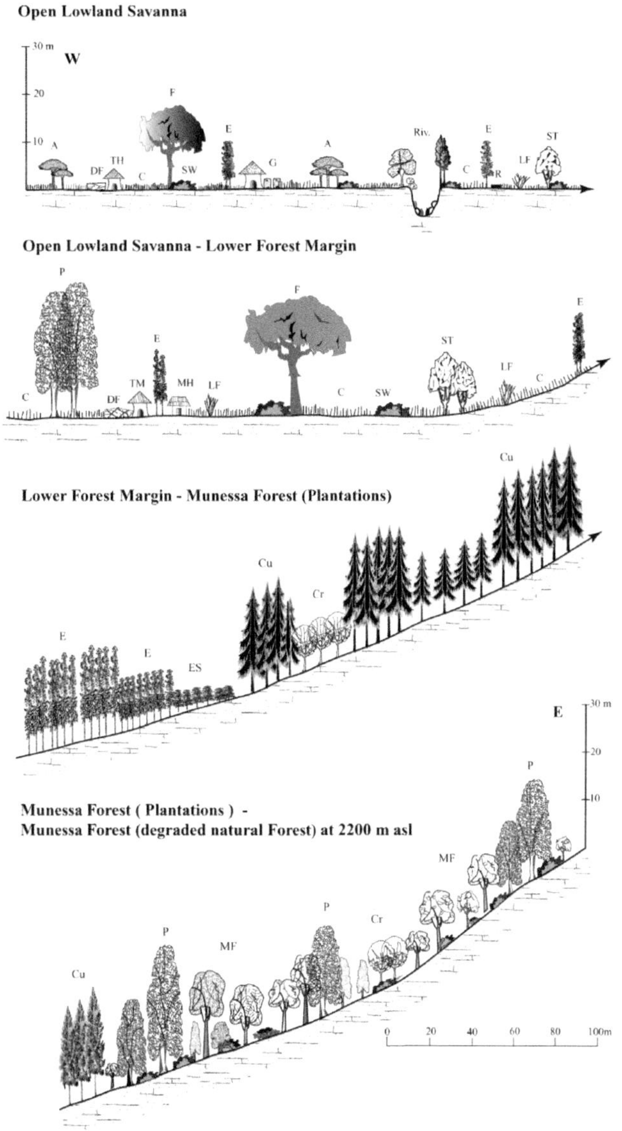

Fig. 3. Transect "Lower forest margin". TH – trad. house (grass roof), MH – modern house (tin roof), G – granary, R – earth road, DF – dead fence, LF – living fence, HG – homegarden, W – water place, C – crops, A – *Acacia* woodland fragments, F – *Ficus spec.*, P – *Podocarpus falcatus*, E – *Eucalyptus spec.*, Es – *Eucalpytus* shoots, SW – shrubs and woodlands, Riv – riverine forest, ST – single tree, Cr – *Croton macrostachys*, Cu – *Cupressus lusitanica*, MF– mixed forest

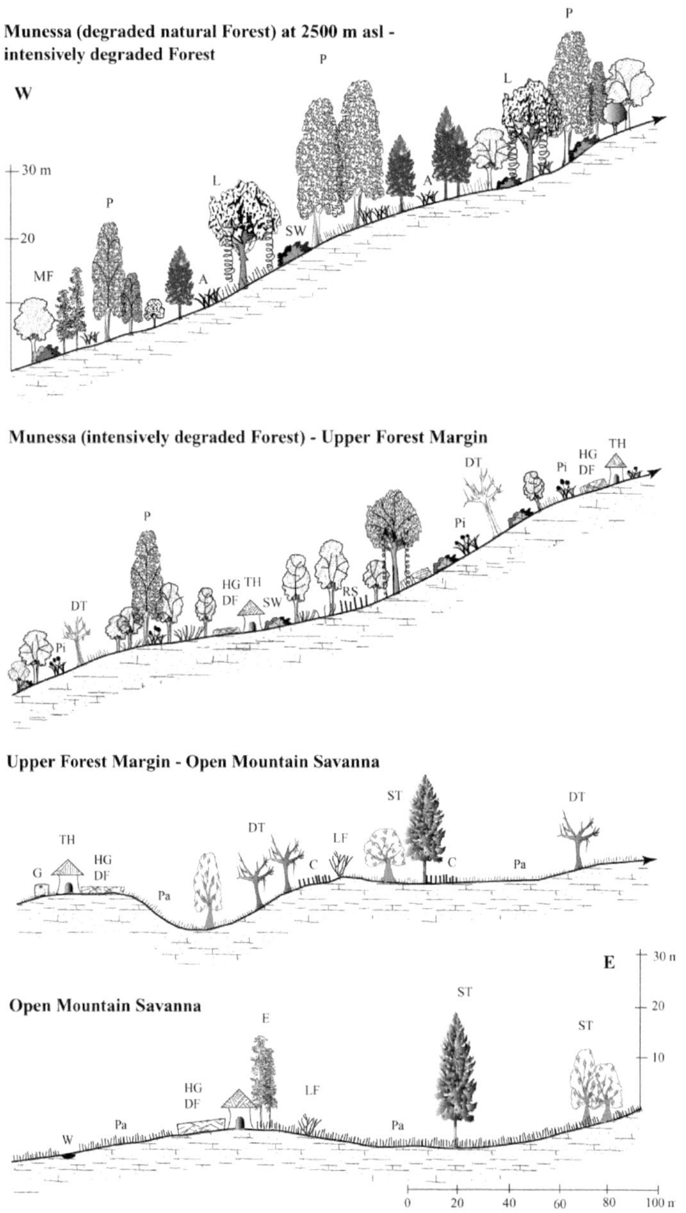

Fig. 4. Transect "Upper forest margin" TH – trad. house (grass roof), G – granary, DF – dead fence, LF – living fence, HG – home garden, W – water place, C – crops, P – *Podocarpus falcatus*, E – *Eucalyptus spec.*, A – *Arundinaria alpina*, L – lianas, Pi – indicator plants for heavy grazing, SW – shrubs and woodlands, DT – dead tree, RS – *Rapanca* shoots, MF – mixed forest

Table 1. Plant species and plant products, sold on Goljota market (1.6.2002) by farmers from the „lower savannah" (ls), the mountain savannah (ms) and merchants from the near-by towns (m). Use is mentioned if not used for preparing meals. According to the season, this list will change, especially concerning the vegetables.

Species Name	coming from			used for
Scientific (English)	ls	ms	m	(other than food)
Allium cepa L. (onion)	X	X		
Allium porum L. (leek)	X			
Allium sativum L. (garlic)	X	X		
Arundo donax L. (spanish reed)	X			baskets
Anethum graveolens L (dill)			X	
Arundinaria alpina K.Schum.	X	X		destilling apparatus
Beta vulgaris L. var. *conditiva* Alef. (garden beet)	X			
Brassica oleracea L. var. *capitata* (common cabbage)	X	X		
Capsicum annuum L. (sweet chillies)	X			
Carica papaya L. (papaya)	X			
Carthamus tinctorius L. (safflower)			X	
Catha edulis (Vahl) Forssk. ex Endl. (khat)			X	mild stimulant
Coffea arabica L. (coffee)			X	
Coriandrum sativum L. (coriander)	X			
Cynodon dactylon L. (stargrass)	X			milk containers
Cyperus papyrus L. (papyrus)	X			coffee-ceremony
Daucus carota L. (carrot)	X	X		
Eleusine floccifolia (Forssk.) Spreng.	X			decorative baskets and plaited saucers
Eleusine coracan (L.) Gaertn. (dagussa)			X	
Ensete ventricosum (Welw.) Cheesm. (ensete)	X	X		
Eragrostis tef (Zuccagni) Trotter (teff)	X			
Guizotia abyssinica (L.f) Cass. (niger seed)			X	
Hordeum vulgare L. (barley)		X		
Ipomea batatas (L.) Poir. (sweet potato)	X			
Nicotiana tabacum L. (tobacco)	X			
Nigella sativa L. (black cumin)	X			
Ocimum basilicum L. (sweet basil)	X			
Ocimum lamifolium Hochst. Ex Benth.		X		
Ocimum urticifolium Roth	X			
Ocimum majorana L.	X			

Table 1. (cont.)

Species Name Scientific (English)	coming from ls	 ms	 m	used for (other than food)
Passiflora edulis Sims. (passion fruit)	X			
Panicum miliaceum L. (true millet)	X			
Pennisetum schimperi	X			part of the milk bottle
Pennisetum glaucum (L.) R.Br. sl. (pearl millet)	X			
Rhamnus prinoides l'Herit	X	X		local alcoholic drink
Ruta chalepensis L.	X			
Saccharum officinarum L. (sugar cane)	X			
Scirpus corymbosus	X			coffee cere-mony
Solanum tuberosum L. (potato)	X	X		
Sorghum durra (Forssk.) Stapf (durra)	X			
Sporobolus spec.	X			brooms
Triticum vulgare Vill. (wheat)	X			
Vicia faba L. (broad bean)	X	X		
Zea mays L. (maize)	X			
Zingiber officinale Rosc. (ginger)			X	

Table 2. Important plant species (excluded cultivated plants) used by the population in the Munessa area for different purposes (according to interviews with people from the lower forest margin and the upper forest margin)

Plant species used at the lower forest margin:

Acacia sieberiana DC	tool handles, firewood
Acacia seyal Del.	important fodder for livestock, dead fences, firewood, charcoal
Acacia tortilis (Forssk.) Hayne	valuable forage, firewood, charcoal
Agave sisalana Perr. Ex Eng	living fences, ropes
Albizia schimperiana Oliv	light timber, tools for agriculture
Aningeria altissima (A.Chev.) Aubrev & Pellegrin	indoor furniture, veneer, high apicultural value
Apodytes dimidiata var. *acutifolia* (A.Rich) Boutique	timber fo carpentry, doors, tool handles, fire-wood r house construction, indoor
Calpurnea aurea (ait.) Benth.	timber for roof construction, and wall of cottages, living fences, apiculture, firewood
Carissa edulis (Forssk.) Vahl	dead fences, roots against aches
Cassipourea malosana (Bak.) Alston	farm tools, plough
Celtis africana Burm. F.	parts for oxplough
Cordia africana Lam.	doors, windows, household tools and furniture (stool, bed, table), mortar and grinder, apicultural value, skin deseases
Croton macrostachys Hochst. ex Del	indoor carpentry, apiculture
Cynodon dactylon L.	roof of huts and houses
Cymbopogon commutatus (Steud.) Stapf.	for parts of milkbottles
Dodonea angustifolia L.f.	wall construction, living fence, firewood
Euclea schimperi (D.C.) Dandy	parts for oxplough
Ekebergia capensis Sparmm.	milk pots and other household utensils
Euphorbia abyssinica L.	roof construction, saddles
Euphorbia candelabrum Frem. ex Kotschy	different boxes, stem for beehives
Euphorbia tirucalli L.	living fences
Fagoropsis angolensis (Engl.) Dale	fruits for treating inflammations
Ficus sur Forssk.	mortars, edible fruits
Ficus sycomorus L.	edible fruits, meeting place
Ficus vasta Forssk.	mortars, meeting place
Maytenus senegalensis (Lam.) Exell.	dead fences
Milletia ferruginea (Hochst.) Bak.	household utensils, apicultural value
Ricinus communis L.	oil to grease the injera-pan, purgative
Syzygium guineense ssp. *afromontanum* F. White	tool handles, charcoal, apiculture
Teclea nobilis Del.	different handles for agricultural tools, parts of oxplough
Typha spec.	to cover roof

Table 2. (cont.)

Plant species used at the lower and the upper forest margin:

Arundinaria alpina K. Schumm.	house construction (walls, roof), leaves for wrapping beehives, queen cages for beekeeping, apparatus for distilling, dead fences
Bersama abyssinica Fresen	roots against ascariasis
Buddleja polystachya Fresen	living fence, important anthelmintic and purgative
Eucalyptus camaldulensis Dehnh.	poles for house construction, fence posts
Eucalyptus globulus Labill.	house construction, charcoal, apicultural value
Cupressus lusitanica L.	house construction (central pole, branches for walls, firewood)
Discopodium penninervium Hochst.	dead and living fences
Maesa lanceolata Forssk.	greasing injera plate, firewood, vermifuge
Nuxia congesta R.Br. ex Fresen	house construction (pole), parts for oxplough, firewood
Olea africana Mill.	valuable timber, central pole for houses, mortar, grinder, tool handles
Podocarpus falcatus (thunb.) R.Br. ex Mirb.	most important timber, central pole for houses, door, doorframes, bed, chair
Prunus africana (Hook f.) Kalkm.	timber for heavy construction (bridges), central pole for houses, handle for axes, mortars
Schefflera abyssinica (Hochst. Ex A.Rich) Harms	high apicultural value

Plant species used at the upper forest margin:

Brucea antidysenteria J.F. Miller	fruits and leaves against fever, diarrhoea and dysentery
Dombeya torrida (J.F.Gmel.) P. Bamps	house construction, farm tools
Echinops ellenbeckii D.Hoffm.	against headache
Elaeodendron buchananii (Loes.) Loes	liana, used as a rope
Erica arborea L.	dead and living fences, firewood, charcoal, high apicultural value
Galineria coffeoides Del.	traditional medicine (gurmit)
Hagenia abyssinica (Bruce) J.F.Gmelin	many parts for house construction, many household tools and furniture, tools for agriculture (lada), firewood, against tapeworm
Hordeum vulgare L.	straw for roofing
Hypericum revolutum Vahl	central pole for houses, parts of oxplough
Jasminum abyssinicum Hochst. ex DC	used as a rope for house construction
Juniperus procera Hochst. ex A.Rich	central pole for houses, firewood
Maytenus gracilipes (Well ex Oliv.) Exell	house construction (walls), firewood
Polyscias fulva (Hiern.) Harms	milk pots, boxes
Rapanea simensis	roof and wall construction of trad. houses
Urera hypselodendron (Hochst. ex A. Rich) Wendel	liana, used as a rope

Obviously the two forest margins are quite different. For these differences at least two main reasons have to be looked at closer. On one hand there are the different natural resources, which are – due to about 1000 m difference in altitude – not the same near the lower and near the upper forest margin. This can already clearly be shown by the different plant species and their uses, listed in Table 2. *Ficus* and *Acacia* tree species can only be found in the lowland savannah, *Hagenia abyssinica, Juniperus procera* and *Hypericum revolutum* trees only in the mountain savannah. The dominating tree species used for the construction of houses, granaries, mortars, grinders, traditional furniture and agricultural equipment differ to a large extent. During the interviews with people from the lowland savannah, only stools made out of *Cordia africana* were offered to sit down on, and at the upper margin those stools were carved out of *Hagenia abyssinica*. There are many more similar examples. They prove that the people in the study area use different resources from the forests for the same purposes. In the same time it becomes clear, that the population in the study area needs the nearby forest resources for their daily life.

On the other hand it is the different economic and political situation, which contributes to the marked differences of the two forest margins. Generally the lower forest margin is very regularly and distinctly shaped due to plantations with exotic tree species. The forest enterprise controls these plantations quite successfully, and new encroachments are minimal. Grazing in the plantations is allowed; people can also take out branches from the fast growing *Eucalyptus* and *Cupressus* trees. However, cutting trees from the adjacent natural forests is not allowed. The forest enterprise has no concept for a sustainable use of the natural forest remnants but tries to protect the remnants of the natural forests, at least for the time being. Furthermore, for almost 20 years primarily *Eucalyptus*-seedlings have been distributed among the population, and training courses for planting and managing have been offered. Today young *Eucalyptus* trees contribute to a remarkable extent to the income of the population in the lowland savannah. Even more important is, that the exotic tree species are fully accepted as a substitute for many native species, especially for the construction of houses, fences and some traditional tools. So the lowland population is "richer" than their neighbours in the mountains. They also have the possibility of getting material for constructing tin roofs and they need no longer lianas from the forest because there are substitutes like plastic ropes and wires, which they can buy at the nearby market. And finally the same is true for medical purposes. There are a large number of medicinal plants for the treatment of different health problems and most of them are still propagated by local healers. But at the weekly market some "modern" medicine is also available. So the plantations of exotic tree species take off some of the pressure from the natural forests at the lower forest margin.

The upper forest margin looks completely different. There are no forest plantations. Therefore, there is also less control and the population is far more inclined to enter the remaining forests, not only to get forest products but also to clear the already heavily degraded forests in order to construct huts inside the forest. Even more land is cleared for homegardens and agricultural fields. For the homegardens a kind of a rotation system exists, a rare contribution to sustainability. During nighttime the younger animals are kept in safely fenced areas. These areas are naturally fertilized with the droppings of the animals and after some time changed into homegardens. But there are no fertilizers or manure for all the other agricultural fields. Only a few years after the clearing the yield is so low, that people try to get new areas for agriculture. The former fields become pastures. That is why cattle breeding is more important in the mountain savannah and in the forests close to the forest margin and why continuous clearing for arable land under the present conditions is necessary. The population living close to the forest margin depends much more on the forest resources, especially on the remaining potential to clear the forests.

The upper forest margin is difficult to identify, because there is no clear border line but a long transitional zone over several hundred meters from the closed forest to the pasture land with single trees, most of them heavily damaged or already dead. The technical equipment for clearing and tree cutting is very basic. Not only axes are used for clearing but also fire is an important tool. Often the cambium of the dead trees is cut off and they all show traces of fire. The long distance to the headquarters of the forest company, where most of the guards are and where the seedlings are produced, explains to a very large extent, that deforestation and forest degradation is much higher at the upper forest margin. In comparison with the lowland savannah, tree planting here is still rare because providing the seedlings and instructing the population in the mountain savannah is more difficult. Furthermore it is almost impossible to control the forest margin in an effective way. Finally, for some people cutting trees and selling the timber to merchants near the asphalted road is the only way to get some more money. Because of a poor harvest in some years, selling agricultural products is hardly possible and the meat, milk and butter produced are mainly for home consumption. The only valuable non-timber product out of the forest is honey. Hunting has been a common practice but presently it is only allowed with an official permission from the regional council. Professional hunters have to pay for such permission. The local population does not benefit from these activities.

There is no doubt that the population in the study area depends very much on the forest resources and that the forest margins look very different from outside according to their natural potential, the needs and activities of the local population and the anthropogenic influences. The present patterns at the lower and the upper forest margin can be understood as an example of a high natural and anthropogenous beta-diversity. However, these patterns are not firm and steady. There is remarkable dynamics and again, the dynamics is different for the two forest margin areas.

In postglacial phases, about 8000 b.p., the mountain forests of Ethiopia were much more extended than they are today, due to the more humid climatic situation at that time (Anhuf et al. 1999). It is also known that still in historical times the mountain forests covered not less than 35 % of the total surface of the country A 20 % coverage was reported for the 19th century (De Vletter 1991). In more recent decades Reusing & Kasberger (2000) recorded on the basis of satellite data only some 6 % for the mid-70s and less than 4 % for the beginning of the 90s of the last century. Only some 50 years ago, first plantations with exotic trees increased the forested area to some extent. Today less than 10% of the still forested area is covered by plantations and experts believe that the decrease of forested areas by human influences is higher than the increase by plantations.

Also in the study area predominantly forested areas in the lowland savannah as well as in the mountain savannah have mainly been converted into farmland for agricultural use or pasture. Here only the already mentioned single large trees and some woodland fragments give evidence of the former forest cover (see Fig. 2). Also the remaining forests have been degraded by selective logging for timber and firewood and by constant grazing. A certain turning point for the development of the forested area in the Munessa-Shashamene area was the change from the military to the present government. The new political situation – including the regional government – and the now successfully managing forest enterprise guaranteed a much better control of the existing rules concerning the use of forest resources by the local population. Taking out dead branches for construction or firewood, as well as grazing, was allowed. Felling trees, neither young nor old ones, exotic or native ones, was forbidden. But often people, looking for firewood made sure to find dead branches also for the next time.

At the lower forest margin the control was and still is quite effective, hence illegal tree felling cannot totally be excluded. The large and highly regular tree plantations, subdivided into geometric patches and only sometimes separated by road corridors, usually form a straight borderline. So the native forests further up are well protected by these plantations. And – as mentioned before – due to some important substitutes, especially the private *Eucalyptus*-plantations near the houses, there is less pressure on the native forests and in total less dynamics.

At the upper forest margin the situation is completely different. Although the area belongs to the forest enterprise, it is not possible to control the human activities in the same way as below, due to 1000 m of altitude difference and 20 km distance from the company headquarters. A very high rate of encroachment is the result. Usually this encroachment starts with selective logging of trees for timber and firewood and of *Arundinaria alpina*, an important plant for the construction of houses and fences. Sometimes a rapid re-establishment of pioneer trees and shrubs follows but intensive grazing leads to first gaps. In the last few years a growing number of houses have been illegally constructed with small homegardens accompanying these houses. In many cases agricultural fields follow and enlarge the gaps. Axes and fire are the common tools, which allow these sequences. According to Forman (1997), this degradation series starts with the process of "perforation", followed by "fragmentation", "shrinkage" and finally "attrition", the total disappearance of patches with trees. The resulting picture in the mountain savannah is pasture with single trees, dead and alive.

Of course, this land transformation does not proceed at a constant rate and not at the same intensity everywhere. Also the different phases do not have the same impact on the basic ecological habitat characteristics and on species richness. Fragmentation especially usually leads to isolation and the separation of plant and animal populations. The number of generalists and edge-species seems to increase, that of the forest-species and the rare species with special habitat requests decrease. But above all, those species with an important economic value such as

Cordia africana in the lowland savannah and the once dominating *Hagenia abyssinica* in the mountain savannah are today endangered. In Lakicha, the only village in this part of the study area, the older people report that the village was constructed only 15 years ago in the middle of a dense forest. Today Lakicha is almost exclusively surrounded by pasture.

5 Conclusions

Worldwide the forests of the tropical mountains have been transformed to a large extent into cultural landscapes more rapidly than at any time in the past. The Ethiopian forests and especially the forests in the study area give evidence of this again. However, the underlying processes, leading to degradation and finally to the complete loss of the forests, may be different. This becomes clear, when we compare the lower and the upper forest margin of the Munessa forest.

From the foregoing presentation of results and discussions it can be shown that there is no general model concerning human influences and their consequences in forest margins in Ethiopia. These influences depend as well on the natural resources (quantity and quality of available plants) as on the socio-economic situation of the population concerned (type of land-use, access to market) and finally on the local, regional and national political conditions (efficiency of control measures and others).

Today ecologists focus more and more on the problems of conservation and sustainable use of the tropical forests. They learn step by step, that a sound conservation planning and setting priorities for a sustainable land use are only possible, if the processes of degradation and deforestation are better known and understood. Up to now, scientists consider economic, social, political and ecological problems often separately. For an integrated landscape management synthesis is indispensable. The study of the relationships between biological systems and cultural systems is the challenge. There is no better place to follow such an interdisciplinary approach than at the margins of forests.

6 Acknowledgements

We thank

- the German Research Foundation (DFG) for supporting our project,
- EARO, our Ethiopian partner in this research, namely Dr. Demel Teketay,
- our German colleagues in the project, namely Florian Fritzsche,
- Elisabeth Schill, Lynn Woelker and Jan Axmacher for technical assistance,
- the population of the study area, who shared their poor meals and their rich knowledge about the forests with us.

References

Abate S (1994): Land use dynamics, soil degradation and potential for sustainable use in Meta Area, Illubabor Region, Ethiopia. Geographica Bernensia A 13. Universität Bern, Bern

Alemu B (1998): Spatial study of soil nutrient status after natural forest conversion to plantation and secondary forests at Munessa Mountain Forest, Ethiopia. MSc. thesis, Swedish University of Agricultural Sciences, Debre Zeit

Barnes RFW (1990): Deforestation trends in tropical Africa. Afr J Ecol 28: 161–173

Beck E & Müller-Hohenstein K (2001) Analysis of undisturbed and disturbed tropical mountain forest ecosystems in Southern Ecuador – DIE ERDE 132: 1–8

Breitenbach Fv (1963) The Indigenous Trees of Ethiopia – Ethiopien Forestry Ass, Addis Abeba

Chaffey DR (1979): South-West Ethiopia Forest Inventory Project – Land Resources Development Center Surrey, England

De Vletter J (1991): Forest genetic resources of Ethiopia. In: Engels JMM, Hawkes JG & Worede M (eds) Plant genetic resources of Ethiopia. Cambridge University Press, pp 82–100

Edwards S, Demissew S & Hedberg I (eds) (1997) Flora of Ethiopia and Eritrea. Vol. 6. Addis Ababa, Uppsala

Edwards S, Tadesse M & Hedberg, I (eds) (1995) Flora of Ethiopia and Eritrea. Vol 2, Part 2. Addis Ababa, Uppsala

Edwards S, Tadesse M, Demissew S & Hedberg I (eds) (2000) Flora of Ethiopia and Eritrea Vol 2, Part 1. Addis Ababa, Uppsala

Feyera S (1998): Native woody species regeneration under the canopies of tree plantations at Munessa-Shashamene forest project area, Southern Oromia. MSc. thesis, Swedish University of Agricultural Sciences, Faculty of Forestry, Stockholm

Fichtl R & Admasu A (1994): Honeybee Flora of Ethiopia. Weikersheim

Forman RTT (1997) Landmosaics – the ecology of landscapes and regions. Cambridge University Press

Friis I (1986) The forest vegetation of Ethiopia. Acta Univ Ups Symb Bot Ups XXVI(2): 31–47

Friis I (1991) Forests & Forest Trees of Northeast Tropical Africa: their Natural Habitats and Distribution Patterns in Ethiopia, Djibouti and Somalia. Kew Bull Additional Series 15(I–IV): 1–396

Friis I and Tadesse M (1990) The evergreen forests of tropical NE-Africa – Mitt Inst Allgem Bot Hamburg 23A: 249–263

Garwood NC (1989) Tropical soil seed banks: a review. In: Leck MA, Parker VT & Simpson RL (eds) Ecology of soil seed banks. London, pp 149–209

Hedberg I & Edwards S (eds) (1989) Flora of Ethiopia Vol 3. Addis Ababa Asmara Uppsala

Hedberg I & Edwards S (eds) (1995) Flora of Ethiopia and Eritrea Vol 7. Addis Ababa, Uppsala

Kelecha WM (1987) A Glossary of Ethiopian Plant Names. Addis Ababa

Knapp R (1968) Höhere Vegetationseinheiten von Äthiopien, Somalia, Natal, Transvaal, Kapland und einigen Nachbargebieten. Geobot Mitt 36: 1–36

Lemordant D (1971) Contribution à l'ethonobotanique éthiopienne. Journal d'Agriculture Tropicale et de Botanique Appliquée 18: 1–35, 99–179
Lemordant D (1981) Plantes toxiques d'Éthiopie. Fitoterapia LII(3): 99–127
Lisanework N and Mesfin T (1989) An ecological study of the vegetation of the Harenna forest, Bale, Ethiopia – Sinet. Ethiop J Sci 12: 63–93
Miehe G & S (1994) Zur oberen Waldgrenze in tropischen Gebirgen. Phytocoenologia 24: 53–110
Rehm S & Espig G (1991) The Cultivated Plants of the Tropics and Subtropics. Weikersheim
Teketay D & Granström A (1995): Soil seed banks in dry Afromontane forests of Ethiopia. J Veg Sci 6: 777–786
Teketay D & Granström A (1997): Seed viability of Afromontane tree species in forest soils. J Trop Ecol 13: 81–95
Tekle K & Tesfaye B (1998): The role of soil seed banks in the rehabilitation of degraded hill slopes in southern Wello, Ethiopia. In: Tekle K (ed) Ecological rehabilitation of degraded hill slopes in southern Wello, Ethiopia. Uppsala

Forest Resource Use by People in Protected Areas and its Implications for Biodiversity Conservation: The Case of Bandhavgarh National Park in India

Puja Sawhney & Stefanie Engel

1 Introduction

Biodiversity brings many benefits to humans, including direct ones, such as the provision of food, shelter and clothing. Many of these products are traded in the market, and have a recognized economic value. Others are subsistence products, but critical to the livelihood needs of people living in rural areas of developing countries, such as wild foods and bush meat, products that normally fall outside conventional economic analysis. There are also indirect ecological benefits and non-consumptive uses such as recreation (Stocking et al. 1995).

Industrialisation, which paved the way for development and advancement generally, has been achieved by drawing heavily on forest resources. The indiscriminate exploitation of forest resources by developed countries and the continued exploitation by the developing and less developed countries has taken a heavy toll on forest resources, especially in the last century. The impact has been greatest in developing and less developed countries, not only because there is heavier reliance on forest resources for economic and subsistence use by the people but because forests in these countries represent the greatest biodiversity.

Growing environmental consciousness has led to the realisation of the implications of continued destruction of forests, not only ecological but also commercial, and need for forest conservation. The most effective way to protect biodiversity is to protect natural habitats. Zoos and botanical gardens can hope to conserve only a small proportion of the earth's biodiversity, and they also suffer from the disadvantage of removing organisms from interactions with their natural environment, thus preventing many ecological and evolutionary processes. Thus, a great emphasis has been placed on the establishment of national parks (NPs) and nature reserves over the years (Stocking et al. 1995). They have since been recognised as playing a crucial role in conserving biological diversity (Wells and Brandon 1992).

In 1969, the International Union for Nature and Natural Resources (IUCN) introduced the concept of a Protected Area (PA) system as a mechanism of protecting and restoring forests and biodiversity around the world. Following conservation trends in the world, India too adopted the categorisation of PAs laid down by the IUCN in 1978[1]. India has adopted three of the IUCN categories, namely NPs, wildlife sanctuaries (WS) and closed areas. All PAs in India fall in the categories of NPs or WS while the category of closed areas is hardly ever used. The biggest drawback of the categorisation of PAs in these two categories is that it fails to take account of the complexities and magnitude of wildlife conservation in the country and leaves out areas such as sacred groves which have been maintained over a long time. The number of PAs in India has increased from a handful in the 1970s to 586 currently.

The importance of wildlife conservation cannot be argued against. However, dependence of local people, mostly marginalized and tribals, on forests cannot be overlooked either. Protected Areas (PAs), especially in the tropics often coincide with areas of human habitation with a heavy reliance on the PA resources for subsistence and economic use. 200–300 million people in South and South East Asia live in close association with forests (Colchester 1992). 86% of the NPs in South America have people living within and using its resources (Amend and Amend 1992 in Kothari et al. 1997). An Indian national survey of PAs conducted in the late 1980s by the Indian Institute of Public Administration revealed that 69% of the surveyed PAs had human populations (upwards of *3 million*) living *inside* them, and 64% had community rights and leases, or concessions inside them (Kothari et al. 1996). This is not surprising, considering that in India, a country with more than a billion people, most people live in rural areas relying heavily on the forests for meeting their subsistence and economic requirements. Out of the 300 million people in India who live below the poverty line an estimated 200 million are partially or wholly dependent on forest resources for their livelihoods. Among these are the 70 million tribal people living inside or in close vicinity to them. Moreover, a correlation exists between the location of tribal people, forests and areas with a concentration of poverty (World Bank 2001).

Lack of viable alternatives compels people to continue their reliance on forest resources, resulting in a multiplicity of problems for the local people, the forest department and for the wildlife. Knowing the extent and type of reliance of people on the resources of a PA can help in formulating programs for reducing or eliminating the dependence by provision of alternatives. Programs aimed at reducing pressures on the Park without assessing the extent and type of dependence cannot hope to be successful in reducing people's dependence and securing the future of the PA. This paper aims at contributing to such an improved understanding by drawing upon the findings of a field study carried out in 2000–2001 in Bandhavgarh National Park, India.

The remainder of this paper is divided into five parts. Part two briefly describes the study area and data collection. Part three assesses the reliance of the resident population on the Park resources for subsistence as well as economic use. Part four describes the impact of this resource use on the Park. Part five focuses on the need for provision of alternatives and possible solutions for lessening the depend-

[1] In 1978, IUCNs commission on NPs and PAs (CNPPA) published its report on 'categories, objectives and criteria for PAs in which it proposed 10 categories, was further revised in 1992 and now there are six categories of PAs (Kothari et al. 1997).

ence of people on the Park resources as well as strategies for involving people in Park management. Part six concludes this paper.

2 Study Area and Data Collection

Bandhavgarh National Park (BNP) is located in the central state of Madhya Pradesh, India (Fig. 1) covering an area of 448.84 sq. km. It was first declared a National Park in 1968 (denoted hereafter as *old Park area*), and the area of the Park was further extended in 1982 (*extension area*). The areas are shown in Fig. 2. In view of the importance of tiger conservation in India, BNP—which has the highest number of tigers in the world—was given the status of a Tiger Reserve in 1993. The old park area contained one village which was relocated in the early 1970s. The extension area of the Park currently has six villages and one illegal settlement within its boundary.

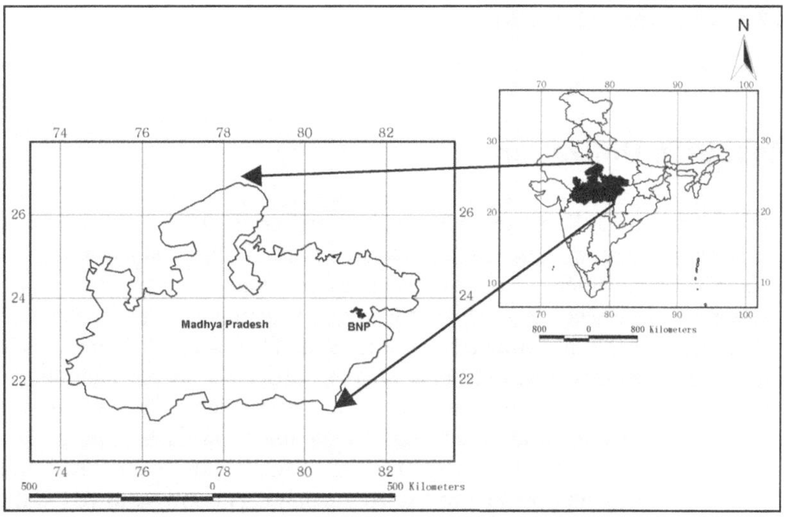

Fig. 1. Location of Bandhavgarh National Park

The data used in this paper was collected during the period from September 2000 to November 2001. After an initial focus group discussion, PRA exercises were conducted in all villages inside BNP. This was followed by a detailed survey of 155 households across all villages as well as a survey of selected Forest Department officials. The surveyed households comprised 31% of the entire Park population.

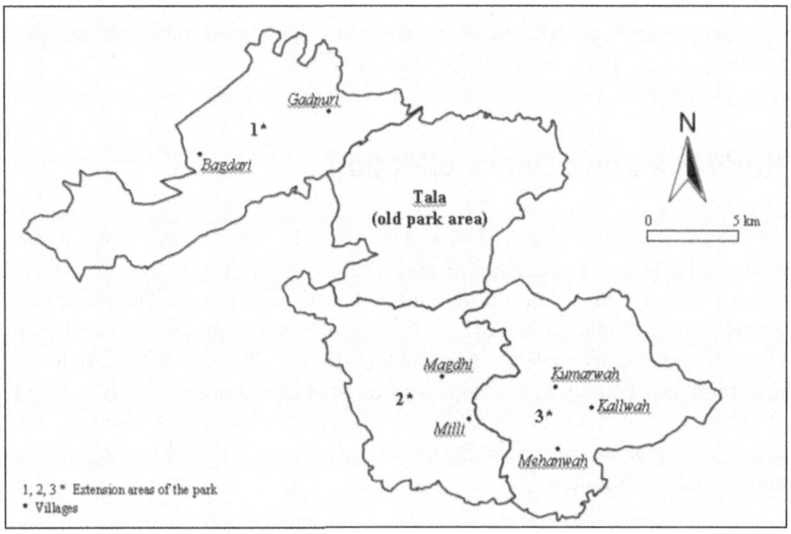

Fig. 2. Old Park area and the extension area of the Park. Courtesy: BNP authorities

3 Resource Utilization

The life of the people living in BNP revolves around the Park resources. It is not only a home for them but also the main source of food, fuel, fodder, medicine and income. The management of the extension area is done in accordance with laws governing NPs in the country. This implies that it is managed under the assumption that all villages will ultimately be relocated. Therefore, all human activities within the boundaries of the extension area are officially illegal, including those for income generation and some subsistence activities such as the cutting of trees for use as fuelwood and building material. Collection of products required for subsistence such as fallen and dead twigs for use as fuelwood, minor forest produce (MFP) for own consumption, and restricted grazing is currently allowed in the extension area of the Park for the inside villages as final notification of the Park is pending. There is a ban on any kind of activity in the old Park area which has already been made public.

Despite these restrictions people continue living in the extension area and depending heavily on the resources of the Park for both subsistence and income generation. Cultivation inside the Park, though illegal by the rules of the NP, is currently tolerated, as the rights of the people have not yet been legally settled. The survey revealed that 96% of the villagers cultivated. Almost all of the agricultural output is used for own consumption or barter. Most people also buy additional food from outside the Park. This is in part a consequence of nearly 50% crop damage due to an increase in wild animal population. Some people reported not being able to buy food grains from outside because of lack of money and their inability to borrow from the money lender. If they had to completely rely on buying food grains from outside, it would affect them adversely. Agricultural production not only sustains them but also provides an income source for some (about two percent). Food grains are also used to pay agricultural labourers.

Besides agriculture, which is the main food source of the villagers, there is a heavy reliance on the Park for the collection of minor forest produce (MFP). The term MFP is a misnomer, as it plays a vital role in the life of the villagers. MFP is an important food source for the people and their livestock and also contributes substantially to household income. The majority of the villagers (97%) collect MFP. All the sampled households collecting MFP also sold them. MFP collection from inside the extension area of the Park was prohibited by the state government in 1982, but was permitted for own consumption purposes again in 1990 in view of the importance of MFP for people living in and near the Park (Sonakia 1993). The different MFPs collected on a regular basis by the villagers are *amla* (*Emblica officinalis*), *char* (*Buchanania lanzan*), *ber* (*Zizyphus mauratiana*), *tendu patta* and *tendu* fruit (*Diospyros melanoxylon*), *mahua* (*Madhuca indica*), mango (*Mangifera indica*), *jamun* (*Syzygium cumini*), *sal seeds* (*Shorea robusta*), *dori* (seed of *Madhuca indica*), honey and medicinal plants. Mushrooms are also collected by people after the rainy season and is an important food supplement.

Though income from sale of MFP is important, its use by the households for own consumption, barter and as feed for livestock is equally important. Of the MFP collected, a minimum of 38% is used by the people for purposes other than sale. Although not quantifiable, if other MFPs collected by people exclusively for own consumption like fruits, nuts, berries, honey, medicinal plants and grasses are included, then the percentage of MFP used for own consumption would be much higher. In real terms, MFP contributes 12% of the household income from all sources (Table1 shows the relative importance of the different activities inside the Park in terms of income and own consumption.). If people were denied the collection of MFP, it would not only dry up an important income source, but would also adversely affect an important food source used as a supplement in times of scarcity.

Wood from the Park is put to different uses by the villagers. For the villagers living inside, fuelwood from the Park constitutes the most important source of fuel. All the villagers use wood as fuelwood, 98% use wood for heating purposes. Most also use it as a crop protection measure. In addition, wood from the Park is used for making agricultural implements, for house construction and repair. For some it is also a source of income, although sale of fuelwood is illegal.

Livestock rearing is the second most common activity of the villagers after agriculture. 90% of the people living inside the Park own livestock. Livestock ownership is not only a status symbol but also represents security for the people as it can be sold in times of hardship. There is nearly complete reliance on park resources for livestock grazing. Of the sampled households owning livestock, 96% exclusively grazed their livestock in the park. Livestock is an important source of income for the households and constitutes a major source of income for the villagers in the park. Livestock rearing is the most important activity for the households in terms of the joint value of own consumption and sales (Table1). Thus, the Park as grazing area for the people's livestock is of crucial importance. Also for the people living in the 34 peripheral villages, the Park is the most important grazing ground.

There is a complete reliance of the villagers on the Park for subsistence. For income generation, working for the forest department (mostly as daily wage employees) constitutes the most important source of income for the people.[2] This is followed by sale of MFP and livestock rearing (Table 1). If, however, the value of own consumption is included, livestock rearing constitutes the most important income source, followed by sale of agricultural produce and sale of MFP.

Most income generating activities of the people take place within the confines of the Park. With only one fifth of the sampled households having family members going outside the Park for employment. Average total household income from all sources in 2000 was $169.81 of which 92% originated from within the Park. In summary, our results show that people living inside BNP rely heavily on Park resources.

Table 1. Contribution from different sources of income to average household income in the park in the year 2000 (US $)

Activity	Income contribution excluding own consumption	Percentage and Rank	Income contribution including own consumption*	Percentage and Rank
Sale of Agricultural Produce	2.0	1.1 (6)**	87.5	19.7 (2)
Sale of MFP	42.5	25.0 (2)	54.2	12.2 (3)
Sale of Fuelwood	1.6	0.9 (7)	7.0	1.5 (7)
Livestock Rearing	35.2	20.7 (3)	206.6	46.5 (1)
Work for the FD	44.1	26.0 (1)	44.1	10.0 (4)
Work Outside the Park	13.0	7.6 (5)	13.0	2.9 (6)
Other Alternative sources of Income	31.5	18.6 (4)	31.5	7.1 (5)
Total Income	169.8		443.9	

Source: Field Survey, 2000–2001.
* The value of own consumption is computed at market prices.
** Numbers in parentheses indicate the rank of different activities in income contribution of households

[2] 41 percent of the surveyed households had members employed by the FD, of which 6 percent were employed on a permanent basis, while 94 percent were daily wage earners.

4 Implications for PA Management

Activities of the people and their livestock within the confines of the Park has implications both for the wildlife and for Park management.

For the collection of the various MFP the villagers move throughout the Park. This causes disturbance to the wildlife. Moreover, consumption of MFP by the people or their livestock is in direct competition with the consumption of MFP by wild animals. Competition amongst the resident human population and the villages on the periphery of the Park has meant that prudent systems of gathering have given way to unsustainable practices and overexploitation. A case in point is the collection of *amla*. In the past, people used to wait for the fruit to ripen before collecting it. Now, as people get good prices and there are hardly any *amla* trees present outside the Park area, people no longer wait for the fruit to ripen but chop down the whole tree to collect the fruit in the process causing wastage and destruction of the trees. Over time this practice might lead to the disappearance of the tree from the Park.

As people are not allowed to cut standing trees, and dead and fallen twigs and branches are not always easily available near the village, people tend to lop branches and let them dry before collecting them for use as fuelwood. Cases of illicit felling are also frequently reported. This is highlighted by the fact that the majority of offences registered (57%) by park authorities during the period from 1995 to 2000 were for illicit felling. The zone of impact of fuelwood collection is extensive and of a high intensity as households from all the villages gather fuelwood regularly from the Park. Illicit felling constitutes a major management problem with the entire Park area affected by it.

Grazing by people's livestock in the Park is in direct competition with the ungulates for forage. As of now, 20% of the Park area is legally open for grazing by livestock from the inside villages as well as from the 34 villages lying on the periphery of the Park, but the area actually grazed is much greater (61%) as people tend to graze their livestock in closed compartments, and grazing by villages not officially allowed to graze is also done. As there are no fenced boundaries, illegal entry of livestock takes place from all around the Park and affects the entire Park area. It is difficult to regulate the number of cattle grazing inside the park as it is nearly impossible for the FD officials to distinguish cattle which is legally allowed to graze inside the Park as opposed to livestock which is not. A ground cover study of the impact of grazing on Park vegetation done by Nagar in 1992 shows the damaging effect of grazing on the designated areas. In the vicinity of four villages inside the Park, 48% of the area near the villages grazed regularly by the livestock were found to be devoid of ground cover as compared to 43% in the areas closed for grazing and 57% in the old park area (Nagar 1992). Over the years there has been a steady increase in the number of livestock in the inside villages and also in villages surrounding the Park. As a consequence the number of cases of illegal grazing has been rising, with more and more livestock grazing in compartments closed for grazing. The likely effects of the increase in number of livestock are overgrazing, decrease in forage availability for the ungulates, soil compaction, weed infestation, changes in floral composition and decimation of soil micro fauna. Moreover, the constant danger of transmission of disease from the livestock to the wildlife is ever present though no cases of serious epidemics have taken place. Lopping of trees for feeding the calves and goats by the villagers poses another problem. Sometimes people, instead of lopping the branches, cut down the whole tree and cause damage to the trees.

Fire also poses a great hazard for the Park. According to Pabla (1998) between 1984–1994, 77 percent of the Park area was subjected to fire with 49 percent burnt a minimum of two to eight times (Pabla 1998). Between 1996–2001, 142 incidences of fire took place in the Park. The potential consequences of large scale forest fires are destruction of habitat, loss of animal life and forage shortage for the ungulates (Nath 2000). When asked during the course of the fieldwork, people categorically stated that most fires were set intentionally, with fires caused due to natural factors a rarity. The most common motive for intentional fire setting is to permit easier collection of MFP especially of *mahua*. Intentional fire setting as revenge against the forest department (FD) cannot be ruled out either. This has been reported to happen when someone gets fined or is stopped from doing something by the FD officials.

Given the presence of people, the high dispersal of settlement, and the easy access to Park resources by the people despite park restrictions, it is difficult for the handful of FD staff in the field to monitor all of people's activities at all times. Moreover, evidence suggests that the peripheral villages contribute significantly to the damages to Park resources. A study conducted by Nath in the late 1990s in BNP of 34 villages (two inside villages and 32 surrounding villages), revealed that all villages without exception have zones of influence extending inside the Park boundary (Nath 2000), even though there is an explicit ban on the use of the Park by the peripheral villages. Of all the forest officials interviewed in the course of our study, only 2 percent stated that more damage is being caused by the people living inside the Park than by those outside the Park. Data pertaining to the number of cases of illicit activities in the Park between 1995–2000 reveals that people from the inside villages were responsible for 25% of the cases, whereas outside villagers were responsible for 37% of the cases.[3]

Apart from restricting the resources available for wildlife populations, human exploitation of the forest for timber, fuelwood and fodder may also degrade the habitat and lower the resource base considerably (Guha 1992: 307).

5 Need for Alternatives

The strong resource dependence of local people on protected area resources is not unique to BNP, but is representative for many other PAs in the country and is a reflection of the situation existing in many PAs around the globe. People living

[3] In one percent of the cases people other than local villagers were involved, while in 37 % of the cases the culprit could not be caught.

within or near forested tracts cannot be expected to easily reduce their resource dependence in view of the sustenance linkages which binds them to their present day lifestyle. The establishments of PAs and the related restrictions on people's behaviour requires sacrifices on the part of the local people who have been using these forests for a long time. Moreover, effective protection of the area from the local people is extremely difficult and costly, leading to ambiguous regulations and low enforcement of existing rules. It is therefore often thought that it is essential to provide the people with viable alternatives before restricting their behaviour. As Wells et al. state

'....it is often neither politically feasible nor ethically justifiable to exclude the poor who have limited access to resources from the parks and reserves without providing them with alternative means of livelihood' (Wells and Brandon 1992:2).

Current efforts, such as the eco-development program, which was initiated by the World Bank, represent one step in this direction. This program aims to reduce people's dependence on PA resources by providing them with better irrigation facilities to improve agricultural productivity, health and transportation facilities, better schooling, as well as alternative income-generating activities (e.g., mushroom cultivation, beekeeping, cottage industry). However, its focus on income generation within the villages might lead to undesired effects, e.g. the attraction of more people into the area or the expansion of agricultural activities. An empirical assessment of the outcomes of current eco-development initiatives would be desirable in this regard.

Moreover, the emphasis of the program is on people living in the periphery of PAs and not on villages within the PA boundary because according to policy stipulation all villages inside a National Park have to be relocated. Despite this policy requirement, eco-development committees have recently been formed in the villages inside BNP, representing the first exception to the rule in India. This appears to reflect the fact that FD officials are realizing the difficulties associated with relocation.

More importantly, it is not clear whether alternative income provision really does reduce people's use of Park resources, since our analysis did not show a significant correlation between the collection of MFPs and alternative sources of income. A more rigorous test through a structured econometric analysis of the hypothesis that alternative income provision reduces pressure on Park resources is needed. Such an analysis was, however, beyond the scope of this paper and is currently being developed as an extension.

Presence of alternatives potentially not only reduces the dependence of the people on the PA, but could also influence people's acceptance of the PA which is often thought to be crucial to ensure the future viability of the PA. In the case of BNP an econometric analysis was done using a logit regression to empirically test for factors which determine people's attitude towards the Park. The variables used in the analysis are outlined in Table 2, and the results are presented in Table 3. The analysis revealed that villagers having alternative sources of income including working for the forest department, working outside the Park and engaging in other alternative activities inside the Park were more likely to be in favor of the Park. Moreover, people who perceived crop damage as a problem due to the Park and those who were unaware of the reasons for the establishment of the Park were found more likely to be against the Park. Female respondents were also less likely to be in favour of the Park. This is intuitive, as they are the ones most involved in fuelwood collection, an activity which has been severely restricted since Park establishment. The results of our model are consistent with a study conducted by Dhaundiyal (1997) on the attitude of the people towards the forests in the forest corridor between Rajaji-Corbett NPs.[4]

Table 2. Variables used in the econometric analysis to measure people's attitude towards the Park

Variable	Description of variable	Expected Sign
ATTITUD	Attitude of people towards the Park (=0 if in favor of the Park, =1 if against the Park or neutral towards the Park)	
TVL	Total value of livestock	–
FEMALE	Whether respondent was female	–
AGE	Age of the respondent	–/+
PPCROPDM	=1 if crop damage as perceived problem due to Park, =0 otherwise	–
RV	=1 if revenue village, =0 if forest village or illegal settlement	–
PCLAND	Per capita landholding	+
NOKNRSPE	=1 if no knowledge of reason for Park establishment, =0 otherwise	–
PCWOPINC	Per capita income excluding income from minor forest produce and livestock	+

[4] The study concluded that households with members employed by the FD are more positive towards the forest, while negative attitudes can be attributed to a result of some recent loss incurred, e.g. crop damage and livestock or human losses due to wildlife (Dhaundiyal 1997).

Table 3. Results of the econometric model

Variable	Coefficient	Std. Error	Significance level
Constant	1.190	0.822	0.148
TVL	0.000	0.000	0.532
FEMALE	–1.145	0.706	0.105
AGE	–0.003	0.017	0.869
PPCROPDM	–1.236	0.061	0.042*
RV	–0.820	0.410	0.046*
PCLAND	0.315	0.267	0.238
NOKNRSPE	–1.738	0.465	0.000***
PCWOPINC	0.000	0.000	0.098

Dependent variable: ATTITUD, n=151.
Percentage of cases predicted correctly: 74%, Chi–squared = 37.46452

Thus, while our analysis supports the idea that access to alternative income sources improves people's attitudes towards the Park, the effect of these attitudes on actual resource use is less clear. Our preliminary analysis showed a positive, but insignificant correlation between agreement with Park establishment and total income from MFP collection.

Relocation of the people is another potential policy option, which is favored by the Forest Department and stipulated by the current conservation policy. However, this option faces serious problems including: i) lack of adequate settlement sites[5], ii) availability of funds for relocation, iii) people's unwillingness to relocate, and iv) lack of political will. But even more important than all these is the fact that the removal of people from inside a PA does not guarantee that the resource use by the masses of people living on the periphery of the PA, having equal if not greater dependence on the resources of the PA, will diminish. This seems highly unlikely. In fact, future increases in population and persistent rural poverty are likely to further enhance the degree of dependence of the peripheral villages on the PA. Forced relocation of people would also constitute a direct violation of an ILO convention to which India is a signatory.

In such a situation it is in the best interest of the PA for the PA managers and the people to join hands. Experience gained from the implementation of Project Tiger has shown that only a people-friendly approach can assure effective management of the PAs on a long term basis (Panwar 1992). This has come about with the realization that putting the needs of wildlife or people above each other will not serve any purpose and will ultimately lead to a worsening of the situation for both. A partnership between local people and the Forest Department would first entail a need for the PA managers to change their perception and move from their previous role of policing to working with the local people in the protection of the PA, rather than viewing people as a management problem.

[5] Although land for relocation exists on paper, in reality there is little land available for relocation of all the people from within the PAs.

Effective participation can only be ensured when the people get a share of the benefits and other measures. Channeling part of the tourism revenue for village development and involving local people in Park management activities are some of the solutions which can go a long way in ensuring peoples participation.

6 Conclusion

The case of BNP illustrates that PA formation—though justifiable in the face of the need for protecting what little biodiversity is left in the world—has often done so at the expense of rural people who are often poor and marginalized and depend heavily on the resources of the PA for meeting their daily needs. In the face of lack of viable alternatives they continue using the resources of the PA to the detriment of both the PA and their own long-run perspectives. On moral, ethical and economic grounds it is frequently not feasible to move the large number of people out of the PA nor is it possible to stop the resource use by the people living on the periphery of the PA who far outweigh the number of people living within the boundaries of the PA. In such a situation alternatives will have to be provided and a partnership between the forest department and the people has to be built to ensure the future viability of the PA.

Our analysis, however, suggests that the provision of alternative sources of income alone may not be sufficient to reduce pressure on Park resources. While access to such alternative income sources does appear to improve people's attitude towards the Park, both attitudes and alternative income provision do not seem to have a significant effect on resource extraction. This suggests that pressure on Park resources can only be reduced when the people get a share of the benefits from conservation. Channeling part of the tourism revenue for village development and involving local people in Park management activities are some of the solutions which can go a long way in this respect. There is no single universal solution for all the PAs. What is universal is the question regarding how serious the governments are towards protecting the remaining biodiversity in their respective countries and in involving the local people who depend on the resources of the PA for their survival in its protection.

The gap between the objective of PAs and the reality of local people's tremendous dependence on Park resources has led to ambiguous regulations and weak enforcement of existing rules. The outcome is a situation where both people and wildlife are suffering from the damages. Moreover, while many current programs seem to put significant emphasis on awareness building and changing people's attitude towards the Park, the actual role of attitudes and perceptions about resource scarcity on people's behaviour remains poorly understood. These aspects are the focus of ongoing research in the present research group.[6]

---

[6] Additional research on these issues is conducted within the research group "The determinants and effects of alternative institutions for natural resource management in developing countries" which is funded by the Robert-Bosch Foundation.

7 Acknowledgements

This paper was prepared for presentation at the STORMA (Stability of Rain Forest Margins) Conference in Bogor, Indonesia, September 29 – October 3, 2002. Funding for this research was provided by the German Academic Exchange Service (DAAD). The research also forms part of an ongoing research group at the Center for Development Research which is facilitated by the Robert-Bosch Foundation. Funding for participation in the conference was also provided by the Robert-Bosch Foundation.

References

Colchester M (1992) Sustaining the forests: the community based approach in South and South East Asia. Discussion paper No.35. UNRISD, Geneva

Dhaundiyal R (1997) Economic assessment of human-forest interrelationship in the forest corridor linking the Rajaji and Corbett National Parks. Unpublished Ph.D. thesis, University of Gwalior

Guha R (ed) (1992) Social Ecology. Oxford University Press, New Delhi

Kothari A et al (eds) (1998) Communities in Conservation: Natural resource management in South and Central Asia. Sage Publication, New Delhi

Kothari A et al (eds) (1997) Building bridges for conservation. IIPA, New Delhi

Kothari A et al (eds) (1996) People and protected areas: towards participatory conservation in India. Sage Publication, New Delhi

Kothari A et al (eds) (1989) Management of National Parks and Sanctuaries in India: a status report. IIPA, New Delhi

Nagar AK (1992) Livestock grazing problems and management options. Unpublished term paper. Wildlife Institute of India, Dehradun

Nath L (2000) Conservation and management of the Tiger, Panthera tigris tigris in, Bandhavgarh National Park, India. Unpublished Ph.D thesis, University of Oxford

Pabla HS (1998) Development of a user friendly wildlife monitoring methodology for protected areas in India. Unpublished Ph.D thesis, Wildlife Institute of India, Dehradun

Panwar HS (1992) Ecodevelopement: An integrated approach to sustainable development for People and PAs in India. Paper presented in the IV World Congress on National Parks and Protected Areas. Caracas Venezuela

Sonakia A (1993) Management Plan for Bandhavgarh National Park for the period of 1993–94 to 2002–03, Vol. 1 Text [Part I & II]

Stocking M, Perkin S, Brown K (1995) 'Coexisting with nature in a developing world' in Morse S, Stocking M (eds.) People and Environment, UCL

World Bank (2001) Final Proceedings, National workshop on community forestry issues, Manesar, 8–9 November 2001

Wells M, Brandon K (1992) People and Parks: linking Protected Area Management with Local communities. World Bank, Washington DC

Land-Use Change, Biodiversity and Ecosystem Functioning in West Kalimantan

Deborah Lawrence

1 Introduction

Shifting cultivation is blamed for 50% of all deforestation in Southeast Asia (FAO 1993). Clearly it has had, and will continue to have, a profound impact on the structure and functioning of rainforests. The most obvious changes that occur are a reduction in tree stature and an increase in the grain size of the landscape. These changes are a consequence of an increase in disturbance frequency (return interval shifts from 100s to 10s of years) and spatial scale of disturbance (patch size increases from 10s or 100s to 10,000s of square meters). Less obvious are the long-term shifts in species composition, species diversity, and forest structure that accompany this altered rainforest disturbance regime. Few have documented long-term responses to shifting cultivation, and fewer have taken the next step to determine the consequences of altering forest structure and organization for critical ecosystem functions such as nutrient cycling and maintenance of soil fertility. Understanding changes in biodiversity and ecosystem function within the context of a dynamic human response to economic and ecological drivers is more complex still. The focus of this paper is to address these issues, in answering two questions about rainforest conversion to shifting cultivation:

- How do centuries of long-fallow shifting cultivation affect tree diversity and soil nutrients?
- How do current patterns of land-use change affect tree diversity and soil nutrients?

2 Methods

To examine long-term effects of shifting cultivation, I controlled for fallow age while sampling natural forest fallows that varied in the number of prior swidden cycles. Details on the methods have already been published (Lawrence and Schlesinger 2001) and will only be discussed briefly here. The vegetation and soils of seven fallows, aged 9–12 years old, were sampled intensively and compared with three undisturbed, primary forest stands nearby. The seven fallows had experienced between one and ten prior cycles of shifting cultivation (slash-and-burn followed by one year of rice and then a 15–30 year fallow period). Site history was determined by interviews; the longest histories were constrained by the age of the village (>200 years) and the average fallow interval (20 years).

Trees > 10 cm dbh were sampled in four $750m^2$ plots per ~1 ha fallow or primary forest location (total 3000 m^2 /site). Trees > 5 cm dbh were sampled in three $100m^2$ quadrats per $750m^2$ plot (total of 1200 m^2 /site), and stems > 50 cm in height and < 5 cm dbh were sampled in three $5m^2$ quadrats per plot (total of 60 m^2 /site). I used regression, dominance-diversity curves, and non-metric multidimensional scaling (NMDS) to analyze changes in the tree community.

Soil cores 0–30 cm deep were systematically drawn from 10 locations per 100 m^2 quadrat, and composited by plot to yield four samples per site. A sequential chemical extraction (Tiessen and Moir 1993) was conducted to determine changes in the distribution and abundance of phosphorus as a function of cultivation history (Lawrence and Schlesinger 2001).

To determine the effects of current land-use change, I sampled the soils, litter and vegetation of the major land-use types in the village-dominated landscape: fallows, fruit gardens and rubber gardens. Trees > 10 cm dbh were sampled in one randomly placed $1000m^2$ plot per site (n=10–11 per type, 32 plots, Lawrence et al 1995, Lawrence and Mogea 1996). I examined species richness and evenness using dominance diversity curves, and compared tree density and basal area as indices of forest structure.

With a team of researchers from Universitas Tanjungpura, soils and standing litter were sampled at seven locations where a fruit garden, a productive rubber garden, and a fallow > 15 years old were adjacent, or < 150 m apart (21 plots, Lawrence et al 2002). Composite soil samples 0–10 cm deep were mixed from nine systematically arrayed cores per site. Soils were air-dried and later analyzed for total carbon and nitrogen (by dry combustion), total and available phosphorus (modified Kjehldahl digestion), and available cations (atomic absorption/emission spectrophotometry). At each of the nine soil sampling points, standing leaf litter on the forest floor (excluding wood > 1.8 cm in diameter) was collected in a 0.5 m^2 area (total 4.5 m^2 per plot). The composited sample was air-dried and weighed . One air-dried subsample was analyzed for the above nutrients, a second was oven-dried to determine moisture content. Analysis of variance (ANOVA) was used to determine differences among land-use types.

3 Results

3.1 Long-Term Effects of Shifting Cultivation

With conversion from primary forest to secondary forest after one cycle of shifting cultivation, tree species richness (> 10 cm dbh) declined by almost 80% (Fig. 1). During subsequent cycles, species loss occurred in the smaller trees (<10 cm dbh), with little change in the richness of large trees. Dominance increased and evenness declined with each cycle (Fig. 2), and a systematic shift in species composition was evident from NMDS analysis (Fig. 3).

In coarse textured (low fertility) soils, total phosphorus in the top 30 cm of soil increased during the first four cycles. In fine-textured (more fertile) soils, no trend was apparent; total phosphorus remained relatively constant for up to 10 cycles of shifting cultivation. For both soil types, shifts in P distribution and availability persisted over 10 cycles, or roughly 200 years. The most important changes were an increase in the proportion of stable, organically-bound P and occluded inorganic P (Lawrence and Schlesinger 2001).

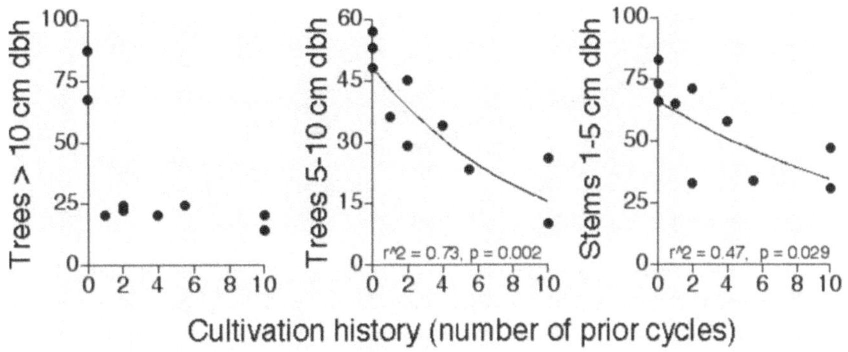

Fig. 1. Decline in tree species richness with repeated shifting cultivation. Richness as a function of the number of previous cultivation cycles for a) trees > 10 cm dbh, b) trees 5–10 cm dbh, and c) saplings > 50 cm in height but < 5 cm dbh. Trees > 10 cm dbh sampled over 3,000 m^2, trees 5–10 cm dbh sampled in 1,200 m^2, and saplings sampled in 60 m^2 per site

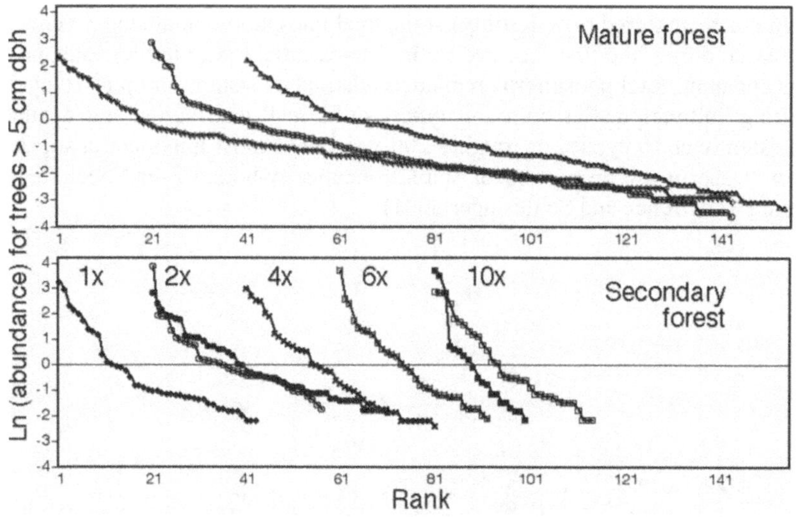

Fig. 2. Dominance diversity curves for trees > 5 cm dbh as a function of cultivation history in a) three mature forests, and b) seven secondary forest, aged 9–12 years old. Each string represents data from one site. The points indicate the abundance of individual tree species, ordered from highest (on the left) to lowest (on the right). Tree abundance is defined as percent basal area. Secondary forest sites are arranged in order of increasing numbers of prior cultivation cycles, from left to right.

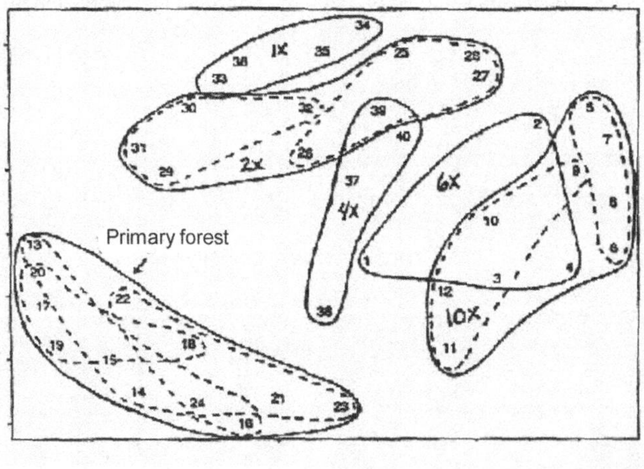

Fig. 3. Graphical representation of results from non-metric multi-dimensional scaling (NMDS) analysis of species composition for trees > 5 cm dbh in sites of varying cultivation history. NMDS performed on four 750m^2 plots per site; plots from a common site encircled with a line. Cultivation history indicated by a number followed by 'x', e.g. 6x, 4x, etc.

3.2 Effects of Current Land-Use Change

In West Kalimantan, the landscape under shifting cultivation is increasingly dominated by rubber gardens rather than naturally regenerating fallows. In 1992, 17% of the landscape was in rubber, and rubber holdings increased on average by ca. 8 % per year from 1991–1995 (Lawrence et al 1998). In 1995 alone, at least 65 % of all rice fields cleared that year were planted to rubber, resulting in the addition of ca. 75 ha of rubber in one year (Lawrence et al 1998). Extrapolating from 1995 to 2003, with a conservative 5% rate of increase per year, 32% of all secondary forests in this landscape should be in rubber. If the rate remained at earlier levels of 8%, 40% of the landscape would be in rubber. Under these scenarios, the proportion of rubber doubled in just over a decade.

The shift from naturally regenerating forest fallows to more intensively managed rubber gardens reduces plot-level species richness by 65%, from a mean of 22 tree species > 10 cm dbh per 1000m^2 in fallows to a mean of 6 in rubber gardens (Fig. 4, from Lawrence and Mogea, 1996). Plot-level species richness was similar between fallows and fruit gardens. Evenness was generally higher in fruit gardens, but the highest evenness was observed in fallows (see the flat shape of the curves in Figs. 5a and 5b).

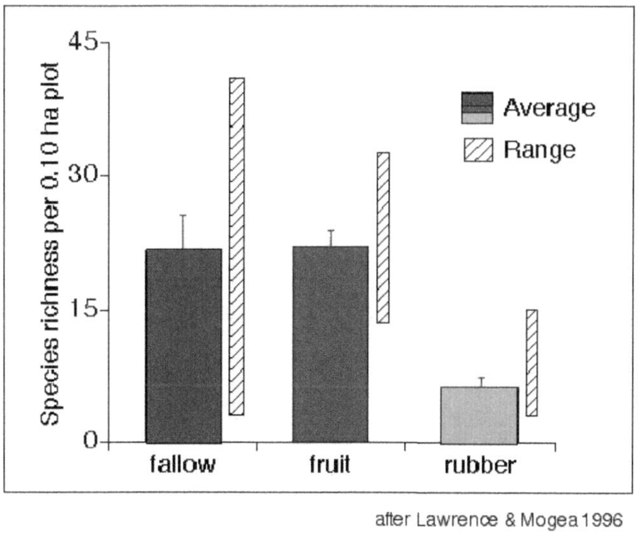

Fig. 4. Species richness as a function of land-use type (n=10–11). Mean, standard error and range of species richness per 1000 m^2 for all trees > 10 cm dbh. Modified from Lawrence and Mogea 1996.

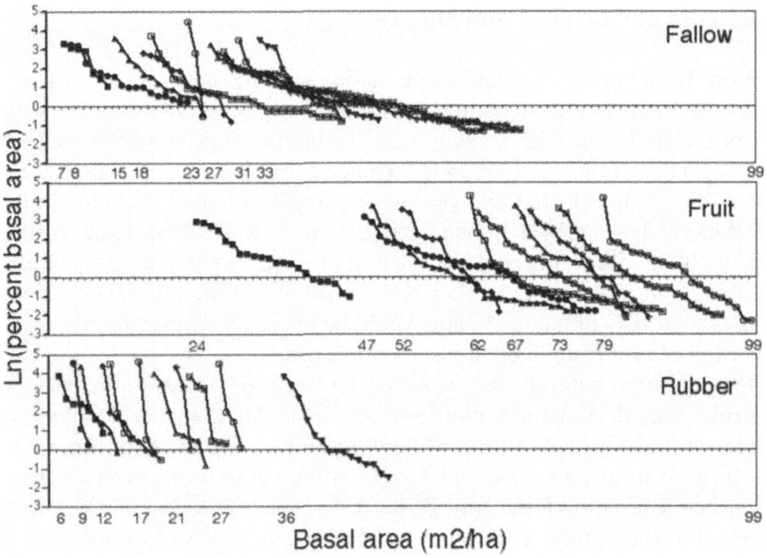

Fig. 5. Dominance diversity curves for trees > 10 cm as a function of land-use type. Sites (n=10–11) from a) naturally regenerating fallows, b) fruit gardens, and c) rubber gardens are arranged left to right in order of increasing stand basal area. See explanation of curves in Fig. 2.

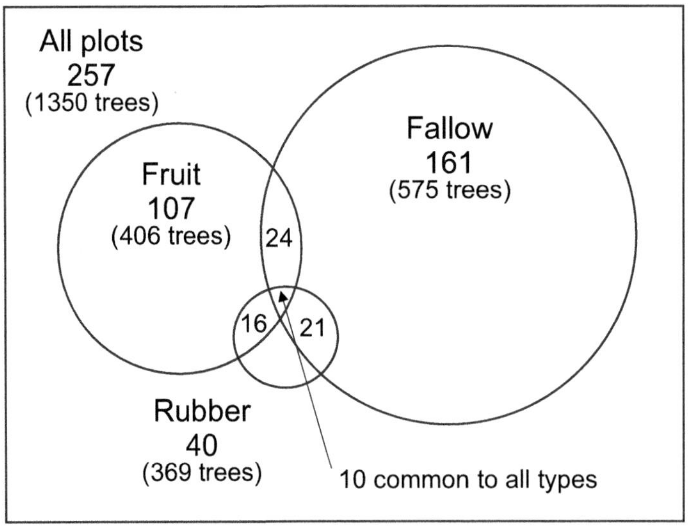

Fig. 6. Overlap in tree diversity among land-use types in the landscape under shifting cultivation. Data on 1,350 trees > 10 cm dbh collected in 32 1,000m^2 plots.

In rubber gardens, increasing dominance by a few species (rubber and several robust, weedy tree species) resulted in homogenization of the tree community at the landscape level as well as a reduction in species richness at the plot level (Fig. 5). Rubber gardens were host to only 40 species out of 257 encountered in 32 1,000m^2 plots across all land-use types (Fig. 6). Two-thirds of these species were found in fallows or fruit gardens as well. Fruit gardens were hotspots of diversity and contributed disproportionately to landscape-level richness. Of the total species pool sampled, 77 (30%) were found exclusively in fruit gardens, yet fruit gardens comprised only 4% of the entire landscape (Lawrence et al 1998). Half of the tree species sampled were found only in fallows (Fig. 6).

The structure of fallows and rubber gardens were quite similar. Basal area ranged from 6 or 7 m^2/ha to 33 or 36 m^2/ha (Fig. 7), in stark contrast to fruit gardens which had a median basal area of 62 m^2/ha. Although total biomass (as indicated by basal area) was probably similar between rubber gardens and fallows of a given age, the distribution of biomass differed between them. Fallows had ca. 50% more stems than rubber gardens, and ca. 25% more stems than fruit gardens (Lawrence et al 2002).

The soils of fruit gardens had significantly higher concentrations of Mg and Ca than fallows and rubber gardens, although there was no significant difference among the three in C, N, or P (Lawrence et al 2002). The Mg content of soils suggested that fruit gardens should have the highest concentration of Mg in their forest floor litter. However, fallow litter had comparable Mg concentration (Fig. 7). The relationship between fallows and fruit gardens in litter K concentration was much as expected, given the abundance of soil K in the two land-use types. Soil and litter [K] were similar, with fallows higher, but not significantly so. Litter [K] for rubber, however, was lower than expected in relation to fallows and fruit gardens, given soil K levels.

Total standing litter mass (with wood < 1.8 cm in diameter) varied little among land-use types, with a mean of 3.6–4.3 Mg/ha (Fig. 8). Litter mass was greatest in fruit gardens, but the only meaningful difference ($p=0.06$) was between fruit and rubber gardens. Litter moisture was significantly higher in fruit gardens than in fallows, with rubber gardens showing intermediate moisture levels.

4 Discussion

One critical conclusion from this research is that management for diversity and soil fertility may be independent in the landscape under shifting cultivation in West Kalimantan. With a 20-year long rotation, fallows showed no sign of decreasing phosphorus in the surface soils that support rice production, and yet tree species diversity was lost with every cycle (Fig. 1). Replacing naturally regenerating fallows with rubber does not currently affect soil fertility (Fig. 7), but greatly reduces both local and landscape tree diversity (Fig. 4).

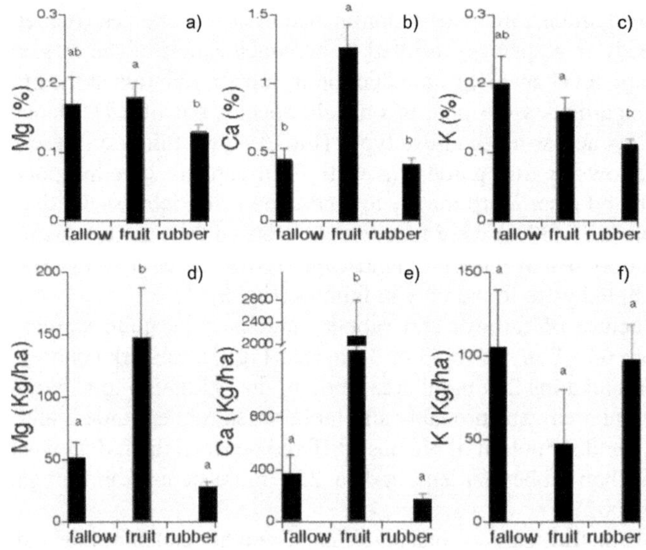

Fig. 7. Cation concentrations in litter (a–c) and total cation stocks in soils 0–30 cm deep (d–f) as a function of land-use type. Mean ± standard error from seven locations where a fallow, fruit garden, and rubber garden were within 150 m of one another. Adapted from Lawrence et al 2002.

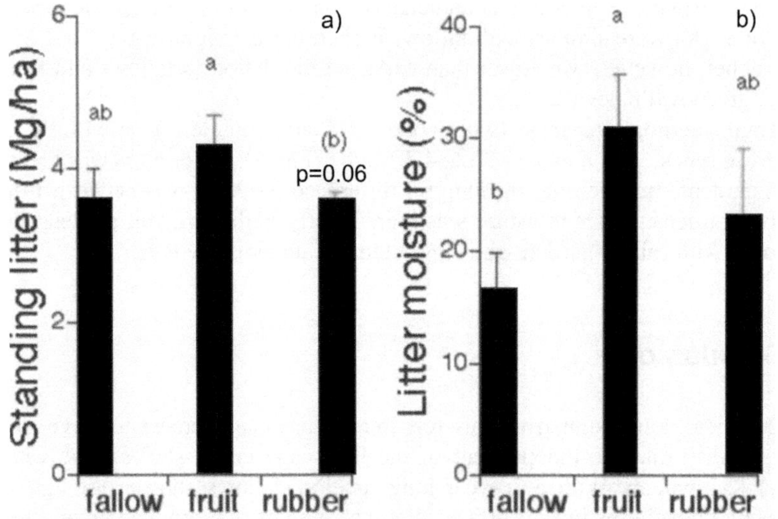

Fig. 8. Variation in the a) mass and b) moisture content of litter on the forest floor under different land-use types. Standing litter was sampled at nine points (0.5 m^2 per point) in each of seven stands per type, as in Fig. 7.

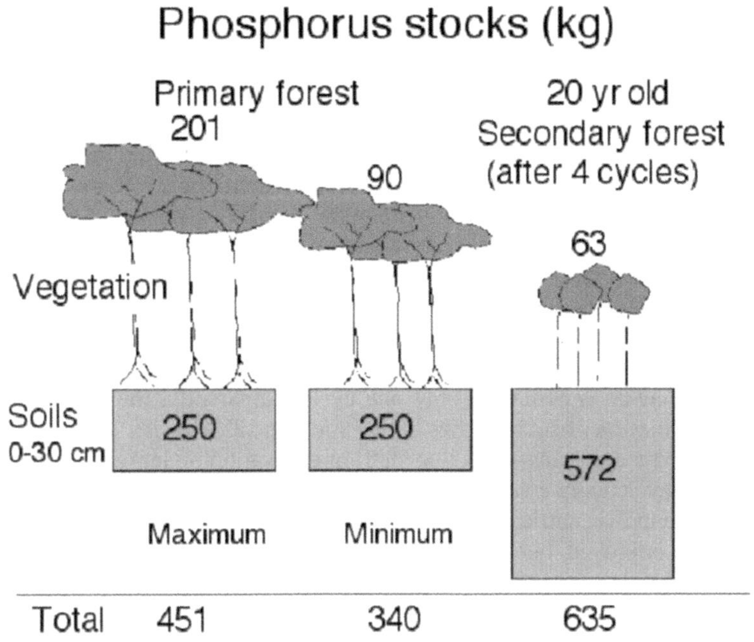

Fig. 9. Phosphorus budgets for primary and 20-year-old secondary forest, including stocks in aboveground biomass and soils 0–30 cm deep. Maximum and minimum P stocks in primary forest represent the quantity estimated in the largest (537 Mg/ha) and smallest (240 Mg/ha) primary forest stands sampled

4.1 Historic Rainforest Conversion

Over the first four cycles, in fact, the increase in total phosphorus of the top 30 cm suggests that the balance is favorable between losses due to harvest and erosion and gains during the fallow period. Lawrence and Schlesinger (2001) showed some evidence that this increase was a result of deep rooting and rapid cycling by secondary forest trees, as suggested by earlier researchers (Nye and Greenland 1960, Ewel 1986, Brown and Lugo 1990). This nutrient-pumping mechanism could account for between 57 and 92% of the phosphorus gained in the top 30 cm, depending on the amount of biomass present in the rainforest prior to the original clearing (maximum and minimum biomass estimated from inventories of mature forest near by, Fig. 9).

In calculating the contribution from nutrient pumping, I have ignored losses to harvest (average ca. 6 kg/ha/1 Mg rice crop) and erosion (unknown). Furthermore, I assume that all primary forest biomass P is transferred first to secondary forest biomass P. This leaves direct inputs to soil P stocks from biomass burning in the range of 27–138 kg/ha. Total soil stocks increased, however, by 322 kg/ha over four cycles. Subtracting biomass burning inputs (27–138) from the observed increase (322) suggests that fallow vegetation pumped between 184 and 295 kg/ha over four cycles of shifting cultivation. In other words, during 80 years of fallow (each period ca. 20 years), a net transfer of 2.3–3.7 kg/ha/yr occurred from below 30 cm to the top 30 cm of soil.

Lawrence and Schlesinger (2001) found that the pattern of primary rainforest clearing was not random, but rather that villagers first cleared the more fertile soils when they established the focal village more than 200 years ago. Only in the last 60–80 years have they brought into cultivation forests growing on sandier, less fertile soils. Therefore, to look beyond the first four cycles to a time period reflecting the entire history of the village (at least 10 cycles), accounting for the effect of original soil fertility was essential. Considering the impact of original fertility, total phosphorus in sites having experienced 10 cycles did not differ significantly from sites having experienced only one cycle. Apparently, the gains in the first four cycles may be offset by losses during subsequent cycles.

An alternative explanation is that shifting cultivation on marginal (sandy, infertile) land may actually enhance the nutrient stocks of those soils. Secondary forest trees may improve nutrient cycling in poor soils. On more fertile land, the soil may not be improved, but nor is it degraded by many cycles of long-fallow shifting cultivation (Lawrence and Schlesinger 2001). Clearly, the success of fallow vegetation in replenishing soil nutrients in West Kalimantan is dependent on a long fallow period. More study is necessary to determine the tolerance of the system to a fore-shortened fallow period. Changes during succession in the stocks of critical nutrients other than phosphorus suggest that a 13-year fallow may be sufficient to replenish soil stocks in West Kalimantan (Kleinman et al 1996).

Although shifting cultivation in West Kalimantan appears to be sustainable in terms of maintaining the potentially limiting nutrient, phosphorus, conservation of biodiversity is not guaranteed. Although biodiversity is high in fallows relative to other land-uses (Figs. 4–6), plot-level species richness is eroding through time (Figs. 1 and 2). A shift in species composition with each swidden-fallow cycle (Fig. 3) suggests that species loss may also be occurring at the landscape scale.

Further work is underway to determine whether local interactions (competition, seed-banking, stem- or root-sprouting) or constraints imposed by the landscape (local and long-distance seed dispersal) are responsible for the steady loss of diversity during 200 years of shifting cultivation. Experimental results from West Kalimantan and a review of literature on tropical seedling growth studies suggest that changing nutrient availability may play a role in the organization of the tree community assembling under the high light conditions of a recently abandoned, large upland rice field (Lawrence 2001, Lawrence 2003). However, distance to primary forest and the richness of neighboring patches also seem to influence the diversity and composition of species (Lawrence 1998). This dispersal constraint has been well documented in the regeneration of pastures in Central and South America (find a reference).

The loss of plant species is likely to be accompanied by the loss of associated species – vertebrates, invertebrates, and microorganisms (Perfecto and Snelling 1996). Thus biodiversity loss is likely to be compounded beyond that documented here, as are the potential ecological effects due to loss of pollinators, seed dispersers, seed predators, herbivores, pathogen-controlling organisms, and mutualistic bacteria and fungi (e.g. Naeem et al, 2000). Biodiversity loss that is apparently neutral (at least with regard to soil fertility in this case) may ultimately have significant effects on broader ecosystem functions, and on human welfare (Naeem et al, in preparation). Despite the maintenance of soil fertility, on long time scales (decades to centuries), other ecological constraints on low-input agriculture may begin to undermine the success of this system of shifting cultivation. As slow erosion of biodiversity under the historic system is replaced by rapid biodiversity loss with land-use change, some of these subtle effects may become more obvious, and change may become more rapid.

4.2 Implications of Current Land-Use Change

Fruit gardens may be considered an alternative fallow management strategy, with the fallow period being much, much longer (>50–100 years). The species are highly selected and managed moderately (sporadic weeding during a particular fruiting episode). In reality, however, fruit gardens tend to be taken out of the rice cultivation cycle, so they are more of an alternative land-use strategy than an alternative fallow management strategy. Basal area was much higher and stem density much lower than in a naturally regenerating, relatively younger fallow, because fruit gardens are allowed to grow (for decades, a century, or more) beyond the normal fallow period. Although similar to fallows in species richness and evenness (Figs. 4 and 5), fruit gardens are dominated by species 'useful' for subsistence, whereas fallows are dominated by less useful species (Lawrence et al 1995).

Rubber gardens are truly an alternative to natural fallows (Dove 1993, but see Lawrence et al 1998). However, rubber requires more intensive management (annual weeding, routine harvesting). The longer rotation period (ca. 40 vs. 20 years) was not reflected in higher mean basal area in this sample. This result may reflect the immaturity of the landscape-wide system of rubber cultivation in the early 1990s. Rubber was introduced in the late 50s and was not immediately adopted by all villagers, so few gardens over 35 years old existed. Lower tree density in rubber gardens is consistent with weeding and stand management. The loss of tree diversity, both in evenness and richness, is also directly related to management intensity and management goals: the production of a single, cash crop.

After several decades, soil organic matter, nitrogen and phosphorus stocks did not differ significantly under the three land-use management options currently pursued by shifting cultivators in West Kalimantan. Soil cation status did differ among land-use types, being highest in fruit gardens and lowest in rubber gardens, with fallows intermediate (Fig. 7). The chemistry of standing litter suggests that in the near future, fallow soils may have better Mg and K status than rubber gardens (Fig. 7), as soil stocks continue to respond to litter inputs.

Greater mass of standing litter on the forest floor of fruit gardens should provide more habitat and more substrate for many elements of the decomposer community. Higher moisture content of the litter in fruit gardens should speed microbial activity. Finally, the quality (cation concentrations) of standing litter in fruit gardens was greater than or equal to that of fallows, and significantly greater than that of rubber gardens (Fig. 7). Thus, decomposition is likely to be most rapid in fruit gardens, suggesting that the rate of nutrient cycling is enhanced under this type of management. Litter mass was similar between rubber gardens and fallows, and lower moisture content in fallow litter was offset by a tendency for higher litter quality. Decomposition rate is not likely to differ between the two.

4.3 Biodiversity and Ecosystem Function in a Changing Landscape

To test the effects of diversity on ecosystem functioning, I compared rubber gardens and fallows. They differ significantly in species richness (Fig. 4) and yet biomass, as inferred from basal area, is similar (Fig. 5). This comparison suggests only a modest benefit of diversity after ca. 20–40 years of alternative forest management. In West Kalimantan, trends in soil nutrients and litter quality suggest that with time, the more diverse natural fallows may prove more efficient at maintaining cations in the soil. At the time of this study, differences between rubber gardens and fallows were small. Ewel and co-workers (1991) also found minimal effects of tree diversity in very young experimental stands. I also compared fruit gardens and fallows to test for effects of forest structure while holding diversity constant. This comparison suggests a strong, positive influence of tree biomass on nutrient cycling (potential decomposition rates) and soil fertility (Mg and Ca stocks).

Biomass, rather than species diversity, may better explain the observed (statistically significant) differences among alternative land-use types in this rainforest landscape. Yet diversity may also play a role. With potential feedbacks between litter nutrient cycling and soil fertility in the future, higher diversity may result in a widening gap between fallows and rubber gardens. The identity of species may also be significant, however, and it may contribute to differences between fallows and fruit gardens as well (see Fig. 6).

If fallow diversity does contribute to maintaining soil fertility, plans for fallow management may bring together the goals of biodiversity conservation and agricultural sustainability at the rainforest margin. This synergism depends on an adequate fallow period for maintaining soil phosphorus (Lawrence and Schlesinger 2001) and requires addressing the slow loss of understory tree diversity (Fig. 1). Although I do not deny the benefits of associated diversity and linked ecosystem services discussed above, further study is necessary before attributing the benefits of higher soil fertility directly to higher diversity in fallows. This topic is well-studied in temperate grasslands (e.g. Tilman 2001, Loreau et al 2003), but is woefully understudied in systems that include long-lived woody plants, including the tropics (but see Ewel et al 1991).

If the fallow period is shortened due to land scarcity or population pressure, mono-specific, but longer-lived rubber gardens may prove to support more efficient nutrient cycles than fallows. Simply because the trees would be allowed to persist much longer (30–40 vs. e.g. 10 years), rubber gardens may retain nutrients better, despite any advantage conferred by the higher diversity of fallows. If fallow length declines, policies to preserve biodiversity and soil fertility would again be disconnected. Even if rubber gardens maintain soil fertility, they are not likely to provide a significant haven for non-rubber tree species or associated animal diversity (Lawrence 1996, Perfecto et al 1995, Perfecto and Snelling 1996).

5 Conclusions

5.1 Historical Rainforest Conversion

- The change from primary forest to secondary forest entails a major, immediate loss of tree diversity. It continues to decline modestly, but steadily, with each subsequent cycle of shifting cultivation.
- Despite a loss of diversity and a radical change in the physical structure of the forest, phosphorus stocks remained intact over 10 cycles (200 years) of shifting cultivation.
- Young, dense forests cycle P (and probably other limiting nutrients) well, but sustaining soil fertility without chemical inputs depends on maintaining a long fallow period.

5.2 Current Land-Use Alternatives

- Conversion to rubber, increasing rapidly in West Kalimantan, results in significant loss of biodiversity at both local and landscape scales.
- Currently, soil fertility is not diminished by a shift to almost mono-specific stands of rubber. However, both fallows and rubber gardens tend to have lower stocks of calcium and magnesium than fruit gardens (older stands with more than twice the biomass).
- Differences in litter nutrient cycling suggest the potential for diminished cation stocks in rubber, in the future. The key question is whether cations or phosphorus are more limiting to productivity in the soils of West Kalimantan, and whether low availability of one may feedback on the availability, uptake and use of the other.

5.3 Biodiversity and Ecosystem Function in a Changing Landscape

- Tree diversity shows modest, if any, effects on current soil fertility, but trends in litter nutrient cycling suggest that higher diversity may positively influence soil fertility in the future. More study is necessary before drawing firm conclusions.

References

Brown S, Lugo AE (1990) Tropical secondary forests. Journal of Tropical Ecology 6: 1–32.
Dove MR (1993) Smallholder rubber and swidden agriculture in Borneo: a sustainable adaptation to the ecology and economy of the tropical forest. Economic Botany 47: 136–147.
FAO (1993) Forest resources assessment 1990. Tropical countries. Food and Agriculture Organization, Rome
Ewel JJ (1986) Designing agricultural ecosystems for the humid tropics. Annual Review of Ecology and Systematics 17: 245–271
Ewel JJ, Mazzarino MJ, Berish CW (1991) Tropical soil fertility changes under monocultures and successional communities of different structure. Ecological Applications 1: 289–302
Kleinman, PJA, Bryant RB, Pimentel D (1996) Assessing ecological sustainability of slash-and-burn agriculture through soil fertility indicators. Agronomy Journal 88: 122–127
Lawrence DC, Leighton M, Peart DR (1995) Availability and extraction of forest products in managed and primary forest around a Dayak village in West Kalimantan, Indonesia. Conservation Biology 9: 76–88
Lawrence DC, Mogea JP (1996) A preliminary analysis of tree diversity in the landscape under shifting cultivation north of Gunung Palung National Park. Tropical Biodiversity 3: 297–319
Lawrence DC, Peart DR, Leighton M (1998) The impact of shifting cultivation on a rainforest landscape in West Kalimantan: spatial and temporal dynamics. Landscape Ecology 13: 135–148
Lawrence D, Schlesinger WH (2001) Changes in the distribution of soil phosphorus during 200 years of shifting cultivation. Ecology 82: 2769–2780
Lawrence D, Astiani D, Syazhaman-Karwur M, Fiorentino I (2002) Alternative fallow management under shifting cultivation: does tree diversity affect soil fertility? In: Cairns MF (ed) Voices from the Forest: Farmer Solutions Towards Improved Fallow Husbandry in Southeast Asia. International Center for Research in Agroforestry, Bogor
Loreau MS, Naeem, Inchausti P (2002) Biodiversity and Ecosystem Functioning: Synthesis and Perspectives. Oxford University Press, Oxford
Naeem S, Engelhardt K, Inchausti P, Lawrence D, Morin P, Perfecto I, Sala O, Wright J, Cerroni M, Costanza R, Jones C. (in preparation) Biodiversity, ecosystem functioning, and ecosystem health: the interface between ecology and policy.
Naeem S, Knops JMH, Tilman D, Howe KM, Kennedy T, Gale S (2000) Plant diversity increases resistance to invasion in the absence of covarying extrinsic factors. Oikos 91: 97–108
Nye PH, Greenland DJ (1960) The Soil Under Shifting Cultivation. Technical Communication No. 51, Commonwealth Bureau of Soils, Harpenden
Perfecto I, Rice R, Greenberg R, Van der Voolt M (1996) Shade coffee as a refuge of biodiversity. Bioscience 46: 598–608

Perfecto I, Snelling R (1995) Biodiversity and tropical ecosystem transformation: ant diversity in the coffee agroecosystem in Costa Rica. Ecological Applications 5: 1084–1097
Tiessen H, Moir JO (1993) Characterization of available P by sequential extraction. In Carter MR (ed) Soil Sampling and Methods of Analysis, Lewis Publishers, Boca Raton, pp 75–86
Tilman D, Reich PB, Knops JMH, Wedin D, Mielke T, Lehman C (2001) Diversity and productivity in a long-term grassland experiment. Science 294: 843–845

Tree Composition in Secondary Forest of Lore Lindu National Park, Central Sulawesi, Indonesia

Ramadhanil Pitopang, Paul J.A. Keßler, S. Robbert Gradstein, Edi Guhardja, Sri S. Tjitrosudirdjo & Harry Wiriadinata

1 Introduction

Indonesia is a country of megadiversity and ranks 5th on the list of the world's richest countries in terms of biological diversity. The rich diversity of Indonesia is evidenced by the rain forests of Lore Lindu National Park (229,177 ha), Central Sulawesi, which may contain about 150 species of trees (> 10 cm dbh) in one hectare at ca. 1000 m elevation (Kessler et al. submitted). More than 2000 species of woody plants have been recorded from Sulawesi in a recent census (Keßler et al. 2002).

The margins of Lore Lindu National Park are characterized by a mosaic of near-primary forests, degraded forests, fallows, forest gardens, and plantations with cocoa, coffee, maize, and paddy rice as the most important crops. Due to the expansion of agricultural activities in the surroundings of Lore Lindu National Park, pressures on the natural resources of the park are increasing especially in areas where villages are located adjacent to the park. These pressures include illegal extraction of wood and rattan, which have considerable impact on the park's natural forests (Achmad 1988). Shifting cultivation in the area around the park has resulted in the development of fallows of different ages. Small patches of secondary forest with a closed canopy usually represent these fallow sites.

In the framework of the interdisciplinary research project STORMA of the German Research Foundation we have undertaken an analysis of the tree diversity in fallows in the region of Lore Lindu National Park. The principal objective of the study was to determine a) the taxonomic composition and structural diversity of the woody flora in the fallows and (b) to compare tree, sapling and seedling diversity of the plots.

2 Material and Methods

The study was carried out near the villages of Alitupu, Wuasa, and Kaduaa in Napu Valley (Lore Utara District, Poso Regency) at an elevation of 900-1100 m. Five plots of 400 m^2 (20 x 20 m) were inventoried, three in young, less than 5 years old fallows and two in somewhat older, 5-10 years old fallows. All species of trees (>10 cm in diameter at breast height = dbh), saplings (4.5-9.9 cm dbh) and seedlings (less than 4.5 cm dbh) in the plots were recorded by species name, abundance (= number of individuals), dbh, and estimated height. Morphospecies were collected with up to 5 duplicates. Processing and identification of the plant specimens was done in the Herbarium Celebense (CEB), Universitas Tadulako, Palu, Indonesia and in Leiden (L). The collected specimens were deposited in several Indonesian herbaria (Herbarium Celebense, Palu; Herbarium Bogoriense, Bogor; Herbarium SEAMEO-BIOTROP, Bogor) with duplicates in the herbaria of the universities of Leiden (L) and Göttingen (GOET). Data recorded were analyzed according to Mueller-Dombois and Ellenberg (1974).

3 Results and Discussion

All species of trees, seedlings, and saplings recorded in the plots are listed in Appendix 1-3. Total number of tree species recorded was 46, saplings 37, and seedlings 32. Number of individuals and taxa per plot, dominant species, and Shannon Diversity Index values are shown in Table 1.

The highest number of trees in young fallows was recorded in the Kaduaa plot, containing 293 tree individuals per hectare belonging to 14 species (14 genera, 9 families). The predominant tree species were *Homalanthus populneus*, *Mallotus barbatus*, and *Macaranga hispida*. *Homalanthus* and *Mallotus* were exclusive to the young fallows, *Macaranga* was also found in the old fallows. In old fallows maximally up to 1350 tree individuals (Wuasa 1 plot) occurred, belonging to 18 different species (18 genera, 11 families). The predominant tree species in old fallows were *Ilex cymosa* and *Artocarpus teysmannii.*

According to Slik (1998), species of *Macaranga* and *Mallotus* (Euphorbiaceae) are the principal elements of secondary vegetation in the Malesian region. Many of them are the first colonizers of disturbed vegetation; others also occur in late secondary vegetation and in the understorey of primary forest.

The highest density of tree saplings was recorded in the plot Wuasa 1 (2500 saplings/ha). Tree saplings included 11 species (11 genera, 6 families), the most common ones being *Calophyllum soulattri, Mallotus barbatus,* and *Canarium hirsutum.* The number of sapling individuals and species recorded per ha was higher than recorded from Kalimantan by Riswan and Kartawinata (1988).

The predominant tree seedlings, finally, are shown in Table 1 (for complete species list see Appendix 3). Similar to saplings, highest density of seedlings was found in the Wuasa 1 plot, with 4200 seedlings/ha belonging to 7 species (6 families, 7 genera). Common species were *Elmerillia ovalis*, *Ficus* sp., *Calophyllum soulattri* (dominant species) *Semecarpus forstenii*, *Trema* sp., *Homalanthus populneus*, *Cananga odorata,* and *Duabanga moluccana*.

Table 1. Tree inventory in secondary forests at Napu Valley of Lore Lindu National Park, Central Sulawesi.

Forest type	Location	Plot size [m^2]	Plant individuals [ha^{-1}]	Families	Genera	Species	Dominant species	Shannon diversity index
Young secondary forest	Alitupu 1	40	275 t	6	8	8	*Homalanthus populneus*	2.25
			1350 s	10	14	14	*Mallotus barbatus*	2.07
			3600 se	9	12	12	*Elmerillia ovalis*	2.20
	Alitupu 2	40	285 t	8	13	13	*Homalanthus populneus*	2.34
			1400s	7	12	12	*Mallotus barbatus*	2.36
			4000 se	9	11	11	*Elmerillia ovalis*	2.40
	Kaduaa	40	293 t	9	14	14	*Homalanthus populneus*	2.00
			1375 s	8	15	15	*Homalanthus populneus*	2.45
			3800 se	10	13	13	*Arenga pinnata*	2.42
Old secondary forest	Wuasa 1	40	1350 t	11	18	18	*Ilex cymosa*	2.38
			2500 s	6	11	11	*Calophyllum soulatri*	2.30
			4200 se	6	7	7	*Calophyllum soulatri*	1.60
	Wuasa 2	40	900 t	17	21	22	*Artocarpus teysmannii*	2.92
			1500 s	9	10	11	*Canarium hirsutum*	2.30
			4100 se	6	7	7	*Semecarpus forstenii*	1.76

t = tree, s = sapling, se = seedling

It appeared that old secondary forest had about 1.5 times more species of trees and 3-4 times more individuals than young secondary forest. Number of individuals of saplings and seedlings was slightly higher in older fallows, but the total number of species was slightly higher in young fallows. Endemism in the fallows was generally low. Only three species, namely *Pigafetta elata* ("Wanga)", *Lithocarpus celebicus* ("Palili") and *Mitrephora celebica* (Annonaceae), are endemic to Sulawesi.

The most important families in both young and old secondary forest in terms of number of species were Euphorbiaceae (9 spp), Palmae (7 spp), Rubiaceae (3 spp), Urticaceae (3 spp), Annonaceae (3 spp), Anacardiaceae, Aquifoliaceae, and Burseraceae (2 spp each) (Table 2). In Kalimantan, Euphorbiaceae may also be the most important family in natural forest (Riswan and Kartawinata 1998). In the study area, most important families in natural forests at similar elevation to the fallow are Lauraceae, Meliaceae, and Euphorbiaceae. The first two mentioned families are absent in the fallows; they help to distinguish natural forest communities from fallows in the study area.

4 Keywords

Fallow, Lore Lindu National Park, secondary forest, species richness, Sulawesi, tree diversity

5 Acknowledgments

We gratefully acknowledge the support of the Rector of Tadulako University, Palu, Sulawesi and the Dean of the Faculty of Agriculture for allowing the first author to carry out this research. We thank Hardianto Mangopo, Fachruddin Lasadam, and Niswan (Herbarium Celebense) for field and laboratory assistance.

References

Achmad A (1998) Vegetation and land cover mapping based on landsat TM Image 1996 Lore Lindu National Park, Central Sulawesi, Indonesia. Final Report. ANZDEC and The Nature Conservancy, Indonesia

Keßler PJA, Bos M, Sierra Daza SEC, Kop A, Willemse LPM, Pitopang R, Gradstein SR (2002) Checklist of woody plants of Sulawesi, Indonesia. Blumea Suppl. 14: 1-160

Kessler M, Keßler PJA, Pitopang R, Bach K, Schmull M, Gradstein SR (submitted) Tree diversity in different land use systems in Central Sulawesi, Indonesia. Submitted to Biodiversity and Conservation

Mueller-Dombois D, Ellenberg H (1974) Aims and methods of vegetation ecology. Wildey, New York

Riswan S, Kartawinata K (1988) A lowland dipterocarp forest 35 years after pepper plantation in East Kalimantan, Indonesia. In: Soemadihardjo S (ed) Some ecological aspects of tropical forests of East Kalimantan. MAB-Indonesia, Indonesia Institute of Sciences (LIPI)

Slik JWF (1998) Key to the taxa of *Macaranga* and *Mallotus* (Euphorbiaceae) of East Kalimantan (Indonesia). Flora Malesiana Bulletin 12(4)

Appendix 1. Tree diversity (DBH > 10 cm) in secondary forest, Napu Valley, Lore Lindu National Park, Central Sulawesi, Indonesia. YSF = young secondary forest; OSF = old secondary forest; + = present; - = absent.

No	Vernacular name (Napu language)	Scientific name	Family	Young fallow			Old fallow	
				YSF 1	YSF 2	YSF 3	OSF 1	OSF 2
1	Andolia	*Cananga odorata* (Lam.) Hook.f. & Thomson	Annonaceae	+	+	+	-	+
2	Belante kuhe	*Mallotus barbatus* Müll.Arg.	Euphorbiaceae	+	+	+	-	-
3	Beranahe	*Acalypha caturus* Blume	Euphorbiaceae		+	+	-	-
4	Balowira	*Pipturus argenteus* (Forst.f.) Wedd.	Urticaceae	-	+	-	-	-
5	Aropi	*Baccaurea tetrandra* (Baill.) Müll.Arg.	Euphorbiaceae	-	-	-	+	+
6	Betau	*Calophyllum soulattri* Burman. f.	Clusiaceae	-	-	-	+	+
7	Bonoh	*Trema orientalis* (L.) Blume	Ulmaceae	+	+	+	-	+
8	Butohulaku	*Canarium hirsutum* Willd.	Burseraceae	-	-	-	+	-
9	Belante	*Homalanthus populneus* (Geiseler) Pax	Euphorbiaceae	+	+	+	-	-
10	Benturu	*Sterculia oblongata* R.Br.	Sterculiaceae	-	-	-	-	+
11	Bure-bure	*Glochidion sp.*	Euphorbiaceae	-	+	+	-	-
12	Dadap	*Erythrina subumbrans* (Hassk.) Merr.	Fabaceae	+	+	+	-	-
13	Dodonga	*Ficus sp.*	Moraceae	-	-	-	-	+
14	Harao mabada	*Areca vestiaria* Giseke	Arecaceae	-	-	-	-	+
15	Harao maeta	*Pinanga caesia* Blume	Arecaceae	-	-	-	+	-
16	Dopi	*Canarium moluccanum* Blume	Burseraceae	-	-	-	-	+
17	Jalata	*Dendrocnide stimulans* (L.f.) Chew	Urticaceae	+	+	+	-	-
18	Kapa	*Ilex cymosa* Blume	Aquifoliaceae	-	-	-	+	+
19	Katindi		Clusiaceae	-	-	-	+	+
20	Kao lira	*Mitrephora celebica* Scheff.	Annonaceae	-	-	-	+	-
21	Kao rano	*Vernonia arborea* Buch.-Ham.	Asteraceae	-	-	-	+	+
22	Kao werani	*Semecarpus forstenii* Blume	Anacardiaceae	-	-	-	-	+
23	Lekotu	*Duabanga moluccana* Blume	Sonneratiaceae	+	-	-	-	-
24	Lowa		Annonaceae	-	-	-	-	+
25	Lebanu	*Timonius sp.*	Rubiaceae	-	-	-	+	-

Appendix 1. (cont.)

No	Vernacular name (Napu language)	Scientific name	Family	Young fallow			Old fallow	
				YSF 1	YSF 2	YSF 3	OSF 1	OSF 2
26	Meapo	*Macaranga hispida* Müll.Arg.	Euphorbiaceae	-	+	+	-	
27	Mbalapapu	*Ficus sp.*	Moraceae	-				
28	Malouba	*Geunsia sp* Blume	Verbenaceae	-				
29	Nunu	*Ficus benjamina* L.	Moraceae	-	-			
30	Mapaleno	*Ilex sp.* Loes.	Aquifoliaceae	-	-	-	-	+
31	Pakabo	*Anthocephalus cadamba* (Roxb.) Miq.	Rubiaceae	+	-	-	-	-
32	Pondo	*Pleomele angustifolia* N.E.Br.	Liliaceae	-	-	-		
33	Tamba	*Buchanania sp.*	Anacardiaceae	-	-	-	+	
34	Tambone	*Glochidion rubrum* Blume	Euphorbiaceae	-	-			
35	Pepolo	*Bischofia javanica* Blume	Euphorbiaceae	+	-			
36	Saguer	*Arenga pinnata* (Wurm.) Merr.	Arecaceae	-	+	+	-	+
37	Siolangi	*Turpinia sphaerocarpa* Hassk.	Staphyleaceae	-	-	+		
38	Walantandu	*Gnetum cuspidatum* Blume	Gnetaceae	-	-			
39	Wone			-	-	-	+	-
40	Wokeh	*Grewia glabra* Blume	Tiliaceae	-	+			
41	Wanga	*Pigafetta elata* (Mart.) H. Wendl.	Arecaceae	-	-	+	-	-
42	Teauru	*Artocarpus teysmannii* Miq.	Moraceae	-	-			
43		*Villebrunea rubescens* (Blume) Miq.	Urticaceae	-	+	+	-	-
44	Uru	*Elmerillia ovalis* (Miq.) Dandy	Magnoliaceae	+	-	-	-	+
45	Poki wana	*Solanum verbascifolium* L.	Solanaceae	-	+			
46		*Antidesma stipulare* Blume	Euphorbiaceae	-	-	-		

Appendix 2. Tree sapling diversity (DBH > 4.5-9.9 cm) in secondary forest, Napu Valley, Lore Lindu National Park, Central Sulawesi. YSF = young secondary forest; OSF = old secondary forest; + = present; - = absent.

No	Vernacular name (Napu language)	Scientific name	Family	Young fallow			Old fallow	
				YSF 1	YSF 2	YSF 3	OSF 1	OSF 2
1		*Arthrophyllum diversifolium* Blume	Araliaceae	+	-	-	-	-
2	Andolia	*Cananga odorata* (Lam.) Hook.f. & Thomson	Annonaceae	+	-	-	-	
3	Bakangkara	*Litsea sp.*	Lauraceae	+	-	-	-	-
4	Belante kuhe	*Mallotus barbatus* Müll.Arg.	Euphorbiaceae	+	+	+	-	-
5	Beranahe	*Acalypha caturus* Blume	Euphorbiaceae	+	+	+	-	-
6	Balowira	*Pipturus argenteus* (Forst.f.) Wedd.	Urticaceae	-	+	-	-	-
7	Aropi	*Baccaurea tetrandra* (Baill.) Müll.Arg.	Euphorbiaceae	-	-	-	+	+
8	Betau	*Calophyllum soulattri* Burman. f.	Clusiaceae	-	-	-	+	
9	Bonoh	*Trema orientalis* (L.) Blume	Ulmaceae	+	+	+	-	
10	Butohulaku	*Canarium hirsutum* Willd.	Burseraceae	-	-	-		+
11	Belante	*Homalanthus populneus* (Geiseler) Pax	Euphorbiaceae	+	+	+	-	-
12	Bure-bure	*Glochidion sp.*	Euphorbiaceae	+	-	-	-	-
13	Jalata	*Dendrocnide stimulans* (L.f.) Chew	Urticaceae	+	+	+	-	-
14	Jambu bol	*Syzygium malaccensis* (L.) Merr. & L.M. Perry	Myrtaceae	-	-	-	-	+
15	Kopi	*Coffea sp.*	Rubiaceae	-	+	+	-	-
16	Kao werani	*Semecarpus forstenii* L.	Anacardiaceae	-	-	-	-	+
17	Lekotu	*Duabanga moluccana* Blume	Sonneratiaceae	+	-	-	-	-
18	Lebanu	*Timonius sp.*	Rubiaceae	-	-	-	+	-
19		*Harpullia sp.*	Sapindaceae	-	-	-	-	+
20	Meapo	*Macaranga hispida* Müll.Arg.	Euphorbiaceae	-	+	+	-	+
21	Mbalapapu	*Ficus sp.*	Moraceae	-	-	-	+	-
22	Mpire	*Caryota mitis* Lour.	Arecaceae	-	-	-	+	-
23	Pakabo	*Anthocephalus cadamba* (Roxb.) Miq.	Rubiaceae	+	-	-	-	-
24	Pondo	*Pleomele angustifolia* N.E.Br.	Liliaceae	-	-	-	-	+
25	Pepolo	*Bischofia javanica* Blume	Euphorbiaceae	+	-	-	-	-

Appendix 2. (cont.)

No	Vernacular name (Napu language)	Scientific name	Family	Young fallow			Old fallow	
				YSF 1	YSF 2	YSF 3	OSF 1	OSF 2
26	Rotan batang	*Calamus zollingerii* Becc.	Arecaceae	-	-	-	+	-
27	Saguer	*Arenga pinnata* (Wurm.) Merr.	Arecaceae	-	+	+	-	+
28	Teauru	*Artocarpus teysmannii* Miq.	Moraceae	-	-	-	+	+
29	Tambekako	*Syzygium sp.*	Myrtaceae	-	-	-	-	+
30	Take	*Arenga undulatifolia* Becc.	Arecaceae	-	-	-	-	+
31	Walantandu	*Gnetum cuspidatum* Blume	Gnetaceae	-	-	-	+	-
32	Wokeh	*Grewia glabra* Blume	Tiliaceae	+	+	+	-	-
33	Wanga	*Pigafetta elata* Becc.	Arecaceae	-	-	+	-	-
34		*Villebrunea rubescens* Blume	Urticaceae	+	+	+	-	-
35	Uru	*Elmerillia ovalis* (Miq.) Dandy	Magnoliaceae	+	+	+	-	-
36	Poki wana	*Solanum verbascifolium* L.	Solanaceae	-	+	+	-	-
37		*Antidesma stipulare* Blume	Euphorbiaceae	-	-	-	+	-

Appendix 3. Tree seedling diversity (DBH <4.5 cm) in secondary forest, Napu Valley, Lore Lindu National Park, Central Sulawesi. YSF = young secondary forest; OSF = old secondary forest; + = present; - = absent.

No	Vernacular name (Napu lang.)	Scientific name	Family	Young fallow			Old fallow	
				YSF 1	YSF 2	YSF 3	OSF 1	OSF 2
1		*Arthrophyllum diversifolium* Blume	Araliaceae	+	-	-	-	-
2	Andolia	*Cananga odorata* (Lam.) Hook.f. & Thomson	Annonaceae	+	+	+	-	+
3	Bangkaraha	*Cryptocarya sp.*	Lauraceae	-	-	-	-	+
4	Beranahe	*Acalypha caturus* Blume	Euphorbiaceae		+	+	-	-
5	Balowira	*Pipturus argenteus* (Forst.f.) Wedd.	Urticaceae	-	+	-	-	-
6	Aropi	*Baccaurea tetrandra* (Baill.) Müll.Arg.	Euphorbiaceae	-	-	-	+	+
7	Betau	*Calophyllum soulattri* Burman. f.	Clusiaceae	-	-	-	+	+
8	Bonoh	*Trema orientalis* (L.) Blume	Ulmaceae	+	+	+	-	+
9	Belante	*Homalanthus populneus* (Geiseler) Pax s	Euphorbiaceae	+	+	+	-	-
10	Jalata	*Dendrocnide stimulans* (L.f.) Chew	Urticaceae	+	+	+	-	-
11		*Siphonodon celastrineus* Griff.	Celastraceae	-	+	-	-	-
12		*Capparis pubiflora* DC.	Capparaceae	-	-	+	-	-
13		*Rubus moluccanus* L.	Rosaceae	-	-	+	-	-
14	Katindi		Clusiaceae	-	-	-	+	+
15	Kopi	*Coffea sp.*	Rubiaceae	-	+	+	-	-
16	Kao werani	*Semecarpus forstenii* L.	Anacardiaceae	-	-	-	-	+
17	Lekotu	*Duabanga moluccana* Blume	Sonneratiaceae	+	-	-	-	-
18	Meapo	*Macaranga hispida* Müll.Arg.	Euphorbiaceae	+	+	+	-	+
19	Mpire	*Caryota mitis* Lour.	Arecaceae	-	-	-	+	-
20	Mbalapapu	*Ficus sp.*	Moraceae	-	-	-	+	+
21	Nunu	*Ficus benjamina* L.	Moraceae	-	-	-	-	+
22	Palili	*Lithocarpus celebicus* (Miq.) Rehder	Fagaceae	-	-	-	-	+
23	Pakabo	*Anthocephalus cadamba* (Roxb.) Miq.	Rubiaceae	+	-	-	-	-
24	Tamba	*Buchanania sp.*	Anacardiaceae	-	-	-	+	-
25	Tambekakao	*Syzygium sp.*	Myrtaceae	-	-	-	-	+

Appendix 3. (cont.)

No	Vernacular name (Napu lang.)	Scientific name	Family	Young fallow			Old fallow	
				YSF 1	YSF 2	YSF 3	OSF 1	OSF 2
26	Tambone	*Glochidion rubrum* Blume	Euphorbiaceae	-	-	-	+	+
27	Saguer	*Arenga pinnata* (Wurm.) Merr.	Arecaceae	+	-	-	-	-
28	Siolangi	*Turpinia sphaerocarpa* Hassk.	Staphyleaceae	-	-	-	+	-
29	Wokeh	*Grewia glabra* Blume	Tiliaceae	-	+	+	-	-
30	Teauru	*Artocarpus teysmannii* Miq.	Moraceae	-	-	-	+	+
31		*Villebrunea rubescens* Blume	Urticaceae	-	+	+	-	-
32	Uru	*Elmerillia ovalis* (Miq.) Dandy	Magnoliaceae	+	-	-	-	+

Effects of Land Use on Butterfly Communities at the Rain Forest Margin: A Case Study from Central Sulawesi

Christian H. Schulze, Ingolf Steffan-Dewenter & Teja Tscharntke

1 Introduction

Many studies focus on the impact of forest disturbance on tropical diversity (e.g. Chung and Maryati 1996; Dahaban et al. 1996; Eggleton et al. 1997; Hamer et al. 1997; Intachat et al. 1997, 1999a; Johns 1992; Lambert 1992; Lawton et al. 1998; Liow et al. 2001; Watt et al. 1997; Willott 1999; Willott et al. 2000). In contrast, agricultural ecosystems and forest plantations have been taken into account only exceptionally (e.g. Chey et al. 1997; Davis et al. 2000; Holloway 1998; Holloway et al. 1992; Intachat et al. 1999b). Published studies of agroforests do not provide comparative data from pristine forests to evaluate the relative importance of agroforestry for maintaining tropical biodiversity (e.g. Klein et al. 2002a, 2002b; Perfecto and Snelling 1995; Perfecto and Vandermeer 1996). Secondary forests and certain land-use systems may be important in preserving at least a fraction of the tropical biodiversity (Hughes et al. 2002; Lugo 1992) of which insects are a major part (Hammond 1992; Stork 1988).

For our study butterflies have been chosen as target organisms because they frequently serve as flagship taxa in biodiversity studies (e.g. Bowman et al. 1990; Lawton et al. 1998) and because of their comparatively well-known biology. This allows a better interpretation of diversity patterns related to human land conversion. We quantified the change of butterfly species richness along a land-use gradient from little disturbed natural forest sites to old and young secondary forest fragments, and, lastly, agroforestry systems and annual cultures.

The fieldwork was conducted in Sulawesi, a region well known for its high extent of endemism (Whitten et al. 1988), which also counts for butterflies (de Jong 1990). Of the 546 butterfly species recorded within the biogeographical region of Sulawesi 43% are stated to be endemic (Holloway et al. 2001).

In this study we address the following questions:

1.) *How is butterfly diversity related to anthropogenic land conversion?* Old secondary forests are known to maintain a high biodiversity (e.g. Beck et al. 2002; Brühl 2001; Intachat et al. 1999a; Schulze 2000), but studies on agroforestry systems, represented in this study by cacao plantations shaded by *Gliricidia sepium* trees, are still rare (Klein et al. 2002b; Perfecto and Snelling 1995; Perfecto and Vandermeer 1996). Studies from the Neotropics showed that agroforestry systems might support a relatively high portion of local biodiversity (e.g. Hughes et al. 2002). Most of the South-East Asian butterflies can be categorized as forest species (e.g. Corbet and Pendlebury 1992; Schulze and Fiedler 1998), so open cultivated habitats, in this study represented by an annual culture, should be of minor importance.

2.) *Are endemic butterflies more affected by land-use activities than widespread species?* Endemic species seem to be more sensitive to human disturbance than widespread species (e.g. Hamer et al. 1997; Hill et al. 1995; Spitzer et al. 1993). Therefore, the number of endemic species should decline more dramatically with increasing land-use intensity than non-endemics.

3.) *Are monophagous butterflies more heavily affected by land use than polyphagous species?* Species more specialized in respect to their food plant requirements respond in a different way to habitat disturbance caused by fragmentation than polyphagous species (Lawton 1995; Steffan-Dewenter and Tscharnkte 2000). Due to the significant change in vegetation composition, food plant specialists should be also more strongly affected by habitat modification.

4.) *Which habitat parameters predict butterfly diversity best?* The diversity of undergrowth vegetation is often related to moth diversity (Beck et al. 2002; Schulze 2000). Does this also hold for butterflies or are they more affected by forest structure determining the light regime of lower vegetation layers (see Schulze et al. 2001)?

2 Study Area and Site Selection

The study area is located in Central Sulawesi (Indonesia) at the eastern margin of the Lore Lindu National Park (area: 229,000ha). The margin of the National Park is characterised by a mosaic of secondary forests, young fallows and land-use systems with cacao, coffee, maize and paddy as the dominating crops. Our study sites were in the northern part of Napu Valley in the vicinity of the villages Kaduwaa, Wuasa, Watumaeta and Alitupu (1100 and 1200m a.s.l., within the lower montane forest zone, Whitten et al. 1988). Four replicate sites have been sampled for each of the five selected habitat types (see Fig. 1), natural forest (NF1–4), old (OSF1–4) and young secondary forest (YSF1–4), agroforestry system (AF1–4) and annual culture (AC1–4). The term "natural forest" was used instead of "primary forest" because human disturbance, in particular illegal selective logging and rattan collecting, already affected the forest at the interior margin of the park. Natural forest sites were characterized by numerous medium-sized trees forming a multi-layered canopy with a height of 25–30m and single big emergent trees of up to 40m. Fragments of old and young secondary forest had a size of 1–3 ha. The upper canopy of old secondary forests reached a height of 20–30m, single emergent trees were up to 35m high. Young secondary forest fragments were ca. 5–6years old with an upper canopy layer 7–8m high. Many large cut tree stems remaining in the plots indicate the recent logging activities. The studied argoforestry systems were represented by ca. 5 yrs old cacao plantation with *Gliricidia sepium* (Leguminosae) as emergent trees. Maize fields have been chosen as annual cultures.

Fig. 1. Studied habitat types in Napu Valley. a: natural forest, b: old secondary forest fragment, c: young secondary forest fragment, d: agroforestry system (cacao plantation shaded by *Gliricidia sepium* trees), e: annual culture (maize field).

3 Methods

3.1 Butterflies

The transect count method proved to be an adequate tool to study butterfly communities in temperate zone habitats (e.g. Erhardt 1985; Pollard and Yates 1993;

Steffan-Dewenter and Tscharntke 2000; Weibull et al. 2000), but also in tropical forests (Schulze and Fiedler 1998; Walpole and Sheldon 1999; Willott et al. 2000). Transect walks have been conducted between May 2001 and January 2002 and were evenly distributed over the whole time period. The sampled transects had a length of 100m and each butterfly count lasted 15 min. Transects have been surveyed just between 9:30 and 15:30 on days without rain and a cloud cover of less than ca. 50%. The sites were visited in a random order. At least 9 transect walks have been conducted per site (but only 7 transects at the old secondary forest OSF3, which was clear-cut before the study could be finished). In case stable estimates of species richness (see next paragraph) could not be reached after 9 transect walks, additional transect surveys have been conducted. At the forest sites 1–2m wide paths have been used, which were cut in advance. All specimens observed in a corridor of 3 m to each side of the transects have been noted. Just unfamiliar species, which could not be identified in the field (like most lycaenids and skippers), were caught and identified later on.

First-order jackknife method described by Colwell and Coddington (1994) was used to estimate the "true" species richness for all sites. This predictor already performed well in other studies (Boulinier et al. 1998; Hughes et al. 2002).

Species have been categorized as endemic, when they are restricted to the Sulawesi region (a biogeographical subregion of Wallacea, see Coates et al. 1997). Because the exact distribution of Lycaenidae and Hesperiidae is still not known for all species, analyses of endemic species took into account just the families Papilionidae, Pieridae and Nymphalidae (following e.g. Spitzer et al. 1993). Information on the geographical distribution was provided by Vane-Wright and de Jong (in prep.).

3.2 Food Plants

Information on larval food plants was extracted from Vane-Wright and de Jong (in prep.). Species for which no information was available have been excluded from the analysis. Butterflies have been defined as monophagous when they are restricted to one plant family. All other species have been categorized as polyphagous.

3.3 Habitat Parameters

(1) Understorey plants were collected in ten 1m^2 squares along each transect. The squares, one every 10m, were situated in a distance of ca. 5m beside the transect line alternating at both sides of the transect. All vascular plants less than 1.30m high were sorted to morphospecies (Beattie and Oliver 1994). As measurement for plant diversity, Fisher's alpha was calculated for the total sample by the computer program provided by Colwell (2000).

(2) The percentage of shade at ground level was estimated for all ten 1m^2 plots per site between 11:30 and 12:30.

(3) The number of trees higher than 1.5m was counted within four 5x5m^2 plots situated beside each transect, one plot every 20m changing between both sides of the transect. Tree density is the mean tree number per 25m^2 (mean of four plots).

(4) The vertical structure of tropical rainforests can be described as number of distinct, though intergrading, vegetation layers (e.g. Whitmore 1984; Pomeroy and Service 1986) gradually modulating certain biotic (e.g. Parker 1995; Terborgh 1985) and abiotic parameters (e.g. Kiltie 1993; Parker 1995; Smith et al. 1992; Yoda 1992). The number of vegetation layers (including herb, understorey and all tree layers) was estimated at every study site for a corridor of ca. 30m beside the transect as indicator for vertical habitat heterogeneity. In case the number of emergent trees was less than 5 along the 100m transect, it was defined that they do not form an own vegetation layer.

(5) Percent canopy cover was estimated at each site for a corridor of ca. 30m beside the transect line.

3.4 Statistics

The program Statistica 5.5 (StatSoft 1999) was used to perform all statistical calculations. Arithmetic means + one standard deviation are given, unless mentioned otherwise. Spearman´s rank correlation was used if data were not normally distributed, otherwise Pearson correlation was calculated to analyse relationships between two variables (Sachs 1992). The computed ANOVA was always of a one-way type. Tukey´s Honest Significant Difference-Test was used for multiple comparisons of means. Estimated numbers of species were calculated with the computer program 'EstimateS Version 6.0b1' (Colwell 2000) by randomising the ranking of the individual transect counts 50 times (see Colwell and Coddington 1994).

4 Results

4.1 Butterfly Abundance and Species Richness

The number of butterfly specimens observed per transect count differed significantly between habitat types (ANOVA: $F_{4,15}$=8.06, p<0.01). The mean abundance of butterflies per transect walk was highest at natural forest and old secondary forest sites, lowest at young secondary forest sites and annual cultures, and intermediate at agroforestry systems (Fig. 2a).

During the total of 212 transect walks 82 butterfly species were recorded. The species totals estimated by Jack1 for single sites differed significantly between habitat types (ANOVA: $F_{4,15}$=7.74, p<0.01). Butterfly species richness steadily decreased from both old-grown forest types (NF and OSF) to younger secondary forests, agroforestry systems and maize field sites (Fig. 2b).

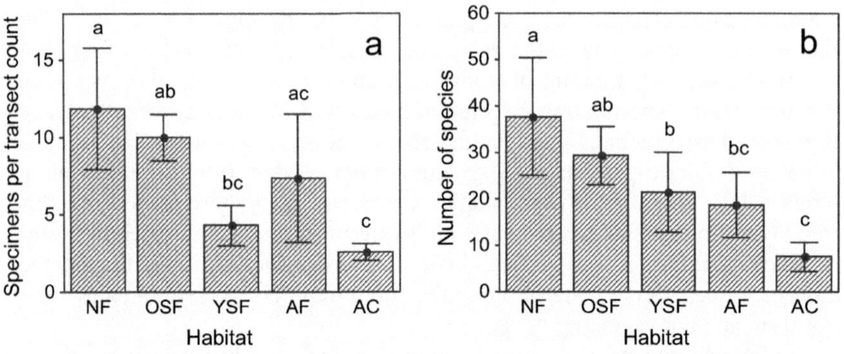

Fig. 2. Mean number butterfly specimens counted per transect walk (a) and mean species numbers estimated by first-order jackknife method (b) for the five sampled habitat types. Different letters indicate significant differences. The ranking of the habitats on the x-axis is related to habitat heterogeneity, which is highest in the old-grown forests (NF, OSF) and lowest in the maize fields (AC). NF=natural forest, OSF=old secondary forest, YSF=young secondary forest, AF=agroforestry system, AC=annual culture.

4.2 Endemic vs. Non-Endemic Species

The estimated species richness (Jack1) of endemics as well as of non-endemics was significantly related to habitat type (ANOVA; non-endemic species: $F_{4,15}$=3.12, p<0.05; endemic species: $F_{4,15}$=7.41, p<0.005). However, the effect on endemic butterflies was much stronger than on non-endemics. The estimated diversity of non-endemic butterflies did not differ significantly between the three forest types (NF, OSF, YSF) and the agroforestry system, but the annual culture had a much reduced diversity (Fig. 3a). In contrast, species richness of endemics of the natural forest sites differed significantly from both types of disturbed forest (OSF, YSF) and the land-use systems (AF, AC). Furthermore, a continuous decline of the mean estimated species total was found with increasing habitat homogeneity from old to young secondary forests, the agroforestry system and the annual culture (Fig. 3b).

4.3 Monophagous vs. Polyphagous Species

Both, the number of monophagous as well as polyphagous species, were significantly affected by habitat type (ANOVA; monophagous spp.: $F_{4,15}$=6.18, p<0.01; polyphagous spp.: $F_{4,15}$=4.65, p<0.05). However, differences of species richness between all forest types (including the agroforestry system) did not reach a sig-

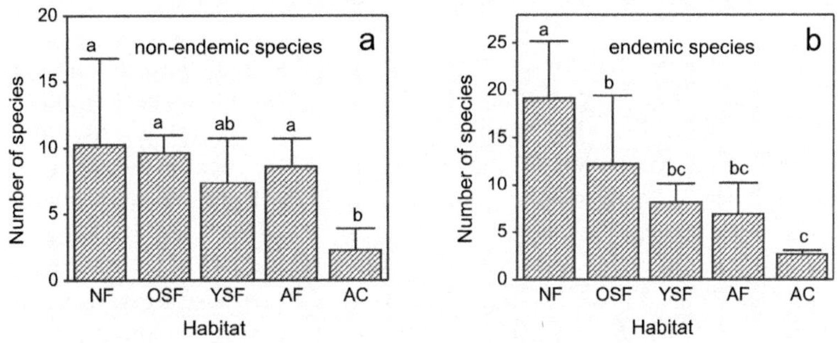

Fig. 3. Mean number of total non-endemic (a) and endemic (b) butterfly species estimated by first-order jackknife for the five sampled habitat types. Different letters indicate significant differences. The ranking of the habitats on the x-axis is related to habitat heterogeneity, which is highest in the old-grown forests (NF, OSF) and lowest in the maize fields (AC). For site abbreviations see Fig. 2.

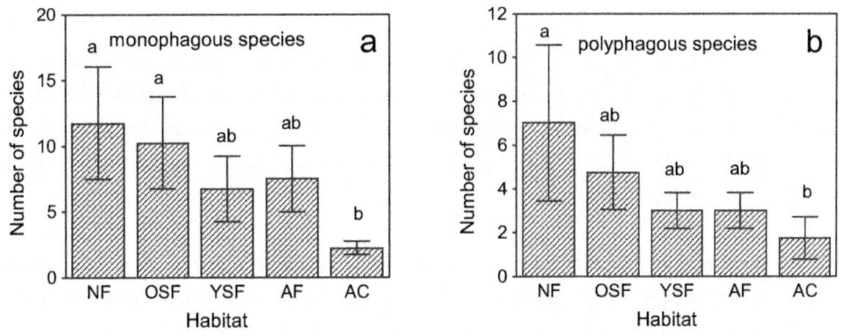

Fig. 4. Mean number of total monophagous (a) and polyphagous (b) butterfly species estimated by first-order jackknife for the five sampled habitat types. Different letters indicate significant differences. The ranking of the habitats on the x-axis is related to habitat heterogeneity, which is highest in the old-grown forests (NF, OSF) and lowest in the maize fields (AC). For site abbreviations see Fig. 2.

nificant level. Just the number of butterfly species recorded for the maize fields differed significantly from the natural forest and, in the case of polyphagous species, additionally from the old secondary forest (Fig. 4). The numbers of recorded monophagous and polyphagous species per site were significantly correlated (Pearson: r=0.74, N=20, p<0.001). Additionally, no significant difference of the number of species recorded per habitat type could be found between both groups (Pearson χ^2 test: χ^2_{16df}=20.00, n.s.).

4.4 Butterfly Species Richness and Habitat Structure

All measured habitat parameters were significantly related to habitat type (see Fig. 5). In particular, the percentage of shade at ground showed a highly significant decrease with declining habitat heterogeneity (Fig. 5b). In addition, of all measured habitat parameters it showed the strongest relationship with butterfly species richness. The number of estimated species decreased significantly with declining percentage of shade at ground (Fig. 6). However, because shade at ground and canopy cover were significantly correlated with all other habitat parameters (see Table 1), we excluded both from the stepwise forward multiple regression (r_{mult}=0.762, $F_{2,16}$=11.04, p<0.001, n=19) performed for the 19 transect sites for which data on all habitat variables were available. The number of vegetation layers (partial correlation coefficient ρ=0.568, p=0.014) as well as the plant diversity of the herb layer (ρ=0.477, p<0.045) had a significant explanatory power for the estimated butterfly species richness, while tree density (ρ=0.041, p=0.875) did not enter into the multiple regression model. The number of vegetation layers decreased significantly with increasing forest disturbance and land-use intensity (Fig. 5d), thereby paralleling the pattern of butterfly diversity. Accordingly, the butterfly species richness increased with increasing number of vegetation layers (Spearman: r=0.720, p<0.001, n=20, Fig. 7a). Also plant diversity of the herb layer showed a significant decline with increasing disturbance and habitat modification, although highest herb layer diversity was found for the old secondary forest sites (Fig. 5a). The relationship between butterfly species richness and plant diversity can be also described by a simple linear regression model (Pearson: r=0.636, p<0.01; Fig. 7b).

Table 1. Pearson (r_P) or Spearman correlation coefficients (r_S) and level of significance (p) for pairwise comparisons of all measured habitat parameters. n=20 except for correlations with tree density (n=19). Significant correlations are marked in bold.

	Shade at ground [%]	Tree density	Vegetation layers	Canopy cover [%]
Fisher's α for herb diversity	r_P=0.65, p=0.002	r_P=0.40, p=0.090	r_S=0.49, p=0.028	r_P=0.45, p=0.046
Shade at ground [%]	—	r_P=0.59, p=0.008	r_S=0.79, p<0.001	r_P=0.89, p<0.001
Tree density	—	—	r_S=0.31, p=0.198	r_P=0.65, p=0.003
Vegetation layers	—	—	—	r_S=0.44, p=0.051

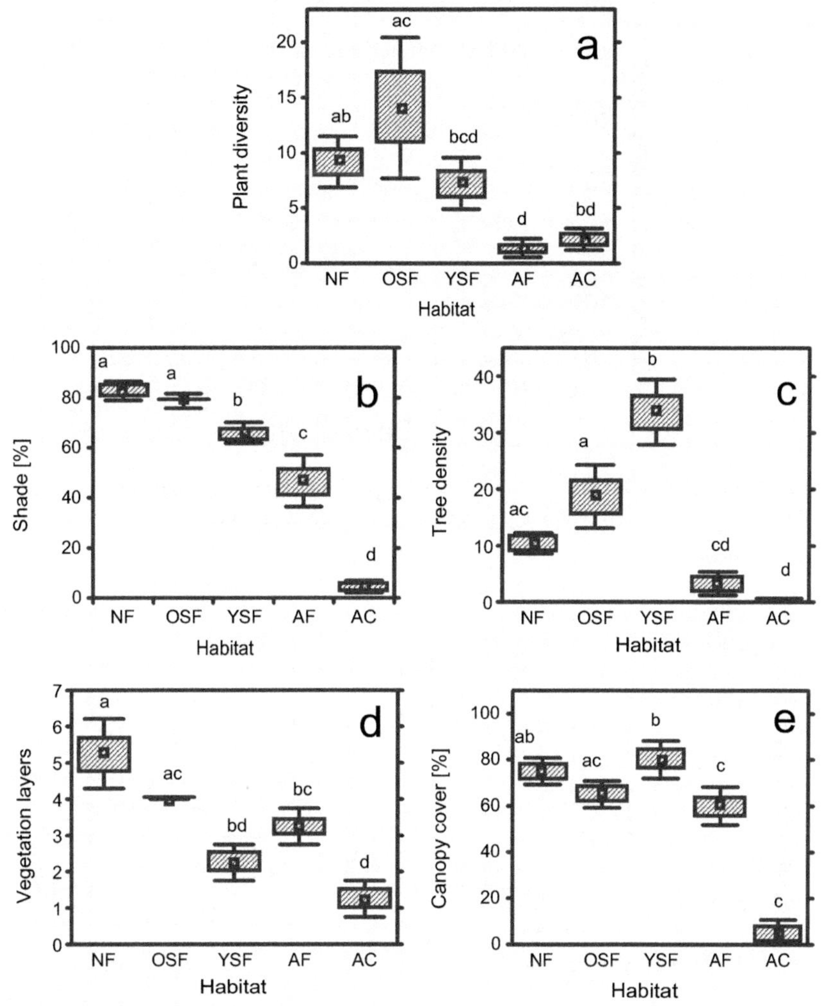

Fig. 5. Change of (**a**) plant diversity of herb layer (quantified as Fisher's alpha) (ANOVA: $F_{4,15}$=10.20, p<0.001), (**b**) percentage of shade at ground ($F_{4,15}$=133.54, p<0.0001), (**c**) tree density ($F_{4,15}$=47.29, p<0.0001), (**d**) number of vegetation layers ($F_{4,15}$=77.79, p<0.0001), and (**e**) canopy cover ($F_{4,15}$=28.65, p<0.0001) across the studied habitat types. The box-whisker-plots present means as well as their standard errors (box) and standard deviations (whisker). Different letters indicate significant differences. For site abbreviations see Fig. 2.

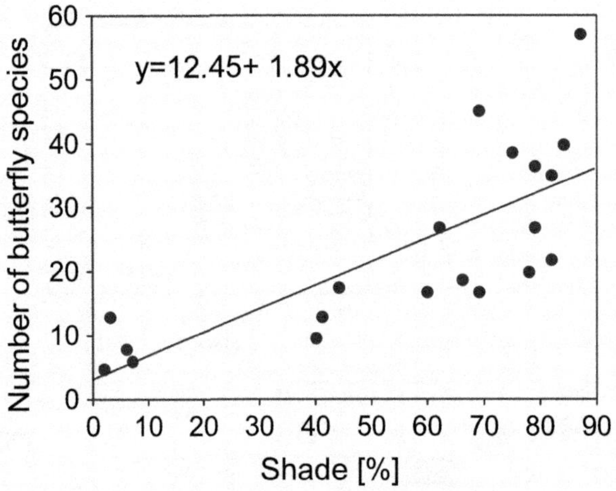

Fig. 6. Relationships between total butterfly species richness (estimated by first-order jack-knife method, Colwell 2000) and percentage of shade at ground.

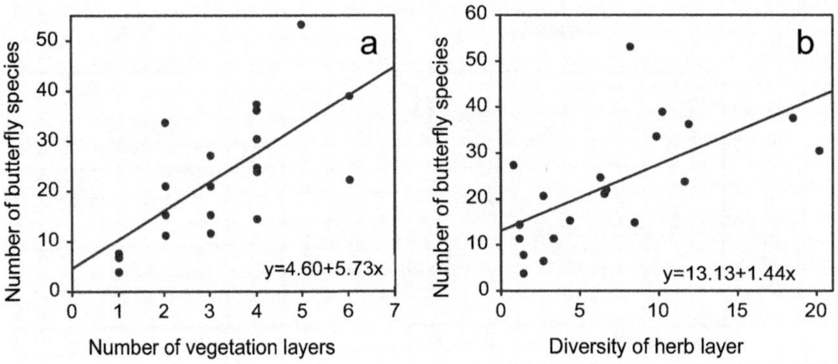

Fig. 7. Relationships between total butterfly species richness (estimated by first-order jacknife method, Colwell 2000) and plant diversity of the herb layer (a) as well as the number of vegetation layers per site (b).

5 Discussion

5.1 Butterfly Diversity vs. Land Use

Studies on various Lepidoptera taxa (Beck and Schulze 2000; Bowman et al. 1990; Intachat et al. 1999a; Lewis 1999; Schulze 2000; Willott 1999) have shown that secondary forests do not necessarily have a significantly lower diversity than primary forests. This pattern was supported by our study. Butterfly richness did not differ between natural forest sites and old secondary forest fragments. Although lower species numbers were recorded for young secondary forests and cacao agroforestry systems, they still reached about 50% of the richness at natural forest sites. Chey et al. (1997) found a remarkable moth diversity in Malaysian tree plantations, while a study by Beck and Schulze (2000) failed to find significant differences between old-grown and 15–30 years old forests with respect to the diversity of fruit-feeding nymphalid butterflies.

Latin American cacao plantations may maintain a relatively high number of species, which are typical for natural forests (see review of Rice and Greenberg 2000). However, cacao plantations in our study area lack woody understorey plants and all kind of canopy climbers, thereby excluding butterfly species, which depend on such plants as food resource. Additionally, cacao plantations provided just a very homogeneous canopy cover, which may not be adequate for species depending on special canopy microhabitats. Nevertheless, a relatively high number of butterfly species was found in these agroforestry systems. However, when taking into account just the butterflies species recorded from natural forest sites, this assemblage of `forest species´ showed a dramatic decrease not just towards the agroforestry systems (loss of 70% of the `forest species´ from natural forest to agroforest) but also towards all other anthropogenically manipulated habitat types including the secondary forest fragments (data not shown). Rice and Greenberg (2000) already emphasized that true forest specialists are often missing in cacao plantations. The species richness documented for our agroforestry systems is still comparatively high because the butterfly community of this habitat was characterized by a remaining fraction of forest species (like *Ideopsis vitrea* or *Cethosia myrina*) and, additionally, a group of species, which can be defined as species of young fallows or open cultivated land (like *Ypthima nynias* or *Junonia atlites*) (Schulze et al. *unpubl. data*).

5.2 Effect of Forest Disturbance and Land Use on Endemic Species

Impact of land use on species richness is particularly interesting for endemic species when regions with a high extent of endemism like Sulawesi are studied (see Whitten et al. 1988). The unique biogeographical position of this island and its high portion of endemic species are already well documented, also for Lepidoptera (Holloway 1987; Holloway and Jardin 1968; 1990; Holloway et al. 2001). Of the

Papilionidae, Pieridae and Nymphalidae butterflies recorded at our study sites in Napu Valley, remarkable 56% of the species can be considered to be endemic to the region of Sulawesi.

Species characterized by a small distribution are missing with a higher likelihood in disturbed areas, perhaps, because they are more specialized than more widespread species (e.g. Hamer et al. 1997; Hill et al. 1995; Spitzer et al. 1993; Thomas 1992). Such a pattern holds also for the butterfly fauna of our study sites. Species richness of non-endemics did not change with forest type, whereas richness of endemics continuously decreased with increasing intensity of forest disturbance and land-use activity. In particular, the natural forest proved to be significantly more diverse in endemics than any other studied habitat type.

5.3 Butterfly Species Richness vs. Habitat Parameters

The percentage of shade at ground level was the best predictor for butterfly species richness. This habitat parameter may act as a simple tool for quantifying rapidly the extent of forest disturbance and land-use intensity and its impact on butterfly diversity. However, due to its strong correlation with all other habitat parameters, shade at ground was excluded from a multiple regression analyzing the relationship between habitat parameters and butterfly species richness. A strong relationship was found between the number of vegetation layers and butterfly species richness. This implies that the vertical habitat structure may play an important role for maintaining butterfly species richness.

The majority of South East Asian butterfly species can be defined as host plant specialists (Fiedler 1998). Therefore, higher plant diversity should increase the number of species, which can be expected at a single site. This was confirmed by our data and indicates a similar response of butterflies to changes of plant diversity as documented for moths (Beck et al. 2002; Schulze 2000). Remarkably, no differences were found between monophagous and polyphagous butterfly species.

5.4 Conservation Aspects

Efforts to promote conservation in the tropics have focused on two approaches: protection of little disturbed high-priority areas and promotion of sustainable forest management. Although a strict protection of pristine forest is irrefutably the most effective conservation strategy, protected areas currently comprise only 8% of tropical forests (Rice et al. 2001). While the expansion of protected areas is often politically unfeasible, protected forests increasingly change their nature due to illegal logging activities. In the margin of the Lore Lindu National Park logging and rattan harvesting caused already a mosaic of secondary forests and agriculturally used sites. Sulawesi belongs to the 25 worldwide most important biodiversity hotspots (Myers et al. 2000), and the inclusion of disturbed forests and little intensive land-use systems should be considered in landscape management. Our data from Central Sulawesi suggest that secondary forests and, to a certain extent, even agroforestry systems could help to maintain butterfly diversity. However, different taxonomic groups do not necessarily respond in the same ways to forest disturbance and land use (Lawton et al. 1998). Therefore, additional taxa have to be studied, before general recommendations can be given to decision makers on how to maintain a maximum of biodiversity on a landscape level.

6 Acknowledgements

We are grateful to Damayanti Buchori, Marion Engelhardt, Veronika Fuest, Gerhard Gerold, Edi Guhardja, Andriyono Kilat Adhi, Diah Ratnadewi, Elim Somba, Silvia Werner, Widia Wati and Rina Yusuf of the German-Indonesian Research Project STORMA (Stability of Tropical Rainforest Margins) for their great support and help. Special thanks are due to all STORMA drivers who were extremely patient most of the time and most helpful in all kind of situations. Dorthe Veddeler kindly provided data on habitat parameters of some secondary forest sites. In particular, we like to thank Paul J.A. Kessler for providing substantial help in sorting our understorey plant samples. Furthermore, we thank all the Indonesian smallholders in Napu Valley, who assisted us in the field and gave permission to conduct research on their land. We are grateful to DIKTI, department of the Ministry of Education in Jakarta, the Governor of Central Sulawesi and Balai Taman Nasional Lore Lindu, who kindly provided all necessary permissions to conduct our fieldwork. The research project was funded by the Deutsche Forschungsgemeinschaft (SFB 552).

7 Keywords

Biodiversity, endemics, rainforest, secondary forest, agroforestry, annual crops, conservation, habitat structure, plant diversity, shading

References

Ackery PR (1988) Hostplants and classification: a review of nymphalid butterflies. Biological J Linn Soc 33: 95–203

Beattie AJ, Oliver I (1994) Taxonomic minimalism. Trends Ecol Evol 9: 488–490

Beck J, Schulze CH (2000) Diversity of fruit-feeding butterflies (Nymphalidae) along a gradient of tropical rainforest succession in Borneo with some remarks on the problem of `pseudoreplicates´. Trans lepid Soc Japan 51: 89–98

Beck J, Schulze CH, Linsenmair KE, Fiedler K (2002) From forest to farmland: diversity of geometer moths along two habitat gradients on Borneo. J Trop Ecol 18: 33–51

Boulinier T, Nichols JD, Sauer FR, Hines JE, Pollock KH (1998) Estimating species richness: the importance of heterogeneity in species detectability. Ecology 79: 1018–1028

Bowman DMJS, Woinarski JCZ, Sands DPA, Wells A, McShane VJ (1990) Slash-and-burn agriculture in the wet coastal lowlands of Papua New Guinea: response of birds, butterflies and reptiles. J Biogeography 17: 227–239
Brühl CA (2001) Leaf litter ant communities in tropical lowland rain forests in Sabah, Malaysia: effects of forest disturbance and fragmentation. Ph.D. thesis, University of Würzburg
Chey VK, Holloway JD, Speight MR (1997) Diversity of moths in forest plantations and natural forests in Sabah. Bull Entomol Res 87: 371–385
Chung AY, Mohamed M (1996) A comparative study of the ant fauna in a primary and secondary forest in Sabah, Malaysia. In: Edwards DS, Booth WE, Choy SC (eds), Tropical rainforest research – current issues. Kluwer Academic Publishers, Dodrecht, pp 357–366
Coates BJ, Bishop KD, Gardner D (1997). A field guide to the birds of Wallacea. Dove Publications, Alderley
Colwell RK (2000) EstimateS – Statistical estimation of species richness and shared species from samples. Version 6.0b1. http:/viceroy.eeb.uconn.edu/ estimates
Colwell RK, Coddington JA (1994) Estimating terrestrial biodiversity through extrapolation. Phil Trans Roy Soc Lond (B) 345: 101–118
Corbet AS, Pendlebury HM (1992) The Butterflies of the Malay Peninsula. Malayan Nature Society, Kuala Lumpur
Dahaban Z, Nordin M, Bennett EL (1996) Immediate effects on wildlife of selective logging in a hill dipterocarp forest in Sarawak: mammals. In: Edwards DS, Booth WE, Choy SC (eds) Tropical rainforest research – current issues. Kluwer Academic Publishers, Dodrecht, pp 341–346
Davis AJ, Huijbregts J, Krikken J (2000) The role of local and regional processes in shaping dung beetle communities in tropical forest plantations in Borneo. Global Ecology and Biogeography 9: 281–292
De Jong R (1990) Some aspects of the biogeography of the Hesperiidae (Lepidoptera, Rhopalocera) of Sulawesi. In: Knight WJ, Holloway JD (eds) Insects and the rain forests of South East Asia (Wallacea). The Royal Entomological Society of London, London, pp 35–42
Eggleton P, Homathevi R, Jeeva D, Jones DT, Davies RG, Maryati M (1997) The species richness and composition of termites (Isoptera) in primary and regenerating lowland dipterocarp forest in Sabah, East Malaysia. Ecotropica 3: 119–128
Erhardt A (1985) Diurnal Lepidoptera: sensitive indicators of cultivated and abandoned grassland. J Appl Ecol 22: 849–861
Fiedler K. (1998) Diet breadth and host plant diversity of tropical- versus temperate-zone herbivores: South East Asian and West Palaearctic butterflies as a case study. Ecol Entomol 23: 285–297
Hamer KC, Hill JK, Lace LA, Langhan AM (1997) Ecological and biogeographical effects of forest disturbance on tropical butterflies of Sumba, Indonesia. J Biogeography 24: 67–75
Hill JK, Hamer KC, Lace LA, Banham WMT (1995) Effects of selective logging on tropical forest butterflies on Buru, Indonesia. J Appl Ecol 32: 754–760
Holloway JD (1987) Lepidoptera patterns involving Sulawesi: What do they indicate of past geography? In: Whitmore TC (ed) Biogeographical evolution of the Malay Archipelago. Clarendon Press, Oxford, pp 103–118

Holloway JD (1990) Sulawesi biogeography – discussion and summing up. In: Knight WJ, Holloway JD (eds) Insects and the rain forests of South East Asia (Wallacea). The Royal Entomological Society of London, London, pp 95–102

Holloway JD (1998) The impact of traditional and modern cultivation practices, including forestry, on Lepidoptera diversity in Malaysia and Indonesia. In: Newbery DM, Prins HHT, Brown ND (eds) Dynamics of tropical communities. 37th Symposium of the British Ecological Society, London, pp 567–597

Holloway JD, Jardine N (1968) Two approaches to zoogeography: a study based on the distributions of butterflies, birds and bats in the Indo-Australian area. Proc Linn Soc 179: 153–188

Holloway JD, Kibby G, Peggie D (2001) The Families of Malesian Moths and Butterflies. Brill, Leiden

Holloway, JD, Kirk-Spring AH, Chey VK (1992) The response of some rain forest insect groups to logging and conversion to plantation. Phil Trans Roy Soc Lond (B) 335: 425–436

Hughes JB, Daily GC, Ehrlich PR (2002) Conservation of tropical forest birds in countryside habitats. Ecol Letters 5: 121–129

Intachat J, Holloway JD (2000) Is there stratification in diversity or preferred flight height of geometer moths in Malaysian lowland tropical forest? Biodiv Cons 9: 1417–1439

Intachat J, Holloway JD, Speight MR (1997) The effect of different forest management practices on geometroid moth populations and their diversity in Peninsular Malaysia. J Trop For Sci 9: 411–430

Intachat J, Holloway JD, Speight MR (1999a) The impact of logging on geometroid moth populations and their diversity in lowland forests of Peninsular Malaysia. J Trop For Sci 11: 61–76

Intachat J, Chey VK, Holloway JD, Speight MR (1999b) The impact of forest plantation development on geometroid moth population and diversity in Malaysia. J Trop For Sci 11: 329–336

Johns AD (1992) Vertebrate responses to selective logging: implications for the design of logging systems. Phil Trans Roy Soc Lond (B) 335: 437–442

Kessler PJA, Bos MM, Sierra Daza SEC, Kop A, Willemse LPM, Pitopang R, Gradstein SR (in prep.) Checklist of woody plants of Sulawesi, Indonesia.

Klein A-M, Buchori D, Steffan-Dewenter I, Tscharntke T (2002a) Effects of land-use intensity in tropical agroforestry systems on coffee flower-visiting and trap-nesting bees and wasps. Cons Biol 16: 1003–1014

Klein A-M, Steffan-Dewenter I, Tscharntke T (2002b) Predator-prey ratios on cocoa along a land-use gradient in Indonesia. Biodiv Cons 11: 683–693

Lambert FR (1992) The consequences of selective logging for Bornean lowland forest birds. Phil Trans Roy Soc Lond (B) 335: 443–457

Lawton JH (1995) Population dynamic principles. In: Lawton JH, May RM (eds) Extinction rates. Oxford University Press, Oxford, pp 147–163

Lawton JH, Bignell DE, Bolton B, Bloemers GF, Eggleton P, Hammond PM, Hodda M, Holt RD, Larsen TB, Mawdsley NA, Stork NE, Srivastava DS, Watt AD (1998) Biodiversity inventories, indicator taxa and effects of habitat modification in tropical forest. Nature 391: 72–76

Lewis OT (1999) Effects of experimental selective logging on tropical butterflies. Cons Biol 15: 389–400

Liow LH, Sodhi NS, Elmqvist T (2001) Bee diversity along a disturbance gradient in tropical lowland forest of south-east Asia. J Appl Ecol 38: 180–192
Moreno CE, Halffter G (2001) On the measure of sampling effort used in species accumulation curves. J Appl Ecol 38: 487–490
Myers N, Mittermeier RA, Mittermeier CG, da Fonseca GAB, Kent J (2000) Biodiversity hotspots for conservation priorities. Nature 403: 853–858
Perfecto I, Snelling R (1995) Biodiversity and the transformation of a tropical agroecosystem: ants in coffee plantations. Ecol Appl 5: 1084–1097
Perfecto I, Vandermeer J (1996) Microclimatic changes and the indirect loss of ant diversity in a tropical agroecosystem. Oecologia 108: 577–582
Pollard E, Yates TJ (1993) Monitoring butterflies for ecology and conservation. Conservation Biology Series, Vol. 1. Chapman & Hall, London
Rice RA, Greenberg R (2000) Cacao cultivation and the conservation of biological diversity. Ambio 29: 167–173
Rice RE, Sugal CA, Ratay SM, da Fonseca GAB (2001) Sustainable forest management: A review of conventional wisdom. Adv Appl Biodiv Sci 3: 1–29
Sachs L (1992) Angewandte Statistik, 7th edn. Springer, Berlin.
Schulze CH (2000) Effects of Anthropogenic Disturbance on the Diversity of Herbivores – An Analysis of Moth Species Assemblages Along Habitat Gradients in East Malaysia. Ph.D. thesis, University of Bayreuth [in German]
Schulze CH, Fiedler K (1998) Habitat preference and flight activity of Morphinae butterflies in a Bornean rain forest, with a note on sound production by adult *Zeuxidia* (Lepidoptera: Nymphalidae). Mal Nat J 52: 163–176
Schulze CH, Linsenmair KE, Fiedler K (2001) Understorey versus canopy: patterns of vertical stratification and diversity among Lepidoptera in a Bornean rain forest. Plant Ecol 153: 133–152
Southwood TRE (1978). Ecological Methods, 2nd edn. Chapman & Hall, London
Spitzer K, Novotný V, Tonner M, Leps J (1993) Habitat preferences, distribution and seasonality of the butterflies (Lepidoptera, Papilionoidea) in a montane tropical rain forest, Vietnam. J Biogeography 20: 109–121
STATSOFT (1999) Statistica for Windows 5.5. StatSoft Inc., Tulsa
Steffan-Dewenter I, Tscharntke T (2000) Butterfly community structure in fragmented habitats. Ecol Letters 3: 449–456
Taylor LR (1978) Bates, Williams, Hutchinson – a variety of diversities. In: Mound LA, Waloff N (eds) Diversity of insect faunas. Sym Roy Entomol Soc Lond 9: 1–18
Thomas CD (1992) Habitat use and geographic ranges of butterflies from the wet lowlands of Costa Rica. Biol Cons 55: 269–281
Vane-Wright RI, de Jong R (in prep.) The butterflies of Sulawesi: annotated checklist for a critical island fauna
Walpole MJ, Sheldon IR (1999) Sampling butterflies in tropical rainforest: an evaluation of a transect walk method. Biol Cons 87: 85–91
Watt AD, Stork NE, McBeath C, Lawson GL (1997) Impact of forest management on insect abundance and damage in a lowland tropical forest in southern Cameroon. J Appl Ecol 34: 985–998
Weibull AC, Bengtsson J, Nohlgren E (2000) Diversity of butterflies in the agricultural landscape: the role of farming system and landscape heterogeneity. Ecography 23: 743–750

Whitten AJ, Mustafa M, Henderson GS (1988) The ecology of Sulawesi. Gadjah Mada University Press, Gadjah Mada

Willott SJ (1999) The effects of selective logging on the distribution of moths in a Bornean rainforest. Phil Trans Roy Soc Lond (B) 354: 1783–1790

Willott SJ, Lim DC, Compton SG, Sutton SL (2000) Effects of selective logging on the butterflies of a Bornean rainforest. Cons Biol 14: 1055–1065

Losing Ground but Still Doing Well - *Tarsius dianae* in Human-Altered Rainforests of Central Sulawesi, Indonesia

Stefan Merker, Indra Yustian & Michael Mühlenberg

1 Introduction

Sulawesi, one of the largest islands of the Malay Archipelago, provides living space for a unique fauna and flora. The island is characterized by vast numbers of endemic plant and animal species originating from the Australian as well as the Oriental region. The destruction and degradation of rain forest habitats pose serious threats to wildlife around the tropical world. On Sulawesi, recent population growth, resettlements, and social unrest have severely added to this problem, hence many of the island's endemic species face a highly unknown future.

Among the potentially endangered animals are the tarsiers (*Tarsius* spp.). These small nocturnal primates are not only unique in their phylogenetic origin and their appearance, the tarsiers' entanglement in Indonesia's myths, their distribution throughout Sulawesi, and their potential for eco-tourism make them – especially when considering the lack of big, charismatic animals on Sulawesi – potential flagship species for raising local awareness for nature.

Detailed field studies on its ecology and behaviour are needed to assess the conservation status of the genus *Tarsius* and to propose useful measures for securing its long-term survival.

Although the number of publications on tarsiers has grown immensely over the past decade (Crompton and Andau 1986, 1987; Dagosto and Gebo 1998; Fogden 1974; Gursky 1994, 1995, 1998a, b, 2000a, b, c; Leksono et al. 1997; MacKinnon and MacKinnon 1980; Merker and Mühlenberg 2000; Niemitz 1979, 1984a; Niemitz et al. 1991; Nietsch 1993, 1998, 1999; Nietsch and Kopp 1998; Nietsch and Niemitz 1992; Shekelle 1997; Tremble et al. 1993; Yustian et al. *in prep*), little emphasis has been given to conservation issues.

The focal species of this study is the Dian's tarsier *Tarsius dianae* Niemitz et al. Because it has just been described one decade ago (Niemitz et al. 1991), still little is known on the biology of this Central Sulawesi endemic. Tremble et al. (1993) were the first to investigate activity budgets, substrate use, and home range sizes. Merker and Mühlenberg (2000) compared population densities of *T. dianae* in habitats differently influenced by man and presented evidence that tarsiers are adaptable to specific kinds of human land-use, but do not tolerate others.

The study, parts of which are presented here, addresses parameters such as population densities, home range sizes, nightly path length, and group sizes of

T. dianae in habitats along a gradient of anthropogenic disturbance. Here, population densities will be focused on.

The research was designed to answer the following questions. To what extent do tarsiers tolerate human activities in the forest? If disturbed, what are their long-term reactions? Do they live "better" or "worse" in disturbed habitats than in pristine forest?

This study was conducted in the same area, partly on the same plots as done by Merker and Mühlenberg (2000). Hence, population densities of tarsiers on the same forest patches can be compared between 1998 and 2001, the years of data collection, and possible changes can be detected.

2 Materials and Methods

2.1 Study Objects

Tarsiers are small nocturnal primates inhabiting the islands of South-East Asia. Currently, six species are recognized, four of which live on Sulawesi and its adjacent islands (Groves 1998). With only 100-140g body weight and 12-13cm body length, tarsiers belong to the smallest primates on earth. Their scaly tail is about 25cm long, and – in the Sulawesi tarsiers – is characterized by a long tuft with long hair in the distal part of the tail. On Sulawesi, tarsiers exhibit a variety of colours of their usually thick fur, ranging from yellowish-brown to dark grey. Huge eyes, big ears, very long fingers and toes with enlarged terminal pads, and a strongly elongated tarsus are all prominent features of the genus *Tarsius*. They live in small family groups of up to 8 individuals (Gursky 1998a; MacKinnon and MacKinnon 1980). At dawn, when they retire to their common family sleeping sites, the adult and subadult Sulawesi tarsiers perform morning duets. These vocalizations are meant for territorrial advertising, the strengthening of social bonds, and mate attraction (MacKinnon and MacKinnon 1980; Niemitz 1984e, Nietsch 1999; Nietsch and Kopp 1998). The calls carry about 150m and are easily audible to the human ear. Tarsiers mark objects in their territory with urine or gland secretions (Niemitz 1984d). Fresh marking sites are easily recognizable by a very distinct smell. This is a simple way to check for the presence of these secretive nocturnal primates in a given area. Tarsiers exclusively feed on live animal prey (Niemitz 1984b, Nietsch 1993) with crickets, grasshoppers, moths, and other insects as their favourite captures.

Tarsius dianae which was described in 1991 by Niemitz et al. is the dominant species in Central Sulawesi (Nietsch and Kopp 1998). Until now, no sympatric tarsier taxa have been documented.

2.2 Study Site and Methodology

All data on *T. dianae* presented here were obtained in the vicinity of Kamarora, a ranger post at the north-eastern boundary of Lore Lindu National Park, Central Sulawesi, Indonesia (1°11'50" S, 120°08'15" E).
In 1998, four plots of natural forest with different forms of human land-use were investigated, in 2001, five patches of forest along a gradient of human disturbance were focused on. All plots were between 10 and 30ha in size.

To localize tarsier sleeping trees in a given area, we recorded all audible duet songs the animals performed every morning around dawn. By triangulating the positions of the call sources and multiple repetition of this procedure, tarsiers could be traced to their sleeping sites mostly comprised of strangler figs, dense shrubs, or bamboo stands. Mapping all sleeping sites in the investigated areas and measuring the distances between the groups resulted in relative population density estimates. To ensure comparability, the population densities given by Merker and Mühlenberg (2000) for 1998 were re-calculated using the same distance-to-density conversion scheme as in 2001.

Testing for statistically significant differences was done using the Wilcoxon signed-rank test and the Mann-Whitney U-test (Sokal and Rohlf 1995). For correlations, Pearson's r was computed (Lamprecht 1992). All tests are two-tailed.

In this study, we quantified anthropogenic disturbances using indicators suggested by Bynum (1999) for lowland forest of Lore Lindu National Park. In each habitat, we established 16 randomly distributed 10x10m plots and measured the number of artificial stumps, exotic plant species, rattan palms (*Calamus* spp.) less than 3m in length and rattans more than 3m in length. The rattans were later pooled for the calculation and the visual presentation (Fig. 1). Epiphytes less than 5m above the forest floor were counted as well as trees with a dbh of more than 50cm. The upper limit of the mossline was also measured or estimated. Integrating these parameters, a disturbance index D_i was developed:

$$D_i = 1/n \Sigma d_{ji} \quad (1)$$

D_i = disturbance index for habitat i
n = number of recorded disturbance parameters
j = disturbance parameter (1 ... n), e.g. stumps
d_{ji} = relative disturbance index for parameter j in habitat i, e.g. 0.2
v_{ji} = value for parameter j in habitat i, e.g. 6 stumps or 6 big trees
$v_{j\,max}$ = maximum value for parameter j over all habitats, e.g. 12 stumps
$v_{j\,min}$ = minimum value for parameter j over all habitats, e.g. 3 big trees

For parameters where the maximum value means maximum disturbance (e.g. stumps), d_{ji} is

$$d_{ji} = v_{ji} / v_{j\,max} \quad (2)$$

If $v_{j\,min} = v_{j\,max} = 0$, then $d_{ji} \stackrel{def}{=} 0$ (3)

For parameters where the minimum value means maximum disturbance (e.g. big trees), d_{ji} is

$$d_{ji} = v_{j\,min} / v_{ji} \qquad \text{(if } v_{j\,min} > 0\text{)} \qquad (4)$$

$$d_{ji} = (v_{j\,max} - v_{ji}) / v_{j\,max} \qquad \text{(if } v_{j\,min} = 0\text{)} \qquad (5)$$

If $v_{j\,min} = v_{j\,max} = 0$, then $d_{ji} \stackrel{def}{=} 1$ (6)

The disturbance index D_i – in the form described above – is suitable to compare the relative degrees of disturbance of plots *within* a study. If it is to be used for a comparison *between* studies, all parameter values for all studies have to be known, for the disturbance indices D_i cannot be obtained independently and then compared. For that, $v_{j\,min}$ and $v_{j\,max}$ constitute the absolute minimum and maximum values for all studies.

3 Results

3.1 Disturbance Index D_i for Five Habitats in 2001

In 2001, instead of describing a land-use form as in 1998, a disturbance index D_i was developed reflecting the degree of human intervention. Five habitats were characterized this way (Fig. 1 and Fig. 2). These areas were chosen along a gradient of disturbance estimated by the observers, with primary forest and heavily disturbed habitat at the extremes. The quantitative analysis using D_i shows a slightly different picture. Here, Habitat 3 (H3) is relatively more disturbed than Habitat 4 (H4), although by looking at it, one would have rated H4 to be affected more severely. This is mostly due to a high number of exotic plants, namely coffee and cocoa, in H3, which is an area with many small forest gardens.

Depletion in rattan palms is also more pronounced in H3. The number of big trees and the number of artificial (branch and tree) stumps resulting from logging or trail-cutting follow a gradient steadily decreasing from H1 to H5, hence the relative disturbance indices d_{ji} for these parameters increase. Fig. 2 depicts the disturbance index D_i for each of the five areas in question. Although, due to the calculation method and also due to some old stumps found in this habitat, the primary forest plot has an $D_1 > 0$, this area is refered to as "undisturbed". Disturbance is most severe in Habitat 5. For all measured parameters j, the relative disturbance indices $d_{j5} = 1$, i.e. anthropogenic influences are at their maximum compared to the other four habitats. This results in $D_5 = 1$.

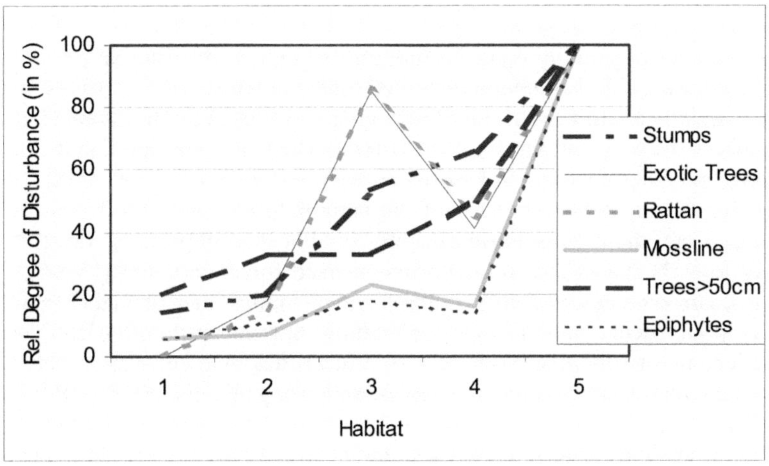

Fig. 1. Relative degree of disturbance in five different forest habitats of Lore Lindu National Park, Sulawesi. Separately for each disturbance parameter, the value in relation to the maximum value found among all habitats is shown (= d_{ji}*100%).

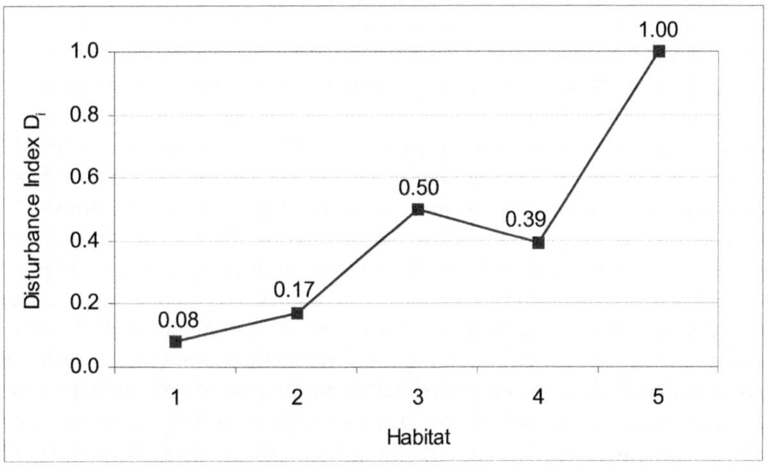

Fig. 2. Disturbance index D_i for each of five different forest habitats of Lore Lindu National Park, Sulawesi. The index has been calculated from the number of stumps, exotic trees, rattan palms, epiphytes, big trees, and the height of the mossline. $D_i = 0$ stands for pristine conditions, $D_i = 1$ means maximum disturbance compared to the other study habitats.

3.2 Population Densities along a Gradient of Human Disturbance

Looking at tarsier population densities along this gradient of disturbance, a clear trend is visible (Fig. 3). While in undisturbed forest, a density of 5.7 groups/10 ha could be found, there are significantly fewer tarsier groups in all investigated areas that are affected by human land-use (**U-test). There are no significant differences between tarsier population densities in the areas with slight or medium disturbance. On the plantations outside of the natural forest, only 1.4 groups/10ha were found. Here, the density is significantly smaller than in primary forest**, in secondary forest*, and in the area with interspersed forest gardens* (U-test). On plantations, tarsiers sleep in dense shrubs, bamboo stands, and remnant strangler figs. They prefer dense undergrowth for hunting, but also use coffee and cocoa plants as locomotor supports. There is a significant negative correlation between tarsier population density and the intensity of anthropogenic disturbance (Pearson, $r = -0.89$*).

We could not find tarsiers in paddy fields or in tea, clove, corn and coconut monocultures. Nevertheless, some groups manage to stay very close to human settlements or at roadside shrubs and bamboos.

3.3 Population Densities in 1998 and in 2001

Population densities of *T. dianae*, obtained in 2001 using the same methodology as in 1998 are generally lower than three years before (Table 1). In 1998, 10.5 tarsier groups/10ha were found in primary forest, in this study, only 5.7 groups/10 ha could be recorded. However, due to the conversion of the primary forest plot of 1998 into plantations, another pristine forest was chosen for the study in 2001, hence the data are not directly comparable. While in 1998, the primary forest as well as the other three habitats were located at an elevation of 700-750m asl, in 2001, no pristine or virtually undisturbed forest patch of at least 10ha in size could be found below the elevation of 950m asl. The chosen primary forest study plot of 2001 is situated at an elevation of 950-1,020m asl. While secondary forest and plantations outside of the natural forest were only investigated in 2001, small-scale logging and small (<0.5ha) forest gardens played a major role in both studies. Population densities in two parts of the forest that were already screened in 1998 were also measured in 2001, hence direct evidence for habitat change can be drawn from the comparison of these data. In 1998, a forest patch characterized by small-scale logging contained on average 4.3 tarsier groups per 10ha. In 2001, the logging had almost ceased, and a number of interspersed forest gardens (coffee and cocoa) could be found. With 3.8 family groups per 10ha, the population density was not significantly different (Wilcoxon) from three years before. However, if this kind of habitat, namely forest with small plantations, is compared with the same habitat type studied in 1998, a dramatic decrease in tarsier population density is apparent (Table 1 and Fig. 4).

Table 1. Relative population densities of *T.dianae* in several habitats of Lore Lindu NP in the years 1998[1] and 2001.[2]

Habitat type	Population density				
	1998			2001	
Primary forest	10.5 gr./10ha	(n=7)		5.7 gr./10ha	(n=6)
Secondary forest				3.6 gr./10ha	(n=6)
Forest with small plantations	12.0 gr./10ha	(n=7)	↗	3.8 gr./10ha	(n=6)
Forest with small-scale logging	4.3 gr./10ha	(n=8)	↙		
Forest with logging and plantations	6.6 gr./10ha	(n=8)	* ↔	3.3 gr./10ha	(n=5)
Plantations outside of the natural forest				1.4 gr./10ha	(n=5)

[1] Density data re-calculated from Merker and Mühlenberg (2000) using the same distance-to-density conversion scheme as in 2001 (this study).
[2] Arrows indicate specific study patches investigated in both years. In one case, the type of land-use changed between 1998 and 2001.

One more specific forest patch – "forest with logging and plantations" – was looked at in both data collection periods. Ongoing logging and conversion of the forest into plantations resulted in a significant decrease of the tarsier population density in this area (*Wilcoxon). While in 1998, on average 6.6 groups could be found per 10ha plot, in 2001 the density had decreased to 3.3 groups/10ha. This change went along with visible degradation of the forest. Fig. 5 shows a map of this area.

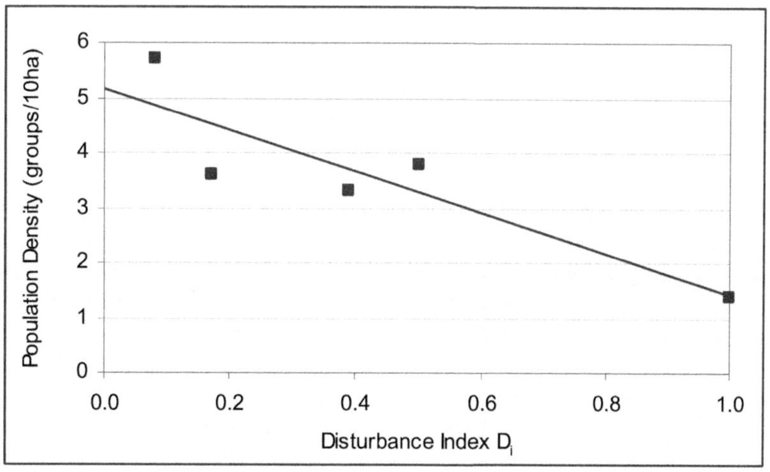

Fig. 3. Relative population densities of *T. dianae* in five different forest habitats of Lore Lindu National Park, Sulawesi. The disturbance index for each area has been calculated from the number of stumps, exotic trees, rattan palms, epiphytes, big trees, and the height of the mossline. $D_i = 1.0$ means maximum disturbance.

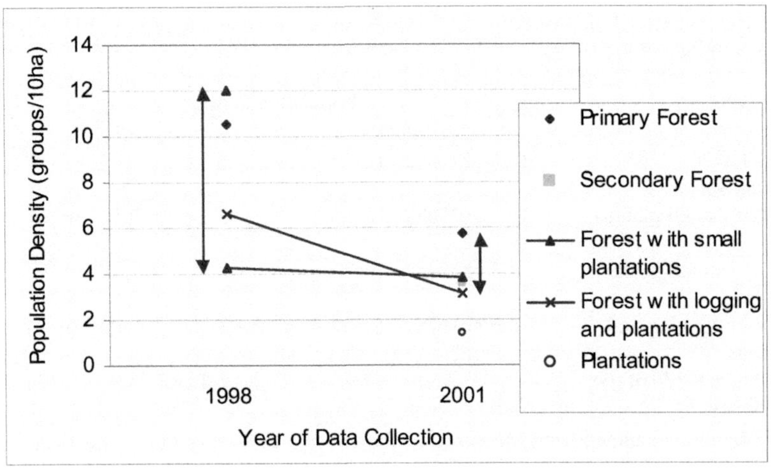

Fig. 4. Comparison of relative population densities of *T. dianae* in several habitats of Lore Lindu NP between the years 1998 and 2001. Connected data points reflect specific plots investigated in both periods. Arrows indicate the range of tarsier densities in habitat types focused on in both studies. Habitat descriptions given were valid in 2001 and are not entirely consistent with 1998 (see table 1).

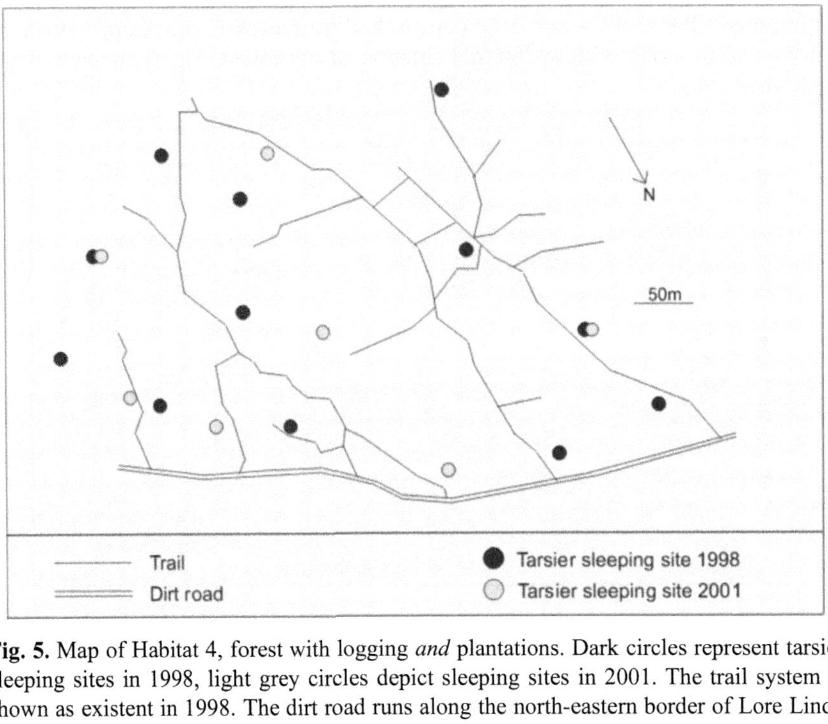

Fig. 5. Map of Habitat 4, forest with logging *and* plantations. Dark circles represent tarsier sleeping sites in 1998, light grey circles depict sleeping sites in 2001. The trail system is shown as existent in 1998. The dirt road runs along the north-eastern border of Lore Lindu NP.

4 Discussion

4.1 Usefulness of the Disturbance Index D_i

The disturbance index D_i provides a possibility to quantify anthropogenic influences and to compare these between habitats. This way, interval/measurement data instead of nominal or ranked variables can be produced, and these can be correlated with the variates in question. This would also be possible by using just one parameter, e.g. the number of stumps. However, calculating D_i gives a more comprehensive notion of anthropogenic effects to the forest.

The index D_i as presented here has been designed to compare relative disturbances of plots of one study (data collection in 2001). Because of different minimum and maximum parameter values for different studies, it is not possible to calculate D_i and then directly compare it with the disturbance indices of a different data set. Instead, data of the different studies have to be pooled producing new extreme values, and D_i have to be calculated anew. As obtaining the complete data sets of other researchers is – in most cases – very difficult, this limits the use of D_i in comparing *between* studies. Additionally, different forest types are characterized by different natural conditions dependent of climate, elevation, topography, geology etc. Indirect indicators for disturbance, e.g. the number of big trees, rattans, and epiphytes, or the upper limit of the mossline are no useful variables to compare montane with lowland forests, or lowland rainforest with mangrove forest. Only direct disturbance indicators such as the number of stumps or exotic trees should be used when comparing different types of forest.

This study uses the "forest integrity indicators" suggested by Bynum (1999). Using statistical analyses, he identified the seven variables mentioned above to detect anthropogenic disturbance. To facilitate comparison, the same variables should be employed in future studies.

4.2 Population Densities along a Gradient of Human Disturbance

There is a strong negative correlation between population densities of *T. dianae* and the intensity of human disturbance of the forest. Highest densities occur in undisturbed primary forest, the lowest number of tarsier groups per area we found in plantations, i.e. heavily disturbed habitat. This finding does not fully confirm the results of Merker and Mühlenberg (2000) who detected the highest population density of *T. dianae* in a slightly disturbed area, i.e. in forest with interspersed plantations (see also 4.3). However, they also state that primary forest harbours almost as many groups of tarsiers as the habitat mentioned above. The density estimates agree with the numbers given by Gursky (1998a) for *T. spectrum* in Tangkoko Nature Reserve in North Sulawesi. She found 5.6 groups/10ha of mixed forest but did not differentiate between habitat types. MacKinnon and MacKinnon (1980) estimated densities of *T. spectrum* in several localities and for a variety of habitats. Listening to the duet songs the tarsiers performed every morning, they found between 3 and 10 individuals per ha. With an average of about 3 members per family group (Gursky 1998a, unpublished data), these numbers correspond to between 10 and 30 groups per 10ha of forest. The latter number is exceptionally high and has been given for thorn scrub at sea level. Whether this habitat type is as "good" as it seems or whether the differences in population densities given by Gursky and by MacKinnon and MacKinnon are due to different methodology remains unclear.

Niemitz (1979) gives a density of 80 individuals per km^2 for *T. bancanus* in Borneo. Assuming a group size of 3, this would mean an average of 2.7 groups/10ha. Considering the different sociality of this tarsier species, comparisons with the Sulawesi tarsiers are difficult, though this number is well within the range of the densities given for *T. dianae* in this study.

Looking at the parameter of population density, we would rate primary forest to provide the best living conditions for tarsiers in Lore Lindu National Park. However, home ranges of female *T. dianae* in Kamarora are smallest in slightly disturbed forest indicating optimum conditions in this habitat (Merker and Mühlenberg 2002, unpublished data). Whether a patch of forest is "better" or "worse" for tarsiers than others cannot be drawn from one single parameter alone. Instead, a set of variables would be needed to estimate habitat quality more reliably. Nevertheless, the significant negative correlation between tarsier density and human disturbance is unmistakable and constitutes a strong reason for conserving what's left of undisturbed forest.

Coffee and cocoa plantations outside of the forest only provide poor living conditions but might serve as stepping stones for the colonization of other forest patches or for the recolonization of a logged and re-grown area. It is crucial for the survival of tarsiers on plantations that there are potential sleeping sites which give shelter from view, predators, and weather, that there is a high abundance of prey items, i.e. large insects, and that the locomotor supports are not spaced too far apart (>2m). In a "clean" plantation without uncultivated patches and with intensive use of pesticides, no tarsiers can survive.

For the reasons named above, paddy fields, corn, clove, tea and coconut monocultures do not permanently harbour tarsiers.

4.3 Population Densities in 1998 and in 2001

While in 1998, the population density in primary forest was an estimated 10.5 groups/10ha, in 2001, only 5.7 groups/10ha were found in pristine habitat. This is probably due to several factors: 1. There is variability between different plots even of the same habitat type. Demographic factors causing fluctuations in sleeping tree distances might play a major role. 2. The primary forest plot of 1998 is situated at 700m asl, the plot of 2001 at 950m asl. According to Whitmore (1984, 1990), one patch falls into the zone of lowland rainforest, the other one belongs to lower montane forest. Availability of prey or locomotor supports might influence the density of tarsiers. However, considering the appearance of both plots, and taking into account that insect abundance estimates in 1998 and 2001 revealed no significant differences (unpublished data), we do not believe the slight contrast in elevation to be of great importance to tarsiers. 3. One more cause for variety might arise from the distinction which forest has been disturbed and which has not been affected by human activities. For instance, the undisturbed study plot of 1998 was characterized by a high number of bamboo stands, none of which could be found on the patch of 2001. Bamboos are favoured by fire and other disturbances (Whitmore 1984). Hence, although there has not been any detectable current or recent disturbance in 1998, and bamboos are not a direct indicator of disturbance, the designation as pristine forest might have been imprecise.

While on one plot investigated in both studies, the population density of *T. dianae* did not change significantly over the course of three years, another one was characterized by a dramatic decrease in tarsier numbers. Although logging and agroforestry occured in both data collection periods, the intensity of human land-use was much higher in 2001, and a high proportion of the forest had already been converted into small cash-crop plantations. Ongoing logging was also severe. After the data acquisition in early 2001 had been completed, this patch was almost clear-cut, and virtually no tarsiers remained in the area. Here, the threshold of severe effects of human disturbance had been crossed, while in other areas, tarsiers were able to cope with a variety of human activities in the forest.

Generally, population densities decline in the study area. Possibly, the year 1998 was characterized by exeptionally good conditions for tarsiers, because so many groups could settle in a small area. On the other hand, these animals are very territorial, long-lived, and raise at maximum one offspring per adult female per year, so major fluctuations between one year and another cannot be expected. Considering the unaltered number of members per family group (unpublished data), no fission or fusion processes were apparent. A high mortality or pronounced emigration seem to have caused the drastic decline in tarsier densities over the last few years. This goes along with recent degradation of the forest and a huge increase in human population density along the margin of Lore Lindu National Park. In particular, a new concrete road leading from Tongoa to Kamarora Field Station attracted many new human settlers to that area. The negative impact of this infrastructural improvement on the park's biodiversity is unmistakeable.

5 Conclusion

Although population sizes decline, future prospects are not too gloomy for *T. dianae* because this species is adaptable to widespread human uses of the forest. Moreover, under certain conditions, cash crop plantations can play a vital role for these small primates, giving additional hope that they may be saved from extinction. However, the area has to be forested to some extent. Tarsiers need sleeping sites, dense undergrowth as locomotor supports, and a high abundance of insects as food source. Controlling the illegal logging in the study area is the single most important action to be taken in order to greatly enhance the chance of survival of this fascinating species. The results of these studies are being included in management plans of several Indonesian authorities.

6 Acknowledgements

The research has been supported by grants from the German Academic Exchange Service (HSP III) and the German National Merit Foundation (to S.M.). We thank the Indonesian Institute of Sciences LIPI as well as BTNLL, Palu for research permits. Further thanks are due to Jatna Supriatna and Noviar Andayani of the Biology Study Program, University of Indonesia, Jakarta for the official sponsorship of this study. We greatly acknowledge the help of Supraptini and Ikin Mansjoer, IPB, Bogor, Ecil Tamalagi and his family in Palu, and Ibu Yulisan in Kamarora. Nothing could have been done without the dedicated work of our field assistants Sapri, Thony, Yulisan, Papa Ece, Herman and Leo.

References

Bynum DZ (1999) Assessment and monitoring of anthropogenic disturbance in Lore Lindu National Park, Central Sulawesi, Indonesia. Trop Biodivers 6(1&2): 43-57

Crompton R, Andau P (1986) Locomotion and habitat utilization in free-ranging *Tarsius bancanus*: a preliminary report. Primates 27: 337-355

Crompton R, Andau P (1987) Ranging, activity rhythms, and sociality in free-ranging *Tarsius bancanus*: a preliminary report. Int J Primatol 8: 43-71

Dagosto M, Gebo D (1998) A preliminary study of the Philippine Tarsier (*Tarsius syrichta*) in Leyte. Am J Phys Anthropol (Suppl) 26: 73

Fogden MPL (1974) A preliminary field study of the Western Tarsier, *Tarsius bancanus* Horsfield. In: Martin RD, Doyle GA, Walker AC (eds) Prosimian biology. Duckworth, London: 151-165

Groves C (1998) Systematics of tarsiers and lorises. Primates 39(1): 13-27

Gursky SL (1994) Infant care in the Spectral Tarsier (*Tarsius spectrum*), Sulawesi, Indonesia. Int J Primatol 15(6): 843-853

Gursky SL (1995) Group size and composition in the Spectral Tarsier, *Tarsius spectrum*, Sulawesi, Indonesia: implications for social organization. Trop Biodivers 3(1): 57-62

Gursky SL (1998a) Conservation status of the spectral tarsier, *Tarsius spectrum*: population density and home range size. Folia Primatol 69(suppl 1): 191-203

Gursky SL (1998b) Effects of radio transmitter weight on a small nocturnal primate. Am J Primatol 46: 145-155

Gursky SL (2000a) Effect of seasonality on the behaviour of an insectivorous primate, *Tarsius spectrum*. Int J Primatol 21(3): 477-495

Gursky SL (2000b) Sociality in the Spectral Tarsier, *Tarsius spectrum*. Am J Primatol 51: 89-101

Gursky SL (2000c) Allocare in a nocturnal primate: data on the Spectral Tarsier, *Tarsius spectrum*. Folia Primatol 71: 39-54

Lamprecht J (1992) Biologische Forschung: Von der Planung bis zur Publikation. Parey, Berlin Hamburg

Leksono SM, Masala Y, Shekelle M (1997) Tarsiers and agriculture: thoughts on an integrated management plan. Sulawesi Primate Newsletter 4(2): 11-13

MacKinnon JR, MacKinnon KS (1980) The behaviour of wild spectral tarsiers. Int J Primatol 1(4): 361-379

Merker S, Mühlenberg M (2000) Traditional land-use and tarsiers – Human influences on population densities of *Tarsius dianae*. Folia Primatol 71: 426-428

Merker S, Mühlenberg M (2002) Endangered or adaptable? - *Tarsius dianae* in man-altered rainforests of Central Sulawesi, Indonesia. Abstracts. The XIXth Congress of the International Primatological Society (IPS), Beijing, China: 21

Niemitz C (1979) Outline of the behavior of *Tarsius bancanus*. In: Doyle GA and Martin RD (eds) The study of prosimian behavior. Academic Press, New York, pp 631-660

Niemitz C (1984a) Biology of tarsiers. Gustav Fischer, Stuttgart New York

Niemitz C (1984b) Synecological relationships and feeding behaviour of the genus Tarsius. In: Niemitz C (ed) Biology of tarsiers. Gustav Fischer, Stuttgart New York, pp 59-75

Niemitz C (1984c) Activity rhythms and use of space in semi-wild Bornean Tarsiers, with remarks on wild Spectral Tarsiers. In: Niemitz C (ed) Biology of tarsiers. Gustav Fischer, Stuttgart New York, pp 85-115

Niemitz C (1984d) An investigation and review of the territorial behaviour and social organisation of the genus Tarsius. In: Niemitz C (ed) Biology of tarsiers. Gustav Fischer, Stuttgart New York, pp 117-127

Niemitz C (1984e) Vocal communication of two tarsier species (*Tarsius bancanus* and *Tarsius spectrum*). In: Niemitz C (ed) Biology of tarsiers. Gustav Fischer, Stuttgart New York, pp 129-141

Niemitz C, Nietsch A, Warter S, Rumpler Y (1991) *Tarsius dianae*: a new primate species from Central Sulawesi (Indonesia). Folia Primatol 56: 105-116

Nietsch A (1993) Beiträge zur Biologie von *Tarsius spectrum* in Sulawesi. Ph.D. dissertation, Free University Berlin

Nietsch A (1999) Duet vocalizations among different populations of Sulawesi Tarsiers. Int J Prim 20(4): 567-582

Nietsch A, Kopp M-L (1998) Role of vocalization in species differentiation of Sulawesi tarsiers. Folia Primatol 69(suppl 1): 371-378

Nietsch A, Niemitz C (1992) Indication for facultative polygamy in free-ranging *Tarsius spectrum*, supported by morphometric data. International Primatological Society Abstracts. International Primatological Society, Strasbourg, p 318

Shekelle M, Leksono SM, Ichwan LLS, Masala Y (1997) The natural history of the tarsiers of north and central Sulawesi. Sulawesi Primate Newsletter 4(2): 4-11

Sokal RR, Rohlf FJ (1995) Biometry, 3rd edn. Freeman and Co., New York

Tremble M, Muskita Y, Supriatna J (1993) Field observations of *Tarsius dianae* at Lore Lindu National Park, Central Sulawesi, Indonesia. Trop Biodivers 1(2): 67-76

Whitmore TC (1984) Tropical rain forests of the Far East, 2nd edn. Clarendon Press, Oxford

Whitmore TC (1990) An introduction to tropical rain forests. Clarendon Press, Oxford

Yustian I, Merker S, Supriatna J, Andayani N (in prep.) Relative population densities of *Tarsius dianae* in man-influenced habitats of Lore Lindu National Park, Central Sulawesi, Indonesia

Home Range, Diet and Behaviour of the Tonkean Macaque (*Macaca tonkeana*) in Lore Lindu National Park, Sulawesi

Anna R. Pombo, Matthias Waltert, S. Supraptini Mansjoer, Ani Mardiastuti & Michael Mühlenberg

1 Introduction

According to Fooden (1969, 1980) there are 19 species in the genus *Macaca*, of which seven occur endemically on Sulawesi, an area covering less than 2% of the total generic range (Albrecht 1978 cited in Bynum et al. 1997). Although there are already several field studies on *Macaca nigra, Macaca nigrescen,* and *Macaca maura* (MacKinnon 1980, Sugardjito et al. 1989, Supriatna 1991, Kohlhaas 1993, Kinnaird and O'Brien 1995, 1996, Reed et al. 1997, Rosenbaum et al. 1998, Okamoto and Matsumura 2002), only little has been done on the behaviour and ecology of the Tonkean macaque *Macaca tonkeana*. The Tonkean macaque (*Macaca tonkeana*) is a Central Sulawesi endemic and considered at lower risk/near threatened (Lr/nt) by IUCN categories (Hilton-Taylor 2000). Sulawesi Macaques are found in lowland and hill forests but only uncommon at elevations above 1500 m (Sarasin & Sarasin 1905, cited in Whitten, 1987). For *Macaca tonkeana*, the primary conservation issue at this time appears to be the ability of populations to persist in highland protected areas, since habitat encroachment, hunting and eradication as crop raiders appear to pose problems at moderate elevations (Bynum *et al.* 1999). The Lore Lindu National Park in Central Sulawesi is located in the centre of the species' range and comprises generally mountainous terrain with over 90% of the park area located above 1000 m (Wirawan 1981).

In this paper we present data from this area on the species' home range, diet and behaviour. During four months of fieldwork at lower montane elevations, two neighbouring groups were studied, one being situated in flat but anthropogenically disturbed forest and one in hilly near-primary forest. The main objectives were to identify and describe (1) group size and composition, (2) diet, (3) home range and daily path lengths, and (4) activity patterns.

We assumed that habitat quality is positively correlated with group size and that – since animal movement is energetically expensive – it is negatively related to daily path lengths and home range size. We also assumed that the proportion of time that animals spend with different activities should be closely related to variability in resource abundance and hence habitat quality (Terborgh 1983; Robinson 1986; Kinnaird 1990 in O'Brien and Kinnaird 1997).

2 Methods

2.1 Study Site

This study was conducted in the Lore Lindu National Park (LLNP), Central Sulawesi [01° 25' S, 120 ° 18' E] (Fig. 1). The 229,000 ha comprising the Park range from c. 200 to 2,610 m above sea level (Watling 1983). Annual precipitation lies between 2,500 and 3,500 mm (Wirawan 1981 in Watling 1983). A 120 ha study area near the village of Wuasa in the Napu valley was established ranging in elevation from 1000 to 1500 m a.s.l. The study area was made accessible by parallel trails at 100 m distance and all trails were marked at 50 m intervals. A map was drawn from the study area on which 0.25 ha blocks were delineated. The forested area was characterised mainly by two different vegetation types: (a) an almost flat area with a mosaic of disturbance regimes including secondary forest and mixed coffee/cocoa gardens under natural forest tree cover and (b) a near-primary hilly area with steep slopes without gardening activities but frequent rattan collection and hunting. Annual cultures, mainly maize fields or beans dominated the farmland bordering the forest.

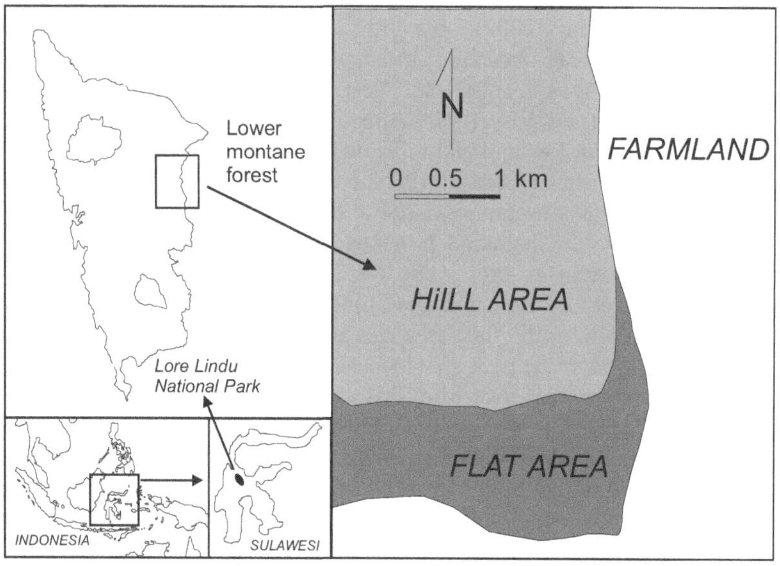

Fig. 1. Location of Sulawesi, Lore Lindu National Park and habitats of the study area. The hilly area (pale grey) was covered by largely undisturbed forest while the flat area (darker grey) was moderately disturbed by farming and tree felling.

2.2 Data Collection

During a preliminary survey, six separate macaque groups and two solitary individuals were located in or near the study area, the groups ranging from *ca.* 10 to 25 individuals (x = 16 individuals). Two neighbouring groups that were located near the forest margin were selected for data collection. Both groups were habituated to human observers.

After a habituation period of four weeks (November 2001), each group was followed for four months and 5 days per week from 06.00 to 18.00 hrs and accompanied for 20-30 min more if necessary to map previously unknown sleeping trees. Every 0.5 hr each group's location was determined by assigning the group centre to a 0.25 ha block being part of a 50x50 m grid system that was measured from the ground. The daily path lengths for each group was also measured using this grid. Behavioural data were collected via scan samples (Altman, 1974; Martin and Bateson 1986; Paterson 1992) on 3 days/week and ad libitum, and inter-group interactions were also recorded. Activities were recorded during a 10 min period at 30 min interval. When feeding, food items were collected using samples of leaves and fruit and later identified at the Herbarium Celebensis in Palu. Four classes of activities were recognised:

- Moving: locomotion including walking, running, climbing and jumping
- Feeding: reaching for, picking up, manipulating, masticating, or placing food in mouth, as well as manipulating the contents of a cheek pouch.
- Resting: body stationary, usually sitting or lying down.
- Social: playing, grooming, sexual and aggressive behaviour.

3 Results

3.1 Group Size and Composition

The two groups observed differed in size: the large group (group A) consisted of 25 and the small group (group B) of 14 individuals. The male-female ratio was 1:1.2 for group A and 1:1.3 for group B. Both groups showed a complete sex and age composition (Table 1) with two adult females being pregnant in group A. In this group, a new infant was born during the fieldwork after which three members left the group. Group A was composed of 44% adults, 44% juveniles, and 12% infants while group B was composed of 50% adults, 43% juveniles, and 7% infants.

3.2 Home Range and Daily Path Lengths

The two groups were observed for a total of 85 days. Fig. 3 shows that the home range of the larger group was almost completely within the level forest area that was disturbed by small forest gardens and secondary forest, while the smaller group's home range was situated mainly within the hilly but undisturbed forest. The frequency of use of 0.25 ha quadrats varied from 1-30 scores per quadrat during the study. The home ranges of both groups were overlapping slightly in the transition between the flat and the hilly area. The sizes of the home ranges estimated from measurements along the transect grid revealed that the home range of the large group was smaller with 44.5 ha than the one of the small group which covered 99.1 ha. However, the horizontal projection of the home ranges using the GPS measurements did not reveal such a large difference in home range size between the groups indicating the extreme hillside character of the small group's home range (Fig.2). The average path length of the large group was 707m (±249m) and was significantly higher with a mean of 1500m (±292) in the small group (Wilcoxon test, Z=-6.93, p<0.001).

3.3 Diet

During the study, 46 plant species were recorded providing food for *Macaca tonkeana* (Table 2). Fruits (71.7%) were the most preferred food item and Moraceae were the most important family constituting 15.6% of all species.

The macaques exclusively fed on ripe fruits and young leaves. Only the soft tissue of fruits was used, and usually parts of the fruit fell down to the ground. Also a small proportion of other plant parts such as flowers and stems and also bird eggs and insects were taken.

3.4 Activity Budget and Behaviour

The percentage of time spent moving was approximately 1.8 times higher in the smaller than the larger group (see below). The large group (A) spent most of its time resting (51%), followed by socialising (19.9%), moving (17.7%) and feeding (11.4%), while the small group (B) spent most of its time resting (35%), followed by moving (31%), socialising (14.7%), and feeding (19.3%) (Fig. 3). The time budgets differed significantly between the groups (Crosstables, χ^2=1086, d.f.=3, p<0.001).

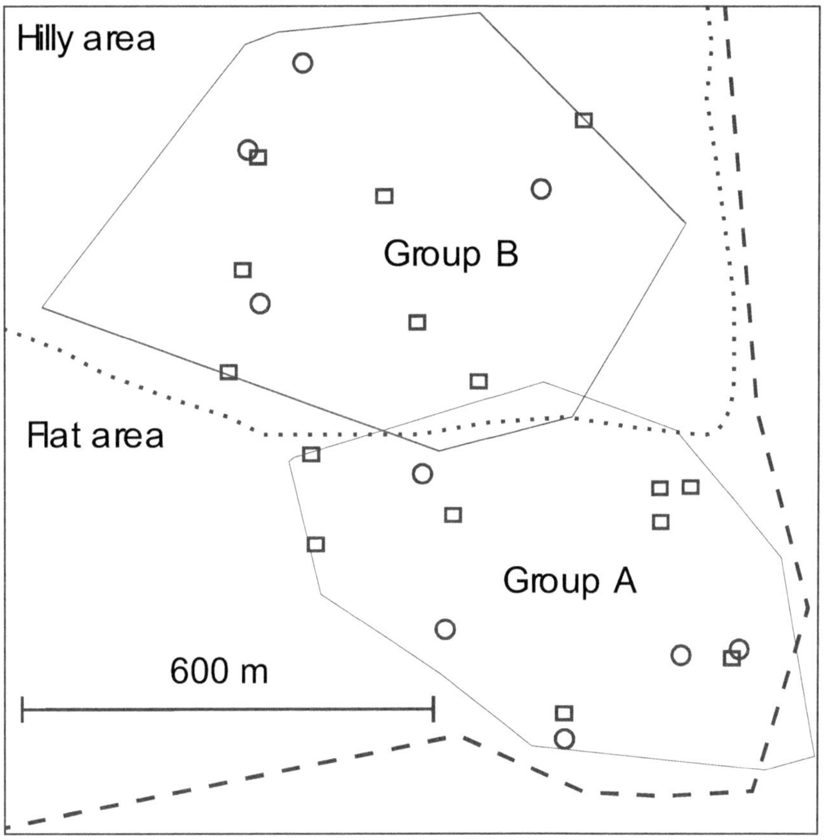

Fig. 2. Minimum convex polygons of home ranges of two groups of Tonkean Macaques, as well as forest border (dashed), delimitation of hilly area (dotted), and location of sleeping (squares) and feeding trees (circles).

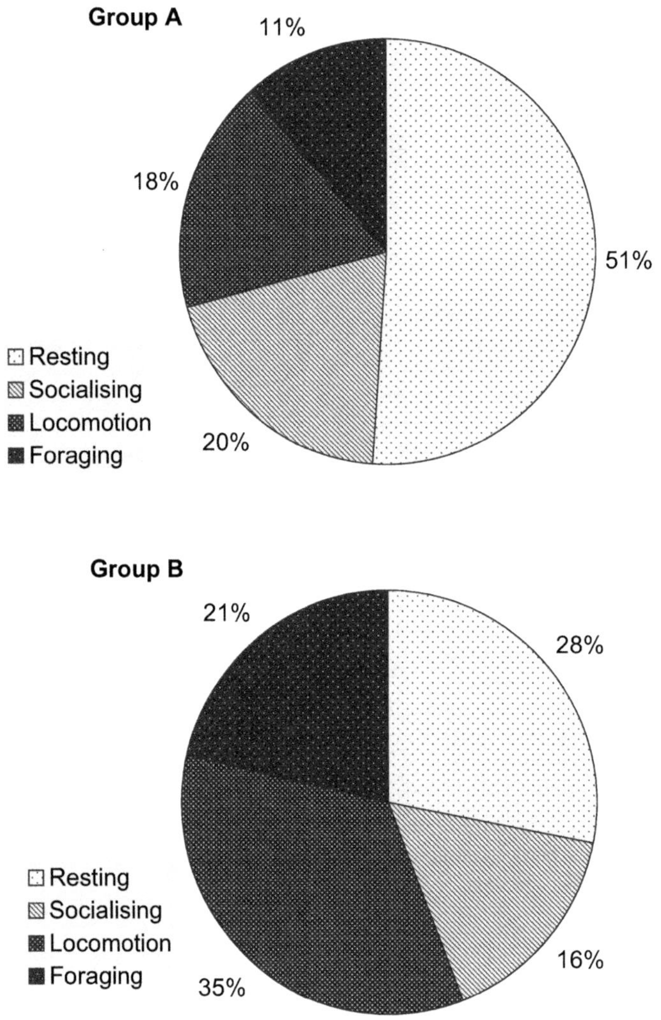

Fig. 3. Activity budgets for two *Macaca tonkeana* groups at Lore Lindu National Park during 41 days of scan sampling. Group A (6657 observations) consisted of 25 individuals while Group B (7380 observations) consisted of 14 individuals.

Table 1. Age and sex composition of two groups of Tonkean macaques *Macaca tonkeana* at Lore Lindu National Park, Central Sulawesi

Group	Adult		Juvenile		Infant		Total
	M	F	M	F	M	F	
A	5	6	4	7	3	-	25
B	3	4	4	2	-	1	14

Table 2. List of plant species and plant parts consumed by *Macaca tonkeana* at Lore Lindu National Park, Central Sulawesi.

No	Local Name	Scientific Name	Family	Parts consumed
1	Mbalahapu (1)	*Ficus variegata*	Moraceae	L Fr
2	Pangarehea	*Elastostema sp.*	Urticaceae	St
3	Mbalahapu (2)	*Ficus sp.*	Moraceae	L
4	Weli-weli	*Photos sp.*	Araceae	St
5	Nuhu	*Ficus sp.*	Moraceae	L Fr
6	Pondo	*Pleomele angustifolia*	Liliaceae	L
7	Bangkakara (1)	*Cryptocarya sp.*	Lauraceae	St
8	Bangkakara (2)	*Litsea sp.*	Lauraceae	Fr
9	Tambekakau	*Eugenia sp.*	Myrtaceae	L Fl Fr
10	Nunu tea	*Ficus sp.*	Moraceae	L Fr
11	Tohiti	*Calamus sp.*	Palmaceae	Fr St
12	Aropi	*Antidesma tetrandum*	Euphorbiaceae	Fr
13	Tiro-tiro	*Elaeocarpus sp.*	Elaeocarpaceae	Fr
14	Betau	*Calophyllum sp.*	Guttiferae	Fr
15	Andolia	*Canangium odoratum*	Annonaceae	L Fr
16	Kaulele	*Rapanea sp.*	Myrsinaceae	Fr
17	Pepolo	*Biscoffia javanica*	Euphorbiceae	Fr
18	Lekatu	*Duabanga mollucana*	Sonneratiaceae	L
19	Warani	*Semecarpus sp.*	Anacardiaceae	L
20	Tea uru	*Artocapus teijsmanii*	Moraceae	Fr
21	Wala	-	Leguminosae	L
22	Kumowatu	*Guiou sp.*	Sapindaceae	L Fl
23	Nunu	*Ficus caulocarpa*	Moraceae	L Fr
24	Nunu Manetetawena	*Ficus sp.*	Moraceae	L Fr
25	Taite	*Dysoxylum sp.*	Meliaceae	L Fr
26	Wale nunu	*Ficus sp.*	Moraceae	Fr
27	Pisang monyet	*Musa sp.*	-	L Fr L
28	Harao	*Areca vestiaria*	Arecaceae	Fr
29	Harao maeta	*Arenga sp*	Arecaceae	Fr
30	Mangkulobi	-	-	Fr

Abbreviations: Fr - fruit, L - leaf, Fl – flower, St – stem

Table 2 (cont.)

No	Local Name	Scientific Name	Family	Parts consumed
31	Kolombio	-	-	Fr
32	Aren	*Arenga pinnata*	Arecaceae	Fr
33	Lebanu	*Nauclea cyrtopoda*	Rubiaceae	L
34	Pala hutan	*Knema cinerea*	Myristicaceae	Fr
35	-	*Memecylon gibbosum*	Melastomataceae	L Fr
36	Pangi	*Pangium edule*	Flacourtiaceae	Fr
37	Kedondong hutan	*Spondia pinnata*	Anacardiaceae	L Fr
38	Anggrek	-	Orchidaceae	Fl L
39	Pondang	*Pandanus sp.*	Pandanaceae	L
40	Manggis hutan (1)	*Garcinia sp.*	Clusiaceae	Fl Fr
41	Manggis hutan (2)	*Garcinia celebica*	Clusiaceae	Fl Fr
42	Jamur	*Fungus*	-	
43	Liana (1)	*Alangium sp.*	Alangiaceae	L Fr
44	Kayu telor	*Alstonia ranvolfia*	Apocynaceae	Fl Fr
45	Kanari	*Canarium sp.*	Burceraceae	Fr
46	Jambu batu	*Psidium guajava*	Myrtaceae	L Fr

4 Discussion

4.1 Group Size and Composition

Macaques (*Macaca* spp.) usually live in large, stable groups. *Macaca nigra* groups can contain 30-90 individuals, also *Macaca fascicularis* can reach 100 individuals (Kohlhaas 1993). In such large groups, the females usually outnumber males 4 to 1, with the society being centred around female families, while males move away from their natal groups (Matsumura 1998). The groups of *Macaca tonkeana* we found were only between 10-30 individuals. With a male-female ratio of 1:1.2 to 1:1.3 we expected our groups to have the capability to develop into bigger associations, but the fact that three group members left group A to form a new group suggests that larger groups might be less frequent in the species.

Although much has been written about factors affecting group size in primates, this subject is still much in debate. Factors proposed to favour grouping in primates include protection from predators, defense of resources, improved foraging efficiency, and improved care-provision opportunities (Dunbar 1988 in Kohlhaas 1993). On Sulawesi, where predation must be relatively low and fruit availability – at least at lower altitudes - is usually high, larger groups than those observed by us can be expected. However, it is yet unclear whether fruit availability at submontane elevations of Lore Lindu is comparable to the lowland areas so far studied within the ranges of *Macaca nigra*.

4.2 Home Range and Daily Path Length

Since forest primates usually use food resources that are heterogeneously distributed in space and time, the search for food is a crucial part of their life, affecting also other behavioural patterns. The balance between the time and energy required for moving between food resources and the food's nutritional quality therefore also affects primate daily path lengths (Strier 1987). In *Macaca nigrescens*, daily path lengths of 850 m per day were recorded (Bismark 1982, cited in Whitten et al. 1987), and a group of *M. nigra* covered approximately 2 miles. The daily path lengths of the groups of *Macaca tonkeana* in our study were also within this range.

Ranging patterns and home range use of our *Macaca tonkeana* groups were almost certainly linked, respectively, to the spatial distribution of natural feeding trees or the availability of crops at the forest border. Although food availability was not measured in this study, it seems likely that it differed between the two habitats studied. The small group was living on a much steeper terrain that was probably less productive than the flat forest margin where the larger group was living. Despite moderate disturbance, this group probably had access to a higher number of feeding trees and could also use farm crops at the forest border including nutritious ones such as maize, papaya, banana and cocoa.

4.3 Diet

Sulawesi macaques are predominantly frugivorous. E.g. *Macaca nigra* in North Sulawesi uses fruits in 66% of all feeding bouts and consumes more than 145 species (O'Brian et al. 1997). In this species, the Moraceae family accounted for 30.4 % of all fruit records, with all but one being *Ficus* species. *Dracontomelum dao* (Anacardiaceae) was the most frequently eaten fruit species and responsible for continuous food supply during the year. Invertebrate food is also important and believed to influence routes taken between fruit trees (O'Brian et al. 1997).

Due to the availability of a herbarium in Palu, we were also able to determine the diet *Macaca tonkeana* although still many tree species on Sulawesi are yet undescribed. Our observations show that the species actually exploits a wide variety of foods. Given the relatively short period of our study, the diet of the species at our study site was composed of relatively many plant species (46). As in *M. nigra* (see above), Moraceae was the most important family and fruits, mostly ripe, were the main class of food eaten. Our data are therefore comparable with what is known from other Sulawesi macaques (O'Brien and Kinnaird 1997; Kohlhaas 1993).

4.4 Activity Budget and Behaviour

The amount of time spent with various activities is a reflection of how an animal balances its energy budget with the demands of life. A monkey that can easily obtain food can spend more time resting and grooming than feeding and moving.

Group differences in activity budgets can often be attributed to gross habitat differences. Both macaque groups studied spent most of their time in trees and less on the ground which appeared to be in contrast to *Macaca nigra* with which the first author is also familiar. It is still unclear whether this behaviour is a diagnostic character for *Macaca tonkeana* but it could be related to thicker ground cover in our – submontane – study area compared to the sparse understorey of the Tangkoko Reserve where *Macaca nigra* has been studied. Another potential explanation could be uncompleted habituation: indeed, long-term studies and ecotourism activities have made *M. nigra* certainly much more habituated to humans.

The larger group spent most of its time resting and socialising and less time moving and feeding while the smaller group spent more time moving and feeding than resting and socializing. The various macaque species often have very different time budgets to cope with environmental circumstances. It is possible that those with reduced resting and social activities are ecologically stressed (Dunbar 1992 in Kohlhaas 1993).

Other environmental factors than resource availability have probably affected several activities in *Macaca tonkeana*. We observed that high rainfall, storm and high daytime temperature increased the resting time of both groups. This has also been reported by Rinaldi (1993) who found the Javan Gibbon to take a rest during rainfall and storm and also Kohlhaas (1993) reports high daytime temperatures to increase resting time and decrease social activities.

Due to group-living, *Macaca tonkeana* individuals have many social contacts everyday. Most social activities during the day were occupied with grooming. Adult females were most likely to groom others of any age or sex class, a behaviour which serves not only for cleaning purposes but also reduces aggression and tightens the group members bond. Thierry et al. (1990) found that in a captive colony of *Macaca tonkeana*, adult females groomed and received most often and that adult female-adult female grooming bouts were usually of especially long duration and carried out facing each other. In contrast, adult males more often groomed on the back and for shorter times.

5 Conclusion

- *Macaca tonkeana* lives in multimale-multifemale groups, showing a complete age and sex composition and a variable group size.
- Out of two groups observed, the larger group had a smaller home range (44.5 ha) probably containing more abundant food resources compared with the

smaller group (91.5 ha) which home range was larger probably containing a lower feeding tree density.

- *Macaca tonkeana* is primarily frugivorous, its diet consisted mainly of fruits with Moraceae being the most important family.
- *Macaca tonkeana* is active during the day (diurnal) and spends most of its time in trees.
- The two groups studied showed differences in their activity budget, with the larger group spending more time with social activities.

6 Recommendations

- Since there are human-monkey conflicts (crop raiding of maize), it may be possible to focus near the forest border on production of crops that are less attractive to macaques (rice, beans).
- Since Lore Lindu National Park is an important protected area in Central Sulawesi and the major refuge for *Macaca tonkeana*, there is a need to carry out further research concerning the species' ability to survive in montane forest areas.
- A field primate research station and/or captive breeding program for Sulawesi macaques is recommended on a long-term scale for monitoring, research and captive breeding.
- On a short-term scale, efficient protection of Sulawesi's protected areas by a combination of increased law enforcement, environmental awareness campaigns and village agreements is urgently needed, especially in areas such as Lore Lindu where both hunting and forest loss are major problems.

7 Acknowledgments

This work was funded by the German Research Council (DFG) through the STORMA project, an Indonesian-German research collaboration. We thank the German co-ordinator and all researchers and Ir. Banjar Laban, Ms head of Balai Taman Nasional Lore Lindu (BTNLL) for their help during our time in Sulawesi. We warmly thank to Prof. Drh. Dondin Sajuthi Ph.D., head of Primate Research Center (PSSP) and all its staff. We especially thank Drh. Ikin Mansjoer, MSc., Dr. Randy Kyes, Jon Riley and all staff of the Wildlife Conservation Society (WCS-Sulawesi IP) as well as of The Nature Conservancy Sulawesi, Herbarium Celebense, as well as the Pekurehua and the Jambata Society, for their help and co-operation.

None of this would have been possible without assistance and dedication of our field assistants Man, Eda, Nimus, Pius and our Cook Johan. We thank Papa Lin and Papa Vito for local names of plants and especially Amat and Icha who were very patient and helpful accompanying me on the field which made the work easier. Thank you very much and God bless you.

References

Altman J (1974) Observational Study of Behavior Sampling Methods. Allec Laboratory of Animal Behavior. University of Chicago, Illionis. (USA)

Anonymous (2002) Lore Lindu National Park. http://www.users.bart.nl/edcolijn/lindu.html [09.07.2002]

Bynum E, Bynum DZ, Supriatna J (1997) Confirmation of the hybrid zone between wild populations of *Macaca tonkeana* and *Macaca hecki* in Central Sulawesi, Indonesia. American Journal of Primatology 43: 181-209

Bynum EL, Kohlhaas AK, Pramono AH (1999) Conservation status of Sulawesi Macaques. Tropical Biodiversity 6: 123-144

Fooden J (1969) Taxonomy and Evolution of The Monkeys of Celebes. Bibliotheca Primatologica

Fooden J (1980) Classification and distribution of living macaques (*Macaca* Lacepede, 1799). In: Lindburg DG (ed) The macaques: studies in ecology, behaviour, and evolution. Van Nostrand Reinhold, New York

Hilton-Taylor C (compiler) (2000) 2000 IUCN Red List of Threatened Species. IUCN, Gland (Switzerland) Cambridge (UK)

Kohlhaas AN (1993) Behavior and Ecology of *Macaca nigrescens*: behavioral and Social Responses to the Environment and Fruit Availability. Ph.D. thesis, University of Colorado

Lee RJ (1997) The Impact of Hunting and Habitat Disturbance on the Population and Behavioral Ecology of the Crested Black Macacaque (*Macaca nigra*). Ph.D. thesis, University of Oregon.

Martin P, Bateson P (1986) Measuring Behavior 2nd edn. Cambridge University Press, Cambridge

MacKinnon KS (1986) The Conservation Status of Nonhuman Primates in Indonesia in Primates: the Road to Self Sustaining Populations. Springer, New York

Matsumura S (1998) Relaxed dominance relations among female moor macaques (*Macaca maurus*) in their natural habitat, south Sulawesi, Indonesia. Folia Primatologica 69: 346-356

O'Brien TG, Kinnaird MF (1997) Behavior, Diet and movements of the Sulawesi Crested Black macaque (*Macaca nigra*). International Journal of Primatology 18: 321-349

Okamoto K, Matsumura S (2002) Intergroup Encounters in Wild Moor Macaques (*Macaca maurus*). Primates 43: 119-126

Paterson JD (1992) Primate Behavior: an Exercise Work Book. Waveland Press Inc.

Reed C, O'Brien TG, Kinnaird MF (1997) Male Social Behavior and Dominance Hierarchy in the Sulawesi Crested Black Macaque (*Macaca nigra*). International Journal of Primatology 18: 247-260

Rinaldi D (1999) Food Preferences and Habitat Utilization of Javan Gibbon (*Hylobates molloch* Audebert) in Ujung Kulon National Park. West Java, Indonesia. MSc Thesis, Georg August University Göttingen [In German]

Rosenbaum B, O'Brien TG, Kinnaird MF, Supriatna J (1998) Population of Sulawesi Crested Black Macaques on Bacan and Sulawesi, Indonesia; Effect of Habitat Disturbance and Hunting. American Journal of Primatology 44: 89-106

Supriatna J, Froechlich JW, Erwin JM, Southwick CH (1992) Population, habitat and conservation status of *Macaca maurus*, *Macaca tonkeana* and their putative hybrids. Tropical Biodiversity 1: 31-48

Strier KB (1987) Primate Behavioral Ecology. Allyn and Bacon (USA)

Thierry B, Gauthier C, Peignot P (1990) Social Grooming in Tonkean Macaques (*Macaca tonkeana*). International Journal of Primatology 11: 357-375

Wirawan N (1981) Ecological Survey of the Proposed Lore Lindu National Park, Central Sulawesi. Hasanuddin University, Ujung Pandang.

Predicting Losses of Bird Species from Deforestation in Central Sulawesi

Matthias Waltert, Maike Langkau, Miet Maertens, Michael Härtel, Stefan Erasmi & Michael Mühlenberg

1 Introduction

Tropical deforestation and forest fragmentation are probably the most serious threats to biodiversity (see Turner 1996) and it has been theoretically stated that even the largest protected areas in the tropics might be too small to sustain populations of all species of the original system (Terborgh 1999). But species loss in forest fragments is a complex process and appears often only after considerable time lags, especially in vertebrates (Brooks et al. 1999b). Therefore, empirical evidence for such extinctions can only be obtained from areas with a long deforestation history and long-known faunal composition (e.g. van Balen 1999). Such empirical data are scarce but are essential in order to convince land use managers of the long-term effects of forest loss on biodiversity. Species area models, however, are a valuable tool in the prediction of tropical vertebrate species loss (van Balen 1999; Brooks et al. 1997, 1999a, c, 2002; Cowlishaw 1999).

The Sulawesi region (Sulawesi plus its satellite islands) in Indonesia is regarded as one of the most important centres of endemism in the world. It supports 97 endemic bird species of which 16 belong to distinctive endemic genera, most of which are found in rainforest (Coates et al. 1997). Due to past deforestation the extent of lowland forest below the foot/hill boundary in Sulawesi is already insignificant (FWI/GWF 2002). Yet our observations in Central Sulawesi indicate even more recent deforestation of remaining hill forest. Much has been converted to intensively used agricultural land which supports very few indigenous bird species (see Coates et al. 1997; Waltert et al. 2003). Forest gardens and secondary forest habitats could potentially play a major role for the conservation of the island's birds, since many species can frequently be found in such habitats (see Schulze et al. 2003 and this volume; Waltert et al. 2003). However, these secondary habitat types are also subject to land conversion and their definitive value for bird conservation has still to be investigated (Pangau et al. 2003).

In this paper we analyse the extent of lowland/hill forest cover in the Lore Lindu National Park in Central Sulawesi and compile information on breeding status, endemism and altitudinal ranges of its bird species. Using this data and the species-area relationship we predict bird species losses from the present deforestation regime. The intention of this paper is to add information to the discussion on land use management in the study area.

2 Study Area

The Lore Lindu National Park was created in 1993 (see TNC/BTNLL 2002). It is 2,180 km² in size and ranges from c. 200 to 2,610 m above sea level. It is largely mountainous with c.70 % of its area between 1,000 and 1,500 m (Watling 1983). Annual precipitation lies between 2,500 and 3,500 mm (Wirawan 1981 in Watling 1983). The Park harbours populations of Sulawesi's many endemic mammals (e.g. the Anoa *Bubalus sp.* or the Babirusa *Babyrousa babirussa*) and is home to some 230 bird species, including a remarkable proportion (68 %) of the 97 Sulawesi region endemics. It is also being considered crucial for the conservation of the global population of the Maleo *Macrocephalon maleo*, a megapode which nests communally on solar-heated beaches and geothermally heated forest soils (Butchart and Baker 2000). The area was designated as UNESCO Biosphere Reserve and World Heritage Site, is a Centre of Plant Diversity (BPD), and is part of the Sulawesi endemic bird area (EBA), as well as the Global 200 Ecoregion Sulawesi/Wallacea (Olsen and Dinerstein 1998).

Human population density around the Park varies from 125 persons per km² to less than 13 persons per km² in different sub-districts and the area was characterized by a comparatively high population growth of 2.4 % per year for the period between 1980 and 2001 (Birner et al. 2002). Some villages are located in two enclaves inside the Park. The Park management strategy can be described as collaborative management (co-management) and the biosphere concept is included in the management plan (TNC/BTNLL 2002).

The present management strategy includes agreements of the National Park administration and several non-governmental organisations with a large number of the villages surrounding the Park. These agreements acknowledge indigenous rights such as the use of non-timber-forest products within the Park and are relevant to many lowland areas situated at its borders. So far, only in a few village agreements wildlife management issues were considered. A special case of deforestation of hill forest occurred in 2001. The action was initially intended to serve as political protest of villagers against government authorities but evidently went out of control affecting probably 60 km² of forest so far. This shows that awareness creation regarding the relevance of submontane forest cover is extremely important. Local stakeholders and even environment-oriented NGOs do not seem to have sufficient knowledge about the relationships described in this paper.

3 Methods

3.1 Bird Community and Altitudinal Range Data

The bird species composition of Lore Lindu National Park is well-documented (see Coates et al. 1997; TNC/BTNLL 2002). We used the species list of Nur Mallo and Ma'dika (1999) with one addition (Mauro 2001) although we are aware of additional, but unconfirmed records (TNC/BTNLL 2002). Each species was classified according to status, geographic range and habitat for which data from White and Bruce (1986), MacKinnon and Phillips (1993) and Coates et al. (1997) were compiled. Information on altitudinal ranges has been taken from Coates et al. (1997) who refer to Central Sulawesi for most of the species. Each one of the resident forest-dependent species was assigned to occurrence in one or more of six elevational distribution classes (0–400m, 400–800m, 800–1,200m, 1,200–1,600m, 1,600–2,000m and 2,000–2,400m). This presence/absence data were used to calculate species similarity between elevation classes with the help of the incidence-based Sörensen index (Colwell 2000). Dissimilarity (1.0 - Sörensen index) between elevation classes was then used in a single-linkage cluster-analysis (Ward method) to analyse sequential turnover. Analyses were made using software of Colwell (2000) and Statsoft (1995).

3.2 Digital Elevation and Remote Sensing Data

Waltert et al. (2003) showed that modernized agricultural systems such as *Gliricidia*-shaded cocoa/coffee plantations and annual cultures (maize) did not support any forest-dependent bird species in the Lore Lindu region, whereas secondary forests were still used by a high percentage of species. We therefore included secondary growth and areas with traditional forest gardens in our analyses of closed forest cover (see Fig. 1). For the calculation of species extinctions the original extent of (closed) forest cover is essential. Since remote sensing data were only available for the period 1999-2001, we assumed that the extent of the original closed forest was equivalent to the sum of present closed forest and areas for which it can be assumed that they once *were* closed forest. Therefore, based on the analyses of the Landsat image from 1999 and 2001, another land use class, open forest, was included in our analysis. This classification represents woodland-like habitats with tree-crowns being clearly spaced as it is the case in areas that were recently deforested for farming of annual crops such as maize (not depicted in Fig. 1). We used digitised topographic maps scaled 1:50,000 and two Landsat ETM images from 1999 and 2001 to analyse the extent of these two classes at different altitudinal bands within the study area.

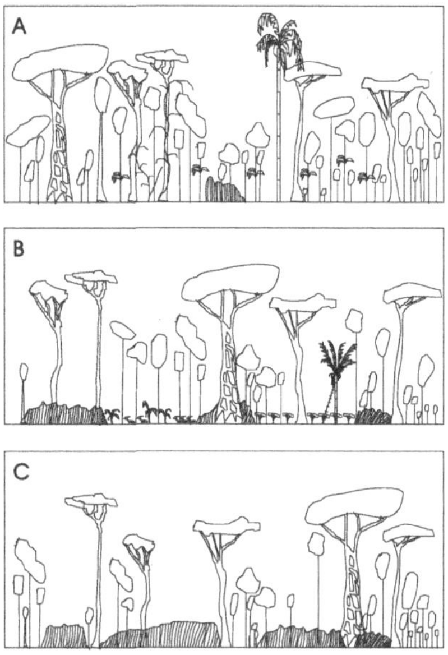

Fig. 1. Field sketches from forest habitats in northern and eastern Lore Lindu National Park (800–1,100 m) showing variation of forest types that were included in the classification of closed forest. A natural forest, B forest garden, C logged forest.

3.3 Predicting Extinctions

We used the inverse species-area relationship (Preston 1962) to assess the number of lowland forest-dependent bird species threatened from deforestation. This procedure has been proven to be valid since it produced results that were in line with assessments of mammal and bird threat status made by conservation organisations and were consistent both in neotropical and SE Asian tropical rainforest (Brooks and Balmford 1996; Brooks et al. 1999a, b). We used the formula $S=cA^z$, where S = number of species, A = area and c and z are constants. Re-arranging this relationship, one can calculate the number of species to go extinct in a fragment (or nested subset) of a given size: Dividing $S_{surviving}=cA_{surviving}{}^z$ by $S_{original}=cA_{original}{}^z$ gives $S_{surviving}/S_{original}=(A_{surviving}/A_{original})^z$. Because $S_{extinct}=S_{original}-S_{surviving}$, one can estimate $S_{extinct}=S_{original}-[S_{original}(A_{surviving}/A_{original})^z]$.

Rosenzweig (1995) reviewed empirical work on species-area patterns and showed that the slope of the species-area curve (z) differs between nested subsets of habitat, real islands, small forest fragments and areas with different biogeographical history. Since there are no empirical data on the species-area relationship available for Sulawesi, we chose the traditional value of z = 0.25 (Preston 1962) which has been used for archipelagos of islands as well as for larger forest fragments within islands (Brooks et al., 1997, 1999). To obtain information on the robustness of our estimate we also used two "extreme" values of z (0.34

and 0.17) to calculate $S_{extinct}$. While a high value such as z = 0.34 is almost certainly overestimating species loss, the low value of z = 0.17 underestimates threat (see also discussions of z values in van Balen 1999 and Brooks et al. 1999b).

The species-area relationship will of course only predict the extinction of species "endemic" to the area of interest (i.e. lowland forest specialists), since species with larger "ranges" (altitudinal generalists) could survive in adjacent mountainous regions even if the lowland was completely destroyed. We therefore used the lowland forest-dependent species for our calculations.

Since bird species composition was so far available only for the whole Park area, not separate areas of lowland forest, we treated the lowland forest area within the Park as a single unit. In our estimations we also have assumed that lowland/hill forest outside the study area will become unavailable and that remaining areas will become isolated patches surrounded either by cultivation or montane forest.

4 Results

4.1 Habitat, Geographic Range and Altitudinal Distribution of Bird Species

A total of 228 bird species have been recorded in Lore Lindu National Park to date (Appendix). Of these, 94 species may be considered non-forest species living mainly in wetlands and open habitats. All but three of the remaining 134 forest-dependent species are breeding residents.

Of the 131 resident forest-dependent species, 69 species occur at higher elevations (altitudinal generalists and montane species). Of these, 75% (52 species) are endemic. 62 (or 47%) of the 131 resident forest-dependent species are associated with elevations below 1,200 m (hereafter called "lowland forest specialists"). Out of these 62 lowland forest specialists, again nearly one half, 30 species, are endemic at least to the Sulawesi region. For predictions of bird species loss (see below) we therefore used $S_{original}$ = 62. The detailed grouping of the resident forest bird species according to altitudinal occurrence is summarised in Fig. 2.

An analysis of resident forest species turnover between altitudes shows that the avifauna of Lore Lindu National Park changes rapidly between 1,200 m and 1,400 m asl. At these altitudes most lowland/hill forest species disappear and montane species appear with increasing elevations (Fig. 3).

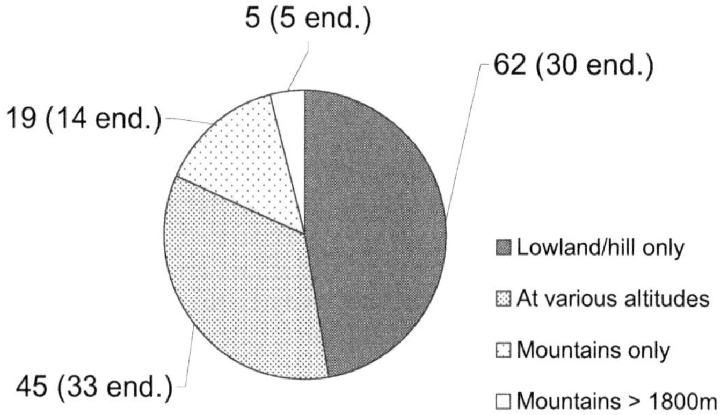

Fig. 2. Habitat characteristics of the 131 resident forest bird species of Lore Lindu National Park, Central Sulawesi in respect to altitudinal specialisation (based on information in Coates et al. 1997). The respective numbers of Sulawesi region endemics are shown in brackets. "Lowland/hill only" represents species found only below 1,200m, "mountains only" those found only above 1,200 m.

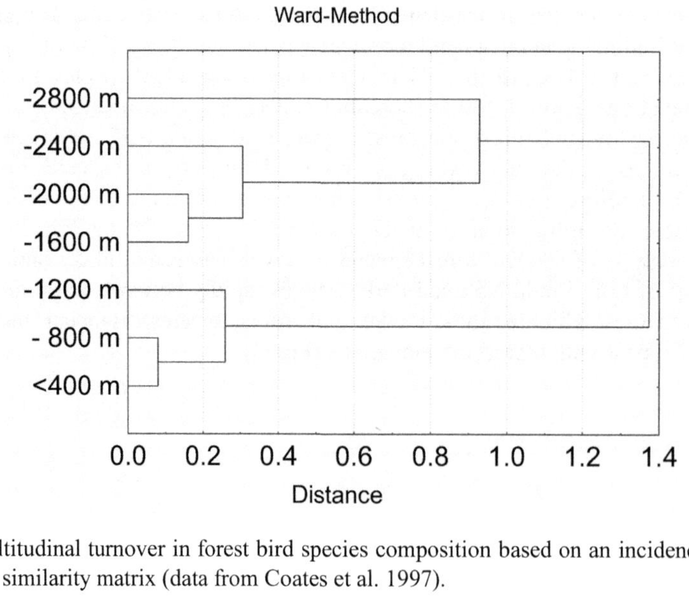

Fig. 3. Altitudinal turnover in forest bird species composition based on an incidence-based Sörensen similarity matrix (data from Coates et al. 1997).

This indicates that many lowland specialists have an upper altitudinal limit at about 1,200m but also reflects the fact that several species are not even found above 800m. Another sharp change occurs between 2,400 and 2,800 m , where many lower montane species have their upper altitudinal limit but where there is still a distinct assemblage of endemic species (Fig. 3).

4.2 Extent of Closed Forest below 1,200 m

According to the digital elevation data, the total surface of the Park area below 1,200 m is about 785 km². In 1999, 774 km² of this area (98.5 %) was still classified as closed forest (forest types shown in Fig. 1) and most of the remaining area, 9 km² (or 1 %), as open forest. In August 2001, closed forest appeared to be reduced to about 628 km² (80 %) leaving 105 km² (13 %) open forest and 52 km² (7%) of perennial crops, grassland and annual crops. This marked change from closed forest to open habitats is mainly visible along the northern border of the National Park (Fig. 4).

In August 2001, closed forest in the lowlands was composed of the following main areas: the already highly disturbed northern Park area, the western Park border area, the south-eastern Park border area and the central (Lindu) area. These strips of land have a rather unfavourable shape and are located along the National Park border. The patches along the south-eastern Park border are very narrow strips of land that are already discontinuous and cover only a few sq. kilometres.

Fig. 4. Lore Lindu National Park (with enclaves Lindu and Besoa) and closed (grey) and open (black) forest cover in areas below 1,200 m in 1999 (left) and August 2001 (right).

4.3 Predicting Species Losses

Since a total of 62 forest-dependent species (including 30 endemics) are confined to Lore Lindu's lowland/hill forest (see above), the Park could potentially lose 45% of its resident avifauna and 34% of its Sulawesi region endemics if this forest is cleared completely. Fig. 5 shows the number of species remaining according to the prediction of the species-area-relationship at various amounts of remaining forest cover. The relationship predicts that if lowland forest within the Park was reduced to approximately 200 km² it would lose between 12 and 22 (19–35 %) forest specialist species. An area reduction down to 100 km² would lead to severe extinctions in the remaining patches (18–30 species).

The analysis of the present situation (August 2001) shows that species loss in the range of 3–8 % is already to be expected (see Fig. 5). Recent observations on deforestation on the ground made after August 2001 indicate the disappearance of appr. 60km² more closed forest in the northern Park area within a six-months period. This would increase the predicted loss to 5–10 % and shows that a single event such as the one observed can already contribute to bird species extinctions. Fig. 6 shows a linear projection of change in lowland closed forest cover in our study area until 2009. If the present trend continues, closed lowland forest could be cleared completely by 2011.

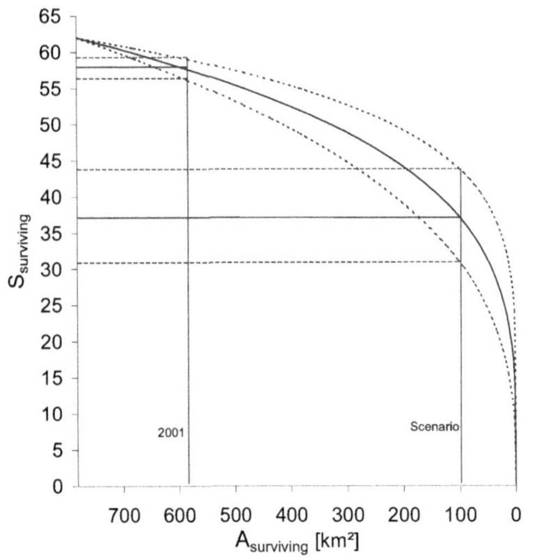

Fig. 5. Relationship between lowland specialist bird species remaining ($S_{surviving}$) and remaining closed forest cover within lowland/hill areas of Lore Lindu National Park, Central Sulawesi. Dotted curves represent calculations using z=0.17 or z=0.34, respectively.

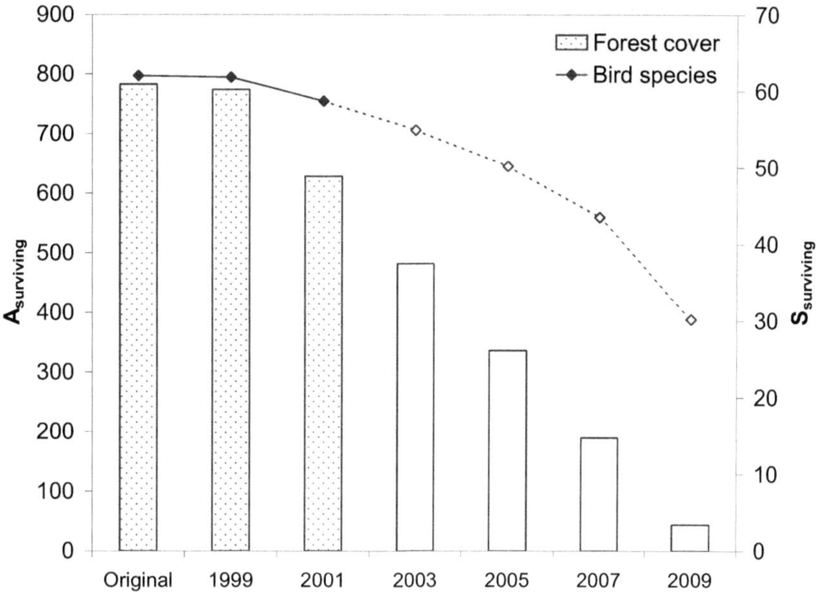

Fig. 6. Development of closed lowland forest cover ($A_{surviving}$) and predicted lowland specialist bird species remaining ($S_{surviving}$) according to the species-area-relationship, using remote sensing data (filled columns) and a linear projection of it (empty columns).

5 Discussion

5.1 Predicting Species Loss from Deforestation

We here present a pattern of bird species extinction from deforestation in the Lore Lindu National Park using the inverse species-area-relationship (Preston 1962). This method has already been applied for the birds of the Atlantic forest in Brazil (Brooks et al. 1999c) and for birds and mammals in insular Southeast Asia (Brooks et al. 1999a). The method has been criticised because - in contrast to its predictions - even the small and highly fragmented Brasilian Atlantic forest did not lose a single endemic bird species until the beginning of the 21st century (Brown and Brown 1992; Budiansky 1994). However, Brooks et al. (1997, 1999a) have demonstrated that the numbers of species considered being threatened by IUCN (Baillie and Groombridge 1996) and Birdlife International (Collar et al. 1994) in the respective regions correspond well with the numbers of species to go extinct predicted by the species-area-relationship. Brooks et al. (1999b, c) therefore conclude that the time-lags involved in population declines of forest birds have prevented species from becoming extinct. This conclusion is rather strong since both alternative explanations, a potential pre-adaptation to deforestation (Brown and Brown 1992) and historical extinctions unnoticed by science, would have lead Brooks et al. 1999a to overestimate the number of threatened species.

This congruence between the "bottom-up" approach of population vulnerability estimation (Collar et al. 1994) and the "top-down" estimation of species extinction (Brooks et al. 1999a) should therefore be a good argument to use predictions from the species-area relationship also for the management of protected areas.

5.2 Empirical Evidence of Bird Extinctions after Fragmentation

The long-term effects of forest fragmentation on the SE Asian rainforest avifauna have recently been assessed by Lambert and Collar (2002) who mainly summarise information on bird species extinction from Singapore (Corlett and Turner 1997) and Java (van Balen 1999). It was estimated that only reserves of 200,000 ha (2,000 km^2) or more will contain all resident lowland forest bird species. None of the 19 lowland forest patches studied by van Balen (1999) including the two largest which contained 50,000 ha and 28,600 ha held the full complement of resident lowland forest bird species. The largest fragment supported 52 out of 65 forest species, suggesting that 24% had become locally extinct. Sixteen forest patches between 6 ha and 10,000 ha contained less than 50% of the forest species. Another prominent example is the 86 ha Bogor Botanical Gardens in Java. After isolation of 60 years, this parkland had lost a third of its original 62 resident bird species, and others were almost extinct. Most of the surviving species (40 out of 42) were habitat generalists, also occurring in the surrounding countryside (Diamond et al. 1987). Similarly, after about 50 years of isolation, the Cibodas Botanical Gardens (*c.* 200 ha) lost 112 (58%) bird species (MacKinnon 1991). A few more empirical examples of forest fragmentation and bird extinctions are available from other continents, as summarised in Appendix 2.

5.3 Which Species are Most Threatened?

It seems clear that not all species are similarly prone to forest fragmentation but that certain ecological characteristics are linked to species' population viability in fragments. From the comparison of the species composition in the 19 forest fragments studied by van Balen (1999), Lambert and Collar (2002) listed the species that are most at risk from fragmentation in the Sundaic region. Perhaps surprisingly many of these threatened birds do not have large territorial requirements. Also in the Bogor Botanical Gardens, some small (<20g) and many medium-sized birds (20–49 g) were lost (Diamond et al. 1987).

This is understandable when looking at the mechanisms driving the extinction of birds in rainforest fragments. There are those that are linked to small populations (e.g. demographic and environmental stochasticity) but also others such as natural population density, dispersal abilities, increased predation (e.g. through mesopredator release), nest parasitism, competitive interactions or alterations in microclimate from edge effects. One of these factors, increased nest predation could be important for the narrow strips of lowland/hill forest in Lore Lindu. An increased predation rate along forest edges and in disturbed forest has been inferred from artificial nest predation experiments for other sites in Southeast Asia (Cooper and Francis 1998, Wong et al. 1998) and also in Lore Lindu National Park (Pangau et al. 2003).

In Lore Lindu we would expect those forest-dependent species to be most threatened that show a combination of the following characteristics (1) altitudinal specialisation to below 800m, (2) large body size, and (3) spatio-temporally varying food resources. Candidates are, for example, the hornbills *Rhyticerus cassidix* and *Penelopides exharatus,* fruit pigeons such as *Ducula aenea, Treron vernans, Treron griseicauda, Ptilinopus subgularis*. The Yellow-crested Cockatoo *Cacatua sulfurea* has suffered from overexploitation (pet trade) and is probably already extinct in Lore Lindu (Coates et al. 1997), and the Maleo *Macrocephalon maleo* has only a few stable nesting grounds (Butchart and Baker 2000). Insectivorous birds that are mainly found below 800m are the pittas *Pitta sordida, P. erythrogaster,* the flycatcher *Ficedula rufigula*, the thrush *Zoothera erythronota* or the kingfishers *Actenoides monachus, Cittura cyanotis* and *Ceyx fallax.*

6 Conclusion

Lowland and hill forest in the Lore Lindu National Park are directly important for the conservation of roughly half the number of its resident bird species. These areas are, however, narrow strips of land that are situated along the Park borders. Much of this forest seems to have been cleared between 1999 and 2001. It is to be expected that these activities and potentially ongoing deforestation will threaten a significant proportion of the Park's biodiversity. Our attempt to use the species-area-relationship for the description of potential bird species loss in Lore Lindu should be used to estimate the extent of these effects.

Park management concepts that do not include some considerations of the total area available for its species assemblages will inevitably fail since even undisturbed forest areas lose species after fragmentation. We therefore strongly recommend that (1) the largest patches of lowland and hill forest be secured in the current condition, (2) maximum connectivity between these patches is maintained, and (3) the possibility to maintain traditional uses of the forest margins instead of the establishment of artificial cacao/coffee plantations be investigated. The examples from Java show that even well-protected patches of 10,000–20,000 ha that today support the full array of the avifauna are likely to lose bird species within one or two decades. Thus if the entire spectrum of Lore Lindu's biological diversity is to be secured for the long-term, only immediate intervention will be sufficient.

Presently, in Lore Lindu National Park the effects of land use on biodiversity and the factors driving deforestation are studied (www.storma.de). Information on forest vertebrate diversity in secondary systems based on mere presence/absence data should be interpreted with caution since there are considerable relaxation times involved after area reduction and habitat change and many species found there now certainly still need adjacent undisturbed forest for their persistence.

7 Acknowledgements

The study was part of the Collaborative Research among Georg-August University of Göttingen, University of Kassel, Bogor Agricultural University and Tadulako University Palu, entitled "Stability of Rainforest Margin Areas in Indonesia (STORMA). STORMA is financed by the German Research Council (DFG). We are grateful for support of our field work to Pak Banjar (Balai Taman Nasional Lore Lindu, Palu). The authors especially acknowledge the work of the coordinating teams in Göttingen, Bogor and Palu, and thank Heleen Fermon, Andreas Mench and Stefan Merker (Göttingen), as well as Drs. Elijohnnadi and Duncan Neville (both Palu) for various ways of support. Special thanks go to Richard Noske for his comments on an earlier version of this manuscript.

8 Keywords

Birds, Sulawesi, fragmentation, extinction, conservation, endemism, remote sensing

References

Baillie J, Groombridge B (1996) 1996 IUCN Red List of Threatened Animals. The IUCN Species Survival Commission, Gland (Switzerland)

Balen B van (1999) Birds on fragmented islands: persistence in the forests of Java and Bali. Tropical Resource Management Papers No. 30. Wageningen University and Research Centre, Wageningen (Netherlands)

Birner R, Waltert M, Mappatoba M, Maertens M. Zeller M & Mühlenberg M (2003) Declining mammal and bird populations in Central Sulawesi, Indonesia – results of a village-level survey using recall methods (submitted).

Brooks TM, Balmford A (1996) Atlantic forest extinctions. Nature 380: 115

Brooks TM, Pimm SL, Collar N J (1997) The extent of deforestation predicts the number of threatened birds in insular south-east Asia. Conservation Biology 11: 382–394

Brooks TM, Pimm SL, Kapos V, Ravilious C (1999a) Threat from deforestation to montane and lowland forest birds and mammals in insular South-east Asia. Journal of Animal Ecology 68: 1061–1078

Brooks TM, Pimm SL, Oyugi JO (1999b) Time lag between deforestation and bird species extinction in tropical forest fragments. Conservation Biology 13: 1140–1150

Brooks TM, Tobias J, Balmford A (1999c) Deforestation and bird extinctions in the Atlantic forest. Animal Conservation 2: 211–222
Brooks TM, Mittermeier RA, Mittermeier CG, da Fonseca GAB, Rylands AB, Konstant WR, Flick P, Pilgrim J, Oldfield S, Magin G., Hilton-Taylor C (2002) Habitat loss and extinction in the hotspots of biodiversity. Conservation Biology 16: 909–923
Brown KS, Brown KG (1992) Habitat alteration and species loss in Brazilian forests. In: Whitmore TC, Sayer JA (eds) Tropical deforestation and species extinction. Chapman and Hall, London, pp 119–142
Budiansky S (1994) Extinction or miscalculation? Nature 370: 105
Butchart SHM, Baker GC (2000) Priority sites for conservation of maleos (*Macrocephalon maleo*) in central Sulawesi. Biological Conservation 94: 79–91
Castelletta M, Sodhi NS, Subaraj R (2000) Heavy extinctions of forest avifauna in Singapore: lessons for biodiversity conservation in South-East Asia. Conservation Biology 14: 1870–1880
Coates B, Bishop KD, Gardner D (1997) A guide to the birds of Wallacea: Sulawesi, the Moluccas and Lesser Sunda Islands, Indonesia. Dove Publications, Alderley Queensland
Collar NJ, Crosby MJ, Stattersfield AJ (1994) Birds to watch 2. Birdlife Conservation Series No. 4. Birdlife International, Cambridge
Colwell RK (2000) EstimateS – Statistical estimation of species richness and shared species from samples. Version 6.0b1. http://viceroy.eeb.uconn.edu/estimates
Cooper DS, Francis CM (1998) Nest predation in a Malaysian lowland rain forest. Biological Conservation 85: 199–202
Corlett RT, Turner IM (1997) Long-Term Survival in Tropical Forest Remnants in Singapore and Hong Kong. In: Laurance WF, Bierregard, ROJ (eds) Tropical forest remnants: ecology, management and conservation of fragmented communities. University of Chicago Press, Chicago, pp 333–345
Cowlishaw G (1999) Predicting the pattern of decline of African primate diversity: an extinction debt from historical deforestation. Conservation Biology 5: 1183–1193
FWI/GWF (2002) The state of the forest: Indonesia. Forest Watch Indonesia, Bogor and Global Forest Watch, Washington DC
Hoogerwerf A (1948) De avifauna van Tjibodas en omgeving. Buitenzorg, Koninkljike plantentuin van Indonesie
MacKinnon J (1991) Field guide to the birds of Java and Bali. Gadja Madah University Press, Yogjakarta
MacKinnon J, Phillips K, van Balen B (2000) Burung-burung di Sumatera, Jawa, Bali dan Kalimantan. LIPI /Birdlife - Indonesia Programme, Bogor
Mauro I (2001) Cinnabar Hawk Owl *Ninox ios* at Lore Lindu National Park, Central Sulawesi, Indonesia. Forktail 17: 118–119
Nur Mallo F, Ma'dika BP (1999) A checklist of the birds of Lore Lindu National Park, Central Sulawesi, Indonesia. The Nature Conservancy, Palu (Indonesia)
Olsen DM, Dinerstein E (1998) The Global 200: a representation approach to conserving Earth's distinctive ecoregions. World Wildlife Fund, Washington DC
Pangau M, Waltert M, Mühlenberg M (2003) Artificial nest predation in forest margin habitats of Sulawesi, Indonesia (submitted)
Preston FW (1962) The canonical distribution of commonness and rarity. Ecology 43: 185–215, 410-432

Robinson WD (1999) Long-term changes in the Avifauna of Barro Colorado Island, Panama, a Tropical Forest Isolate. Conservation Biology 13: 85–97

Robertson JMY, van Schaik CP (2001) Causal factors underlying the dramatic decline of the Sumatran orang-utan. Oryx 35: 26–38

Rosenzweig ML (1995) Species diversity in space and time. Cambridge University Press, Cambridge

Schulze CH, Waltert M, Kessler PJA, Pitopang R, Shahabuddin, Veddeler D, Steffan-Dewenter I, Mühlenberg M, Gradstein SR, Tscharntke T (2003) Biodiversity indicator taxa of tropical land-use systems: comparing plants, birds and insects. *Ecological Applications* (in press).

StatSoft (1995) STATISTICA for Windows, vol I-V. StatSoft, Tulsa.

Terborgh J (1999) Requiem for Nature. Island Press, Washington DC

TNC/BTNLL (2002) Lore Lindu National Park. Draft Management Plan 2002–2027. Directorat Jenderal Perlindungan Hutan dan Konservasi Alam and The Nature Conservancy, Palu (Indonesia)

Turner I (1996) Species loss in fragments of tropical rain forest: a review of the evidence. Journal of Tropical Ecology 33: 200–209

Waltert M, Mardiastuti A, Mühlenberg M (2003) Effects of land use on bird diversity in Sulawesi, Indonesia (submitted).

Wong TCM, Sodhi NS, Turner IM (1998) Artificial nest and seed predation experiments in tropical lowland rainforest remnants of Singapore. Biological Conservation 85: 97–104

Appendix 1. Status, geographic range and altitudinal ranges of Lore Lindu bird species as in column code (after White and Bruce 1986, Coates et al. 1997, Nur Mallo and Ma'dika 1999, Mauro 2001, additional information from MacKinnon and Phillips 1993). Code contains information on the species status (B - breeding resident, V – visitor), geographic range (Si - endemic to the Sulawesi island, Sr - endemic to the Sulawesi sub-region, W - endemic to Wallacea, X - distributional range extending Wallacea), and altitudinal range (L - lowland specialist species up to 1,200 m, G – generalist lowland/montane species up to 2500 m, M – montane species confined to above 1,200 m, U – upper montane species confined to above 1800 m). Marked/unmarked boxes indicate presence/absence at respective altitudinal bands, grey-scale boxes indicate infrequent occurrence. Asterisks * in code mark genera endemic to Wallacea, superscript ^{x} mark open-land species.

Family	Species	Code	Presence/absence at altitude (m)						
			0–400	4–800	8–1200	12–1600	16–2000	20–2400	24–2800
Podicipedidae	*Tachybaptus ruficollis*^{x}	BXG	■	■	■	■	░		
Phalacrocoracidae	*Phalacrocorax melanoleucos*^{x}	BXL	■	░					
Anhingidae	*Anhinga melanogaster*^{x}	BXL	■	■	░				
Pelecanidae	*Pelecanus conspicillatus*^{x}	VXG	■	■					
Ardeidae	*Ardea purpurea*^{x}	BXG	■	■	■	■			
	Egretta alba^x	BXG	■	■	■	■			
	Egretta intermedia^x	BXL	■	■	■				
	Egretta garzetta^x	BXG	■	■	■	■			
	Bubulcus ibis^x	BXL	■	■	■	░			
	Ardeola speciosa^x	BXL	■	■	■	░			
	Butorides striatus^x	BXG	■	■	■	░			
	Nycticorax nycticorax^x	BXL	■	■	■				
	Ixobrychus sinensis^x	BXL	■	■					
	Ixobrychus cinnamomeus^x	BXL	■	■	░				
	Ixobrychus flavicollis	BXL	■	■					
Ciconiidae	*Ciconia episcopus*^{x}	BXL	■	■	■				
Threskiornithidae	*Plegadis facinellus*^{x}	BXL	■	■	■	░			
Accipitridae	*Aviceda jerdoni*	BXL	░	■	░				
	Spilornis rufipectus	BSrG	■	■	■	■	■	░	
	Milvus migrans^x	BXL	■	■	░				

Appendix 1 (cont.)

Family	Species	Code	Presence/absence at altitude (m)						
			0–400	4–800	8–1200	12–1600	16–2000	20–2400	24–2800
Accipitridae	*Haliaetus leucogaster*[x]	BXG	■	■	■	■			
	Pernis celebensis	BXG	■	■	■	■	■		
	Haliastur indicus[x]	BXG	■	■	■	■	■	■	
	Hieraetus kienerii	BXG	■	■	■	■	■	■	
	Ictinaetus malayensis	BXG	■	■	■	■	■	■	
	Butastur liventer[x]	BXL	■	■	■				
	Elanus caeruleus[x]	BXL	■	■	▒				
	Accipiter griseiceps	BSrG	■	■	■	■	■	▒	
	Accipiter trinotatus	BSrG	■	■	■	■	■		
	Accipiter nanus	BSiG		▒	■	■	■		
	Accipiter rhodogaster	BSrG	■	■	■	■	■	▒	
	Circaetus assimilis[x]	BXL	■	■	▒				
	Macheiramphus alcinus	BXL	■	■	▒				
	Ichthyophaga humilis	BXL	■	■	■	▒			
	Spizaetus lanceolatus	BSrG	■	■	■	■	■	▒	
Falconidae	*Falco moluccensis*[x]	BXG	■	■	■	■	■	■	
	Falco severus[x]	BXG	■	■	■	■	■		
	Falco peregrinus[x]	BXG	■	■	■	■	■	■	▒
Dendrocygnidae	*Dendrocygna guttata*[x]	BXL	■						
	Dendrocygna arcuata[x]	BXL	■	■	▒				
Anatidae	*Anas gibberifrons*[x]	BXG	■	■	■	■	▒		
	Anas superciliosa[x]	BXG	■	■	■	■			
	Aythia australis[x]	VXL	■	■	■				
Megapodidae	*Megapodius cumingii*	BXG	■	■	■	■	■		
	Macrocephalon maleo	BSi*L	■	■	■	▒			
Phasianidae	*Coturnix chinensis*[x]	BXL	■	■	■	▒			
	Gallus gallus	BXL	■	■	▒				
Turnicidae	*Turnix suscitator*[x]	BXL	■	■	▒				
Rallidae	*Gallirallis striatus*[x]	BXL	■	■	▒				

Appendix 1 (cont.)

Family	Species	Code	Presence/absence at altitude (m)						
			0–400	4–800	8–1200	12–1600	16–2000	20–2400	24–2800
Rallidae	*Gallinula tenebrosa*[x]	BXL	■	■	░				
	Gallirallus torquatus[x]	BXL	■	■	░				
	Aramidopsis plateni	BSi*L	■	■	■	░			
	Gymnocrex rosenbergii	BSrL	■	■	■	░			
	Poliolimnas cinerea[x]	BXL	■						
	Amaurornis isabellinus[x]	BSiL	■	■	■	░			
	Amaurornis phoenicurus[x]	BXL	■	░					
	Gallinula chloropus[x]	BXL	■	■	░				
	Gallirallus phillippensis[x]	BXL	■	■	░				
	Gallicrex cinerea[x]	VXL	■	■	■				
	Porphyrio porphyrio[x]	BXL	■	■	░				
Jacanidae	*Irrediparra gallinacea*[x]	BXL	■	■	░				
Recurvirostridae	*Himantopus himantopus*[x]	BXL	■	■	░				
Charadriidae	*Pluvialis fulva*[x]	VXL	■	■	░				
	Charadrius dubius[x]	VXL	■						
	Numenius phaeopus[x]	VXL	■						
	Tringa nebularia[x]	VXL	■	░					
	Tringa glareola[x]	VXL	■	■	░				
	Actitis hypoleucos[x]	VXL	■	■	░				
	Gallinago megala[x]	VXL	■	■	░				
	Scolopax celebensis	BsiM			░	■	■	░	
	Chlidonias hybridus[x]	VXL	■	■	░				
Columbidae	*Columba livia*[x]	BXL	■	■	■	░			
	Streptopelia tranquebarica[x]	BXL	■						
	Streptopelia chinensis[x]	BXG	■	■	■	■			
	Macropygia amboinensis	BXG	■	■	■	■	■		
	Turacoena manadensis	BSrL	■	■					
	Chalcophaps indica	BXL	░						
	Chalcophaps stephani	BXL	■	░					
	Gallicolumba tristigmata	BsiG	■	■	■	■	■		

Appendix 1 (cont.)

Family	Species	Code	Presence/absence at altitude (m) 0–400	4–800	8–1200	12–1600	16–2000	20–2400	24–2800
Columbidae	*Treron vernans*	BXL	▒						
	Treron griseicauda	BXL	■	■	▒				
	Ptilinopus fischeri	BsiU			▒	▒	■	■	■
	Ptilinopus subgularis	BSrL	■	■					
	Ptilinopus superbus	BWG	■	■	■	■	▒		
	Ptilinopus melanospila	BWG	■	■	■	■			
	Ducula forsteni	BSrG			■	■			
	Ducula radiata	BsiM				■	■	▒	
	Ducula aenea	BXL	■	■	▒				
	Ducula luctuosa	BSrL	■	■					
	Cryptophaps poecilorrhoa	Bsi*M				▒	■	▒	
Psittacidae	*Trichoglossus ornatus*	BSrL	■	■	▒				
	Trichoglossus flavoviridis	BSrG			■	■	■	■	
	Cacatua sulphurea	BWL	■	■					
	Prioniturus platurus	BSrG	▒	▒	■	■	■	▒	
	Tanygnathus sumatranus	BXL	■	▒	▒				
	Loriculus stigmatus	BSrL	■	■	■				
	Loriculus exilis	BsiL	■	■	▒				
Cuculidae	*Cuculus crassirostris*	BsiL		▒	■	▒			
	Cuculus fugax[x]	VXL	■	▒					
	Cuculus saturatus	VXL	■	▒					
	Cacomantis merulinus	BXL	■	■	■				
	Cacomantis sepulcralis	BXG	■	■	■	■	■	■	▒
	Chrysococcyx minutillus	BXL	■	■					
	Chrysococcyx russatus	BXL	■	■					
	Surniculus lugubris	BXL	■	■	■				
	Eudynamis melanorhyncha	BSrG	■	■	■	■			
	Eudynamis scolopacea[x]	BXL	■	■	▒				
	Scythrops novaehollandiae[x]	BXL	■	■	▒				
	Phaenicophaeus calyorhynchus	BSrG	■	■	■	■			
	Centropus bengalensis[x]	BXL	■	■	▒				
	Centropus celebensis	BSrL	■	■	▒				

Appendix 1 (cont.)

Family	Species	Code	Presence/absence at altitude (m) 0–400	4–800	8–1200	12–1600	16–2000	20–2400	24–2800
Tytonidae	*Tyto rosenbergi*[x]	BSrL	■	■	■				
	Tyto inexpectata	BsiL	░	■	■	░			
Strigidae	*Otus manadensis*	BSrG	■	■	■	■	■		
	Ninox punctulata	BsiL	■	■	░				
	Ninox ochracea	BSrG	■	■	■	■	░		
	Ninox ios	B?SrM				■			
Caprimulgidae	*Eurostopodus diabolicus*	BsiG	■	■	■	■	░		
	Eurostopodus macrotis	BXG	■	■					
	Caprimulgus celebensis[x]	BSrL	■						
	Caprimulgus affinis[x]	BXL	■	■					
Apodidae	*Collocalia fuciphaga*	BXL	■	■	░	░			
	Collocalia vanikorensis	BXL	■	░					
	Collocalia infuscata	BSrG	■	■	■	■	■	■	░
	Collocalia esculenta[x]	BXG	■	■	■	■	■	■	■
	Hirundapus caudacutus	VXG	■	■	■	■	■	■	■
	Hirundapus celebensis	BXG	░	■	■	■	■		
	Apus pacificus[x]	VXL	■	■	░				
	Hemiprocne longipennis	BXL	■	■	■	░			
Halcyonidae	*Actenoides monachus*	BSrL	■	■	░				
	Actenoides princeps	BsiM			░	■	■		
	Cittura cyanotis	Bsi*L	■	■	░				
	Halcyon melanorhyncha[x]	BsiL	■	░					
	Halcyon chloris[x]	BXG	■	■	■	■	░		
Alcedinidae	*Ceyx fallax*	BSrL	■	■	░				
	Alcedo meninting	BXL	■						
	Alcedo atthis	BXL	■						
Meropidae	*Merops superciliosus*	BXL	■	■					
	Merops ornatus	VXL	■	■	░				
	Meropogon forsteni	Bsi*G	■	■	■	■	■	░	
Coraciidae	*Coracias temminckii*	BSrL	■	■	■				
	Eurystomus orientalis	B?XL	■	░					
Bucerotidae	*Penelopides exharatus*	BSrL	■	░					
	Rhyticeros cassidix	BSrL	■	■	░				

Appendix 1 (cont.)

Family	Species	Code	Presence/absence at altitude (m) 0–400	4–800	8–1200	12–1600	16–2000	20–2400	24–2800
Picidae	*Dendrocopus temminckii*	BSrG			■	■	■	■	
	Mulleripicus fulvus	BSrG	■	■	■	■	■	▒	
Pittidae	*Pitta erythrogaster*	BXL	■	■					
	Pitta sordida	BXL	■	▒					
	Pitta elegans	BWL	■	■	■				
Hirundinidae	*Hirundo rustica*[x]	VXL	■	■	▒				
	Hirundo tahitica[x]	BXL	■	■	▒				
Motacillidae	*Motacilla flava*[x]	VXG	■	■	■	■	■		
	Motacilla cinerea[x]	VXG	■	■	■	■	■		
Campephagidae	*Coracina temminckii*	BSiG		▒	■	■	■	▒	
	Coracina leucopygia	BSrL	■	■					
	Coracina abbotti	BSiM					■	▒	
	Coracina tenuirostris	BXL	■						
	Coracina morio	BSiG	■	■	■	■	▒		
	Lalage leucopygialis	BSrL	■	■	▒				
	Lalage sueurii[x]	BXG	■	■	■	■			
Dicruridae	*Dicrurus montanus*	BSiG		▒	■	■	▒		
	Dicrurus hottentottus	BXL	■	■	■	▒			
Oriolidae	*Oriolus chinensis*	BXL	■	■	■	▒			
Corvidae	*Corvus enca*[x]	BXG	■	■	■	■			
	Corvus typicus	BSrG	■	■	■	■			
Timaliidae	*Trichastoma celebensis*	BSrG	■	■	■	■	▒		
	Malia grata	BSi*M				■	■	■	
	Geomalia heinrichi	BSi*U					▒	■	▒
Turdidae	*Heinrichia calligyna*	BSi*M				▒	■	■	▒
	Zoothera erythronota	BSrL	■	■	▒				
	Cataponera turdoides	BSi*M				■	■	■	
	Saxicola caprata[x]	BXG		▒	■	■	▒		
Sylviidae	*Gerygone sulphurea*	BXG	■	■	■	■	■	▒	
	Bradypterus castaneiceps	BWM			▒	■	■	■	▒
	Locustella fasciolata[x]	VXL	■	■	▒				

Appendix 1 (cont.)

Family	Species	Code	Presence/absence at altitude (m)						
			0–400	4–800	8–1200	12–1600	16–2000	20–2400	24–2800
Sylviidae	*Acrocephalus stentoreus*^{x}	BXL	■	■					
	Acrocephalus orientalis^x	VXL	■						
	Orthotomus cucullatus	BXM					■	▒	
	Phylloscopus trivirgatus	BXG		▒	■	■	■	■	
	Phylloscopus sarasinorum	BsiG		▒	■	■	■	■	
	Megalurus timorensis^x	BXL	■	■	■	▒			
	Cisticola juncidis^x	BXG	■	■	■	■			
	Cisticola exilis^x	BXL	■	■	■	▒			
Muscicapidae	*Muscicapa griseisticta*^{x}	BXG	■	■	■	■	■		
	Muscicapa dauurica	BXL	■	▒					
	Eumyias panayensis	BXM				■	■	■	
	Ficedula mugimaki	VXM			▒	■	▒		
	Ficedula hyperythra	BXM			▒	■	■	■	
	Ficedula rufigula	BsiL	■	▒					
	Ficedula westermanni	BXG			■	■	■	■	▒
	Cyornis hoevelli	BsiM				▒	■	▒	
	Cyornis omissus	BSrL		▒	■	▒			
Monarchidae	*Hypothymis azurea*	BXL	■	■	■				
	Rhipidura teysmanni	BSrG		▒	▒	■	■	▒	
Petroicidae	*Culicicapa helianthea*	BXL		▒	■	▒			
Pachycephalidae	*Hylocitrea bonensis*	Bsi*U				▒	▒	■	
	Coracornis ravenis	Bsi*M				▒	■	▒	
	Pachycephala sulfuriventer	BsiG			■	■	■		
Artamidae	*Artamus leucorhynchus*^{x}	BXL	■	■	■	▒			
	Artamus monachus	BSrG	■	■	■	■	■		
Sturnidae	*Aplonis minor*	BXL	■	■	▒				
	Aplonis panayensis	BXL	■	▒					
	Basilornis celebensis	BSrL	■	■	▒				
	Streptocitta albicollis	Bsi*L	■	■	■				
	Enodes erythrophris	Dsi*M			▒	■	■	▒	
	Scissirostrum dubium	Bsi*L	■	■	▒				
Meliphagidae	*Myza celebensis*	Bsi*M			▒	■	▒		

Appendix 1 (cont.)

Family	Species	Code	Presence/absence at altitude (m)						
			0–400	4–800	8–1200	12–1600	16–2000	20–2400	24–2800
	Myzomela sanguinolenta	Bsi*G		▒	■	■	■	■	
	Myza sarasinorum	Bsi*U				▒	■	■	■
Nectariniidae	*Anthreptes malacensis*[x]	BXL	■	■	▒				
	Nectarinia aspasia	BXL	■	■					
	Nectarinia jugularis[x]	BXL	■	■					
	Aethopyga siparaja	BXL	■	■	■	▒			
Dicaeidae	*Dicaeum aureolimbatum*	BSrG	■	■	■	■	■		
	Dicaeum nehrkorni	BsiG		▒	■	■	■	■	▒
	Dicaeum celebicum	BSrL	■	■	▒				
Zosteropidae	*Zosterops montanus*	BXM			▒	■	■	■	▒
	Zosterops chloris	BXL	■	■	▒				
	Zosterops atrifrons[x]	BXL	▒	■	■	▒			
	Lophozosterops squamiceps[x]	BSiM				▒	■	▒	
Passeridae	*Passer montanus*[x]	BXL	■	■	■	▒			
Estrildidae	*Erythrura hyperythra*	BXG		▒	■	■	■	▒	
	Erythura trichroa	BXM				▒	■	▒	
	Lonchura molucca [x]	BWL	■	■	▒				
	Lonchura punctulata[x]	BXL	■	■	▒				
	Lonchura malacca[x]	BXL	■	■					
	Lonchura pallida[x]	BSrL	■	■					
Fringillidae	*Serinus estherae*	BSiU					▒	■	

Appendix 2. Size, age since isolation and numbers (percentages) of bird species that went extinct since isolation in different rainforest areas. Data compiled from (Kattan et al. 1994, Robinson 1999, van Balen et al. 1988, Hoogerwerf 1948, MacKinnon 1991, Castelletta et al. 2000)

	Size (ha)	Age (years)	No. spp. extinct
Java forest patches	40	?	80%
Bogor Botanical Gardens	~100	42	89 (64%)
Cibodas Botanical Gardens	~200	~50	112 (58%)
San Antonio cloud forest	700	50	40 (31%)
Barro Colorado island	1,600	85	65 (30%)
Singapore's last forest	2,000	75	138 (67%)
Forest patch C Java	10,000	?	25 %

Chapter 3:

Water and Nutrient Cycles, Land-Use Systems and Modelling

The Effects of Rainforest Conversion on Water Balance, Water Yield and Seasonal Flows in a Small Tropical Catchment in Central Sulawesi, Indonesia

Alexander Kleinhans & Gerhard Gerold

1 Introduction

The ongoing debate over the causes of deforestation in Indonesia views smallholder production and their growing number as a main cause of deforestation in Indonesia. With the common terminology 'smallholder conversion' a broad range of conversion strategies like the shifting cultivation-forest pioneer continuum, smallholder tree crop production and transmigration activities is summarized (Sunderlin and Resosudarmo, 1996). However, after 'smallholder conversion' the created landscape along the rainforest margin area is usually characterized by a patchwork of different land use types in ever smaller patches undergoing a gradual change from forest dominated areas via annual crops to perennial plantation interspersed by secondary forest, pasture and annual crops. Since the Indonesian economic crisis in 1997 land clearing increased dramatically for establishing export tree crops. Indonesia is now the third largest producer of cocoa (Sunderlin, 2000, Maertens et al, 2002) with Central Sulawesi as one of the main Indonesian production areas. The question then arises: to what extent do these conversion activities affect the hydrological behaviour of the affected areas ?

In general, the role of forests in watershed hydrology and the effect of deforestation in the humid tropics on the hydrological response has been widely researched with respect to logging activities. With a few exceptions catchment experiments generally indicate increased runoff from 'short-crop vegetation' with lower leaf area index and vegetation coverage compared with forested areas. Conversion impacts on seasonal flows are more difficult to predict. Different site-specific, often competing, processes and causes can exist and the impact of land use change is therefore difficult to predict for a particular site (see Bruijnzeel 1990; Bosch and Hewlett 1982; Calder 2000).

Within the framework of 'the stability of rainforest margins area project' (STORMA) in Indonesia an experimental catchment along the rainforest margin of the Lore Lindu National Park in Central Sulawesi has been designed in 2001 to generate scientific information about the influence of the unique transition process from natural forest to smallholder agroforestry on the hydrological cycle.

The focus of this paper is to introduce the research area (Figure 1) with its long-term research possibilities and give first hydrological characterizations and preliminary results of recent conversion effects on annual water yield and seasonal hydrological behaviour of the research area.

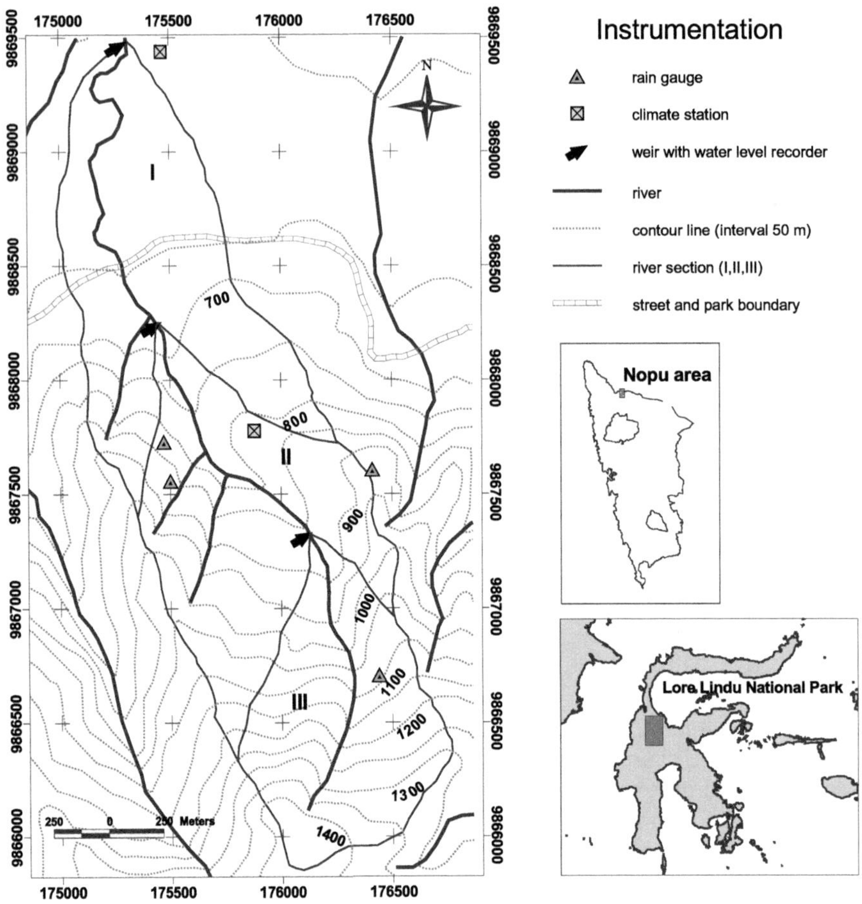

Fig. 1. Location of the research area along the border of the Lore Lindu National Park in Central Sulawesi, Indonesia with topography and instrumentation of the experimental Nopu catchment

2 Methodology

2.1 General Site Description

The greater study area of the STORMA project is located in the province of Central Sulawesi along and in the borders of the Lore Lindu National Park (ca 229000 ha), a mountainous natural forest reserve reaching up to 2500 m a.s.l. with rice paddies, agroforestry and annual crops in the valleys outside the National Park and remarkable forest conversion rates along the forest boundaries.

Located just south of the equator the research area has a humid tropical climate with mean annual temperatures between 25 and 26 °C at sea level, high humidity and a mean annual precipitation >2500 mm. Because of the mountainous research area, rainfall distribution is very heterogeneous. The Köppen classification describes the climate as tropical rainy (Chang and Lau, 1993). Figure 2 shows at the example of the Lake Lindu meteorological station (2371 mm mean annual precipitation) that there is no distinct dry season at the vicinity of the research area however a slightly bimodal rainfall distribution can be found due the annual movement of the Inter-Tropical Convergence Zone (ITCZ) with greater monthly rainfall from March till June and a secondary rainy period from November till December (Berlage 1949, Whitten et al 1987). Rainfall distribution is also modified by topography and long-term rainfall data of the vicinity of the National Park is scarce.

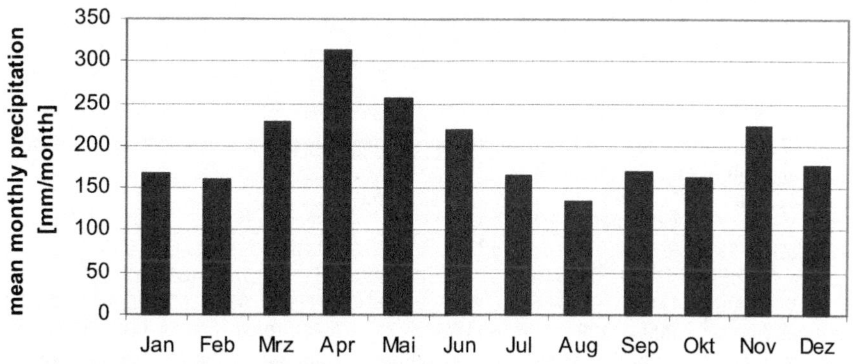

Fig. 2. Mean monthly rainfall (1931-1941) at the Lake Lindu meteorological station (1°18' S, 120°05' E; 1000 m a.s.l.) after Berlage (1949)

2.2 Experimental Watershed with Catchment Sections

The Nopu catchment (1° 11′ S, 120° 05′ E) is situated at the border of the Lore Lindu National Park in Central Sulawesi, Indonesia. The mountainous catchment has an area of approximately 2.45 km² and altitudes range between 600 and 1400 m a.s.l. (Figure 1).

There are three dominant landforms in the catchment (Table 1). In the lowland are floodplains with terraces and dissected rivers located followed by footslopes with a wide river valley and gentle slopes (700-800 m a.s.l). At higher elevations the mountainous upland with a deep incised valley and steep slopes is situated.

The research area belongs to the sub-montane rainforest of Central Sulawesi and ten years ago the complete area was forest covered. Land use changes fast in the former forested catchment. Since approximately 10 years smallholder established cocoa plantations in the lowland of the catchment where the hamlet Nopu is located along the National Park boundary. Between 650 and 950 m a.s.l. smallholder are creating an upward moving transition zone of quickly disappearing tropical forest which is replaced by patches of slash and burn agriculture, pasture, secondary forest and agroforestry. There, the hamlets' increasing population is establishing their new farm plots. Above this zone primary forest still remains quite undisturbed. Using this land use type distribution three catchment sections with predominant land use types can be separated (I agroforestry, II transition zone, III natural forest). Dominant soil types in the research area are Cambisols. The predominant texture is sandy loam on the slopes adjoining the small riparian zone at the valley bottom. Deep soils developed on gentle slopes and medium soils on steep slopes. Catchment section II and III are homogeneous units with respect to topography, soil substrate and hydrological behaviour before the forest conversion.

Table 1. Catchment and catchment section characteristics of the experimental Nopu catchment in Central Sulawesi, Indonesia

parameter	catchment	section I	section II	section III
size (ha)	245	47	116	82
geographic position of the outlet	51 M 0175405 UTM 9869412	51 M 0175405 UTM 9869412	51 M 0175405 UTM 9868365	51 M 0176212 UTM 9867344
altitude (m)	600 - 1420	600-750	750-900	900-1420
predominant land use	natural forest, cocoa, annual crops	cocoa, fallow	annual crops, secondary forest, fallow	natural forest

2.3 Experimental Design and Data Analyses

Along the rainforest margins suitable natural forest headwater catchments are not anymore available. Therefore an experimental design with a classical paired catchment approach is not practicable due to the absence of a forested reference catchment with quite similar catchment pattern as the disturbed catchment. Therefore a modified paired catchment approach has been used by dividing the research catchment in three river sections (Table 1) with the natural forest headwater section (III) serving as a reference for the disturbed river sections (II transition zone, I agroforestry).

Weirs serve as divide for the different river sections. The first weir is located at the outlet of the catchment, the second weir below the transition zone and the third weir at the outlet of the natural forest head water catchment. The following equation allows to allocate measured runoff (Q) at the weirs to the runoff of unmeasured river sections by using the section area (A) as distributor:

$$Q_{weir2}*A_{weir2} = Q_{section\ II}*A_{section\ II} + Q_{weir3}*A_{weir3} \quad (1)$$

This design allows the observation of the influence of land use changes on water yield and water balance on the land use section scale.

At the outlet of all three river sections double rectangular concrete notch weirs crested with thin sharpened metal plates have been constructed in front of a concrete stilling basin. Weir dimensions are adjusted to catchment size. The water level behind the notch is measured by water level recorder (SEBA Hydrometrie GmbH Kaufbeuren, Germany). Water level is logged at 10-min intervals. Runoff was calculated by the stage-discharge relationship of the rectangular weir, additionally the calculated stage-discharge relationship have been validated by direct runoff measurements using the salt dilution method. Runoff is continuously measured since November 2001.

To quantify the precipitation and meteorological inputs four automatic rain gauges and two automatic weather stations (AWS) are operating in the catchment. The AWS were manufactured by Theodor Friedrichs GmbH Germany and measure rainfall (0.2 mm per tip), temperature, relative humidity, soil temperature, wind speed and direction and global radiation. Data are logged at 5-min intervals and downloaded regularly. The four automatic rain gauges (Theodor Friedrichs GmbH Germany) are collecting data at 10-min intervals (0.1 mm per tip). These six, continuously recording tipping bucket rain gauges, operate to measure the temporal and special distribution of rainfall.

The experimental catchment has been instrumented since October 2001. Data from 2002 have been used for a descriptive analyses of the annual, seasonal and event based precipitation and runoff behaviour in the natural forest and transition section of the research area.

With this data the volumetric water balance per unit area can be calculated as

$$P = ET + Q + SS \quad (2)$$

where P is precipitation, ET evapotranspiration, Q runoff and SS soil storage. The comparison of the water balance equations of the transition and natural forest section serve as a mean for the change of water yield due to forest conversion. Differences in the catchment section runoff time series and flow duration curves indicate the effects of conversion on extreme flows and seasonal patterns.

Catchment section I in the lowland is more complex with respect to the water divide and higher leakages due to an existing water supply system for the hamlet

Nopu and higher transmission losses in the permeable riverbed of the floodplain. The use of the volumetric water balance equation for the river section I would require the integration of additional data. Therefore, this study focuses on the comparison of natural forest and transition section.

On the base of the digital elevation model derived from the topographical map 1:50,000 the catchment section boundaries and sizes have been separated (Table 1) by using the topographical analysis module of the WASIM-ETH water balance model. Area precipitation is calculated using the inverse distance weight method of WASIM-ETH (Schulla, 1997).

3 Results and Discussion

3.1 Precipitation

During the year 2002 the total area precipitation in the Nopu catchment was about 2500 mm, mean point rainfall (6 stations) was 2430 mm. Comparing our measured monthly rainfall distribution (Figure 3) with data from earlier years at the Lake Lindu station (Berlage 1949) rainfall characteristics of the experimental catchment appear quite similar to the long-term average. In 2002 precipitation (P>0.2 mm/day) occurred every 1.7 days. Rainfall intensities are high and rainfall durations usually less than two hours. Maximum daily point rainfall was 128 mm. Maximum hourly point precipitation was 66 mm and 10 minutes maximum was 25 mm. Because of the mountainous research area spatial variability of rainfall is high. More than ten day periods without rainfall occurred four times in 2002 (February, Mai, June, August) The lowland (600-700 m a.s.l) received considerable less precipitation in comparison to higher elevations. When examining separate storms maximum values were also observed in the lowland.

3.2 Runoff

Data on runoff from the catchment and catchment sections can be seen in Table 2. There are remarkable differences between the two different river sections although river section II and III are regarded as homogeneous with respect to topographical properties and clearly defined water divides.

Seasonal discharge of the research area (Figure 3) follows the precipitation seasonality. In 2002 there is a steady increase in streamflow from February to mid-June followed by low monthly runoff during the rest of the year. The secondary wet season in November and December results in no significant increase of runoff on a monthly scale. Highest monthly rainfall amounts in March did not lead to the highest water yield, but in the recharge of the depleted soil moisture storage. Only after soil storage reached a higher charged state, with soil moisture contents near field capacity (37 - 46 vol % soil moisture content), high monthly precipitation results in high discharge. In contrast to the main wet season, the secondary wet season was not able to recharge soil moisture storage. During drier months precipitation with lower intensities mainly feeds soil moisture storage, which seems to be important to maintain baseflow and ET during days without rainfall.

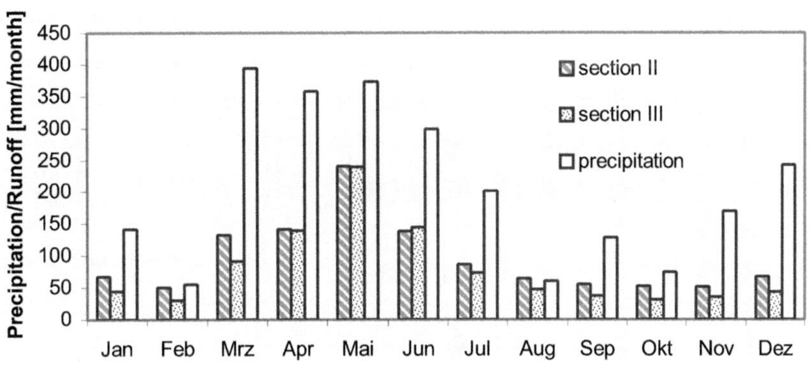

Fig. 3. Monthly area precipitation with monthly runoff at the natural forest section (III) and transition zone section (II) in the experimental Nopu catchment, Central Sulawesi, Indonesia

Daily discharge demonstrates a fast precipitation-runoff reaction (Figure 4). However reaction is largely dependent on preceding soil moisture storage and rainfall intensities. In November a daily area precipitation of 86 mm results only in a moderate flood. On the other hand, under saturated soil moisture conditions considerably less precipitation results in the highest runoff of the year 2002.

Comparing runoff behaviour at the different river sections on a daily temporal scale we observe the highest peak runoff at the transition zone. The reasons are probably found in the decreasing infiltration capacities after forest conversion due to dispersion and sealing of the unprotected soils by heavy rainfall as reviewed in Bonell and Balek (1993), although the unambiguous conclusion of an increasing stormflow due to land clearing couldn't be drawn on the base of a one year time series. During the measurement period the transition zone showed a flashier precipitation-runoff behaviour with often higher peaks after a dry period but also faster decay shortly after the runoff event. This could be seen as an indication that the amount of water leaving the transition section immediately as stormflow is augmented. Delayed stormflow shortly after precipitation seems to be higher in the natural forest due to the higher storage capacities of the canopy. Higher evapotranspiration associated with faster depletion of the soil moisture storage under natural forests during periods without sufficient rainfall finally result in a diminished dry season flow in the natural forest in comparison to the partly cleared transition section. The comparison of the flow duration curves of the different catchment sections clearly demonstrates these differences under low flow conditions (Figure 5).

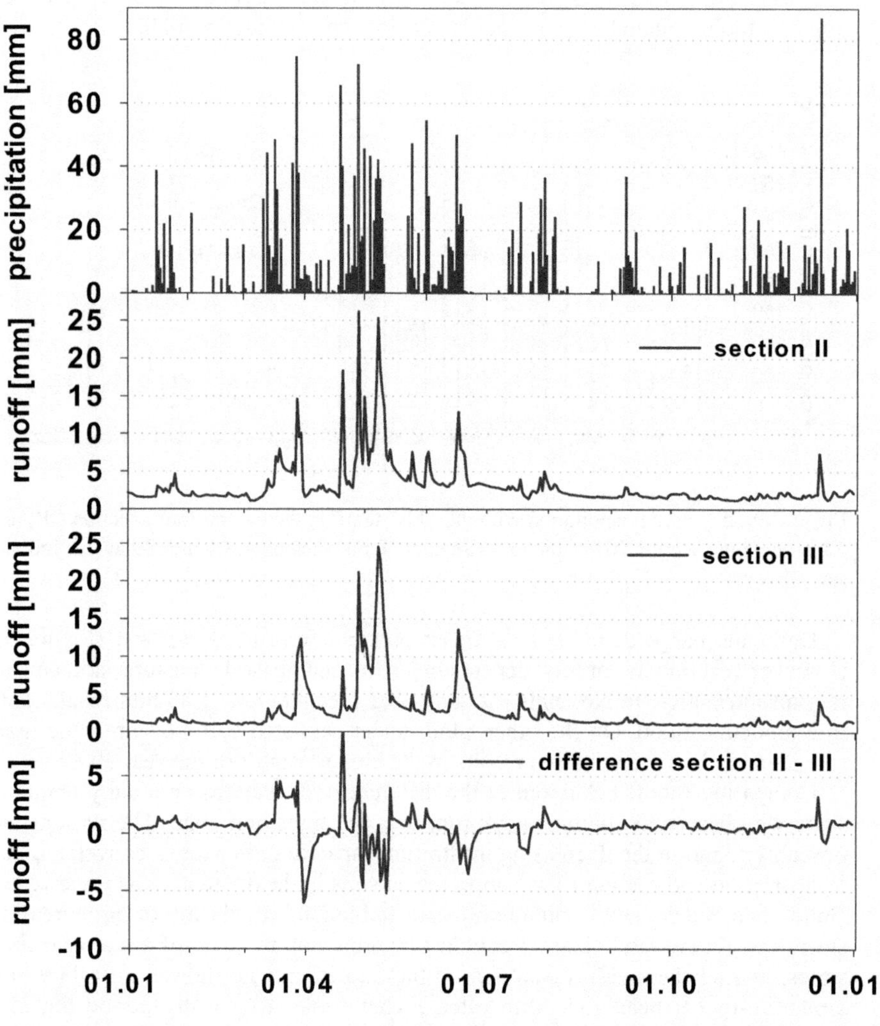

Fig. 4. Daily precipitation and discharge dynamics in the natural forest section (III) and the transition section (II) with differences between the sections in the experimental Nopu catchment in Central Sulawesi, Indonesia

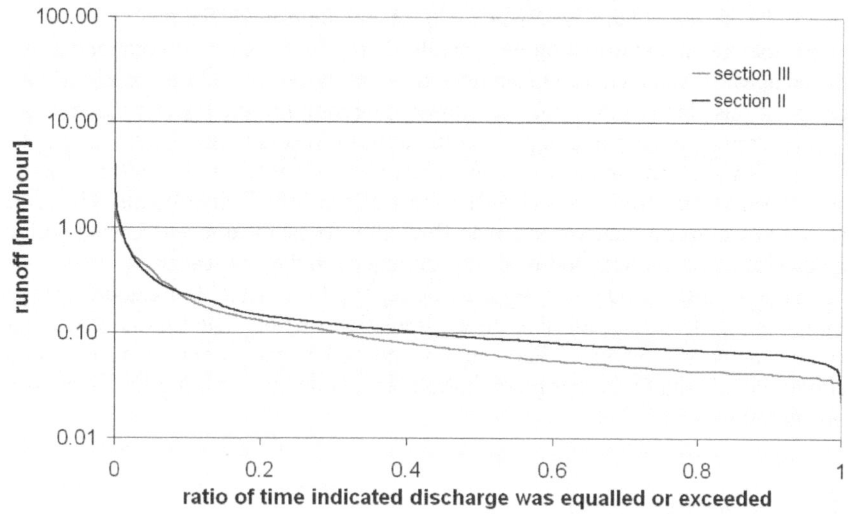

Fig. 5. Hourly flow duration curve of all river sections for the year 2002 in the experimental Nopu catchment in Central Sulawesi, Indonesia

3.3 Water Yield Changes

Water yield is defined as the discharge of a stream at a particular cross section as calculated over a specified time (Benham 1974). Results from the Nopu catchment show differences in yearly and monthly water yield between the natural forest river section and the disturbed transition section. Because of the homogeneous river sections we assume that the water yield differences are mainly caused by land use change. In total, yearly differences summarize up to 190 mm (Table 2).

The effects of land use change on water yield has been one focus at the hydrological decade (1965-1974). Hibbert (1967) summarized 39 studies of forest clearings, mainly situated in North America. He shows that the upper limit of yield increase is 4.5 mm/y for each percentage of reduction of forest cover. However most treatments produced less than 2.5 mm/y increase per percentage forest cover change and the magnitude of treatment response varied considerably. The yield increases almost invariably begin to decline soon after treatment and establishment of a new vegetation cover. According to Hibbert (1967) water yield increases in the field experiments mainly during the growing season, however most reviewed sites are located in temperate climate.

Using this relationship a percentage of converted land in the transition section of the Nopu catchment of 42-76% can be calculated. These results seem to be quite reasonable and correlate with results from land use change mapping in the catchment area and satellite image analyses. However, due to the different degrees of disturbance in the different patches of the transition section the direct comparison of the degree of land use change with other studies is difficult.

Beside the annual effect on water yield there is also seasonal component in the differences of water yield. Differences in monthly water yield are highest during drier seasons (January-March). Reasons can be found in the lower evapotranspiration on converted land in comparison to natural forest and the higher water uptake by the root zone in natural forest as reported by Hodnett et al. (1995). These results consist with Bruijnzeels (1990) review of controlled experiments. The review showed that largest increase in streamflow after forest clearance occurred with the delayed flow component and is most pronounced in the dry season.

In the wettest period of the measurement period water yields are equal or higher in the natural forest section than in the transition section. The reasons can be presumed in the saturation of interception and soil water storage capacities under conditions of slightly higher precipitations inputs due to the altitudinal increase of precipitation.

3.4 Evapotranspiration

Evapotranspiration (ET), as calculated from the water balance for the transition section II and the natural forest section III for the year 2002 was 1360 mm and 1570 mm. or respectively 3.7 mm/day and 4.3 mm/day.

Compared with other studies like the more recent and comprehensive study by Malmer (1993), who investigated the influence of different logging techniques on the water yield of different rainforest catchments in Sabah (Malaysia) these results seem quite plausible. There, a mean annual ET of 1540 mm in mature forest and 1253 mm in a mixed area of secondary vegetation and surviving trees was calculated. A survey of tropical forest water use by Bruijnzeel (sited in Schellekens et al. 1999) reveals that highest annual ET (2000-2400 mm) are observed at continental edges and island locations of high rainfall, therefore we can assume high annual evapotranspiration rates in the research area.

Due to scarcity of reliable ET data from tropical montane rainforests the existing climatic data will be used to apply the site specific Penman-Monteith equation (Monteith, 1975) to the research area. The comparison of both results allows the detection of possible leakages in the catchment sections.

Table 2. Water balance of the year 2002 in the experimental Nopu catchment, Indonesia

	weir 2	section II	section III
area (ha)	198	116	82
precipitation (mm)	2520	2510	2530
runoff (mm)	1070	1150	960
rest (ET,SS, D)	1450	1360	1570

where SS is soil storage and DD is deep drainage

4 Conclusion

The Nopu experimental catchment proved to be a useful location for the investigation of hydrological effects of land use changes by smallholders in the humid tropics. The modified paired catchment approach proved to be a suitable and robust tool for hydrological investigations. Although only short time series of measurements in our recently established Nopu research catchment are available, the profound influence of smallholder conversion on the hydrological responses becomes already apparent.

The use of the water balance method to investigate the effects of forest conversion produces plausible results. Land use change in the transition section of the research increased annual water yield by 190 mm, resulting in an decrease of ET. However the determination of the ET is dependant on cumulative errors in the three variables of the water balance equation (Equation 2). Losses by deep drainage or leakages out of the catchment sections can considerably influence the results. Therefore additional, independent methods to obtain the evapotranspiration must be applied to the experimental catchment. In a next step the ET will be calculated with the Penman-Monteith equation and results from eddy-covariance measurements of an adjacent cocoa plantation will be compared with the results of the water balance equation.

The influences of forest conversion on the seasonal behaviour of runoff and stormflow generation need more detailed analyses. However statistical methods like high and low flow analyses need data sets based on longer measurements which are often not available. The application of more process-oriented, physically based water balance models like WASIM-ETH can be a solution to improve the knowledge about the hydrological processes and allows the integration and validation of land use scenarios.

Land use change in the catchment is still continuing, Therefore measurements should continue until a new equilibrium of land uses is attained. Then the complete temporal and spatial process of forest conversion by smallholders from natural forest over a transition period to cocoa plantation can be applied to the modelling process.

5 Acknowledgements

We gratefully acknowledge the German Research Council (DFG) for funding this project and the STORMA organization in Goettingen, Kassel, Bogor and Palu for the management of the project. Thanks to Sance Lipu from University of Tadulako Palu, Indonesia, John F. Salomo and the village Nopu for their continued support in watershed instrumentation and measurement.

References

Benham AD (1974) Determination of the monthly, seasonal, annual, or long-term average yield of a stream. UN Water Resources Series No 34, pp 56-62

Berlage HP (1949) Rainfall in Indonesia: Mean rainfall figures for 4399 rainfall stations in Indonesia, 1879-1941. Verhandelingen van het Koninklijk Magnetisch en Meteorlogisch Observatorium Batavia 37

Bonell M, Balek J (1993) Recent Scientific Developments and Research Needs in Hydrological Processes of the Humid Tropics. In Bonell M, Hufschmidt MM and Gladwell JS (eds) Hydrology and water management in the humid tropics – Hydrological Research Issues and Strategies for Water Management. UNESCO and Cambridge University Press, pp 571-773

Bosch JM, Hewlett JD (1982) A review of catchment experiments to determine the effect of vegetation changes on water yield and evapotranspiration. Journal of Hydrology 55: 3-23

Bruijnzeel LA (1990) Hydrology of moist tropical forests and effects of conversion: a state of knowledge review. IHP-UNESCO, Paris

Calder I (2000) Land use impacts on water resources. In: Land-Water Linkages in Rural Watersheds – Electronic Workshop. Background Paper No. 1. Food and Agriculture Organization of the United Nations (FAO), Rome

Chang JH and Lau LS (1993) Appendix A: Definition of humid tropics. In: Bonell M, Hufschmidt MM and Gladwell JS (eds) Hydrology and water management in the humid tropics – Hydrological Research Issues and Strategies for Water Management. UNESCO and Cambridge University Press, pp 571-773

Hibbert A (1967) Forest treatment effects on water yield. In: Sopper, WE and Lull HW (eds) International Symposium on Forest hydrology. Pergamon Press, Oxford, pp 527-545

Hodnett MG, Pimenel da Silva L, da Rocha HR, Cruz Senna R (1995) Seasonal soil water changes beneath central Amazonian rainforest and pasture. Journal of Hydrology 170: 233-254.

Maertens M, Zeller M, Birner R (2002) Explaining Agricultural Land Use in Villages surrounding the Lore Lindu National Park in Central Sulawesi, Indonesia, STORMA Discussion paper series No. 4, University of Goettingen, and Kassel, Germany and Institut Pertanian Bogor and Universitas Tadulako, Indonesia

Malmer A (1993) Dynamics of hydrology and nutrient losses as response to establishment of forest plantation: a case study on tropical rainforest land in Sabah, Malaysia. Ph.D. thesis, Umea

Monteith JL (1975) Vegetation and the atmosphere, vol. 1: Principles. Academic Press, London

Schellekens J, Scantena FN, Bruijnzeel LA, Wickel AJ (1999) Modelling rainfall interception by lowland tropical rain forest in northeastern Puerto Rico. Journal of Hydrology 225: 168-184

Schulla (1997) Hydrologische Modellierung von Flussgebieten zur Abschätzung der Folgen von Klimaänderungen. Zürcher Geographische Schriften, Heft 69. Zürich

Sunderlin WD (2000) The effect of Indonesia's Economic crisis on small farmers and natural forest cover in the outer islands. Occasional Paper No. 28(E). Center of International Forestry Research (CIFOR), Bogor (Indonesia)

Sunderlin WD and Resosudarmo IAP (1996) Rates and causes of deforestation in Indonesia: Towards a resolution of the ambiguties. Occasional Paper No. 9. Center of International Forestry Research (CIFOR), Bogor (Indonesia)

Whitten T, Henderson GS, Mustafa M (1987) The Ecology of Sulawesi. The Ecology of Indonesia Series Vol VI. Gadjah Mada University Press, Singapore

Water Tenure in Highland Watersheds of Northern Thailand: Tragedy of the Commons or Successful Management of Complexity?

Andreas Neef, Liane Chamsai, Martina Hammer, Apai Wannitpradit, Chapika Sangkapitux, Yeeb Xyooj, Prapinwadee Sirisupluxuna & Wolfram Spreer

1 Introduction

In recent years, Thailand has faced serious water problems, such as water shortages in the dry season, flooding during the rainy season and pollution with agrochemicals and industrial waste. Agriculture is the main user of available water resources and accounts for 70% of the water demand. Whereas traditionally irrigation in Thailand has been supplementary during the rainy season, water is now increasingly used for irrigation during the dry season. However, other sectors are continuously increasing their share in the withdrawal of water resources. The management of water resources is characterized by institutional and legal pluralism, responsibilities being shared among as much as eight different ministries, divided into more than 30 ministerial departments (Sethaputra et al. 2001).

Until recently, the highlands of northern Thailand were characterized by rainfed agriculture for subsistence based on slash and burn practices. Only in the small mountain valleys, wet-rice cultivation was possible. With the opium substitution programs of the Thai government and international development programs and as a response to increasing land scarcity due to extensive reforestation programs, cash crops such as fruits, flowers and vegetables have spread rapidly since the late 1980s. Especially in highland areas with good market access, this development went along with a sharp increase of land under temporary or permanent irrigation. Since then, water conflicts between upstream and downstream water users in many watersheds of northern Thailand have become a widespread phenomenon and an expression of increasing water scarcity (Charoenmuang 1994). Water shortages are aggravated by increasing water demands of non-agricultural sectors, such as tourism and processing industries. The objectives of this paper are (1) to challenge the conventional view that water in the highlands of northern Thailand is an open access resource and therefore subject to mismanagement and overexploitation, (2) to present the complexity of water tenure ranging from purely individualized and family-managed rights to communal and public ownership and use rights, (3) to analyze the effects of interventions from government agencies on customary water management schemes, and (4) to compare different conflict resolution mechanisms with regard to water management.

The paper is the result of an interdisciplinary cooperation between researchers from the field of resource economics, social sciences and agricultural engineering.

The study was done in the framework of a Thai-Vietnamese-German Collaborative Research Program (SFB 564 – The Uplands Program) "Sustainable land use and rural development in mountainous regions of Southeast Asia".

2 Common-Pool Resources and Property Regimes – Theoretical Considerations

2.1 Misconceptions of Common-Pool Resources

The debate about the respective benefits of private, common and government resource tenure regimes has been dominated by a number of misconceptions and confusions. Based on Ostrom (2001)[1] two main sources of these misconceptions can be distinguished: (1) confusions between common property and open access regimes, and (2) differences between common-pool resources (CPR) and common property regimes.

(1) Confusions between Common Property and Open Access Regimes

Gordon (1954) and Hardin (1968) state that under common property regimes natural resources are prone to overexploitation because the costs of negative externalities like pollution of water or overgrazing of pastures are born by the community as a whole whereas the potential benefits accrue to the individual. For many years the general interpretation of these theorems was that communal ownership was the culprit for forest destruction, land degradation and water pollution and that private property or government control was crucial to sustain natural resources (Chalamwong and Feder 1985, Narkwiboonwong et al. 1994). In the meantime, several authors - the most prominent being Ciriacy-Wantrup & Bishop (1975), Bromley (1991), Ostrom (1994) and Bruce (1999) - have made clear distinctions between common property and open access. In common property regimes the members of a clearly demarked group have the right to exclude nonmembers of that group from using the resource (e.g. community forests, community-based irrigation schemes). Open access describes a situation where no one has the legal right to exclude anyone from using a resource (e.g. open seas and fresh air). However, the confusions between common property and open access regimes have proven extremely resistant, especially in the minds of some policy-makers. In Thailand, the perception of water as an open access resource is widespread and the existence of diverse forms of control, ownership and use rights of water resources is widely neglected. These misperceptions are used as arguments for government interventions in the management of water resources.

[1] Actually, Ostrom mentions a third factor (the confusion between the resource and the flow of resource units) which is not discussed here.

(2) Differences between Common Pool Resources (CPR) and Common Property Regimes

This issue has drawn much less attention than the first factor and has been widely neglected in the literature on CPR. Ostrom (2001) points out that common-pool resources share two important attributes: first, it is costly to exclude non-members of a certain group from using the resource either through physical barriers or legal restrictions and, second, the benefits consumed by one individual or subgroup subtract from the benefits available to other individuals or subgroups. The first attribute of common-pool resources is shared with public goods whereas the second attribute CPRs have in common with private goods. In contrast to widespread beliefs that common-pool resources are always under a common property regime, Ostrom emphasizes that CPR may be owned by private individuals or corporations, by communities, or by national, regional or local governments. Hence, automatic linkage between common-pool resources and common property regimes cannot be made. Furthermore, before drawing any conclusions on the efficiency or sustainability of some form of property to be used in connection with common-pool resources, it is important to base one's assessment on solid empirical studies considering the local context rather than on theoretical and general assertions (Ostrom 2001).

2.2 Complexity of Water Rights and Legal Pluralism

As compared with rights over other natural resources, water rights are particularly complex due to the characteristics of water as "a mobile, fluid, and fugitive resource, with a great deal of inherent uncertainty regarding its quantity and location" (Meinzen-Dick and Pradhan 2002, p 17). Supply of and demand for water can vary considerably by time and space. In most areas, multiple stakeholders compete for the use of water. Therefore, water rights are dynamic, flexible and often subject to continuous negotiation (Bruns and Meinzen-Dick 2000). Water rights can be closely linked to land rights, but the mobile character of water also allows for disconnecting water rights from other resource rights.

The concept of legal pluralism describes best the complexity of water rights. This concept acknowledges the co-existence and interaction of different legal systems in the same socio-political context (Woodman 1998, von Benda-Beckmann F and K 2001, Meinzen-Dick and Pradhan 2002). These interactions can both create uncertainties and open up new opportunities for groups and individuals. Legal pluralism can exist at the farm household, the community, and the national level. In this paper we will concentrate on legal pluralism at the community and farm household level.

To date, research on the complexity of local water tenure systems in mountainous regions of northern Thailand remains scant. There is also a lack of empirical evidence on how local water tenure systems change under government influence and market forces. The following case study from a watershed in northern Thailand will shed light on the problems of and potentials for sustainable management of common-pool resources under different tenure systems and varying external factors.

3 Case Study of Mae Sa Watershed, Chiang Mai Province, Northern Thailand

3.1 Material and Methods

Three villages in two subcatchments of the Mae Sa watershed in Chiang Mai province, northern Thailand, were selected for this study (Fig. 1). Mae Sa watershed is located in the outskirts of the northern Thai capital Chiang Mai and feeds into the Ping river, one of the major tributaries of the Chao Prya river. This watershed is intensively used for market-oriented agriculture, mainly fruit, flower and vegetable production. It is also a major tourist destination for day trips from Chiang Mai and comprises several resorts and tourist attractions.

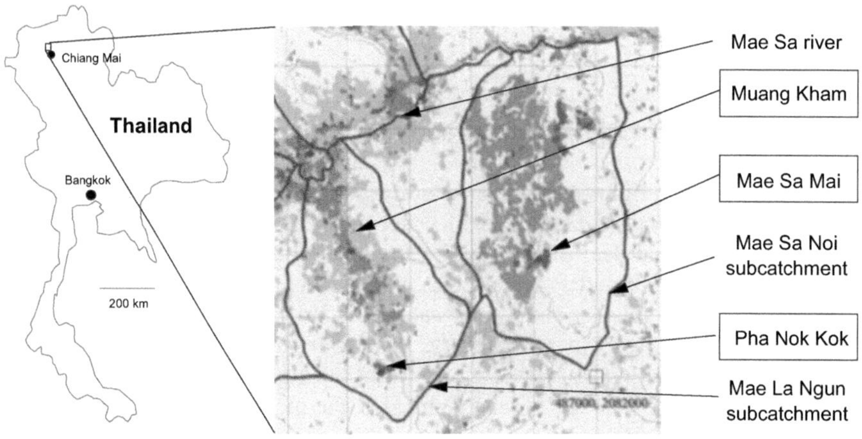

Fig. 1. Map of the study area with location of study villages and borders of subcatchments. Source: Own draft based on GIS-map from C. Riedel 2002

The village of **Mae Sa Mai**, a Hmong ethnic minority community, represents a case with no major upstream-downstream conflicts as the village is the only settlement in the subcatchment. The other two villages are located in a neighboring subcatchment where upstream-downstream land and water conflicts are a common phenomenon: **Pha Nok Kok** is an upstream Hmong settlement located only a few kilometers from the downstream Thai community of **Muang Kham** (Fig. 1). While the two Hmong communities are located in the area of the Doi Suthep-Pui National Park and most of the farm households do not have officially recognized land titles, the major part of the Thai village is outside the borders of the National Park and most of the farm households have official land titles.

In all three villages, open and semistructured interviews with key persons (village headmen, government officials, heads of water management committees and women groups) were conducted from January to August 2002. Researchers also went with selected, knowledgeable villagers to the various weirs, reservoirs and other conveyance systems and asked questions about water management, use and conflicts. Users of different water conveyance systems were asked about practices and regulations with regard to water use. Participant observation was applied in the case of conflict mediation, e.g. when water committees made field trips to areas of water conflicts. In addition, meetings of newly established basin committees were attended. The study was completed by a review of existing research documents and project plans about the study area. Intermediate research results were fed back to the informants for cross-checking and validation of the information obtained during the survey.

3.2 Management of Water Resources at the Community and Farm Household Level

Local Knowledge, Local Management and Local Elites – The Case of Mae Sa Mai

History of land use and resource endowment. The Hmong community of **Mae Sa Mai**, founded in 1965 by farmers in search for better access to water, is located in the Doi Suthep-Pui National Park. The village is the only major settlement in the Mae Sa Noi subcatchment which feeds into the Mae Sa river. Villagers do not have legal land titles as their presence is only tolerated by the National Park authorities. Recent land claims by an adjacent Botanical Garden, a Royal Project operating in the subcatchment and by the Royal Forest Department for reforestation purposes have decreased the available area for agriculture considerably. Land use changed from opium and upland rice production over corn and coffee to its present land use system, consisting mainly of cash crops like fruits and vegetables. The area planted with fruit trees (mainly litchi) has increased from less than one percent in 1974 to over sixty percent in 2002. A farm household survey of 60 farm households done in 1998 in the framework of an interdisciplinary study project under the author's supervision shows that the increase of fruit tree area was the main factor for the expansion of irrigated fields in the village (Fig. 2).

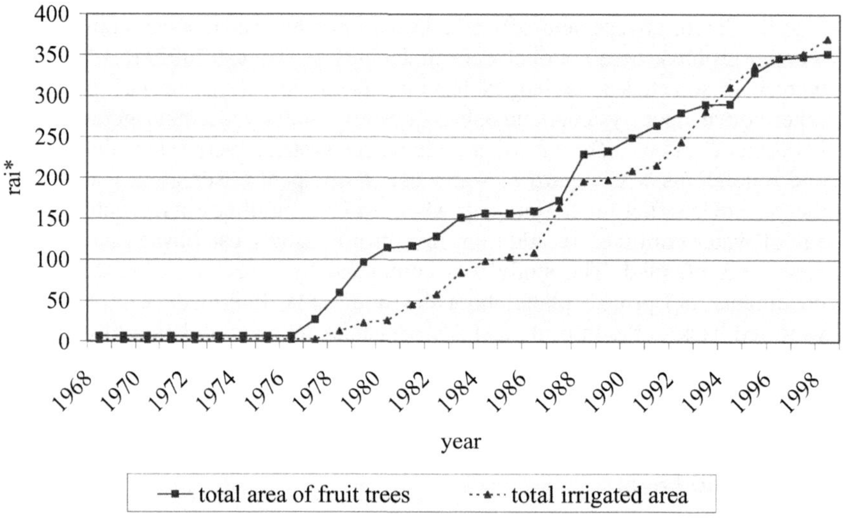

*1 rai = 0.16 ha

Fig. 2. Increase of irrigated area and fruit tree area in Mae Sa Mai from 1968-1998 (based on a representative survey with 60 farm households).
Source: Schiller 1999.

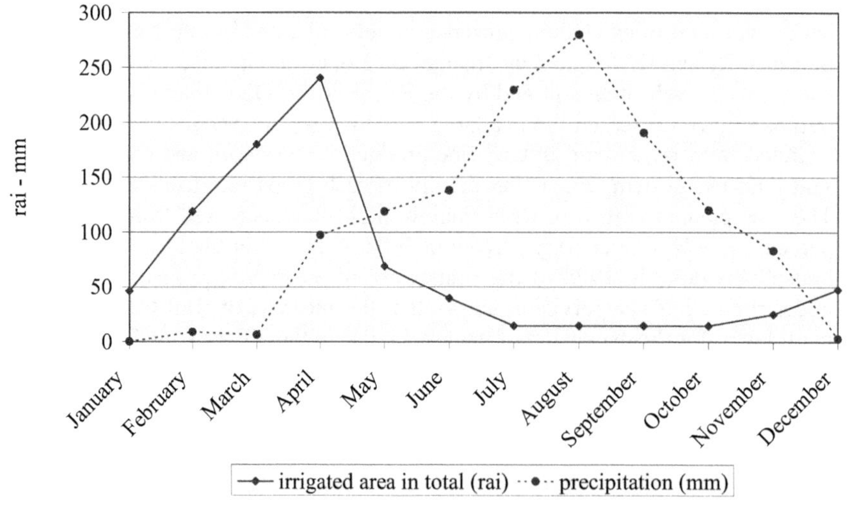

Fig. 3. Distribution of irrigated area in Mae Sa Mai in 1998 (based on a representative survey with 60 farm households).
Source: Schiller 1999.

More than half of the total arable land is now under irrigation by sprinklers. Fruit trees are the largest consumers of irrigation water: they use around 30% more water than triple cropped rice and around 60% more than vegetables (Ringler 2001). As litchi trees usually flower in December and are harvested in May, the most crucial time for irrigation is during the hot and dry season from February to May where water for irrigation is relatively scarce as precipitation is below 10 mm on average in the months from January to March (Fig. 3). Evaporation reaches its maximum in April, while relative humidity comes to its lowest point and temperature as well as wind speed reach their highest levels.

Major water sources in the Mae Sa Noi subcatchment are small creeks which vary strongly in terms of water availability. Some of them have water the whole year round, others only during the rainy season and a short time right afterwards. From February to April, surface water of the various small springs and creeks in the Mae Sa Noi subcatchment is significantly reduced and most conflicts around water allocation occur.

Water for household consumption. Water for household consumption is of high priority in Mae Sa Mai. Hence, irrigation in the hillsides above the village is forbidden to ensure that there is enough water for all households. Due to the size of the village (about 1,700 inhabitants), the village is subdivided into six zones each with their own leader. These zones, which are generally inhabited by members of the same clan, have their individual supply system for household water. A non-representative survey of water consumption of nine households in February 2003 showed that the average per-capita consumption of water was around 40 liters. Thus, we can calculate a total consumption of household water in the water scarce months from February to April of about 6,120 m^3. This amount appears negligible as compared to the 270,000 m^3 of irrigation water consumed by the estimated 54,000 litchi trees grown in the subcatchment[2].

Water management for irrigation. Over the years the farmers have developed a network of water conveyance systems to regulate the water resources of the valley. The whole Mae Sa Noi watershed is organized and divided into different watershed parts and tap-offs. Depending on the location of their fields, the water rights of each farmer for the different sources and the time, when the system was set up, the farmers tap off the water and convey it to their plots by various techniques based on their local knowledge and availability of material. Water rights are part of a resource tenure system adapted to social, cultural, economic and geographical conditions. Major influencing factors are traditions, time, power, locations, amount of water, and capital. Embedded in this socio-technical system, the farmers always know, when, where and how long it is their turn and possibility to use the water.

[2] The calculation is based on measurements with eight-year old litchi trees done in subproject B1 "Water, energy and nutrient efficient production of fruit trees on hillsides of northern Thailand" of The Uplands Program (SFB 564).

Currently, there are three main water tenure systems in the village, (1) individual water use rights, (2) user groups sharing the same pipe or ponds, and (3) user groups sharing the same springs or creeks.

(1) Individual users may be nuclear or extended families having agricultural fields located directly at a water source that can be used exclusively by them. Individual use rights can also exist if farmers drain water directly from streams above their fields with their individual pipes without sharing with other users although they might use the same water source. 40% of the orchard owners enjoy these individual use rights.

(2) Joint management systems exist when user groups share the same pipe or the same pond. For instance, the water would be drained by a pipe from the water source to a pond located in the orchard of one farmer. The pipes of the different users are then directly connected to this pond under different regulations. In Mae Sa Mai, around 20 such groups existed during the survey, each with a particular management system.

(3) Sometimes, user groups allocate water from the same spring or creek under a joint management arrangement. This tenure system is mostly found in higher altitudes of the watershed. In one case, farmers jointly constructed a small weir or reservoir with a concrete outlet at the bottom. This outlet is divided into different compartments to which the pipes of the individual pipe owners are connected (Fig. 4). The user group agreed on the diameter of the pipes that individual users are allowed to install. Size and number of pipes depend on the number of farmers, their social position, their modes of cooperation, the location of the water source, the amount of available water and the distance to the fields.

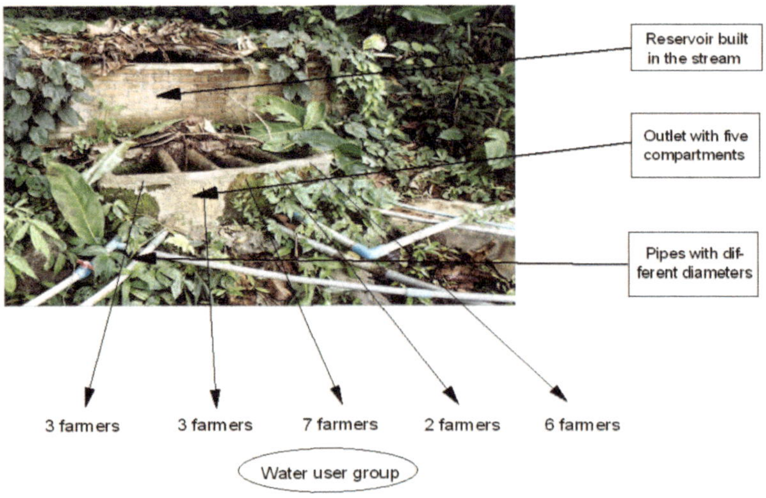

Fig. 4. Water conveyance system under a joint management arrangement

The history of the joint management system shown in Fig. 4 illustrates how these communal systems evolved from formerly individualized systems. Until the mid-1990s, farmers connected their pipes directly to the water source. As inequalities in sharing water occurred, the farmers agreed on constructing a reservoir with a compartmentalized outlet. Pipes with a given size were connected to the five cemented sections to allocate water based on the needs of water for the orchard owners while respecting the customary rights of the first users. The head of the user group comprising 21 members coordinates the water use, mediates in disputes and performs the religious ceremony *sya-hao-dhale*, hold annually at the beginning of the 'irrigation' season. He represents the clan which the majority of the group members belong to. This clan also dominated the decision-making process leading to the amendment of the water management and use regulations. Whereas traditionally cooperation between members of different clans is not very frequent in the Hmong society, the establishment of the user group followed pragmatic economic considerations: farmers sharing the same compartment (between two and seven per compartment) are owners of neighboring orchards and belong to different clans in the village.

In general, water use rights reflect local power relations, with rich and long-established families, lineages and clans controlling the water conveyance systems by tapping water directly from the source and by using pipes with a bigger diameter than others. Members of local elites usually would take as much water as they need and nobody would dare to interfere. Powerful farmers also have the opportunity to take water from different sources and thus are able to diversify risks in an insecure institutional environment (see example in Box 1).

When established, the system was simple and clear, but with the time, changes of power relations, technical innovations and intervention of different outside organizations, it became more and more complex. The basic system was developed and extended step by step, but never completely re-structured under the aspects of current needs and efficiency. Most of the older reservoirs are located above the village close to the water sources of the small creeks feeding into the subcatchment. Until 1984 the villagers settled in a higher altitude but then had to move further downstream in order to save and regenerate the remaining forest in the head watersheds. However, the old water tapping points were kept in operation with the connecting canals and pipes being extended to the new fields. This caused a fundamental change in the land and water tenure system: land rights and access to water – which had previously been closely interlinked – were then separated from each other[3] (cf. Box 1).

[3] Results from a representative survey from 2002 including 40 farm households show that the average distance from the fields to the water sources was more than 600 m.

Box 1: History of water rights at the field level and strategies to cope with water insecurity – The case of Mr. Kasem. Mr. Kasem is a member of an old and respectable Hmong family of Mae Sa Mai. He has plots at different locations in the valley and therefore claims rights to several water sources. After starting the litchi production at his field close to the Queen Sirikit Botanical Garden he was using the water rights of the tap-off situated next to the field. There was also a major pipeline (line H) coming down the whole valley from a reservoir located above the village, which guaranteed the availability of water as a second source.

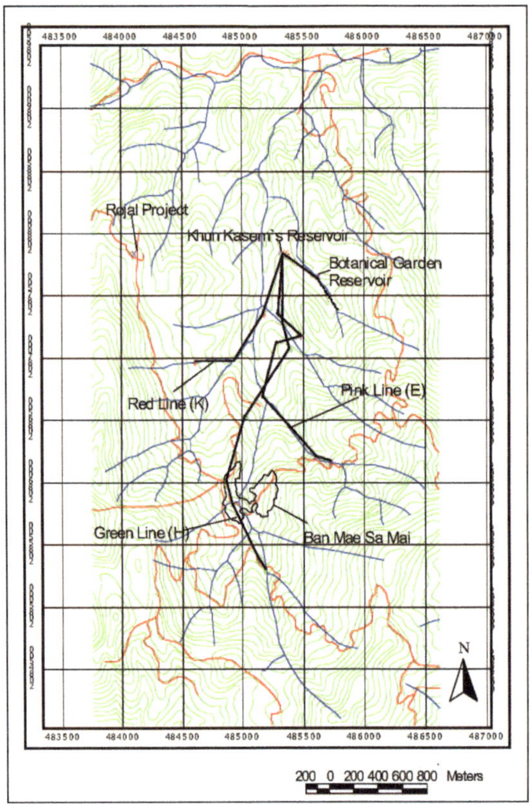

By the time, the farmers needed more and more water, the amount of water reaching his field from this reservoir was decreasing. In addition, the conveyance system got older and its functionality decreased. When a nursery was set up in the Queen Sirikit Botanical Garden, an enormous reservoir was built for this purpose and the Queen Sirikit Botanical Garden got the official permission for the water rights of the source next to Mr. Kasem's litchi field. Thus, he lost this water source. As a consequence he set up a new system to get the needed water supply for his field. The water was taken from a source far away at the other side of the valley (line K). There, each farmer is allowed to tap off water by a standardized PVC-pipe. The Department of Irrigation constructed a new dam for another source at his side of the valley and the big mainline (line E) is passing his field, too. Therefore, he should be in the situation to have two water sources available. However, the pipe of the new dam is broken; hence he has currently only one source as water supply.

Individual Water Access and Community-Based Regulations – The Case of Pha Nok Kok

History of land use and resource endowment. Like Mae Sa Mai, the Hmong village of Pha Nok Kok is located inside the borders of Doi Suthep-Pui National Park in the Mae Sa catchment area. The village was officially recognized in 1984. Its legal status, however, has become ambiguous due to its inclusion into the National Park which was due to an expansion of the park's boundaries in 1996. None of the villagers has documented land rights. Farming systems are dominated by cabbage, litchi and flowers as the main cash crops. Located on a mountain top, villagers of Pha Nok Kok use land on both sides of the mountain, namely the Mae Ta Chang catchment to the south and the Mae La Ngun subcatchment which feeds into the Mae Sa river in the north.

The village has access to six streams, three in the Mae La Ngun subcatchment namely Huai Pha Buu, Huai Don Dua and Huai Muang Chang and three in the Mae Ta Chang catchment called Huai Deen Dok, Huai Hier and Huai Maffai. The water situation in Pha Nok Kok is better than in neighboring Hmong villages to the west of the watershed where water for household consumption has to be fetched directly at the stream because they do not have a water conveyance system. After installation of such a system in Pha Nok Kok drinking water supply is sufficient, with the exception of very dry years. However, the village regularly faces shortage of irrigation water, which is greatest from April to May, just before the litchi harvest. As the scarce water resources in the dry season can only be used for irrigating litchi plantations, it is not possible to grow annual cash crops, such as vegetables, during the dry season. Therefore water, followed by land, can be considered as one of the main limiting factors for agriculture in this area.

Water management. In analyzing water management in Pha Nok Kok village we have to differentiate clearly between irrigation water and water used for household consumption. The first is mainly organized individually, the latter in a more centralized way.

In former times water had been regarded as a free resource without any regulations to control or rationalize its use. Today, water sources for households and consumption are clearly defined and rules and regulations for water use had been set up and summarized in one document (see Table 1). Beside these rules, internal cooperation within the village is perceived as a main aspect enabling resource management at community level. Obviously, the Hmong village of Pha Nok Kok has a more cohesive community structure and therefore more social capital and capacity for collective action than the village of Mae Sa Mai which might be related to the smaller size of the community and a stronger authority of the village headman.

Table 1. Regulations concerning water use in Pha Nok Kok

No.	Regulation
1.	Water from the stream Huai Pha Buu on village elevation is exclusively to be used for household consumption and after 12 a.m. it is not allowed to use big amounts of this water (i.e. laundry, car washing)
2.	Irrigation water can be used from Huai Don Dua (the rest further downstream), Huai Muang Chang, Huai Deen Dok, Huai Hier, Huai Maffai.
3.	Irrigation water: Construction of new water access facilities (i.e. pipes) has been forbidden since 2000.
4.	Valves have to be installed in every plantation, so that the pipes could be closed if the reservoir is filled.
5.	Maximum size of pipes conveying water from the stream to the plantation: 1.5 inches diameter.
6.	At the time of the Songkran Festival [traditional water pouring ceremony], 11-15 April, it is not allowed to use irrigation water at all.
7.	If two or more farmers use a reservoir together, access to irrigation water is regulated through the number of sprinkler and an irrigation schedule.
8.	In the dry season it is not allowed to irrigate the vegetable plantations. (Note: Official, village internal regulation, but reality is different).
9.	All villagers are asked to use water economically.

Source: Own survey 2002

Table 2. Steps towards the establishment of a water conveyance system in Pha Nok Kok

Time	Activity	Implementing agency/person
1970	Construction of a water reservoir made from soil	UN Development Project
1988	Building of a concrete basin at the site of the UN basin	Private investor
n.a.	Construction of water tanks nearby basin	n.a.
n.a.	Construction of more tanks at the same location	Dept. of Social Welfare
2002	Construction of small dam at Don Dua stream	Royal Forest Department
2002	Construction of new tank (~18m^3) nearby old ones	Subdistrict Office

Source: Own survey 2002

Water for household consumption. The year 1970 marked the beginning of the establishment of a village water conveyance system. By that time the UN started with the construction of a village water storage basin facilitating access to water for household consumption. By 1988 a private investor had improved these water storage facilities built by the UN, and more constructions had been added by different government agencies, i.e. the Dept. of Irrigation, the Dept. of Land Development and the Dept. of Social Welfare (Table 2). Currently water from the Pha Boo stream is stored in six tanks after passing some filter device. The villagers themselves are responsible for the maintenance of the whole water conveyance system.

Water for irrigation. According to the village headman, most families have direct access to streams for irrigation. Families with no access would ask others for permission to use water from their water source, e.g. private reservoirs. Sometimes, pipes allocating water from streams towards the plantations, serve more than one party. Like in Mae Sa Mai, villagers who were first to build reservoirs and pipes have first priority in water use, causing an uneven allocation of water. One attempt to equalize the distribution of water among the villagers and, at the same time, to reduce water use was the standardization of pipes used as water allocation facilities. This measure was initiated by the village headman and accepted by the village community.

Access to water depends largely on the time of the settlement, the financial situation of individual farmers, village internal regulations[4], and distance to the water source and is thus a function of economic, social and geographical factors. The individually organized water management seems to lead to less efficiency in water management due to a lack of financial resources. In order to meet the demand for irrigation water and water used for fire protection more water storage facilities (basins, reservoirs, tanks) would be necessary but cannot be constructed because there is no funding. Also automatic tapping-systems regulating water flow into private basins are considered useful but cannot be implemented for the same reason[5].

Commercialization and Individualization – The Case of Muang Kham

History of land use and resource endowment. Muang Kham is a Thai village founded in 1914 and one of the oldest settlements in Mae Sa Watershed. It is located just a few kilometers downstream from the Hmong village of Pha Nok Kok. The major part of the village territory is outside the boundaries of the Doi Suthep-Pui National Park. Hence, most of the farmers have full or partial ownership rights, documented by official titles.

---

[4] Since 2000 the construction of new pipes opening new access to water is forbidden. This is the reason why people with fields very close to water sources might not have access to water.

[5] This automatic tap-system would close the pipe when the reservoir is filled. It would cost around 200 Baht, equivalent to the salary of three days.

Until the end of the 1980s the main crop was paddy rice and it was not possible to grow crops the whole year round, but only during the rainy season, water being the most limiting factor. Since the beginning of the 1990s the Thai community of Muang Kham has experienced a rapid shift from a subsistence-based towards a market-oriented agricultural economy. Today, with the installation of irrigation facilities such as reservoirs and wells, agriculture is highly intensive and commercialized. Throughout the year farmers grow cash crops, mainly flowers, sayote (*Sechium elude*), sweet pepper and strawberries. Paddy rice production is reduced to a few small plots and does not have any economic significance.

Muang Kham has a relatively rich endowment with water sources. The small creeks Huai Don Dua, Huai Pha Boo, Huai Muang Chang, Huai Buak Kiad and Huai Pha Pung contribute their water to the Mae La Ngun stream which flows into the Mae Sa river. Three creeks, namely Huai Don Dua, Huai Pha Boo and Huai Muang, are shared with the upstream Hmong village of Pha Nok Kok.

Irrigation water management. Traditionally, water management in Muang Kham was done on a communal basis as long as paddy rice was the most important crop. The muang-fai[6] system is a traditional form of irrigation system common in many areas of northern Thailand and used for the irrigation of the paddy fields. In the past there were four main irrigation canals operating in Muang Kham. For every canal there was one person responsible, the gae muang, whose duty it was to care for the maintenance of the muang-fai system, to organize the annual ceremony[7] performed at every canal and to mediate between the users. In the muang-fai system, land and water rights are closely linked as only farmers with fields along the canals can get access to water. Today, three of the muang in Muang Kham are still in use and managed by the gae muang and the users, mainly to irrigate vegetable and flower plantations. The Mae Sa river, which feeds one of the canals, is also used directly by farmers who withdraw irrigation water through individual pipes.

In the slope areas, farmers would use water directly from the creeks either by installing pipes that lead towards their fields or by construction of small earthen or concrete basins/ponds, funded privately or by the government. These ponds can be located at the stream or directly at the farmers' fields, with pipes leading towards the plantations. Usually water allocation facilities are shared among family members, only in a few cases used jointly by independent parties across family boundaries. Different farmers using the same creek form water user groups. In some cases, such as the Buak Kiad creek (used by as many as 24 farmers) and the Muang Chang creek, the members of the water user group elaborated a list with rules and regulations concerning water use (box 2). The meeting in which these rules were elaborated was attended only by 16 of the 24 members. According to

---

[6] Translation: "Muang" = small canal; "Fai" = weir

[7] During this ceremony held in the 9th month of the year (according to the northern Thai calendar) offerings are made to the water spirits. The *gae muang* has to collect money from the users to buy these offerings.

the village headman, the others were not interested because they did not use the water at that time.

Another user group consisting of four members built a common reservoir in the field of one farmer. The agreement was that the person who provided the land for the reservoir did not have to contribute to the construction costs. Water is shared equally among the members of the user group. However, the field owner stated that he would prefer a private reservoir as sharing water with others can easily cause problems.

Box 2. Regulations of the water user group of Buak Kiad creek in Muang Kham (as of October 17, 1998)

1. It is forbidden to install additional pipes above the already existing pipes that are used within the water user group.
2. As a member of the water user group I promise not to cheat on or cause damage to other group members.
3. In case I am a member of the water user group and do not practice agriculture, I promise not to use any water from the Buak Kiad stream.
4. We agree to help each other taking care of the maintenance of the water pipes in the area of Buak Kiad stream without consideration who the pipe owner is.
5. In case I violate these rules I agree to pay a fee, which is to be determined by the group, and to refrain from any further water use.
6. In case of violation of the rules this procedure has to be followed:
   1st violation: Admonishment, exclusion from water use for 7 days.
   2nd violation: Fee of 500 Baht, exclusion from water use for 15 days.
   3rd violation: Prohibition of water use.
7. We, the members of the Buak Kiad water user group, are obliged to develop water use (...) in the area of Buak Kiad stream in a joint manner for common benefit.

This document is signed by the 16 members present during this meeting.

Commercialization of agriculture and land which started with the introduction of flowers and cabbage has induced a shift to more individualized systems of water conveyance and irrigation. The trend to individualized water management strategies was accelerated by the introduction of strawberries in 1994/95, as farmers started to construct their private wells in order to ensure a sufficient supply of clean water. From 1998 onwards, more private wells have been built by contract farmers growing sweet peppers under hydroponic conditions for a foreign company. Clean water is a prerequisite for sweet pepper cultivation to prevent diseases like mould and to ensure a high quality of the product.

As communal forms of water allocation continue to exist – even with much lower importance – legal pluralism of water rights has emerged and water management has become far more complex. Legal pluralism exists both at the village level and at the farm household level (see Box 3). Today, three main types of access to water can be observed in the village, (1) direct access to natural water sources, (2) access through *muang-fai* system, and (3) private groundwater wells, the latter continuously increasing its importance.

Like in Mae Sa Mai, distribution of water among the users is often characterized by strong inequalities because it depends on factors like distance between the resource and the fields, the time of settlement (who came first will have long-established and therefore more secure access rights), the location of the field (due to the first-come/first-serve principle upstream users are in a better position) and the financial situation of the farmer (need for investment to built a conveyance system). Thus, many farmers suggested to improve the water management in the village by constructing a community reservoir with the argument that this would equalize access among users.

Box 3. Legal pluralism of water rights (Example of Mr. Narong, farmer in Muang Kham)

Individual right: Mr. Narong uses water of Muang Chang and Mae La Ngun and stores this water in two reservoirs, one funded from government sources (built in 1994) and one from his own budget (built in 1995). In addition, he built a well for drinking water supply around 10 years ago. For his plantations the amount of irrigation water is currently not sufficient. Mr. Narong shares water from his two reservoirs and the well with the family of his brother.

Extended individual right: As a consequence of the construction of the lower reservoir, soil was released into the stream which then moved towards the fields of Mr. Amnuay, a neighbor of Mr. Narong, causing considerable land loss. As a compensation it was agreed that Mr. Amnuay was allowed to use water from one of the reservoirs.

Common right: Mr. Narong has to share his access to Muang Chang stream with four other parties. This was the cause of ongoing conflicts until the construction of a small reservoir, built two years ago nearby the stream upon initiative of the involved parties and with support of the village headman. Its purpose is to make sure that water is equally distributed among the users. It was funded and constructed by the villagers themselves. From this reservoir one pipe per family is installed to lead the water towards the plantations. Unfortunately, one party uses more water than the others (by installing an additional pipe), because they were the first to use water from the Muang Chang stream for irrigation of their former rice fields and therefore believe themselves to be the "owner" of the water. Mr. Narong considers this as very injustice especially because this family/clan already has good access to two other sources of water further downstream. Although the reservoir brought a slight relief for the tensed situation the issue remains an ongoing conflict, emerging every year in the dry season.

Access to drinking water. Like water for agricultural purposes, access to drinking water has undergone a process of individualization. Today, supply of drinking water is organized on a private basis. According to one informant, there are a few exceptional cases like Ban Luang, a subcommunity of the village Muang Kham, where allocation of drinking water is organized on community level. In some cases people use the same water sources for both household consumption and irrigation, sometimes they distinguish between uses. Compared to the village Pha Nok Kok, where one stream is used exclusively for consumption, sources of drinking water are not clearly defined in Muang Kham and can either be streams, wells or bottled water. The latter source is getting more important as many villagers are afraid of pesticide residues in their drinking water.

A drinking water company located at the village center started its production in August 2002. It was built on the private initiative of a villager of Muang Kham and is declared as a communal village project, and therefore receives governmental support. The initiator holds 70% of the shares for the company and 37 villagers of Muang Kham hold shares of 30%. The company plans to sell the water mainly to local customers within the subdistrict. Villagers interviewed during our survey did not seem to be concerned about problems of decreasing water tables and the danger of pesticide residues.

Internal Water Conflicts and Local Institutions for Mediation

Traditionally, water conflicts in the customary system rarely occurred as well-working schedules – although distorted to favor the rich and long-established families – had been developed over the years and the number of people sharing a stream, reservoir or pipe was relatively small and based on social relations between the users. Farmers usually knew exactly with whom they share their water, which amount of water they receive, how the schedule works and what their individual use rights are. Recently, however, water scarcity has become more accentuated. In very dry years, such as during the El-Niño phenomenon, water conflicts can occur more frequently. With the growing age of the litchi trees in the orchards, the water requirement is also steadily increasing. In **Mae Sa Mai**, a water committee was established in 1996 after a period of droughts and a series of smaller conflicts. The major task of the water committee was to mediate in conflicts that could not be solved by the user groups of each individual water source in the sub-catchment. Initially, the committee was composed by one representative of each of the five clans in the villages to ensure objectivity and transparency for all water users. The major contribution of this management group to successfully mediating in past water conflicts was the suggestion of a general reduction of pipe diameters. The current water committee is composed of representatives from three clans only. According to some informants, this committee does not have the authority to solve conflicts within the community. Some villagers even would prefer the involvement of external intermediaries; they stated that it is difficult for local committees to mediate in conflicts in which members of the same clan, close friends and relatives are involved.

Similar conflicts occurred in **Pha Nok Kok** where a water committee was established in 1993. In contrast to the group in Mae Sa Mai it consists of two representatives for each of the six streams of the village. Conflicts are solved by regulations concerning the number of sprinklers and time of irrigation. If the water committee is not able to solve the conflict the case is transferred to the village headman. In the past, potential for conflicts arose also through influential migrants as the village is not far from the urban center of Chiang Mai. In 1988, a Member of Parliament bought land from villagers of Pha Nok Kok. He built a concrete reservoir replacing the basin formerly funded by the United Nations Development

Project (cf. table 1) and asked the villagers for permission to use water for consumption. However, in contrast to the agreement he used the reservoir for irrigation and continuously increased the amount of water from the stream feeding the reservoir. Currently he uses 25% of the water of the Pha Puu stream. The villagers regularly control his water withdrawal to prevent him from claiming an increasing amount of water.

In **Muang Kham**, most internal conflicts are solved within the water user groups for each natural water source, reservoir or *muang-fai* system. In case conflicts cannot be solved at this level, the village headman would act as the main mediator. One recent conflict occurred when a villager claimed water from one stream for himself because he was the first one to use this water source. The construction of private wells and the formulation of regulations for water user groups has considerably reduced the potential for conflicts between villagers.

3.3 External Interventions in Water Management and Upstream-Downstream Water Competition

Influence of Government Institutions and NGOs on Water Tenure, Management and Conflict

Recently, government agencies and national projects have started to exert more influence on the customary system in **Mae Sa Mai** by building bigger reservoirs in the lower parts of the subcatchment. As the number of users varies between 20 and 70 farmers and the conveyance systems are relatively new, the irrigation schedule is difficult to manage and lack transparency for local users. Thus, the management of these newly established systems is often characterized by distrust and misunderstandings. Along the main road to the village eight smaller reservoirs were built by the Department of Irrigation in cooperation with the Royal Project. Initially 20 such reservoirs were planned but budget limitations forced the government agency to reduce the number. The sites of the reservoirs were selected based upon initial surveys on the number of orchard owners in the area and the geographical conditions, such as slope. The owners had to approve the construction which took place in 2001. Paradoxically, two ponds constructed upon the private initiative of two farmers are considered illegal by the government agency following National Park regulations which do not allow any permanent constructions in the fields. Thus, they do not have the rights to receive water from the water conveyance system that feeds into the other eight reservoirs.

In early 2002, the construction of a huge reservoir was completed with support from the Department of Irrigation and the Royal Project Foundation. This reservoir was designated to secure water supply for five of the eight smaller reservoirs. However, long-established families and lineages in the village continue to take water directly from the stream that was supposed to feed into the new reservoir which is running dry. Whereas the construction of the new reservoir could in principle provide more distributional equity by opening up access to water to less influential people in the village, the responsible agencies failed in recognizing local power structures and involving the villagers in planning the reservoir and setting up a workable schedule. The people who currently control the customary water conveyance system are afraid of losing power and influence; other villagers fear that the Royal Project will be the main user of the new reservoir and will not be able to work out an adequate user scheme. Those villagers, however, who are disfavored by inequalities in the customary system, are positive towards a big reservoir and external interventions in water use regulations. They argue that the local water management group does not have enough power to ensure equitable access to water.

Box 4. Forest, trees and water – Local perceptions and scientific evidence
Recent water shortages in the dry season in northern Thailand have raised increasing concerns and sparked the search for explanations. The main culprit is believed to be deforestation reducing water yields and creating a more uneven distribution of rainfalls. This belief has become the main driving force for forest-for-water policies in Thailand. On the other hand, a direct link between deforestation and precipitation has not been proven yet. The same applies for the generally held conviction that forests act as a major water store, absorbing water during the rainy season and releasing it during the dry season. There is evidence, however, that the impact of deforestation on water yields depends strongly on the type of vegetation and land use that replaces the forest (Bruijnzeel 1991, Kaosa-ard 2000).

In the local context of the northern Thai highlands, ethnic minority groups traditionally make a strong connection between forest, trees and water as they believe that there are spirits to live in and protect these natural resource. In Mae Sa Mai, Hmong villagers hold the sya-hao-dhale ceremony each year at the major streams before using water for irrigation or on the occasion of the Hmong New Year in December. The ceremony serves to protect the water sources and to guarantee abundant water for household consumption and agriculture. If someone violates traditional norms, such as cutting down trees around the water source or polluting it, this person would be punished and would have to compensate the spirit as an apology.

Some villagers in Mae Sa Mai and Pha Nok Kok believe in the ability of the banana tree to convey water to the surface. In fact, banana trees are found at most springs and along the creeks downstream. Other farmers, however, try to eradicate banana trees because they regard them as a major competitor for water. The village headman of Pha Nok Kok stated during the survey that a link between forest conservation and water availability does not exist. He argued that water availability in the village was still good when a large share of the forest had already been vanished, but the amount of water decreased several years after. Whereas researchers would interpret this as a time-lag between deforestation and decline of water availability, the village headman and other influential villagers of Pha Nok Kok use this observation as an argument against claims of the Royal Forest Department (RFD) to reconvert agricultural land into forest. The village headman even maintains his unprofitable litchi orchard because of fears that the RFD would claim the land for reforestation after he cut down his trees. The leaders of the women group of Pha Nok Kok, on the other hand, stated that the water situation in the village has improved since the villagers engaged in forest protection measures.

Since the Royal Project is not operating in the village of **Pha Nok Kok**, interventions in water conveyance systems are exclusively initiated by government institutions. Nevertheless, there exists a high degree of institutional pluralism: the Department of Irrigation, the Department of Land Development, the Department of Public Welfare and the Royal Forest Department all have intervened indirectly in the water allocation system by constructing tanks, reservoirs, dams and irrigation channels. However, the actual water management is mainly done by the villagers themselves which reduces the potential for conflict as compared to the case of Mae Sa Mai.

In **Muang Kham**, the Department of Land Development has supported the construction of earthen reservoirs for irrigation purpose since the year 2000. A total of 10 reservoirs have been built on farmers' land and used by individual families. In September 2001 the construction of three community wells had been initiated by the Dept. of Irrigation for the village drinking water supply, but not been completed so far because of budget problems. Farmers fear the possibility of having to pay for the water as well as for the maintenance of these facilities. Contradicting policies of different government agencies is the cause of slow procedures or the complete failure of projects. Sometimes, government agencies working under the same ministry are not able to properly coordinate their activities. A project for building a big reservoir was approved with 10 Mio. Baht by the Department of Irrigation, but could not be implemented because of an intervention of the Royal Forest Department which did not allow the construction at the designated site.

In all three villages, the Royal Forest Department through its policy of reforestation, forest protection and delimitation of forest boundaries indirectly influences the water management system. By engaging in government-supported forest protection activities and by caring for 'community forests' assigned by the Royal Forest Department, the villagers 'negotiate' the security of their land and water rights even if not all of them are convinced of the often-stated strong linkages between forest, trees and water (see Box 4).

Upstream-Downstream Water Disputes and Conflict Resolution Mechanisms

Until recently, there were regular conflicts between the upstream Hmong village of Pha Nok Kok and the downstream Thai village of Muang Kham. These conflicts occur mostly in particularly dry years, such as in 1992/93 and 1999/2000. During a severe drought and a disastrous hailstorm, Thai villagers from Muang Kham visited the village of Pha Nok Kok and urged them to reestablish their traditional ceremony of *dong seng* which is held to pay respect to the sacred tree of the village. Villagers of Muang Kham thought that a relationship exists between the abolition of that ceremony and the natural disasters. Due to the conversion of many Hmong villagers of Pha Nok Kok to Christianity, animist ceremonies to protect sacred trees and water sources were abandoned since the beginning of the 1990s. The villagers of Pha Nok Kok did not show any willingness to revive this tradition and the Thai villagers had to organize their own ceremony. As conflicts between Pha Nok Kok and Muang Kham became more frequent, a joint committee consisting of two representatives per stream and village was set up to solve these conflicts. A local NGO functioned as an external mediator. In 1999/2000, one stream used by farmers from one village fell dry, leading to an acute conflict between the two villages. Villagers of Muang Kham initiated meetings with villagers of Pha Nok Kok where the issue of water distribution was discussed. Regulations were set up that would satisfy both parties, such as a standardization of the diameter of pipes.

The main solution for the increasing conflicts, however, was driven by market forces: the introduction of strawberries and sweet peppers in Muang Kham brought about the construction of private wells for pumping ground water which made the farmers in this village much less reliant on surface water from upstream creeks formerly shared with farmers from Pa Nok Kok (cf. 3.2.3). While the conflicts with Muang Kham gradually decreased in this process, villagers in Pha Nok Kok experienced a much more serious conflict with downstream stakeholders on the other side of the hill, in the Mae Ta Chang watershed. Lowland Thai farmers accusing them of using too much water for litchi production invaded some of their orchards, cut down trees and destroyed water pipes. Neighboring highland villages, such as Ban Doi Pui, were affected by similar problems. The conflict was exacerbated by the fact that several tourist resorts located between the conflicting communities also take a big share of the water.

With the help of a local NGO and financial support from the Thai Research Fund, a committee consisting of representatives of all involved parties, NGO members and scientists from Chiang Mai University was established to work out solutions for this conflict. The Mae Ta Chang Rehabilitation Project was created as an action-research project using a scientific approach to solve problems at the community level. The objectives of this project were to identify the water availability and the course of the different streams, to investigate the actual water requirement and use of water by the different stakeholders and to set up regulations and standards with participation of the parties involved (Institute of Community Rights 2001). In order to supply representatives (two from Pha Nok Kok) for the Mae Ta Chang Water Committee, the Water Committee of Pha Nok Kok, which was not active at that time, had been revived.

At the beginning of the process the discussions were very emotional and the involved parties accused each other of overexploiting water resources. The lowland Thais had the perception that the upland Hmong have no right at all to use water because they do not pay taxes. Moreover, they blamed them for natural catastrophes (i.e. floods during the rainy season and water scarcity in the dry season). On the other hand, the Hmong who prefer to use expensive, but more economical sprinkler systems consider the irrigation system of the lowlanders to be inefficient; the lowland Thais lead water through small gutters towards and through their fruit plantations. According to the Hmong the lowlanders are not able to guarantee prevention from agricultural expansion and the prohibition of land sale, i.e. to resorts, large-scale agriculture and speculators.

After regular meetings and watershed inspections which are made every three months, a real dialogue between the conflicting parties has been initiated. Whereas some committee members are skeptical about the real intentions of the NGO and doubt about their neutral position in the conflict, the village headman of Pha Nok Kok strongly supports this initiative as he thinks that researchers would work more systematically towards finding solutions based on facts and empirical findings rather than on assumptions and false accusations.

4 Conclusions

Results of this case study suggest that the time when water in highland watersheds of northern Thailand could be regarded as an open access resource is past. Today, water use is subject to different tenure regimes, rights and obligations. In the study area, control over and use of water is regulated in diverse ways that involve a bundle of rights, ranging from individualized and exclusive property rights to rights shared in water user groups, rights under communal control and water rights under the public domain. These rights are extremely dynamic and new regulations and tenure institutions evolve at the local level if the need arises and external factors, such as market forces and government interventions, change. However, rules and regulations are not equally respected by all members of the community as local elites often use their powerful positions to increase their share in the control over water resources. These findings are in sharp contrast with common perceptions of policy-makers that all laws and norms at the local level are customary, long-established and uniform and that everybody in the community acts according to these local rules (cf. von Benda-Beckmann, F and K 2001). Results also contradict the romanticizing picture that some NGOs tend to draw about local communities as harmonious and peaceful entities, working towards a common goal of sustainable resource management. Local water tenure regimes strongly reflect power structures and unequal social and economic relations both within local communities and between upstream and downstream water users.

Therefore, any future interventions in these tenure systems should build upon careful assessments of existing rights to water and other natural resources and their embeddedness in the local socio-political, institutional and economic setting. Until today, many interventions from government agencies and national projects in water tenure and management have faced difficulties in enhancing more equitable access to water due to lack of transparency, legitimacy and participation of local stakeholders. Policy-makers are well advised not to ignore the legal complexity of local water tenure regimes. They should refrain from designing uniform laws and regulations that cannot accommodate the needed flexibility at the local level. Local representatives of government agencies should provide more than just technical assistance, such as building water conveyance systems to improve efficiency of water use. They should be enabled to play a stronger role in facilitating changes of inequitable tenure regimes through understanding local social and economic context of the communities they are working with. New partnerships between local communities, government agencies, NGOs and research institutions as described for the case of the Mae Ta Chang Rehabilitation project could serve as an example for future directions towards more equitable water tenure regimes, negotiated water rights and more sustainable water management systems.

5 Acknowledgement

The authors would like to thank the villagers of Mae Sa Mai, Pha Nok Kok and Muang Kham and representatives from local government agencies for sharing their knowledge and perspectives with us. The financial support of the Deutsche Forschungsgemeinschaft, Germany, and the National Research Council of Thailand in carrying out this study is gratefully acknowledged.

References

von Benda-Beckmann F and K (2000) Recognizing water rights. In: Meinzen-Dick RS and Rosegrant M (eds) Overcoming water scarcity and quality constraints. International Food Policy Research Institute (2020 Focus 9), Washington D.C.

Bromley DW (1991) Environment and economy: Property rights and public policy. Basil Blackwell, Cambridge (MA)

Bruce JW (1999) Legal bases for the management of forest resources as common property. FAO Community Forestry Note 14, Rome

Bruijnzeel LA (1991) Hydrological impacts of tropical forest conversion. Nature and Resources, vol 27(2)

Bruns BR, Meinzen-Dick R (eds) (2000) Negotiating water rights. Intermediate Technology Publications, London

Chalamwong Y, Feder G (1985) Land values and land title security in rural Thailand. Discussion Paper. World Bank, Washington D.C.

Charoenmuang T (1994) The governance of water allocation problems in Thailand – Four case studies from the upper northern region. In: Thailand Development Research Institute (ed) Water conflicts. Natural Resources and Environment Program, TDRI and Queen's University, Canada. Bangkok, pp 111-147

Ciriacy-Wantrup SV, Bishop RC (1975) "Common property" as a concept in natural resource policy. Natural Resources Journal, vol 15(4), pp 713-727

Gordon S (1954) The economic theory of a common property resource: The fishery. Journal of Political Economy, vol 62(2), pp 122-142

Hardin G (1968) The tragedy of the commons. Science 162: 1243-1248

Institute for Community Rights (2001) Mae Ta Chang Rehabilitation Study (Hang Dong, Mae Rim, and Muang Districts, Chiang Mai Province). Thai Research Fund, Chiang Mai (in Thai)

Kaosa-ard MS (2000) Ecosystem management in northern Thailand. Center for Sustainable Development Studies, Chiang Mai University, Chiang Mai

Meinzen-Dick R, Pradhan R (2002) Legal pluralism and dynamic property rights. CGIAR Systemwide Program on Collective Action and Property Rights Working Paper No. 22. International Food Policy Research Institute, Washington D.C.

Narkwiboonwong W, Charuchandra C, Aunchaisri B, Satutum P (1994) Land tenure and production structure of agriculture in Thailand. Ministry of Agriculture and Cooperatives, Bangkok

Ostrom E (1990) Governing the commons: the evolution of institutions for collective action. Cambridge University Press, New York

Ostrom E (1994) Neither market nor state: governance of common-pool resources in the twenty-first century. International Food Policy Research Institute (IFPRI), Lecture Series 2, Washington

Ostrom E (2001) The puzzle of counterproductive property rights reforms: A conceptual analysis. In: de Janvry A, Gordillo G, Platteau J-P, Sadoulet E (eds) Access to land, rural poverty and public action. Oxford University Press, Oxford

Schiller S (1999) The dynamics of land tenure and natural resource management under the influence of social organisations of a Hmong village in northern Thailand. MSc thesis, University of Hohenheim, Stuttgart

Sethaputra S, Thanopanuwat S, Kumpa L, Pattanee S (2000) Thailand's water vision: A case study. Office of the National Water Resources Committee, Bangkok

Woodman GR (1998) Ideological combat and social observation: recent debate about legal pluralism. Journal of Legal Pluralism 42: 95-106

Growth and Phosphorus Nutrition of Maize in Pot and Field Experiments as Affected by other Plants Grown in Association or in Rotation

Alexander zu Dreele, Hanadi El Dessougi, Gary Schlosser, Juergen Kroschel & Norbert Claassen

1 Introduction

A viable practice to overcome low availability of P in soils could be the intercropping of P efficient species with inefficient ones, or rotation systems in which crop residues of P efficient species are incorporated into the soil. Efficient plants can alter the chemical mobility and bioavailability of inorganic P in the rhizosphere through their uptake and exudation activities, which modify vital processes governing the P concentration in soil solutions such as adsorption-desorption, through pH changes, organic ligands secretion, affecting concentrations of metal cations which fix P such as Ca, Fe and Al (Gardner et al. 1983, Hinsinger, 2001).

Model calculations showed that the root exudates are concentrated within a short distance of around 0.5 mm from the root surface (Claassen et al. 2001). Thus, a close contact between the roots of inefficient and efficient species must be ensured so that the solubilized P is directly available to the inefficient plants. Hocking and Randall (2001) showed a residual P benefit and proved that the organic-acid secreting species lupine and pigeon pea could access soil P pools not available to the non-secreting species sorghum and wheat. Several studies have shown the beneficial effect of such practices on growth, yield and P nutrition of P inefficient species and some showed higher P uptake by the plants. However, the influence of intercropping or rotating of a P efficient species on the plant uptake components, i.e. whether the enhanced uptake was because of an enhanced root growth or higher influx, is not well understood. Both Horst and Waschkies (1987) and Kamh et al. (1999) reported a higher P uptake of wheat intercropped with white lupine as compared to wheat alone. The enhanced P uptake was partly due to a higher P influx. In both studies a larger soil volume of intercropped wheat plants could have been a factor contributing to the higher P influx besides the improved soil P status, which results from solubilizing fixed P and/or mineralization of organic matter.

Although some experiments were conducted, there is less knowledge about effects and importance on P nutrition of intercropping systems in farmers' practice. In a field experiment on a high P fixing soil in Nigeria Horst et al. (2001) found a positive effect on soil native P uptake of maize if residues of P efficient legumes as Cajanus cajan and Chamaecrista rotundifolia were incorporated into the soil. The residual effect was small compared to the effect of freshly applied P and the uptake of applied P (60 kg ha^{-1}) was not enhanced if crop residues were incorporated.

Information about changes in maize yields in mixed culture with companion crops (cc) are sparingly available. A depression of maize and cc yields is reported by Evans (1960), IRRI (1972) and Dalal (1974). The cc were groundnut, mung beans and pigeon peas, respectively.

The objective of this study was to investigate the effect of P efficient species on the nutrition of maize, which was considered as inefficient, with focus on the P uptake parameters. Therefore three pot experiments and one field experiment were conducted.

2 Materials and Methods

The intercropping of maize was carried out in two separate pot experiments and a subsequent field experiment. Effect of plant residues was tested in a third pot experiment.

2.1 Pot Experiments

For the first and second pot experiments Mitscherlich pots of 6 L capacity were filled with a sandy soil having 1.1% clay, 6.2% silt, 2.23% $C_{org.}$, pH ($CaCl_2$) 5.5, 27 mg kg^{-1} soil CAL extractable P. Nutrients added per kg soil were 50 mg N as NH_4NO_3, 66 mg K as K_2SO_4, 16 mg Mg as $MgSO_4$. Water content of soil was kept to 15% by weight. Three pots per treatment were left unplanted as control. Treatments were replicated 3 times.

In the first pot experiment maize (*Zea mays* L.) cv. Consul was intercropped with white lupine (*Lupinus albus* L.) cv. Bardo, sugar beet (*Beta vulgaris* L.) cv. Sonja or oilseed rape (*Brassica napus* L.) cv. Lizard. In the second experiment maize was intercropped with groundnut (*Arachis hypogaea* L.) cv. Tschad. The companion crops were sown 21 days (groundnut 15 days) before maize, since their growth is slower and it was necessary to ensure sufficient roots that might interact with those of maize.

In order to minimize aerial competition maize was planted on one half of the pot and separated by a plastic board (20 x 20 cm) from the companion crops. Monocropped maize was planted in the same manner, i.e. with a plastic board. In this way any difference in growth between mono and intercropped maize would be due to the below ground interaction.

The treatments were maize intercropped with each of the companion crops at no P fertilization and maize alone with 0 (P0), 33 (P1) and 66 (P2) mg P kg^{-1} soil given as 60% NaH_2PO_4, 40% $Na_2HPO_4 \cdot H_2O$. Plants were harvested 20 and 37 (groundnut 15 and 36) days after sowing (DAS) the maize. This allowed the calculation of the P influx.

For the rotation experiment plastic pots of 3 L capacity were used. The ground plant material of white lupine, sugar beet and oilseed rape was incorporated into the soil where they had been grown in the previous intercropping experiment. Because of the differences in dry matter production and P concentration, incorporated crop residue were equivalent to 58, 34 and 51 mg P pot^{-1} from white lupine, sugar beet and oilseed rape, respectively. The soil was allowed to incubate for 3 weeks with adequate moisture. Then the soil was air-dried, thoroughly mixed, supplied with 50 mg N kg^{-1} (NH_4NO_3) and maize was sown. As control maize was planted on the previously unfertilized treatments of the intercropping experiment without incorporation of the maize residues. Plants were harvested 10 and 25 DAS.

All pots were grown in a growth chamber with a day/night regime 16/8 hours, relative humidity 70% and temperatures of 20/18°C for white lupine, sugar beet, oilseed rape and the rotation experiment and 25/20°C for groundnut. The photosynthetic active radiation during the day time was 250 μE m^{-2} s^{-1}. The number of plants of each species per pot was 6 for the first and 3 for the second harvest. The experiments were carried out in a completely randomized design. Analysis of variance (ANOVA) was conducted using SigmaStat version 2 program.

The following soil parameters were analyzed for the pot experiments: CAL extractable P was determined following Schüller (1969). Soil solution was obtained by a modified displacement technique of Adams (1974). The P concentration in soil solution and the extracted orthophosphate was measured by spectral photometry according to Murphy and Riley (1962). After harvests the total and inorganic water soluble P was determined: soil samples were mixed with water in 1:2 ratio, incubated for 24 hours, shaken for 2 hours and filtered. Samples for total P concentration were wet digested in a concentrated triacid mixture and the pH neutralized with NaOH. Inorganic and total P were then determined according to Murphy and Riley (1962). The difference between the total and inorganic P gave the organic P concentration. The pH was measured in 0.01 M $CaCl_2$ in a 1:2.5 soil:solution ratio.

Shoot dry matter was determined after drying at 105°C till constant weight and the plant material was then finely ground. The P concentration in shoot was determined after wet digestion in a concentrated triacid mixture, by the Molybdenum Vanadate method (Scheffer and Pajenkamp 1952). The roots were separated from the soil by washing them gently over a 0.2 mm sieve. The root length was measured using the method of Tennant (1975). In the first intercropping experiment, at the second harvest root length and soil measurements were done on sub-samples of each pot. To ensure that the sub-samples were representative of the whole pot, 2 soil sectors per pot were cut (ca. 1.5 to 1.7 kg soil) from the center of the pot to the edges opposite of the boundary maize-companion crop. Root counting was done on the mixture of these two samples. Assuming exponential growth of young plants, average P influx was calculated according to Williams (1948).

2.2 Field Experiment

The field experiment was conducted from January – May 2002 in Wanga, Central Sulawesi, Indonesia, on a loamy clay soil having 2 mg P kg $^{-1}$ (CAL P), pH 4.4 ($CaCl_2$) and a high content of sesqioxides. Rainfall during the experimental period was 1290 mm, average temperature was 26°C (max.) and 18°C (min.). The treatments were maize (M) cv. C-7 grown in monoculture at four fertilization levels 1. no nitrogen (N) no phosphorus (P) (N0P0), 2. only N (N1P0), 3. only P (N0P1) and 4. N and P (N1P1). In addition to the monocropped maize, six treatments were established with companion crops (cc) grown in mixed culture with maize. The cc were: groundnut (GN) var. Kelinci, mucuna (Mu, *Mucuna pruriens* L. var. Utilis) and sweet potato (SP, *Ipomoea batatas* L, var. Sukur). Plots intercropped with groundnut or mucuna were fertilized as (N1P0) and (N0P1), whereas plots intercropped with the non-legume sweet potato were fertilized only with (N1P0). The nitrogen fertilized plots received 100 kg N ha^{-1} as urea, given in two equal doses at planting and 30 DAS. Phosphorus was applied at 60 kg P ha^{-1} as superphosphate. All plots received 200 kg K ha^{-1} as KCl. Potassium and P were added 29 days before planting maize.

To obtain information about growth of companion crops in monoculture, a separate block was established. Fertilizer levels were the same as those of monocropped maize. Plant density was 80,000 plants per ha and species. Also for intercropping 80000 plants of maize and 80,000 plants oc cc were used.

Plant material of maize and companion crops was analyzed 25, 70 and 120 DAS for dry weight. Phosphorus concentration was determined after wet digestion in a tri-acid mixture with ICP analyzer. At the same dates, root samples were taken from 0-15 and 15-30 cm soil depth using a root auger, washed over a 0.2 mm sieve and root length was determined after Tennant (1975). Roots of maize and cc were distinguished by color and shape and counted separately. Soil solution was displaced after Adams (1974) at 0, 62 and 110 DAS and analyzed for pH and 2 P fractions: inorganic (P_i) measured by spectral photometry following Murphy and Riley (1962) and P_{total} measured with ICP. Organic P (P_{org}) was calculated by the difference of $P_{total} - P_i$.

The experiment was carried out in a completely randomized block design; analysis of variance (ANOVA) was made using SigmaStat version 2 program.

3 Results and Discussion

3.1 Mixed –Cropping

The mixed cropping pot experiments showed that maize yield was strongly reduced when grown together with white lupine, sugar beet or oilseed rape. The dry matter yield of maize grown with lupine, sugar beet or oilseed rape was about 27, 15 and 13% as compared to the maize grown solely at the same fertilization (data not shown). The reduced growth might be due to competition for nutrients as suggested by Horst et al. (2001) for some cases. Even though the soils were fertilized with other nutrients but P, but since the cc were grown 3 weeks before maize, they might have depleted the soil and the remaining nutrients were not enough to support maize growth. The dry matter yield of maize in monoculture responded to P fertilization indicating that the soil P was strongly limiting growth (Table 1). In contrast, dry matter yield of maize intercropped with groundnut (Table 1) was 3 times higher than maize yield grown alone without P fertilization (P0) and was even 44% more than the yield at the highest P fertilization level (P2) of 66 mg P kg^{-1} soil. Other researchers reported the beneficial effect of intercropping P efficient species with inefficient ones (Horst and Waschkies 1987; Kamh et al. 1999). Groundnut is reported to have a high P uptake efficiency probably because of its ability to mobilize sparingly soluble P reserves (Ae et al. 2001).

The higher dry matter yield of maize grown in mixed culture with groundnut was not because maize could use absorbed P more efficiently for yield production, since P concentration in shoot at all treatments was similar and was even somewhat higher at (P0) fertilization level (Table 1). This was probably because of the lower dry matter production (concentration effect).

Table 1. Dry matter yield, P concentration in dry matter, root length and P influx of maize intercropped with groundnut (M/GN) at 0 (P0), or alone (M) at 0 (P0), 33 (P1) and 66 (P2) mg P kg-1 soil (second pot experiment).

Treatment	Dry matter yield g pot^{-1}	Shoot P conc. mg P(g DM) $^{-1}$	Root length m pot^{-1}	P influx 10^{-14} mol cm^{-1} s^{-1}
M/GN – P0	15.1a	2.1a	1224*a	1.74a
M – P0	4.6c	2.4a	532b	1.08b
M – P1	5.9bc	2.1a	586b	1.33ab
M – P2	10.5ab	1.9a	851ab	1.62a

*Root length calculated from the root shoot ratio of the plants grown at (P2) fertilization level.
Within columns values followed by the same letters are not significantly different (Tukey $p < 0.05$).

The improvement of intercropped maize growth could also not be due to better N nutrition, since sufficient N was fertilized. Furthermore the soil did not contain the specific rhizobium for groundnut (*Bradyrhizobium* sp.); as a consequence no nodules were formed at the roots of groundnut. This indicates that biological N_2 fixation did not contribute to the N nutrition of the plants.

An improved P uptake by the maize through the interaction with groundnut roots was the reason for the observed enhancement growth of maize growth.

It was not possible to separate the roots of the two intercropped species, hence the root length of intercropped maize was calculated from the ratio of root length - shoot dry matter at P2, since here maize had a dry matter yield similar to that of intercropped maize as was done by Troelstra and Berendse (1982). Maize intercropped with groundnut had more than double the root length and a 60% higher P influx than the (P0) monocropped maize (Table 1). Both Horst and Waschkies (1987) and Kamh et al. (1999) attributed the higher P efficiency of wheat intercropped with white lupine to increased P uptake per unit root length. But in both experiments it is not clear whether the higher uptake per unit root length was solely due to a higher P availability as a result of solubilization by root exudates of the efficient species or in addition due to a higher amount of P in the larger available soil volume. In contrast, in this study the roots of intercropped maize had the same soil volume as when grown alone. The increased influx must have been through a higher P availability due to P mobilization by groundnut roots.

To look for an explanation for the high P influx of the intercropped maize the soil was analyzed for different P fractions. Table 2 shows that plant growth decreased the P concentration in soil solution as well as the CAL P. The soil with intercropped maize had the lowest concentration in soil solution already at the start of the measuring period, i.e. first harvest, and only 25% of

This P concentration in solution was found at the second harvest. Nevertheless, the P influx of intercropped maize was 60% higher than the monocropped maize (Table 1) and even somewhat higher than the fertilized maize (M-P2) which had 5 to 10 times higher P in soil solution (Table 2).

This indicates that intercropped maize had higher available P fractions because of the P mobilizing effect of groundnut roots. Several authors reported the capacity of various plant species to mobilize sparingly soluble P through root exudates, for example white lupine (Gardner et al. 1983; Kahm et al. 1999). Hence it might have been that groundnut in our study had the capacity to improve P availability by its root exudates or some other components. Ae et al. (2001) reported a high P uptake by groundnut as compared to other leguminous species and explained the superior P uptake of groundnuts by the solubilizing-ability of root cell-walls-components. Whether the root cell-walls of groundnut were responsible for the observed higher P uptake of maize may be questioned, since one would expect that the mobilizing effect would be restricted to the root surface but not transfer P to another root. Therefore, some other mechanism must have been effective in this experiment. However, if the mobilized P is transferred to another root as in this experiment, the mobilized P was transferred from groundnut to the maize root, one would expect an increase of P in bulk solution. But this was not observed here (Table 2).

Organic P (P_{org}) in soil solution may be used by plants through root phosphatases (Seeling and Jungk 1996; Beißner and Römer 1998). In Table 2 it can be seen that P_{org} was lower in the intercropped than monocropped soil, indicating that the concentration of P_{org} is no explanation for the observed high P influx of the intercropped maize either. Therefore, it is not clear yet by which mechanism groundnut improved the P uptake of maize.

Table 2. Soil solution P inorganic concentration (Pi), CAL P, pH and water soluble inorganic (Pi) and organic (Porg.) phosphorus in soil suspension (soil : water = 1:2) at first and second harvest of maize (M) intercropped with groundnut (M/GN), or grown alone at 0 (P0), 33 (P1) and 66 (P2) mg P kg-1 soil and the unplanted control at P0 fertilization.

Treatment	First harvest			Second harvest		
	P_i	CAL P	pH	P_i	CAL P	pH
	µM	mg (kg soil)$^{-1}$		µM	mg (kg soil)$^{-1}$	
M/GN-P0	0.78a	32.5a	5.7a	0.20a	29.4a	6.3a
M-P0	0.90a	35.3a	5.5ab	0.80a	36.0a	5.6b
M-P1	3.60b	48.8b	5.5ab	1.79a	46.2b	5.7b
M-P2	9.07c	63.0c	5.5ab	4.46b	64.9c	6.4a
Control	1.00a	33.4a	5.4b	1.22a	37.0ab	5.4b

Soil suspension

	First harvest		Second harvest	
	P_i	P_{org}	P_i	P_{org}
	µM		µM	
M/GN-P0	2.13a	0.85a	1.91a	0.91a
M-P0	2.33a	1.04a	2.28a	1.60b
M-P1	4.36b	1.60b	3.60b	1.46b
M-P2	7.18c	1.29a	8.53c	1.87c
Control	1.83a	1.50b	2.02a	2.00c

Within columns values followed by the same letters are not significantly different (Tukey $p < 0.05$).

Dry matter production and yield of maize in the field experiment responded to phosphorus fertilization indicating that the soil P was strongly limiting growth (Fig. 1). At 120 DAS, in the nitrogen fertilized treatment (N1P0), yield of maize grown with the different companion crops (average yield 4.85 t ha^{-1}) was almost similar to the yield of monocropped maize (4.62 t ha^{-1}). That means that the companion crops did not compete with maize for nitrogen, whereas, in the treatment N0P1, yield of the monocropped maize differed from those of intercropped maize (Fig. 1). Maize grown with groundnut yielded 5 t ha^{-1} and yield of maize grown with mucuna was 7.51 t ha^{-1} as compared to the yield of monocropped maize (5.95 t ha^{-1}). This could be due to a mobilization of fertilizer or native P by mucuna, in contrast, groundnut might have competed with maize for fertilizer or native P.

The clay fraction of the soil had an extremely high sesqioxides content, which means a high P fixing capacity (Fox et al. 1971). Phosphorus fixing soils are able to absorb up to 5600 kg ha^{-1} P until they provide satisfactory crop growth (Sanchez and Uehara 1980). The high fixing capacity of the soil is also indicated by the fact that the P_i contents of the soil solution in the P fertilized treatments did not increase in the soil depth 10-30 cm (Table 3). Phosphorus is already fixed in the upper 10 cm. Moreover, P_i levels were very low although with 60 kg P ha^{-1} a relatively high amount of P was given. In general, the concentrations of P_i and $P_{org.}$ decreased, as is shown by the values 140 days after application (Table 3). Although there is no statistical significance between the P levels at one date, it seems that P_i in soil solution of the M/GN plots decreased more slowly between 62 and 110 DAS as compared to the monocropped plots or those intercropped with mucuna or sweet potato. At 93 DAS, on N1P0 treatment, the P_i concentration in M/GN plots was around 6% higher than in monocropped maize plots. On the N0P1 treatment the increase was about 16%. At 110 DAS the increase was about 21% on N1P0 and 172% on N0P1.

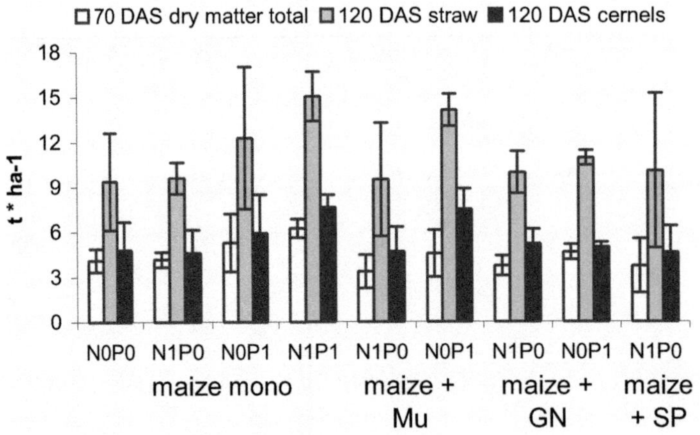

Fig. 1. Dry matter production 70 and 120 DAS and grain yield of maize 120 DAS grown in mono or mixed cropping with mucuna, groundnut or sweet potato at different fertilizer treatments. Vertical bars are standard error of means.

Table 3. Concentration of soil solution inorganic (Pi) and organic (Porg) phosphorus after maize planted mono (maize mono) or intercropped with mucuna (M/Mu), groundnut (M/GN) or sweet potato (M/SP) at different fertilizer treatments at 0, 62 and 110 days after sowing (DAS). P was fertilized 30 days before sowing.

Treatment		depth	0 DAS	62 DAS		110 DAS	
			P_i	P_i	P_{org}	P_i	P_{org}
		(cm)			µM		
maize mono	N0P0	0-10	0.64 ab	1.48 ab	9 ab	0.46 ab	20 a
		10-30	0.56 ab	0.46 ab	12 ab	0.24 a	13 a
	N1P0	0-10	1.03 ab	1.07 ab	13 a	1.18 ab	16 a
		10-30	0.55 ab	0.17 a	19 a	0.48 ab	10 a
	N0P1	0-10	2.47 ab	1.48 ab	7 ab	0.71 ab	18 a
		10-30	0.41 a	0.41 a	14 a	0.32 ab	15 a
	N1P1	0-10	1.89 ab	2.14 ab	13 ab	1.28 ab	17 a
		10-30	0.31 a	0.49 ab	9 ab	0.37 ab	12 a
M/Mu	N1P0	0-10	0.72 ab	2.46 ab	9 ab	0.37 ab	16 a
		10-30	0.53 ab	0.45 a	18 a	0.14 a	12 a
	N0P1	0-10	2.95 ab	4.32 b	16 a	1.09 ab	15 a
		10-30	0.62 ab	1.02 ab	6 ab	0.48 ab	16 a
M/GN	N1P0	0-10	0.49 a	1.14 ab	9 ab	1.43 ab	13 a
		10-30	0.38 a	0.40 a	17 a	0.27 a	16 a
	N0P1	0-10	5.32 b	2.39 ab	1 b	1.93 b	16 a
		10-30	0.46 a	0.27 a	12 ab	0.41 ab	10 a
M/SP	N1P0	0-10	0.95 ab	1.88 ab	13 ab	0.56 ab	21 a
		10-30	0.53 ab	0.57 ab	13 a	0.35 ab	11 a

Within columns values followed by the same letters are not significantly different (Tukey $p < 0.05$).

Intercropping with mucuna also increased the P_i concentration of soil solution. But here the highest increase was at 62 DAS. The P_i concentrations in the P fertilized and P unfertilized plots were 2 to 3 times higher than those on monocropped ones. The increase of P_i concentration in soil solution was detected in both soil depths (0-10 cm and 10-30 cm). However, at 110 DAS the increase in P_i concentrations was negligible or even negative (Table 3). Sweet potato increased the P_i concentration in soil solution by 75% and 235% in the 0-10 and 10-30 cm soil depths, respectively, at 62 DAS. However, P_i concentrations at 110 DAS were lower than on the monocropped plots (Table 3). It seems that at earlier stages of their growth, mucuna and sweet potato were able to change P concentrations in soil solution. Groundnut showed similar effects but at later stages.

Maize is reported to be inefficient in P uptake in earlier growth stages but increases its efficiency in later stages (Bhadoria et al., 2001). Consequently, an effect of companion crops on P nutrition of maize can only result in higher P uptake if it occurred in early growth stages.

In the mono and mixed cropping systems maize showed the highest P influx between 25-70 DAS (Fig. 2). P influx of monocropped maize responded to fertilization. The highest P influx occurred in the N1P1 treatment. This shows that also nitrogen can affect uptake of fertilizer and native P. Between 25-70 DAS P influx in mixed cropped treatments was almost the same as P influx in monocropped ones. Between 70-120 DAS P influx, especially of maize intercropped with mucuna, was higher than that of monocropped maize (Fig. 2). Intercropping with mucuna increased the P influx on the N0P1 treatment. Also intercropping with groundnut or sweet potato seemed to increase P influx of maize. The higher P influx of mixed cropped maize was due to the higher P_i concentrations in soil solution (Table 3). In intercropping treatments, mucuna seemed to increase P_i concentrations in earlier growth stages more effectively as compared to groundnut or sweet potato. This can have positive effects on maize growth and P nutrition.

The yield and biomass production of monocropped companion crops were around 2 to 3 times higher than those in intercropped treatments (data not shown). This could have been due to above ground competition for light in intercropped treatments.

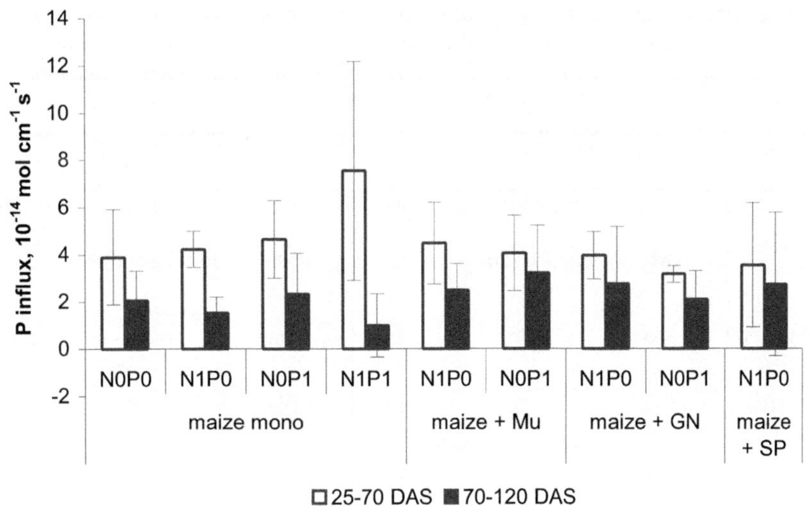

Fig. 2. Phosphorus influx of maize 25-70 and 70-120 DAS grown in mono or mixed cropping with mucuna, groundnut or sweet potato at different fertilizer treatments. Vertical bars are standard error of means.

3.2 Rotation

Table 4 shows that only white lupine residue substantially increased the total water soluble P as compared to the maize grown after maize and the unplanted control, mainly because of a higher organic P fraction. The increase in (P_i) due to crop residue incorporation was between 45-90% as compared to maize grown after maize, but, except for the residue of white lupine, the increase was negligible as compared to the unplanted soil (Table 4). Also it can be seen that for all treatments the organic P fraction decreased from first to second harvests of maize.

The P concentrations in soil solution (Table 4) were lower under maize grown after oilseed rape or sugar beet than under maize after white lupine, although not statistically significant, and all were significantly lower than under maize after maize and unplanted soil. The P concentration in soil solution from maize after oilseed rape and sugar beet residue was around 40% of the P concentration in soil under maize grown after maize and after white lupine residues and only about 14% of that from unplanted control. Also the CAL extractable P fraction was lower under maize after crop residues than under maize without residue and unplanted control (Table 4).

Table 4. Soil solution P inorganic concentration (Pi), CAL P and pH and water soluble inorganic (Pi) and organic (Porg.) phosphorus in soil suspension (soil : water = 1:2) of a sandy soil at the first and second harvest of maize grown after incorporation of crop residues of white lupine (M/L), sugar beet (M/SB), oilseed rape (M/R) and without incorporation of crop residue at 0 (P0) mg P kg-1 soil and the unplanted control (P0).

Treatment	First harvest			Second harvest		
	P_i	CAL P	pH	P_i	CAL P	pH
	µM	mg (kg soil)$^{-1}$		µM	mg (kg soil)$^{-1}$	
M/L-P0	0.66a	40.7a	6.45a	0.45ab	30.1bc	6.10b
M/SB-P0	0.59a	34.7a	6.53a	0.19a	26.2ab	6.97a
M/R-P0	0.46a	38.1a	6.49a	0.20 a	24.5a	6.84a
M-P0	0.96b	37.6a	5.48b	0.54 b	33.6c	5.86bc
P0	1.12b	45.6a	5.54b	1.36c	29.3abc	5.26c

Soil suspension

	First harvest		Second harvest	
	P_i	P_{org}	P_i	P_{org}
	µM		µM	
M/L-P0	2.75a	5.59a	1.96a	1.27b
M/SB-P0	2.10ab	3.22b	1.44b	0.91b
M/R-P0	2.04ab	3.56b	1.15c	1.06b
M-P0	1.41b	3.03b	1.39b	0.92b
P0	1.76b	3.51b	1.41b	2.72a

Within columns values followed by the same letters are not significantly different (Tukey $p < 0.05$).

Oilseed rape and sugar beet crop residue increased the pH as compared to the crop residues of white lupine, maize grown after maize and unplanted control (Table 4). Hafner et al. (1993) found that crop residue application to an acidic soil increased soil pH and base saturation and decreased exchangeable Al and Al saturation. The author explained that the decrease in exchangeable Al, which strongly adsorbs P, will improve the solubility of P and increase the P concentration in soil solution. Horst et al. (2001) attributed such an increase in pH after crop residue incorporation into soil to nitrate uptake . Nitrate fed plants excrete OH^-/HCO_3^-, thus alkalinize the rhizosphere and enhance bioavailability of inorganic soil P by desorption of P ions from metal oxides through ligand exchange reactions (Parfitt 1979).

Incorporating the crop residues of the P efficient species had a positive effect on the growth of subsequent maize (Fig. 3). Crop residues of oilseed rape more than doubled maize yield, those of sugar beet increased maize yield by a factor of 1.6, whereas white lupine residues had no effect on maize yield (Fig. 3). These results are in partial agreement with several reports from the literature which showed a residual P effect of incorporating crop residues of various P efficient species on the growth and yield of subsequent crops (e.g. Hafner et al. 1993; Horst et al. 2001). No beneficial effect of white lupine on maize growth and yield was found in this study in contrast to several studies (e.g. Horst and Waschkies 1987; Kamh et al. 1999; Hocking and Randall 2001), even though the P amount returned in lupine residue (ca. 20 mg P kg^{-1} soil) was higher as compared to oilseed rape (17 mg P kg^{-1} soil) and sugar beet (12 mg P kg^{-1} soil). This shows that besides the quantity of P returned by the crop residues, also their quality is of importance, as was already observed by Singh and Jones (1976) and Friesen et al. (1997).

The P concentration of maize shoots did not vary much (Figure 3) but total P uptake was strongly affected by the preceding crop, being up to 3 times higher after oilseed rape, while white lupine had no effect on P uptake of maize (data not shown).

The factors that determine nutrient uptake, in this case P, are the size of the root system and the influx (rate of uptake per unit root length). Because of its low mobility P transport to the roots is mainly by diffusion (Barber 1995).

Accordingly an extensive root system with fine roots is beneficial for accessing larger soil volumes and could result in more efficient uptake. Presence of root hairs increases the root surface area and shortens diffusion distances. Such adaptation mechanisms are reported for plants grown under P deficiency (Föhse et al. 1991). Figure 4 shows that crop residues strongly affected the P influx. Maize grown after incorporating oilseed rape residues showed an 11 times higher P influx than maize after maize. Sugar beet residues increased the P influx by a factor of 4, while the residues of white lupine had no effect on P influx.

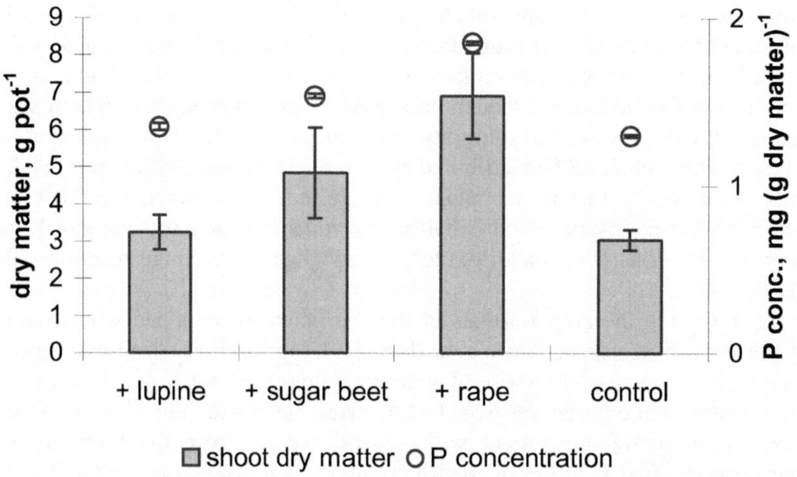

Fig. 3. Dry matter production and P concentration of maize grown after incorporation of crop residues of lupine, sugar beet, rape and without incorporation of crop residues at no P fertilization (control). Vertical bars are standard error of means.

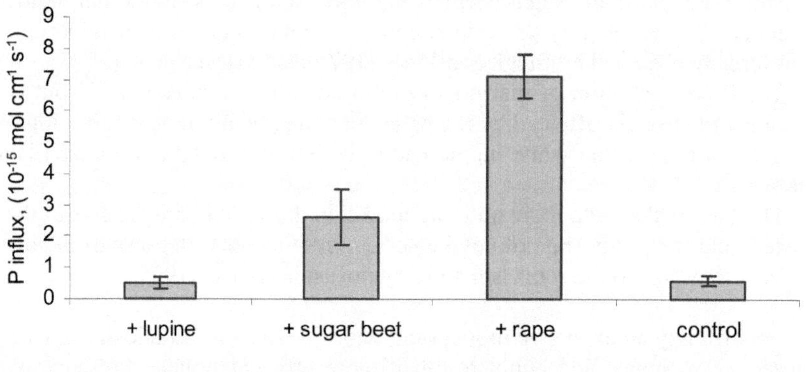

Fig. 4. Phosphorus influx of maize grown after incorporation of crop residues of white lupine, sugar beet, rape and without incorporation of crop residues at no P fertilization (control). Vertical bars are standard error of means.

What caused the large effect on P influx? The P uptake is the rate of P intake into the root and has to be equal to the rate of transport to the root. Intake into and transport to the root are generally accepted to occur in the soil liquid phase, i.e. the soil solution. Therefore, after oilseed rape residue incorporation, where the P influx of maize was highest, a strong increase in P soil solution concentration would have been expected, but the reverse happened (Table 4), i.e. the P concentration was the lowest, at both harvest dates, while after white lupine with the lowest P influx, soil solution was the highest. For CAL-P the trend was similar but not so strong. Table 4 shows the organic P (P_{org}) in a soil : water suspension of 1:2. After incorporation of oilseed rape residue there was no higher P_{org} than in the other treatments and on the contrary the highest P_{org} was after white lupine where the P influx had been the lowest.

The results of the experiments showed that P uptake of maize either intercropped or grown in rotation after incorporation of crop residues of other species may vary greatly. The main factor of P uptake affected by the treatments was the P influx. The soil parameters measured could not explain the differences encountered, on the contrary a high influx was related to the lowest P values in soil.

Root hairs, which were not measured in this experiment, may have a large effect on P uptake per unit root length (Föhse et al. 1991). Plants can have different root hair formation depending on environmental conditions (Jungk, 2001). Under equal conditions, the formation and morphology of root hairs seems to be regulated by nutrient supply, especially nitrate and phosphate (Föhse and Jungk 1983). Jungk (2001) concluded that mineral nutrient supply, e.g. P, is a factor that activates physiological reactions leading to increased root hair formation. Michael (1990) proposed that ethylene plays a central role in such reactions, hence it might have been that crop residue of oilseed rape had an effect on this process and resulted in more root hair formation.

References

Adams F (1974) Soil solution. In: Carson EW (ed) The Plant Root and its Environment. University Press of Virginia, Charlottesville

Ae N, Shen R, Otani T (2001) The significance of the root cell wall in phosphorus uptake. In: Ae N, Arihara J, Okada K, Srinivasan A (eds) Plant nutrient acquisition new perspectives. Springer, Tokyo, pp 251-275

Barber SA (1995) Soil nutrient bioavailability: a mechanistic approach. 2nd edn. Wiley, New York

Beißner L, Römer W (1998) Der Einfluß von P-Ernährung und pH auf die Phosphataseaktivität von Zuckerrübenwurzeln. J. Plant Nutr. Soil Sci 162: 83-88

Bhadoria PBS, Singh S, Claassen N (2001) Phosphorus efficiency of wheat, maize and groundnut grown in low phosphorus supplying soil. In: Horst WJ et al. (eds) Plant Nutrition – food security and sustainability of agro-ecosystems. Kluwer, Dordrecht, pp 530-531

Claassen N, Steingrobe B, Syring KM (2001) A mechanistic model to describe the effect of complexing root exudates on transport and uptake of soil nutrients. In: Horst WJ et al. (eds). Plant nutrition- Food security and sustainability of agro-ecosystems through basic and applied research. Kluwer, Dordrecht, pp 600-601

Dalal RC (1974) Effects of intercropping maize with pigeon peas on grain yields and nutrient uptake. Exptal. Agr 10: 219-224

Evans AC (1960) Studies of intercropping: I. Maize or sorghum with groundnuts. East Afr Agr For J. 26: 1-10

Friesen DK, Rao IM, Oberson A, Sanz JI (1997) Phosphorus acquisition and recycling in crop and pasture system in low fertility soils. In: Ando T et al.(eds) Plant nutrition for sustainable food production and environment. Kluwer, Dordrecht, pp 493-498

Föhse D, Jungk A (1983) Influence of phosphate and nitrate supply on root hair formation of rape, spinach and tomato plants. Plant and Soil 74: 359-368

Föhse D, Claassen N, Jungk A (1991) Phosphorus efficiency of plants: II. Significance of root radius, root hairs and cation-anion balance for phosphorus influx in 7 plant species. Plant and soil 110: 101-109

Fox RL, Hasan SM, Jones RC (1971) Phosphate and sulfate sorption by Latosols. Proc. Int. Symp Soil Fert Eval 1: 857-864

Gardner WK, Barber DG, Parberry DG (1983) The acquisition of phosphorus by Lubinus albus L: III. The probable mechanism by which phosphorus movement in the soil/root interface is enhanced. Plant and Soil 70: 107-124

Hafner H, George E, Bationo A, Marschner H (1993) Effect of crop residue on root growth and phosphorous acquisition of pearl millet in an acid sandy soil in Niger. Plant and Soil 150: 117-127

Hinsinger P (2001) Bioavailability of soil inorganic P in the rhizosphere as affected by root-induced chemical changes: a review. Plant and Soil 237: 173-195

Hocking PJ, Randall PJ (2001) Better growth and phosphorus nutrition of sorghum and wheat following organic acid secreting crops. In: Horst WJ et al. (eds). Plant Nutrition-Food security and sustainability of agro-ecosystems through basic and applied research. Kluwer, Dordrecht, pp 548-549

Horst WJ, Kamh M, Jibrin JM, Chude VO (2001) Agronomic measures for increasing P availability to crops. Plant and Soil 237: 211-223

IRRI (1972) Multiple cropping: Annual Report. International Rice research Institute, Los Baňos (Philippines), pp 21-35

Jungk A (2001) Root hairs and the acquisition of plant nutrients from soil. J. Plant Nutr. Soil Sci. 164: 121-129

Kamh M, Horst WJ, Amer F, Mostafa H, and Maier P (1999) Mobilization of soil and fertilizer phosphate by cover crops. Plant and Soil 211: 19-27

Michael G (1990) Vorstellungen über die Regulation der Wurzelhaarbildung. Kali-Briefe (Bünthof) 20: 411-429

Murphy J, Riley JP (1962) A modified single solution method for the determination of phosphat in natural waters. Analytica Chimica Acta 27: 31-36

Parfitt RL (1979) The availability of P from phosphate-geothite bridging complexes. Desorption and uptake by ryegrass. Plant and Soil 53: 55-65

Sanchez PA, Uehara G (1980) Management Considerations for Acid Soils with high Phosphorus fixation Capacity. In: ASA-CSSA-SSSA, The Role of Phosphorus in Agriculture

Scheffer F, Pajenkamp H (1952) Phosphatbestimmung in Pflanzenaschen nach der Molybdän-Vanadin-Methode. Z. Pflanzenernähr. Düngung Bodenk. 56: 2-8

Schüller H (1969) Die CAL-Methode, eine neue Methode zur Bestimmung des pflanzenverfügbaren Phosphats im Boden. Z. Pflanzenernäh. Bodenk. 123: 48-63

Seeling B, Jungk A (1996) Utilization of organic phosphorus in Calcium chloride extracts of soil by barley plants and hydrolysis by acid alkaline phosphatases. Plant and Soil 178: 179-184

Singh BB, Johnes JP (1976) Phosphorus sorption and desorption characteristic of soils as affected by organic residues. J Soil Sci Soc Am 40: 389-394

Tennant D (1975) A test of a modified line intersect method of estimating root length. J. Ecol 63: 995-1001

Troelstra SR, Berendse F (1982) Root CEC determinations to establish root biomass of two plant species grown in mixtures. Plant and Soil 64: 277-281

Williams RF (1948) The effect of phosphorus supply on the rates of uptake of phosphorus and nitrogen upon certain aspects of phosphorus metabolism in gramineous plants. Aust J Sci Res(B) 1: 333-361

Adaptability Analysis and Risk Assessment of N Fertilizer Application to Maize in the Napu Valley of Central Sulawesi, Indonesia

Jürgen Anthofer, Garry Schlosser, Jürgen Kroschel & Norbert Claassen

1 Introduction

Throughout the developing world, many development projects promote various technologies to address the decline in soil fertility. The majority of them rely on research results which are often not reproducible in different agro-ecological zones because of differences in soil properties, rainfall regime, economic framework, availability of fertilizers and other inputs, and management practices. Limited economic resources, coupled with the focus on extension instead on research and the short life span of development projects are major reasons not to conduct intensive adaptation trials for potential innovations before promoting them to farmers. Consequently, many extension projects have failed because they have not been able to address the real problems adequately in different project environments with their own specific characteristics. Therefore, there is a need to apply suitable methodologies to quickly assess the potential of technologies to be introduced under specific local conditions and to assess their risk of failure.

The concept of adaptability analysis (Hildebrand and Russell 1996), better known as stability analysis (Yates and Cochran 1938; Finlay and Wilkinson 1963; Hildebrand 1984), addresses these concerns by relating treatment yields to a location mean yield of all treatments tested. This approach originates from plant breeding programs in their search for methodologies for the adaptability of varieties to different environmental conditions. Finlay and Wilkinson (1963) considered varieties with regression slope of $b_i = 1$ as average stable or adapted to different environments. Eberhard and Russell (1966) defined an adapted variety or a variety with maximum stability as one with a high mean yield, $b_i = 1$, and a small deviation from regression. According to Piepho (1998) $b_i = 1$ is neither a necessary nor a sufficient condition for one system to outyield the other in all environments because in addition to b_i the mean yield itself is another important parameter.

Despite these controversial discussions about the definition of "stability" the regression approach provides a useful tool to identify such technologies or cropping systems which show a below or above average response to local farmers' practice under different environmental conditions, as measured by the location mean. It also helps to identify whether a technology is more effective under poor or under superior environmental conditions when compared with a control treatment. Therefore, the stability of yield is an important aspect when judging the value of one technology relative to others. So far, analysis of stability and adaptability has been mainly employed to multilocation trials of crop cultivars.

There is also an increasing interest in the assessment of risk (Piepho 1998, 1996; Hildebrand and Russell 1996; Eskridge 1990; Mead et al. 1986). It is important for researchers and extension programs to know the risk that a certain technology may not achieve a certain yield level or obtain an expected yield difference above the usual farmers' yield. In addition, it is important to address the high yield variability usually found within small scale farming systems.

In the Napu valley of Central Sulawesi, maize (*Zea mays* L.) is an important crop in small scale farming systems. Productivity may be enhanced either by the application of mineral fertilizer or through rotations or intercropping with legumes. The objective of the present study was to test the efficacy of N based mineral fertilizers on maize yield and economy within the diverse farming environments in comparison to the current farmers' practice. Therefore, different methods of adaptability analysis and risk assessments were applied in order to

- identify whether various N based fertilizer treatments are able to improve yields and are economically justified
- reveal whether these treatments are more effective in farming environments of low or high production levels
- estimate the risk for a randomly chosen farming environment to obtain a yield which falls below a critical yield level
- determine the probability that the various fertilizer treatments will outperform the unfertilized control based on crop yields alone and under consideration of marginal costs of the fertilizer application.

2 Materials and Methods

2.1 On-Farm Experimentation

In the period January 20th to March 6th 2002, an on-farm fertilizer trial was established on 32 farmers' maize fields in the villages Wuasa, Wanga and Rompo, in Central Sulawesi, Indonesia. Maize fields were selected according to the interest of farmers to participate in the assessment, and if they planned to start planting maize within the next two weeks or had already planted maize within the previous two weeks.

The fertilizer trial was located within the total field but was careful to avoid larger trees and stumps. The four treatments comprised: (1) 60 kg N; (2) 60 kg N + 60 kg K; (3) 60 kg NPK; and (4) Control (farmers' practice) without fertilizer. Urea was used as the N fertilizer source while muriate of potash (KCl) was the K source. The compound fertilizer contained 15% N, P and K. Each treatment plot had an area between 150 and 250 m^2. In treatment (1) N was split into two dressings: one at planting and the second after six weeks. Fertilizer of the other treatments were applied during planting and all fertilizers were broadcast.

The experimental design was a completely randomized block design with each farmer's location representing a block in which each treatment was implemented once.

With few exceptions maize was planted in rows but spacing varied considerably among farmers. Maize varieties differed as well with most farmers using local or replanted hybrid varieties. Weeding was either done with a hoe or herbicides. Type, frequency and time of weed control was decided by the farmers. Fertilizer was provided by the STORMA ("Stability of Rainforest Margins in Indonesia") project while all other inputs including maize seeds and labor were provided by the farmers. Management of the plot was completely left to the farmer.

Maize was harvested at physiological maturity. Within the inner 100 m^2 of the treatment plots maize cobs were removed and weighed with an electronic scale. Ten randomly chosen cobs were weighed, seeds removed and weighed again to calculate the seed-husk ratio. Subsamples of seeds were weighed in the field, dried at 70°C for 72 h and dry weight was recorded. All yields presented are based on dry matter.

2.2 Parameters

Maize grain yield was the major parameter with economic variables also being considered. Since the main interest of the trial was to estimate treatment differences between fertilized and unfertilized plots, only the return and cost factors differing among treatments were taken into account. These were: grain yield (kg ha^{-1}), maize grain price = 611 Rupia (Rp.) kg^{-1}, 100 kg NPK (15-15-15) fertilizer = 190,000 Rp.; 100 kg muriate of potash (KCL) = 205,000 Rp.; 100 kg Urea = 114,000 Rp.; daily wage for farm labour = 15,000 Rp. d^{-1} (= 6 h); time for each fertilizer application: 1.67 days ha^{-1}. This information was used to calculate marginal costs which were expressed in maize grain yield.

2.3 ANOVA and Linear Contrasts

Yield and net return data were subjected to ANOVA with the GLM procedure of the SAS system (SAS 1992). Mean separation was done with the PDIFF option of the LSMEANS statement. Contrasts were used to compare each fertilizer treatment separately with the control.

2.4 Adaptability Analysis

Yield performances y_{ij} were related for each fertilizer and control treatment on the mean of both treatments at a farmer's location with $x_j = (y_{fj} + y_{cj}) / 2$, where y_{fj} is the performance of the fertilizer treatment at the j-th farmer's location, y_{cj} is the performance of the control treatment at the j-th farmer's location and x_j is the mean yield of the two treatments at a farmer's location. The regression model can be written as

$$y_{ij} = \mu_i + \beta_i \mu_j + d_{ij}, \tag{1}$$

where β_i is a regression coefficient corresponding to the i-th treatment, μ_j is an effect of the j-th farmer's location and d_{ij} is a random deviation from the regression line (Piepho 1998).

Three different scenarios are possible by comparing two treatments with each other (Fig. 1):

1. $\Delta y_A < \Delta y_B \wedge x_A < x_B$,
   where $\Delta y_A = y_{A2} - y_{A1}$ and $\Delta y_B = y_{2B} - y_{1B}$ is the yield difference at location A and B between treatment 1 and 2, respectively, and x_A and x_B is the mean yield of both treatments at location A and B, respectively;
2. $\Delta y_A > \Delta y_B \wedge x_A < x_B$;
3. $\Delta y_A = \Delta y_B \wedge x_A < x_B$.

In scenario (1) treatment *2* shows an above average response to improved environmental conditions, while the opposite holds true for treatment *1*. In scenario (2) the situation is reversed and in scenario (3) both treatments show an average response to different environmental conditions.

The different scenarios can be measured by the regression parameters β_i:

1. $b_2 > 1 \wedge b_1 < 1$,
   where b_1 and b_2 are estimations of β_1 and β_2 for the regression of treatment *1* and treatment *2* on the environmental mean, respectively;
2. $b_2 < 1 \wedge b_1 > 1$,
3. $b_1 = b_2 = 1$.

Whether b_1 is significantly different from b_2 was tested by regression analysis of the location yield difference on the location mean yield with the following model:

$$y_{i\Delta} = \mu_{\Delta i} + \beta_\Delta \mu_j + d_{\Delta j}, \tag{2}$$

where β_Δ is a regression coefficient corresponding to the treatment difference Δ, $\mu_{\Delta j}$ is a difference effect of the j-th farmer's location and $d_{\Delta j}$ is a random deviation from the regression line. The three scenarios correspond to the following estimation of β_Δ:

1. $b_D > 0$,
   where b_D is an estimation of β_Δ;
2. $b_D < 0$;
3. $b_D = 0$.

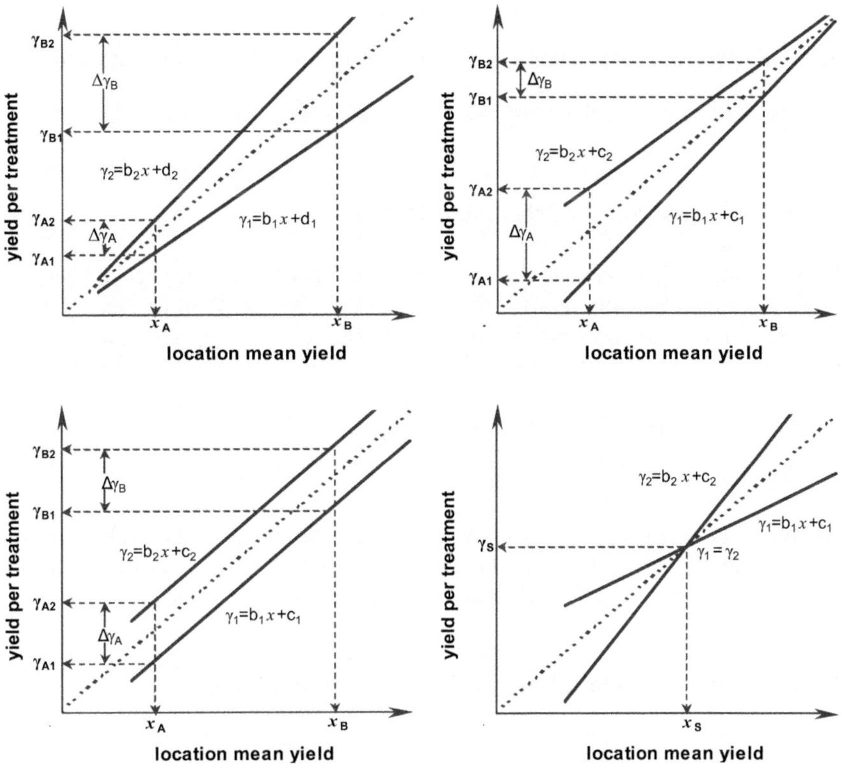

Fig. 1. Basic concept of adaptability analysis and scenarios: Separate regressions of yields or economic variables of treatment 1 and treatment 2 on the mean yield of the two; scenario 1 (above left): above-average response of treatment 2 to improved environmental conditions; scenario 2 (above right): below-average response of treatment 2 to improved environmental conditions; scenario 3 (below left): average response to environmental changes; scenario 4 (below right): the two regression lines are crossing each other; the crossing point depicts the point at which the treatment with the lower yield switches into one with a higher yield or *vice versa*.

The significance and the confidence interval for b_D were calculated.

Adaptability analysis based on yields was combined with economic parameters to assess risk since the economic superiority of one technology over another is the major single parameter for farmers to adopt it or not. For that purpose the marginal costs between two practices to be compared, which can be expressed in yield, were considered by subtracting the yield of the fertilizer treatment by the grain equivalent of the marginal costs.

If the two regression lines are crossing each other within the data range and the slopes are significantly different from each other the two regression equations are set equal to identify the point at which the treatment with the lower yield switches into one with a higher yield or *vice versa*:

$$a_1x + b_1 = a_2x + b_2 \tag{3}$$

To address marginal costs the equation can be extended by a marginal yield:

$$a_1x + b_1 = a_2x + b_2 + \Delta y \tag{4}$$

2.5 Risk Assessment of Mean Grain Yield

Probabilities of risk associated with mean maize grain yield were calculated separately for each treatment and for the treatment difference. Risk of mean yield can be quantified by determining the lowest expected yield or income at a given level of probability using confidence intervals. Average yield and yield difference uncertainties or risk levels were calculated as the lower confidence limit (LC) of the mean yields according to the formula:

$$LC = \overline{Y} - t_{df=n-1} \cdot s_d \cdot n^{-0.5}, \tag{5}$$

where, $\overline{Y}$ is the mean yield or the mean yield difference, t is the probability level from a one-tailed t table, s_d is the standard deviation of the mean, and n is the number of observations (Hildebrandt and Russell 1996).

2.6 Probability of Yield Falling Below a Critical Level

Where normality of data can be assumed, the probability that the yield of a technology will fall below a critical level λ in a randomly chosen environment j can be described by

$$\Pr(y_j < \lambda) = \Phi[(\lambda - \mu)\,\sigma^{-1}], \tag{6}$$

where Φ is the cumulative distribution function of the standard normal distribution, μ is the mean and σ is the standard deviation of the technology (Eskridge 1990). Where normality of data are not given, other distributions may be considered like the triangular (Anderson and Dillon 1992) or the log-normal (Piepho 1996).

To get information about the relative risk of the two systems compared with the risk of treatment *1*, defined as $\Pr(y_{1j} < \lambda)$, was plotted against the risk of treatment *2*, defined as $\Pr(y_{2j} < \lambda)$, for several values of λ.

2.7 Probability of One Treatment Outperforming Another One

It may be difficult to determine a critical level λ. In addition, new technologies are competing with existing ones, which calls for a direct comparison of any new technology with a control treatment. When normality of yields can be assumed, the probability of one treatment outperforming another one can be calculated by:

$$\Pr(D_j > 0) = \Phi\,[\delta\,\sigma_D^{-1}], \quad (7)$$

where D_j is the yield difference between the two treatments in a randomly chosen location, Φ is the cumulative distribution function of the standard normal distribution, $\delta = \mu_1 - \mu_2$, μ_1 and μ_2 are the mean yields of the two treatments and σ_D is the standard deviation of the difference D_j (Eskridge and Mumm 1992).

If the two treatments, compared with each other, require different resources and inputs, this approach might over or underestimate the probability. As with the calculation of the probability of falling below a critical total yield level, it is possible to calculate the probability of falling below a critical yield difference which needs to be surpassed before one treatment becomes economically attractive compared with another:

$$\Pr(D_j < \lambda_D) = \Phi[(\lambda_D - \mu_D)\,\sigma_D^{-1}], \quad (8)$$

where D_j is the yield difference between the two treatments in a randomly chosen location, λ_D is the critical yield difference, Φ is the cumulative distribution function of the standard normal distribution, μ_D is the mean difference and σ_D is the standard deviation of the difference.

3 Results

3.1 ANOVA and Linear Contrasts

Maize grain yield data subjected to ANOVA revealed significant differences between treatments. Maize grain yield after application of NPK, N+K and N was 2,173, 2,438 and 2,326 kg ha^{-1}, respectively, compared with 1,907 kg ha^{-1} of the unfertilized control. The yield difference between the N+K treatment and the control of 531 kg ha^{-1} was the only difference which was significant at $P<0.05$ (Table 1).

After subtracting marginal costs due to fertilizer application from gross return, the computed net return of the NPK and N+K treatment with 575,915 and 753,843 Rp. ha^{-1} were significantly lower than the net return of the control with 1,209,550 Rp. ha^{-1}. The N treatment with a net return of 1,306,971 Rp. ha^{-1} was the only fertilizer treatment with a higher return than the control but the difference was not significant (Table 2).

3.2 Adaptability Analysis

While the mean yield of the NPK treatment was slightly higher than the control, the two regression lines were almost parallel to each other with a similar slope (Fig. 2). Corresponding analyses of the N+K and N treatments showed above average responses to improved environmental conditions compared with the control.

Table 1. Differences in maize grain yield between selected treatments

Comparison	Difference (kg ha^{-1})	Probability
NPK vs. Control	266	0.244
N+K vs. Control	531	0.025
N vs. Control	419	0.074

Table 2. Differences in net return between selected treatments

Comparison	Difference (Rp.)	Probability
NPK vs. Control	-633,635	<0.001
N+K vs. Control	-455,707	0.003
N vs. Control	97,421	0.491

However, the regression of the yield differences of each of the fertilizer treatments to the control (Fig. 3) was not significant. Therefore, the slopes in each graph of Fig. 2 were not significantly different from each other. However, maize grain yields of the control treatment were positively correlated to plant available P (R^2 = 0.36**; Fig. 4) but the yield difference between fertilizer treatments and the control was not affected by available P. Soil pH and number of year s of cultivation to yield differences were also not significant.

3.3 Risk Assessment of Mean Grain Yields

Maize yields ranged between 485-2,824 kg ha^{-1}, 560-3,403 kg ha^{-1}, 381-2,629 kg ha^{-1} and 409-2,627 kg ha^{-1} in the NPK, N+K, N and control treatment, respectively. The variability as indicated by the standard deviation (s_d) and the confidence interval of the mean (CI) was larger in the fertilizer treatments than in the control treatment.

The objective of the risk assessment based on mean yields is to provide information and potential tools for research and extension programs whether or not to recommend a specific technology to farmers in a given target area. Using kg ha^{-1} grain yield as the evaluation criterion, the risk that the mean yield of the fertilizer treatment falls below the mean yield of the unfertilized control was 12.0%, 0.5%, and 8.6% for the NPK, N+K and N treatment, respectively (Fig. 5). The probability that on average the mean difference between fertilizer treatment and control would fall below 0 was 3.2%, 0.6% and 2.5% for the NPK, N+K and N treatment, respectively.

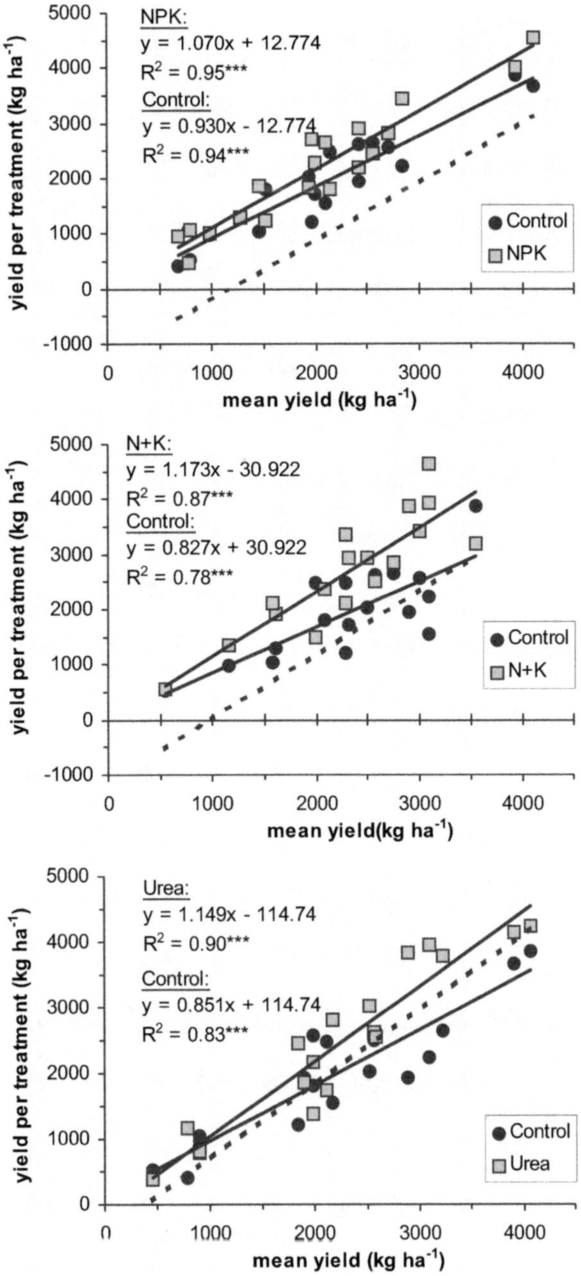

Fig. 2. Adaptability analysis comparing maize fertilized with 60 kg NPK ha^{-1} (above), with 60 kg N+60 kg K ha^{-1} (middle), and with 60 kg N ha^{-1} (below) with the respective unfertilized control. The dotted line represents the fertilizer treatment adjusted by marginal costs.

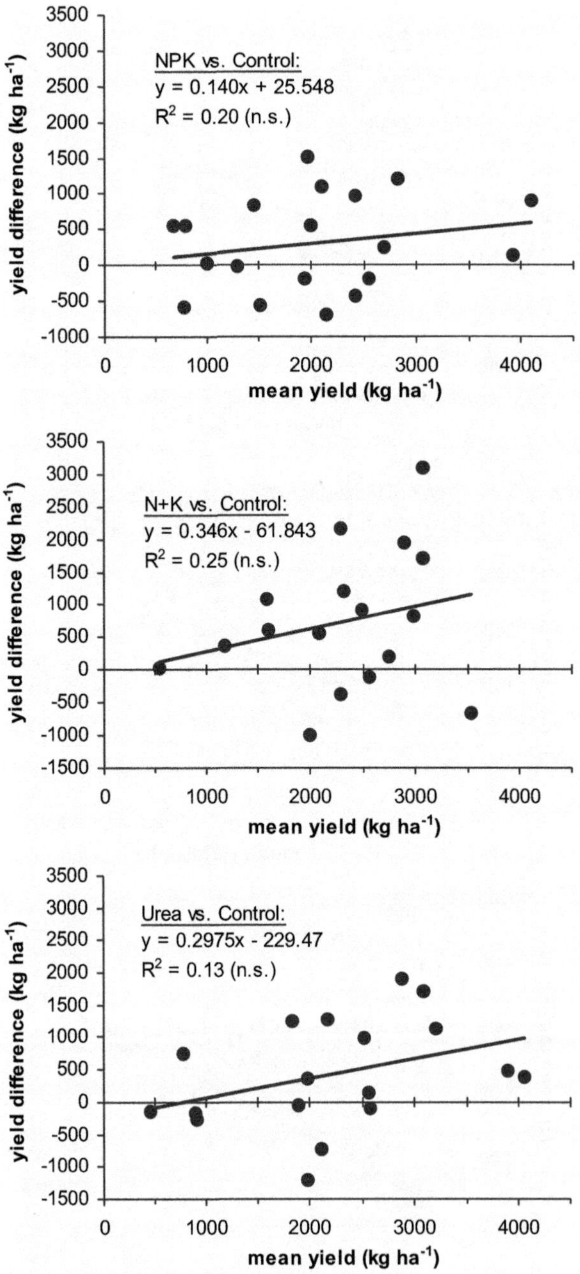

Fig. 3. Regression of maize yield difference between maize fertilized with 60 kg NPK ha^{-1} (above), with 60 kg N+K ha^{-1} (middle), and with 60 kg N ha^{-1} (below) and unfertilized control on location mean yield of the respective two treatments.

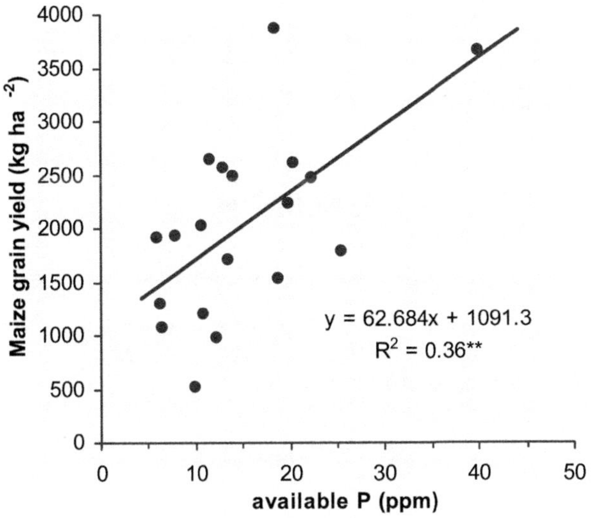

Fig. 4. Unfertilized maize grain yields as influenced by plant available P contents of soils at different locations.

3.4 Probability of Yield Falling Below a Critical Level

With maize yield as the criterion, NPK and N+K had lower risk regardless of the critical level (Fig. 6). The same trend could be detected for the N treatment but at low critical levels the control treatment presented less risk.

To readjust the analysis by economic parameters the marginal costs for the fertilizer treatments were transformed to break even in maize yield differences. The marginal maize yields which had to be subtracted from the fertilizer yields were 1,285 kg ha^{-1}, 1,129 kg ha^{-1}, and 324 kg ha^{-1} of the NPK, N+K and N treatment, respectively. In contrast to the comparison of yields alone, the control entailed less economic risk than the NPK and N+K treatment regardless of the threshold level if marginal costs are considered (Fig. 6). For N application the critical yield level determined whether N application of the unfertilized control is less or more risky.

Based on maize yield as the evaluation criterion, NPK and N+K application entailed less risk than the control regardless of the critical level (Fig. 7). The situation was reversed when marginal costs for fertilizer application were considered. The risk of N application to fall below a critical level was lower than for the control at all critical levels except at very low values but the consideration of fertilizer marginal costs implied a clear S-shaped curve. Where the critical level of net return is higher than 1,000,000 Rp. ha^{-1} the N treatment is less risk prone than the control while the opposite is the case below 1,000,000 Rp. ha^{-1}.

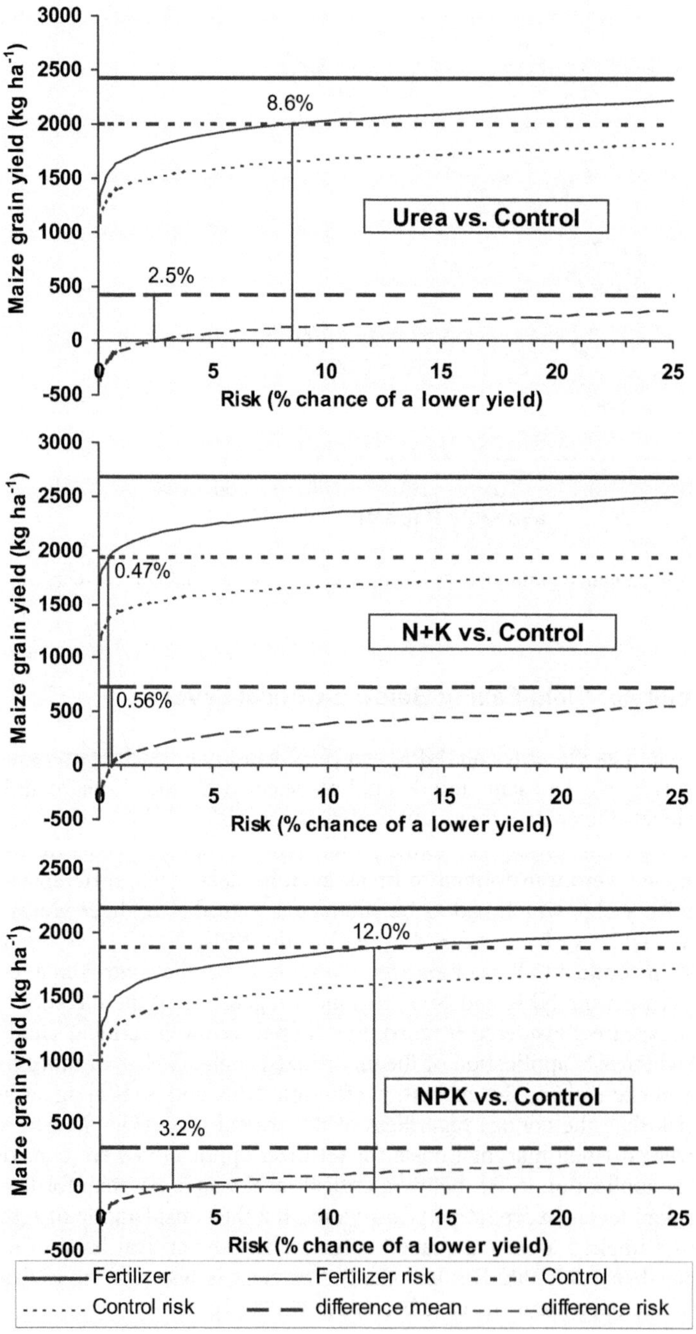

Fig. 5. Risk for fertilizer and unfertilized control treatments and their respective yield difference to change their mean values to a lower level.

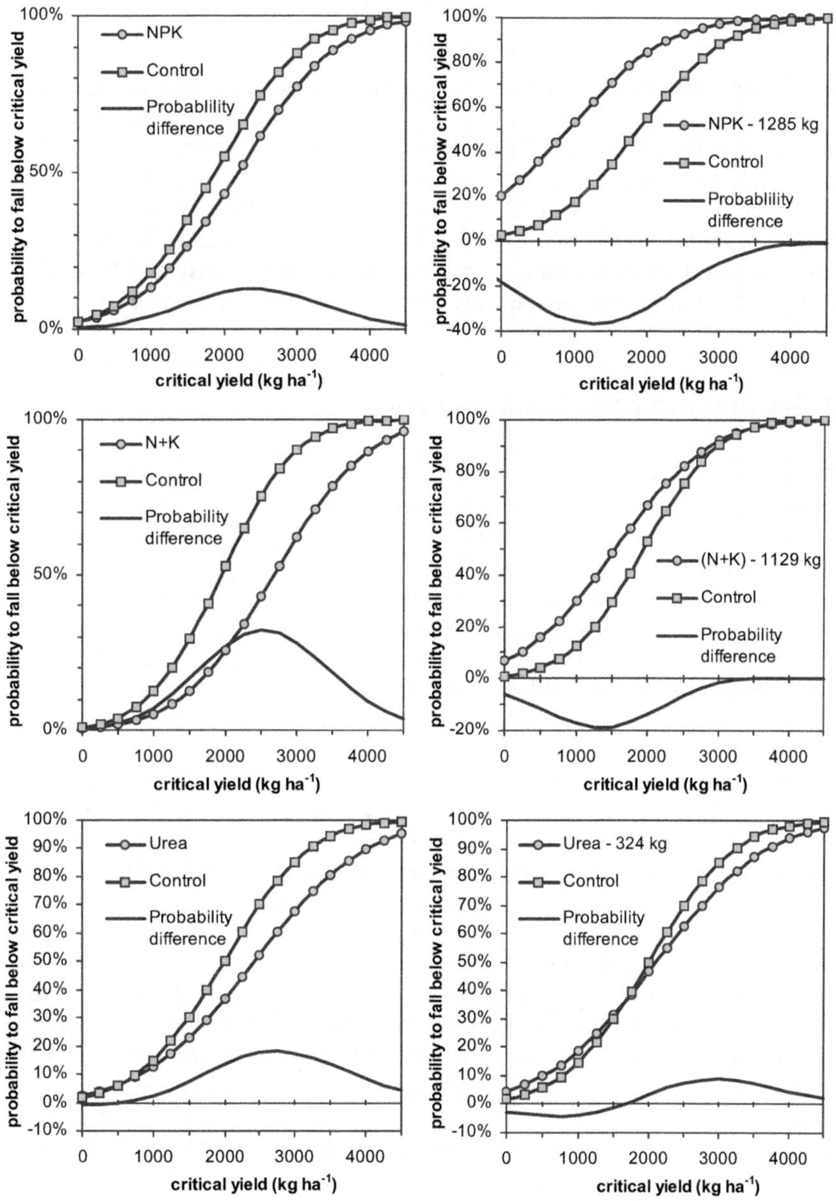

Fig. 6. Probability that maize yields after NPK application (top), after N+K application (middle), after N application (bottom), and of unfertilized control will fall below a critical yield for various critical yield levels based on yields alone (left) and under consideration of marginal costs for fertilizer application (right).

3.5 Probability of One Technology Outperforming Another Technology

One of the main interests of the study was the comparison of the various fertilizer treatments with an unfertilized control. The probability that NPK, N+K, and N would outperform the control in a randomly selected environment based on maize grain yield alone was 68%, 76% and 70%, respectively. This is equivalent to the risk of NPK, N+K and N of 32%, 24% and 30% of falling below a yield difference compared with the control of 0 (Fig. 8). However, considering the marginal costs for fertilizer application, the probability that NPK, N+K and N would outperform the control decreased to only 8%, 36% and 55%, respectively.

4 Discussion and Conclusions

The results of this on-farm experimentation revealed that average maize yield increased through the application of N based mineral fertilizers. However, the yield variation varied immensely. Therefore, the differences in yield compared with the unfertilized control were not significant. The regression model of the adaptability analysis showed an above average response with improved environmental conditions for the NPK and N+K treatment but the slope differences were not significant. Hence, other unexplained factors may have contributed to the yield variation. After application of the fertilizer a dry spell lasting about three weeks followed. Two to three weeks before harvest heavy rains occurred resulting in flooding of parts of the experimental plots. For that reason, N recovery by maize is expected to be low. Efficiency of utilization of N fertilizer in a multiple cropping trial in Lampung Province, Indonesia, was also low with 32-40% for dry season maize (Sisworo et al. 1990). Low effects of N fertilizers on maize are not uncommon. In field trials on a Regosol in East Timor, maize yield was not significantly affected by application of N, K and trace element fertilizers but responded well to P (Gunarto 1992). In this study, maize yields of unfertilized plots were also significantly correlated to soil P content (Olsen).

The response of the N treatment was lower than that of N+K. Besides the fact that K was also added in this treatment, the total amount of fertilizer was applied at planting while the N treatment was the only treatment in which the rate was split in two equal parts. Nihayati and Damhury (1996) obtained highest yields of sweet maize with a 50:50 split application 14 and 44 days after sowing. The current trial followed a similar application sequence which might have negatively affected yield considering the adverse weather conditions during that time.

NPK fertilization showed the smallest difference compared with the control. Analysis of a random sample of the fertilizer revealed that it did not contain any nutrients. However, at some locations the fertilizer appeared to be effective, suggesting that only parts of the applied NPK fertilizer contained no nutrients. Therefore, these results have to be considered with caution but should not be excluded from a risk analysis because they represent the uncertainties farmers are facing.

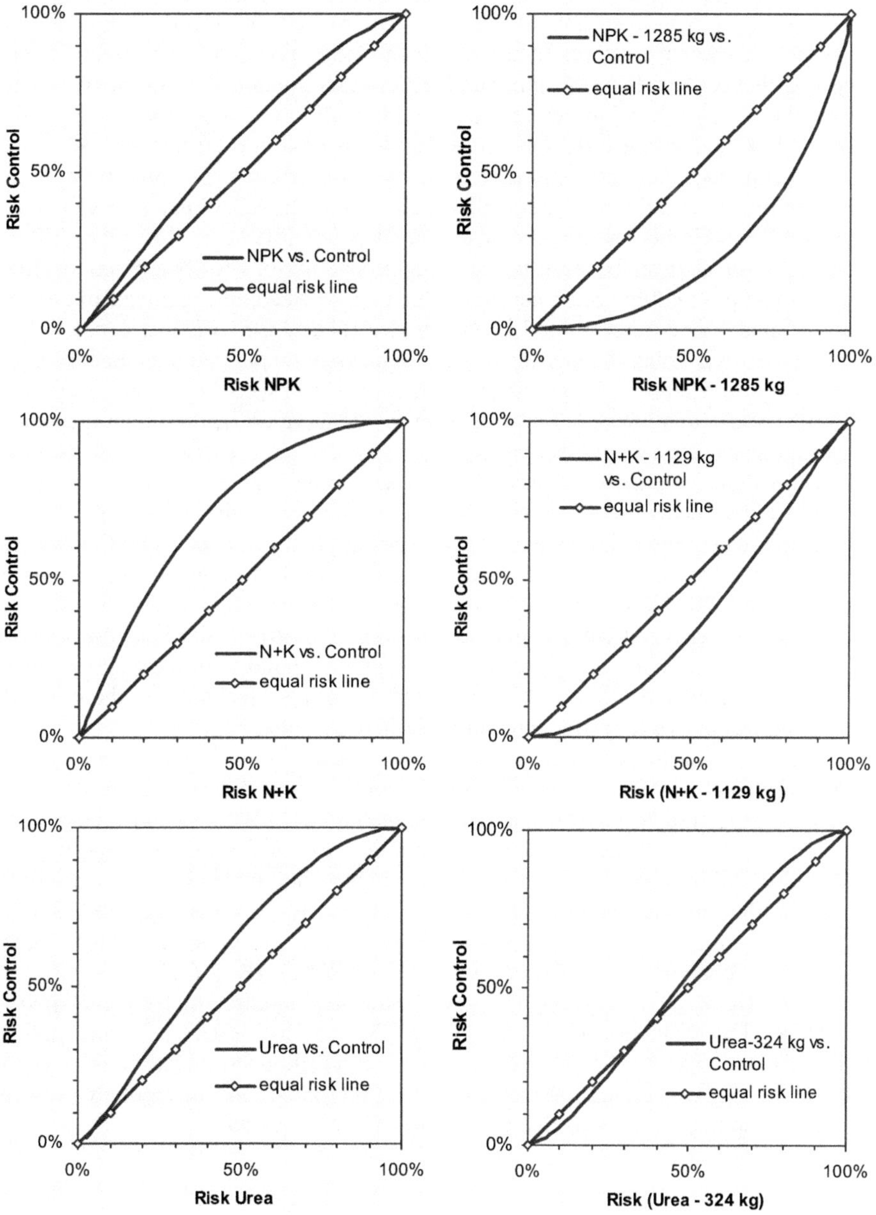

Fig. 7. Risk of maize yields after fertilizer application falling below a critical yield based on maize yield alone (left) and under consideration of marginal costs for fertilizer application (right). Data points represent the same critical yield of both compared treatments.

NPK fertilization showed the smallest difference compared with the control. Analysis of a random sample of the fertilizer revealed that it did not contain any nutrients. However, at some locations the fertilizer appeared to be effective, suggesting that only parts of the applied NPK fertilizer contained no nutrients. Therefore, these results have to be considered with caution but should not be excluded from a risk analysis because they represent the uncertainties farmers are facing.

Weed management strongly determined yield performance. Some farmers did not apply any weed control after planting which resulted in very low yields and yield differences between treatments. Wherever possible such plots were still included in the analysis because farmers might have reasons for this practice. However, some plots could not be harvested due to poor weed management. In farming households where other crops like paddy rice (*Oryza sativa* L.) or cocoa (*Theobroma cacao* L.) play the key role maize might only be planted to bridge excesses in terms of available land and labor. At the time when weeding is necessary these farmers prefer to allocate labor to other more lucrative enterprises than maize and accept the small maize yield.

The various risk analyses showed that based on the maize yield level alone, the fertilizer treatments were less risky for the farmers with the exception of the N treatment where the risk depended on the critical yield level. The situation was reversed when marginal costs of fertilizer application were considered. The probability that NPK, N+K and N in a randomly selected farming environment outperforms the unfertilized control reduced to only 7%, 36% and 55%, respectively. Since farmers do not only expect a break-even compared with their current practice but to achieve a considerable surplus by adopting a technology, there is little economic incentive for farmers to apply fertilizers.

Other options to add N to the soil-plant system is the inclusion of legumes (Noordwijk et al. 1995). However, the N contribution of food legumes is often small. Removal of N in cowpea (*Vigna unguiculata* (L.) Walp.) and soybean (*Glycine max* (L.) Merr.) seed exceeded the amount derived from fixation (Sisworo et al. 1990). Alley cropping systems for annual crops like maize have an agronomic potential but are very labor intensive and sensitive to poor management (Dvorák 1996). Rotations with leguminous cover crops show promising results. Maize yields in rotations with *Mucuna pruriens* (Stickm.) DC., *Calopogonium mucunoides* Desv. and *C. caeruleum* (Benth.) Sauvalle increased maize yields by 0.5-1.0 t ha^{-1}. Use efficiency of biomass N was 0.8 times that of urea in 2 split applications (Noordwijk et al. 1995). Hence, suitable innovations to increase maize yields in Central Sulawesi could be based either on the integration of legumes as cover (*Mucuna* spp.) or intercrops like groundnut (*Arachis hypogaea* L.).

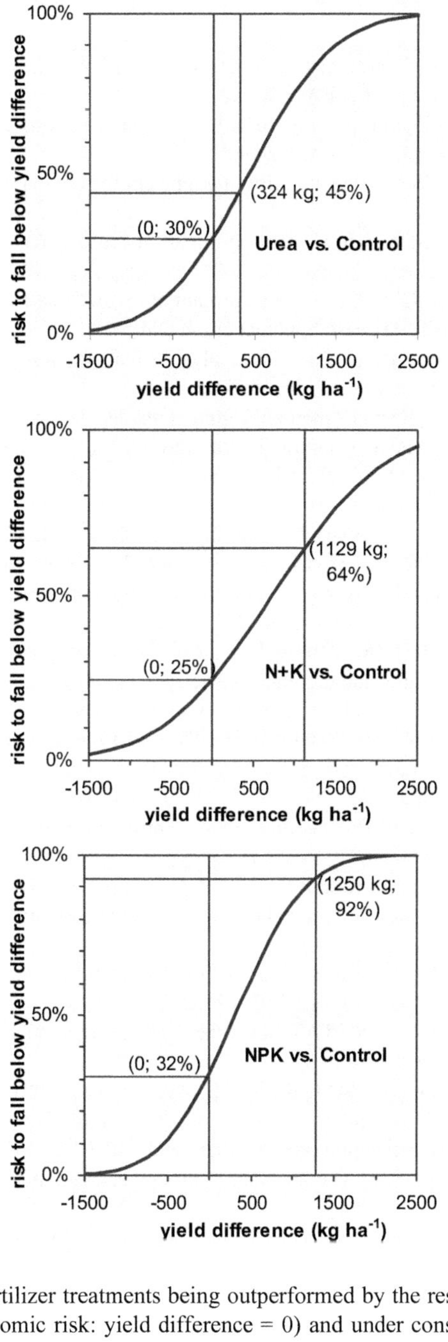

Fig. 8. Risk of the fertilizer treatments being outperformed by the respective control based on yields alone (agonomic risk: yield difference = 0) and under consideration of marginal costs (economic risk).

References

Anderson JR, Dillon JL (1992) Risk analysis in dryland farming systems. Farming Systems Management Series , vol 2, FAO, Rome

Dvorák KA (1996) Adoption potential of alley cropping. Resource and Crop Management Research Monograph 23. IITA, Ibadan

Eberhard SA, Russell WA (1966) Stability parameters for comparing varieties. Crop Sci 6: 36–40

Eskridge KM (1990) Selection of stable cultivars using a safety first rule. Crop Sci 30: 369–374

Eskridge KM, Mumm RF (1992) Choosing plant cultivars based on the probability of outperforming a check. Theoret Appl Genet 84: 894–900

Finlay KW, Wilkinson GN (1963) The analysis of adaptation in a plant breeding programme. Aust J Agric Res 14: 742–754

Gunarto L (1992) Tanggapan tanaman jagung terhadap pemberian N, P, K dan unsure mikro di Regosol-Maliana, Timor. Timur Agrivita 15(2): 65–69

Hildebrand PE (1984) Modified stability analysis of farmer managed on farm trials. Agron J 76: 272–274

Hildebrand PE, Russell JT (1996) Adaptability analysis: a method for the design, analysis and interpretation of on-farm research-extension. Iowa State University Press, Iowa

Mead R, Riley J, Dear K, Singh SP (1986) Stability comparison of intercropping and monocropping systems. Biometrics 42: 253–266

Nihayati E, Damhury S (1996) Pengaruh proporsi dan waktu pemberian urea terhadap pertumbuhan dan produksi jagung manis varietas SD-2. Agrivita 19(2): 51–56

Noordwijk M, Sitompul, SM, Hairiah K, Listyarini E, Syekhfani M, Noordwijk M (1995) Nitrogen supply from rotational or spacially zoned inclusion of Leguminosae for sustainable maize production on an acid soil in Indonesia. In: Noordwijk M, Date RA, Grundon NJ, Rayment GE, Probert ME (eds) Plant-soil interaction at low pH: principles and management. Proceedings of the Third International Symposium on Plant-Soil-Interactions at low pH, Brisbane, Queensland, Australia, 12–16 September 1993. Developments in Plant and Soil Sciences vol. 64, pp 779–784

Piepho H-P (1996) A simplified procedure for comparing the stability of cropping systems. Biometrics 52: 315–320

Piepho H-P (1998) Methods for comparing the yield stability of cropping systems – a review. J Agronomy and Crop Sci 180: 193–213

SAS Institute Inc (1997) SAS/STAT Software. Changes and enhancement through release 6.12. SAS Institute, Cary

Sisworo WH, Mitrosuhardjo MM, Rasjid H, Myers RJK (1990) The relative roles of N fixation, crop residues and soil in supplying N in multiple cropping systems in a humid, tropical upland cropping system. Plant and Soil 121(1): 73–82

Yates F, Cochran WG (1938) The analysis of groups of experiments. J Agric Sci 28: 556–580

Characterisation of Biodiversity in Improved Rubber Agroforests in West-Kalimantan, Indonesia: Real and Potential Uses for Spontaneous Plants

Stéphanie Diaz-Novellon, Eric Penot & Michel Arnaud

1 Introduction: The Improvement of Jungle Rubber

Since the introduction of rubber at the turn of the 20th century smallholders have developed an original complex agroforestry system called jungle rubber, in which non selected young rubber trees (seedlings) are managed extensively alongside secondary forest re-growth.

The issue of improving smallholder rubber productivity at affordable capital investments and levels of inputs while maintaining the environmental benefits of jungle rubber has been addressed by the Smallholder Rubber Agroforestry Project (SRAP, a joint project run by ICRAF, GAPKINDO and CIRAD). In 1995-1996, 27 trials (with a total of 100 plots) were set up in three provinces in Indonesia to assess the possibility of associating clonal rubber with agroforestry practices under smallholder conditions (Penot, 1997). Two RAS types were selected for this study: RAS n° 1 and n° 3. RAS n° 1 is basically improved jungle rubber using clonal planting material (see a description of RAS types in annexe 1). The rubber trees are in competition with spontaneous vegetation in the inter-row, but results show that there are no negative consequences for rubber growth during the immature period. RAS n° 3 was designed for areas infested by *Imperata cylindrica*, with the establishment of shrubby leguminous cover crops and fast-growing tree species in the inter-rows with the aim of shading out weeds. The other type, RAS n° 2, is based on intercropping clonal rubber with various annual and perennial crops, including fruit and timber trees (Penot et al, 1994). In all cases, RAS have a planting density of 550 clonal rubber trees/ha and a variable number of associated fruit, timber or fast growing shade trees (from 92 to 256/ha).

In addition to the RAS experimental plots, "RAS sendiri" (or "endogenous RAS") are rubber agroforests improved by farmers without outside assistance.

The district of Sanggau in the province of West Kalimantan was identified by SRAP as representative of traditional jungle-rubber-based local farming systems that have developed over the last 90 years. The district of Sanggau is located in the central area of the Kapuas river basin, between 1° N and 0°6' S and 09°8' W and 11°33' E. The district covers 18 302 km2, i.e. 13 % of the province. The trial plots described in this study are located in the villages of Embaong, Engkayu, Kopar, and Trimulia (the last being in the transmigration area).

Most soils in the province of West-Kalimantan are acrisoils associated with ferralitic soils. Such soils have relatively good physical characteristics but poor chemical value and become acid. Rubber is widely grown in this area, as it can grow in poor soils. The landscape is dominated by logged-over forest, secondary forest and a mosaic of jungle rubber and fallow with secondary forest re-growth. Large scale logging activities took place from 1950s to the 1980s at the expense of primary forest. At present, forested areas are located in hilly or remote areas and are very limited in extent. Oil palm and *Acacia mangium* plantations developed exponentially in the 1990s increasing the conversion of degraded forest areas into Estates that cultivate perennial crops.

The main objective of this study is to assess existing plant biodiversity in RAS systems compared to that of jungle rubber. The second objective is to review the current uses of certain plants and their market potential.

2 Methodology

The data presented in this paper were collected between August and October 2001 in 4 villages in the West-Kalimantan provinces and included 23 rubber agroforests plots. The nature of the previous vegetation, neighbouring vegetation and soil characteristics were recorded in addition to classical data collected from plots used for on-farm trials (rubber growth. etc).

2.1 Transect Method

In order to qualify existing biodiversity, the "transects" method was used with a sampling size per transect of 1 m x 0.2 m, and 15 replications for each treatment. The measurement criteria for floristic analysis classically used with agroforestry systems are: i) abundance, ii) density, iii) frequency and iv) dominance. In addition, in our study data were collected on the number of species and identity.

This method was implemented to try and cover a wide range of situations. However, the results obtained with this method do not allow a direct comparison of biodiversity between cropping systems (between RAS and jungle rubber) because the initial plots (the transects) are too small. Further analysis merging replications into larger transects should provide information that will enable comparison of systems. Further research is therefore required, both in the field to cover a wider range of situations and in data processing to complete the picture. This paper presents the preliminary results.

2.2 Statistical Analysis: The Use of Correspondence Analysis

Statistical analyses (correspondence analysis)[1] was implemented using "Winstat", a software developed by CIRAD. This analysis allowed us to focus on different effects that influence plant biodiversity in rubber inter-rows under the different systems. The five first axes were taken into account for each analysis. Data were collected in RAS systems as well as in selected fallow plots with different densities of existing vegetation. The plotting was as follows: 23 RAS plots, fallow plots respectively 1,2,3,4 and 5 years old, 6-year-old jungle rubber plots (the same age as RAS plots), some *tembawang* plots *(Tembawang* are fruit and timber agroforestry systems developed by Dayak people) and some secondary forest plots near the study villages. Complete detailed results are available in Diaz Novellon's MSc. Thesis (2001)

The S-Plus software[2] allowed us to implement discriminating analyses to explain variations in variables of diversity as a function of different explanatory variables (a "TREE function"). The selected variables of diversity were i) the number of species per replication, ii) the number of individuals in a replication (all species included) and iii) the ratios of individuals/number of species (N). These variables of diversity were correlated with different explanatory variables such as:

- The village: Kopar, Engkayu, Embaong, Trimulya;
- The farmer (21);
- The type of RAS (RAS 1 or RAS 3);
- The sub-type of RAS (RAS1.1/1.2/3.1/ 3.2/ 3.3/ 3.4)[3];
- The treatment (A, B, C, D, E, F, G, H, I, J);
- The surrounding plot environment (forest, fallow, jungle rubber, oil palm, etc.);
- The type of crop or vegetation before rubber (annual crops, *Imperata cylindrica*, Jungle Rubber, fallow, secondary forest, etc.)

[1] In French: AFC = analyse factorielle des correspondances.

[2] This software is available on the web and is usually used for botanical analysis.

[3] RAS trials are subdivided to study one component per trial in order to keep the procedure very simple and useable by farmers. Treatments are planting density, type of clones, weeding intensity and so on.

3 Assessment of the Biodiversity under Rubber Agroforestry Systems: A Comparison between Jungle Rubber and RAS n°1 & 3

Specific diversity, i.e. the number of species, appears to be higher in jungle rubber than in improved rubber agroforests (RAS).

Graphs 1 and 2 show the main results of the effect of environmental factors and previous vegetation on the biodiversity of each system (graph 2) and a comparison of the density (number of species and replications) between RAS and jungle rubber of the same age (graph 3).

However, in RAS n° 3.4, in which fruit and timber trees are planted in the inter-rows, the biodiversity of a number of species per transect is comparable to that of jungle rubber. A similar result was observed with "RAS sendiri". It thus appears that the different cultivation methods can directly influence the spontaneous diversity of plants in the inter-rows, and in fact, experimental RAS, RAS sendiri and jungle rubber are managed differently, resulting in a significant "farmer effect" .

According to trial protocols discussed each year with project farmers the number of weedings in the inter-rows is higher in the case of RAS. In practice in RAS, weeding is limited to the selective cutting of trees and shrubs that grow taller than young rubber trees, as compared with jungle rubber, where there is no weeding at all during the first few years. Weeding appears to be the main factor that influences plants diversity. When the cutting of spontaneous vegetation of the inter-row is spread out over time, the number of species is higher. On the other hand, the type of rubber trees (clonal or seedlings) does not influence the type, the diversity and the quantity of vegetation. Inter-row biodiversity is therefore more influenced by the farming practices and in particular by the frequency of selective cutting or the number of weedings.
The distribution and biodiversity of RAS plots is shown in table 1.

Table 1. Distribution of plants per biological type for all plots

	Trees	Herbaceous	Lianas on the ground	Lianas on trees	Bamboo rattan	Shrubs
Number of plants	1138	2480	368	128	54	231
Number of species	55	24	7	6	1	3

One important question concerning the comparison between RAS and jungle rubber is whether jungle rubber has a higher specific plant density, i.e. the number of plants per unit area, than that of improved rubber agroforests.

Our results show that jungle rubber does indeed have more individual plants in inter-rows than RAS, although the density of plants is very similar to that of "RAS sendiri". The different agroforestry practices (and in particular the frequency of selective cuttings) explain this difference. In comparison with biodiversity in secondary forest or *tembawang*, the number of species appears to be similar to that of RAS when the ground-level density of species is considerably lower (see further tables). In other words, the difference is mainly quantitative.

What are the most significant factors that explain the variation in biodiversity? Discriminating analyses showed that previous farming practices are a significant factor. An area that had been cultivated for at least 3 years showed higher specific biodiversity than fallow. A possible explanation for this result is that cultivated areas as "open systems" have a more significant seed bank and can collect seeds from surrounding forests or agroforests. Environmental factors probably also influence biodiversity. The presence of jungle rubber in the immediate vicinity induces greater biodiversity. 1 to 5 years of fallow around plots probably also increases diversity. As far as agricultural practices are concerned, the number of selective cuttings per year appears to be the most significant factor influencing plant biodiversity in the inter-row[4].

Previous results on jungle rubber biodiversity that were available to the author (Werner 97, De foresta, 97) as well as a guide book on plant uses (Levang & al, 1991) provided very useful preliminary studies for this paper.

Smallholders' Perception of Plant Biodiversity

It was immediately clear that local populations knew the plant species in their fields and their specific uses perfectly well. During field surveys as well as during interviews with farmers, more than 300 species had to be indexed. The most common uses of spontaneous biodiversity (in forest and agroforests) are in order of importance: health (medicinal plants) food (fruit, vegetables), construction (wood and timber), firewood and others (see table 2).

However, most spontaneous vegetation is not yet used by the local population and is thus available for other "potential uses". Medicinal plants have considerable potential (see table 3): they are not widely used, as farmers prefer "modern" drugs, which are considered to be far more effective against malaria, diarrhoea and other illnesses.

In the case of timber and wood, the most valuable species (see table 4) are becoming scarce in the vicinity (in local forests) in the study area.

Most farmers are interested in integrating certain timber species see table 5) in their agroforests for both housing (construction) and sale.

Prices vary considerably with the timber species, showing that this market is already well developed (table 6).

Some local species are maintained or preserved by replanting or favouring regeneration from natural re-growth in the different types of agroforests (table 7) and have a range of different uses.

---

[4] Detailed results are available in the MSc Thesis by the author, see bibliography (Diaz Novellon, 2001).

Table 2. Potential uses of biodiversity by the Dayak population

Uses	Number of species
Timber for construction, housing	83
Firewood	40
Timber for sale or furniture	2
Fruits	112
Vegetables	68
Medicinal plants	179
Animal food	24
Pulp (for paper)	1
Cosmetics	1
Colouring properties	2
Use as paper	9
Weed control	14
Insecticide	6
Handicraft	66
Latex	4
Oil	7
Fertilisation	14
Spices	55
Others	8

Table 3. Uses of medicinal plants

Diseases cured by local plants	Number of identified species
Coughs	12
Fever	23
Itching	15
Tiredness	11
Malaria	2
Dysentery	1
Throat	13
Toothache	1
Bellyache Belly sickness	44
Burns	9
Headaches	11
Others	11

Table 4. Timber species that are becoming scarce in remaining forests

Local names	Latin names	Village
Belian	*Eusideroxylon zwageri*	All villages
Tapang	*Koompassia excelsa*	Embaong, Kopar
Tekam		All villages
Benkirai	*Shorea sp*	Embaong

Table 4 (cont.)

Meranti	*Shorea spp*	Engkayu, Trimulya
Terenak		All villages
Jeluntung	*Dyera costulata*	Trimulya
Kayu Raya	*Shorea sp.*	Kopar
Majau	*Shorea palembanica*	Embaong
Omang	*Hopea dryobalanoides*	Sanjan, Engkayu
Medang	*Litsea elliptica*	Kopar, Engkayu
Tunam	*Shorea lamellata*	Kopar
Nyatuh	*Palaquium spp*	Engkayu
Owan		Engkayu
Ubah	*Glochidion sp*	Sanjan, Engkayu
Taba	*Aquilaria malaccensis*	Kopar
Keladan	*Dryobalanops beccarii*	Engkayu
Tengkawang	*Shorea macrophylla*	Kopar

Table 5. Timber species preferred by farmers.

Local names	Latin names
Belian	*Eusideroxylon zwageri*
Keladan	*Dryobalanops beccarii*
Tekam	
Ketuat	
Meranti	*Shorea spp*
Terindak	*Shorea senimis*
Tengkawang	*Shorea macrophylla*
Mengkirai	*Trema orientalis*
Mentibu	
Medang	*Litsea elliptica*
Nyatuh	*Palaquium spp*
Oman	*Hopea dryobalanoides*
Owan	
Jonger	*Ploiarium alternifolium*
Taba	*Aquilaria malaccensis*
Tantang	*Buchania sessifolia*

Table 6. Prices for local timber species

Timber species	Latin name	Price in ruplah
Belian	*Eusideroxylon zwageri*	50 000 Rp / board
Raya		3 500 Rp / board
Jonger	*Ploiarium alternifolium*	4 000 Rp / board
Owan		8 000 Rp / board

Table 6 (cont.)

Medang	*Litsea elliptica*	8 000 Rp / board
Paku		5 000 Rp / board
Tapang		20 000 Rp / board
Tengkawang	*Shorea macrophylla*	4000 to 10 000 Rp / board
Tantang	*Buchania sessifolia*	200 000 Rp / m2

Note : 1 US $ = 10500 Rp in July 2001.

Table 7. Spontaneous timber species maintained in local agroforests and their uses

Local names	Latin names	Uses
Leban	*Vitex pinnata*	Timber, wood, spice, medicinal
Medang	*Litsea elliptica*	Timber, latex
Ramboutan	*Nephelium lappaceum*	Fruits, timber
Jengkol	*Pithecellobium jiringa*	Fruits, vegetable, timber, medicinal
Durian	*Durio zibethinus*	Fruits, timber
Pingam	*Artocarpus sp*	Fruits, timber, vegetable
Cempedak	*Artocarpus integra*	Fruits, medicinal, vegetable
Lengsat	*Lansium domesticum*	Fruits, medicinal, handicrafts
Pekawai	*Durio c.f. dulcis*	Fruits
Mentawa	*Artocarpus c.f. anisophyllus*	Fruits
Nyatuh	*Palaquium spp*	Timber, latex
Owan		Timber, handicrafts
Bungkang	*Polyalthia rumpfii*	Timber, spice
Belian	*Eusideroxylon zwageri*	Timber
Ubah	*Glochidion sp*	Timber
Kemenyan	*Styrax benzoin*	Timber, latex, animal food
Tantang	*Buchania sessifolia*	Timber
Bidara	*Nephelium maingayi*	Fruits

Some of these species have been re-introduced in agroforests (see table 8), in particular in "*tembawang*", or are protected when they emerge in natural regrowth in jungle rubber and RAS systems.

Some other species that farmers do not consider suitable in RAS inter-rows were also identified (table 9). These species are in fact still used in that their products are still collected in true forests, but are not specifically selected in agroforests due to the fact that – at least in the farmers' opinions - they may have a negative effect on rubber growth or production.in the farmers' opinions For example, after 20 years of growth the number of durian trees per hectare has to be limited to fewer than 20 to reduce shading when the durian's canopy begins to outgrow that of rubber. Another example is "tengkawang" (Illipe nut tree) which is considered to "drain" soils and therefore limit rubber trees growth (although this observation has not been scientifically confirmed).

Table 8. Local species reintroduced in agroforest

Local names	Latin names	Uses
Jengkol	*Pithecellobium jiringa*	Fruits, vegetable, timber, medicinal
Mangga	*Mangifera indica*	Fruits
Ramboutan	*Nephelium lappaceum*	Fruits, timber
Manggis	*Garcinia mangostana*	Fruits
Durian	*Durio zibethinus*	Fruits, timber
Cempedak	*Artocarpus integra*	Fruits, medicinal, vegetable
Coklat		Cocoa
Kopi		Coffee
Petai	*Parkia speciosa*	Fruits, vegetable
Lengsat	*Lansium domesticum*	Fruits, medicinal, handicraft
Kedupai	*Mischocarpus pentapetalus*	Fruits
Sibau	*Xerospermum norotanum*	Fruits
Mentawa	*Artocarpus anisophyllus*	Fruits
Pekawai	*Durio c.f. dulcis*	Fruits
Melinjo	*Gnetum gnemon*	Fruits, vegetable
Nangka	*Artocarpus heterophyllus*	Fruits
Tengkawang	*Shorea macrophylla*	Fruits, oil, timber
Tekam		Timber, handicraft
Ketuat		Fruits, timber
Tempuih	Baccaurea sp	Fruits
Pisang	Musa spp	Fruits, vegetable, medicinal

Table 9. List of species not chosen for agroforests and their uses

Local names	Latin names	Uses
Belangai	*Eurya nitida*	Timber, medicinal, handicraft
Tucet	*Alstonia angustifolia*	Timber
Plaik	*Alstonia scholaris*	Timber, latex, medicinal, handicraft
Bamboo		Housing, handicraft , various uses
Todoh	*Phrynium capitatum*	Wrapping
Ringkan	*Ficus grossularoides*	Fruits, wrapping, timber
Resak	*Melastoma malabathricum*	Timber, fruits, vegetable, medicinal
Pakis		
Semolang	*Euodia aromatica*	Medicinal, timber
Siyet	*Sceria prupurescens*	Medicinal
Entiup	*Artocarpus sericicarpus*	Fruits, oil, handicraft
Leban	*Vitex pinnata*	Timber, spices, medicinal
Jambu america	*Bellucia axinanthera*	Fruits, wrapping, timber
Alang-Alang	*Imperata cylindrica*	Medicinal
Marade		Timber

Certain species (see table 10) may be selected to limit invasion of *Imperata cylindrica* in young agroforests.

Table 10. Species used to limit Imperata cylindrica (alang²) in young agroforests.

Local names	Latin names	Type of action
Semenput		Shading
Beringing		
Melastoma	*Melastoma malabathricum*	Cover cropping
Coklat		Cover cropping
Nenas	*Ananas comosus*	Root competition
Gmelina	*Gmelina arborea*	Shading
Orok-Orok	*Crotolaria mucronata*	Competition with alang²
Gamal	*Gliricidia sepium*	Shading (limited)
Akacia	*Acacia* mangium	Shading
Albizia	*Albizia sp*	Shading

Table 11 summarises the different species in agroforests, "*tembawang*" and jungle rubber, RAS sendiri, RAS1 and RAS 3 as well as in "house gardens" of the local population *(pekarangan)* used and sold on local markets. It gives an idea of the wide variety of products that have an impact on both the household food supply and on the economy.

4 Conclusion: Market Potential

Some forest products are of obvious economic interest (see table 11). Smallholders try to domesticate some of these species in the inter-rows of their agroforest (RAS and Jungle rubber), by replanting or favouring regeneration from natural re-growth, which has the advantage of involving almost no cost.

Timber species and fruit trees are particularly appreciated when they emerge from forest re-growth as they require no cost for plantation and very little additional labour to maintain. They are also replanted in order to enrich the vegetation in inter-rows.

Fruit trees have the most obvious potential market value , in particular durian (Durio zibethinus) which is already sold everywhere in Indonesia as well as in other countries in Southeast Asia (e.g. Thailand & Malaysia), rambutan and dukus, the latter profiting from a keen demand on the Indonesian market. National markets do not yet appear to be saturated, but export would provide the best market for smallholders, particular in the case of durian, which could easily be exported to neighbouring Malaysia, and north Borneo. The lack of larger organised marketing channels other than the traditional Sino-Indonesian one is a severe obstacle to the expansion of fruit markets and exports.

As a result of the keen demand for timber and wood products such as plywood in consumer countries (Japan, USA, and Europe), a timber shortage may already be expected in the very near future. Smallholders in West-Kalimantan would be well advised to anticipate this trend and include in their agroforests inter-rows species that can be used to supply the demand of the the plywood industry. Some species (Nyatoh in particular /*Palaquium spp.*) have a life-span similar to that of rubber (30 to 40 years). The ultimate life cycle of rubber agroforestry systems might then be completed with the exploitation of timber trees such as Belian (*Eusideroxylon zwageri*),(lifespan 60 years) or meranti (up to 90 years). In this way, old rubber-based agroforests may develop into "*tembawang*". Finally, at the end of rubber life-span, rattan might prove to be a useful crop as indicated by the strong demand for furniture for export.

One important obstacle is the Indonesian legislation on land and tree tenure that needs to be re-examined and adapted to the context of smallholder production whose future potential could be highly significant. At present regulations concerning timber exploitation practically precludes trade in timber from forest or agroforest by smallholders.

Other forest products with future potential are without any doubt medicinal plants. Local sales of these products, although limited, have also gradually declined due to the effectiveness and availability of pharmaceutical products. However, pharmaceutical firms could be interested in several forest and agroforest species from Borneo and perhaps develop research projects that could indirectly benefit local populations. Examples of this type have been already been observed in other countries in Amazonia, as well as in Côte d'Ivoire, where a product to control hypertension was discovered under rubber

Irrespective of how great the future potential of agroforest products is, and even if it is high for fruit, timber, rattan and medicinal plants, most products still remain under-exploited and this represents a great challenge for the very near future. Several contraints remain in terms of market organisation as well also as in terms of official regulations.

References

Diaz-Novellon S, Boutin D (2001) Flora of rubber agroforests in the West Kalimantan Province. (CD rom). SRAP ICRAF-CIRAD, Bogor

Diaz-Novellon S (2001) Caractérisation de la biodiversité spontanée des agroforêts à hévéa en Indonésie: utilisation et potentialités des espèces utiles. CNEARC, MSc « Agronomie et Innovations en milieu rural ». CNEARC, Montpellier

De Foresta H (1997) Smallholder rubber plantations viewed through forest ecologist glasses: an example from South Sumatra. ICRAF/SRAP workshop on RAS (Rubber Agroforetsry Systems), Bogor

Levang P and d Foresta H (1991) Economic plants of Indonesia: a Latin, Indonesian, French and English dictionary of 728 species. Bogor, Orstom - Seameo/Biotrop

Penot E (1997) Introduction to SRAP methodology and concepts: summary of the preliminary results. The ICRAF/SRAP Workshop on RAS (Rubber Agroforestry Systems), September 1997. ICRAF, Bogor

Penot E (1999) Rubber Agroforestry Systems (R.A.S.) methodology and main results: technical report. CIRAD/ICRAF, project paper. Montpellier

Werner S (1997) Biodiversity of jungle rubber in West-Kalimantan. ICRAF/SRAP workshop on RAS (Rubber Agoforestry Systems), Bogor

Annex 1: Description of the Various RAS Cropping System

RAS 1: An Improved Jungle Rubber

The first system (RAS 1) is similar to the current jungle rubber system, in which unselected rubber seedlings are replaced by clones selected for their potential promising adaptation. These clones must be able to compete with the natural secondary forest growth; various planting densities and weeding protocols are tested. The biodiversity is very similar to that of jungle rubber, being itself close to that of secondary forest of the same age. (H. de Foresta, 1997). This system is similar to that originally tested by the Dutch in the 1930s and called "jungle weeding" (Dijkman, 1950) after estates planters recognize that with the same unselected planting material, yields of jungle rubber were very comparable to that of estates with high weeding and intensive management. This proves that rubber, at least unselected rubber (seedlings), could grow and produce well in a complex agroforestry system with no major incidence on production. Clonal rubber requires definitely more weeding and maintenance that seedlings. The idea of RAS 1 is to reintroduce clones in a jungle rubber environment and to assess the minimum maintenance required for rubber to grow well. The main hypotheses derives from the precedent statement : rubber production, to a certain extend, is not affected by associated trees. In RAS 1, associated trees are not selected. These trees are those that grow naturally. However, farmers do a selection at opening (between 8 and 15 years after planting).

One main question might be the following " With an early opening (at 5/6 years old compared to 8/15 years with jungle rubber)), and consequently an earlier canopy closure : will biodiversity composition and evolution be different ?, can we therefore expect the same type of biodiversity ?

RAS 2: An Intensified Rubber Agroforest with Intercrops

The second, RAS 2, is a complex agroforestry system in which rubber (550/ha) and perennial timber and fruit trees (92 to 152 trees/ha) are established after slashing and burning. It is very intensive, with annual crops being intercropped during the first 3 or 4 years, with emphasis on improved upland varieties of rice, with various amounts of fertilization as well as dry season cropping such as groundnut.

In that case, the number and type of trees associated with rubber are deliberately chosen at planting time.

Several combination of planting densities (92, 140 and 250 trees/ha) and of selected species are being tested according to a tree typology, in particular : rambutan, durian, petai and tengkawang. Biodiversity is limited to the planted species as well as those which naturally regenerate and are eventually selected by farmers.

The Various Types of RAS 2

In RAS 2.1: Each field is divided in 5 plots (rubber + X): rubber alone, Durian, Rambutan, Durian + Rambutan, farmer's Mix (a combination of various fruit and timber trees)

In RAS 2.2: fruit and timber trees are planted at 3 planting densities: 92, 140 and 150 trees/ha with rubber 550 trees/ha with a maximum of 30 big trees such Durian or timber trees. These planting densities are within the range of what farmers already do in Sanjan for instance, without noticeable change in rubber production.

In RAS 2.5: rubber is combined with cinnamon (in the Jambi province) Cinnamon is collected at the 8th year. Cinnamon at 1000 trees /ha is planted with rubber 550 trees/ha in RAS 2.5.

RAS 3: A Rubber Agroforest Adapted to Imperata Grassland

The third system, RAS 3, is a complex agroforestry system with rubber and other trees planted with a frame similar to that of RAS 2; the difference is that it is established on degraded lands covered by *Imperata cylindrica,* or in area where *Imperata* is a major threat.

Imperata increase rubber immature up to 8/9 years (G Wibawa, 1995). Labour or cash for controlling *Imperata* with herbicide are the main constraints. In RAS 3, annual crops, generally rice, are grown the first year only, with non vine cover crops grown immediately after rice harvesting *(Mucuna, Flemingia, Crotalaria, Setaria and Chromolaena*) with a selection of trees : fruit and timber trees, multipurpose trees *(wingbean, Gliricidia..*) and fast growing trees as pulpwood for shading *(Paraserianthes falcataria, Acacia mangium or crassicarpa and Gmelina arborea* have been selected in West Kalimantan. The objective here is to eliminate the weeding constraint after the first year by providing a favourable environment for rubber and associated trees and consequently suppressing *Imperata* with low labour requirements. Several designs and combinations have been tried in RAS 3 in West Kalimantan.

The Various Types of RAS 3

RAS 3.1 (Planting in 1995): A Preliminary Test

The different treatments where based on the use of the following trees (1 per treatment): Leucena leucocephala, Gliricidia sepium, Caliandra callothyrsus, Ceasalpinia sappam, Chromolena odorata and Flemingia congesta with classical LCC (legume cover crops) as a control. In addition, 250 fruit and timber trees are combined with 550 rubber/ha

RAS 3.2 (Planting in December 1996): Confirmation Test

This trial was established according to the results of RAS 1 trial with the following trees: *Gliricidia s., Gmelina a.,* and the following cover crops: *Chromolena odorata, Crotalaria and Mucuna spp.*

The first control is rubber in monoculture + LCC. The second control is rubber monoculture + *Imperata.* In addition , 250 fruit and timber trees are combined with 550 rubber/ha

RAS 3.3: No Associated Fruit/Timber Trees: Pulp Trees Only

A first set of 4 replications has been planted in January 1996. The selected trees have been the following: Paraserianthes falcatarian Acacia mangium and Gmelina arborea. Rubber is at 550 trees/ha as well as pulp trees in the interrow (3 meters between tres on the pul tree line and 3 meters apart from the rubber line). The covercrop is Flemingia.

RAS 3.4: Pulp Trees + Associated Trees

Same structure as RAS 3.3 with 1 pulp tree out of 3 replaced by associated fruit/timber trees: in that case : the planting densities are for rubber: 550/ha, for pulp trees : 460/ha and for associated trees: 92/ha

Table 11: Species and products already sold on local

Indonesian/local names	Latin names	Sale price	Origin
Pisang / Banana	*Musa spp.*	1 500 Rp /lot	Agroforêst
Pakis piding / ferns		500 Rp / lot	Agroforêst
Kangkong	*Ipomea aquatica*	500 Rp / lot	
Cangkok manis		500 Rp / lot	Agroforêst
Daun kacang/bean leaves		500 Rp / lot	Home garden
Daun ubi / cassava leaves	*Gnetum gnemon*	500 Rp / lot	Agroforest
Bunga pisang/banana flower	*Musa spp*	1 000 Rp / fleur	Agroforêst
Jengkol	*Archidendron pauciflo-rum*	1 000 Rp / Kg	Agroforeesêt
Maram	*Eleiodoxa conferta*	2 000 Rp / Kg	Agroforest
Kacang panjang/bean	*Vigna unguiculata*	2000 Rp / Kg	Home garden
Timun /cumcumber	*Cucumis sativus*	2000 Rp / Kg	Home garden
Bunga jagung/maize flower	*Zea sp*	500 Rp / fleur	Pontianak
Bayam	*Amaranthus hybridus*	500 Rp / lot	Home garden
Petai	*Parkia speciosa*	2 500 Rp / Kg	Agroforest
Labu air / pumpkin	*Lagenaria siceraria*	2 500 Rp / Kg	Home garden
Jahe / gingember	*Zingiber officinale*	2 500 Rp / Kg	Home garden
Kelapa / coco nuts	*Cocos nucifera*	1 000 Rp / fruit	Home garden
Peringgi		4 500 Rp / Kg	Home garden
Kecambah		1 000Rp/portion	
Ubi / cassava	*Manihot esculenta*	2 500 Rp / Kg	Agroforest
Kedondong	*Spondias pinnata*	500 Rp / lot	
Pekawai	*Durio c.f. dulcis*	10 000 Rp /lot	Agroforest
Terong	*Solanum melongens*	5 000 Rp / Kg	Home garden
Cabe / pepper	*Capsicum annuum*	20 000 Rp / Kg	Pontianak
Buncis	*Phaseolus vulgaris*	3 500 Rp / Kg	
Gambas	*Luffa acutangula*	2 000 Rp / Kg	
Jeruk / lemon	*Citrus sp*	3 000 Rp / Kg	Home garden
Nangka / Jacqj fruit	*Artocarpus heterophyllus*	2 500 Rp / Kg	Agroforêst
Kencur	*Kaempferia galanga*	10 000 Rp / Kg	
Kunyit	*Curcuma longa*	5 000 Rp / Kg	Agroforest
Serai	*Cymbopogon nardus*	500 Rp / lot	Agroforest
Keladi	*Colocasia esculenta*	1 000 Rp / lot	Agroforest
Kundur	*Benincasa hispida*	2 500 Rp / Kg	
Asam	*Tamarindus indica*	500 Rp / fruit	Agroforest
Labu siam	*Sechium edule*	2 500 Rp / Kg	Home garden
Pane	*Momordica charantia*	5 000 Rp / Kg	Pontianak
Wartel / carott	*Daucus carota*	9 000 Rp / Kg	Pontianak
Jeruk nipis / lemon	*Citrus aurantifolia*	4 000 Rp / Kg	Pontianak
Kol / cabbage	*Brassica oleraceae*	5 000 Rp / Kg	Pontianak
Kentang / potato	*Solanum tuberosum*	4 500 Rp / Kg	Pontianak
Tomat / tomato	*Lycopersicon esculentum*	6 000 Rp / Kg	Pontianak

Table 11 (cont.)

Bawang merah/red onion	*Allium cepa*	7 000 Rp / Kg	Pontianak
Bawang putih/white onion	*Allium sativum*	7 000 Rp / Kg	Pontianak
Kayu manis/cinnamon	*Cinnamomum burmanii*	2 000 Rp / lot	Agroforest
Nenas / pinepale	*Ananas comosus*	2 000 Rp / fruit	Agroforest
Sawih / cabbage	*Brassica rugosa*	5 000 Rp / Kg	Pontianak
Jambu air	*Syzygium aquaeum*	1 500 Rp / Kg	Home garden
Pepaya / papaya	*Carica papaya*	2 500 Rp / Kg	Home garden
Kenikir	*Cosmos caudatus*	500 Rp / lot	
Lengkuas	*Alpinia galanga*	1 000 Rp / lot	Agroforest
Daun salam / leaves	*Eugenia polyantha*	500 Rp / lot	Agroforest
Daun sop /celery leaves	*Apium graveolens*	1 000 Rp / lot	Home garden
Daun pepaya/ papaya leaves	*Carica pepaya*	500 Rp / lot	Home garden
Mangga	*Mangifera indica*	8 000 Rp / Kg	Agroforest
Petai	*Parkia speciosa*	2 000 Rp / lot	Agroforest
Kacang tanah/peanut		3 000 Rp / Kg	Home garden
Cempedak hutan	*Artocarpus integra*	500 Rp / fruit	Agroforest
Kumis kucing	*Orthosiphon aristatus*	1000 Rp / lot	Home garden

NB : These species 's latin names should be utilized cautiously because of the difficulties of identification and correspondence between vernacular names and scientific names

Graph 1: Effect of environmental factors and previous vegetation on agroforest biodiversity

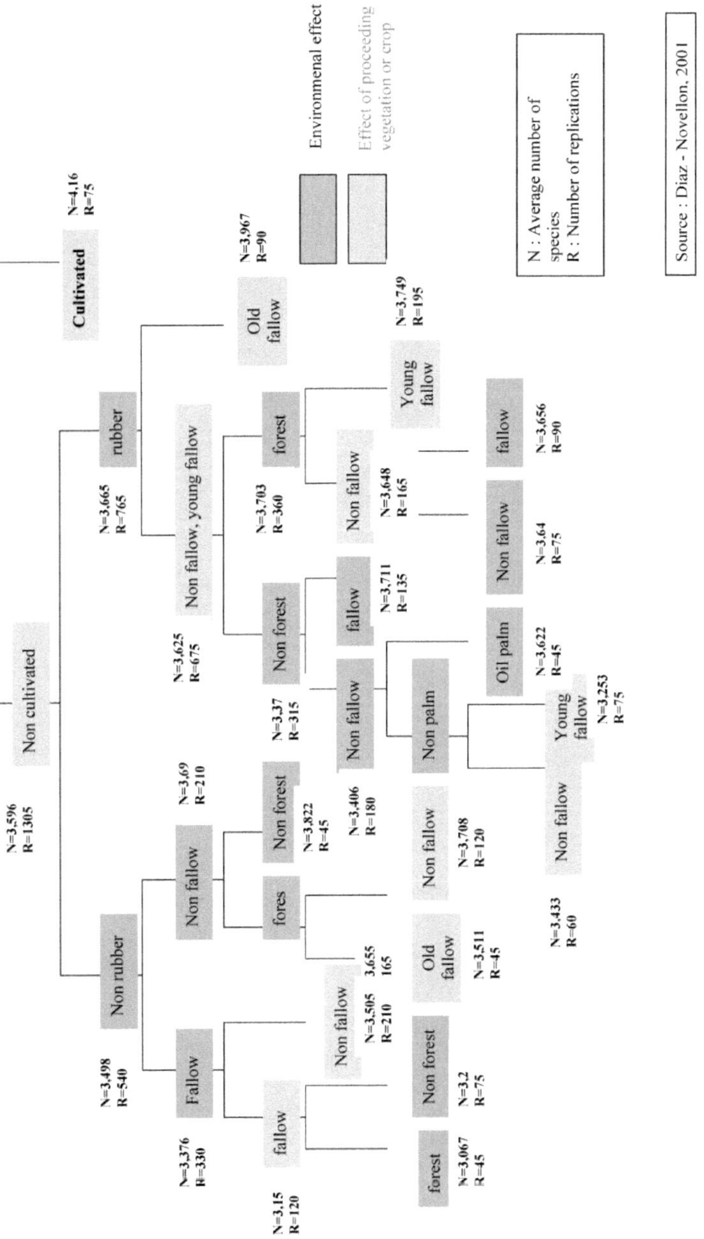

Graph 2: Comparison of density in RAS and jungle rubber at the same age

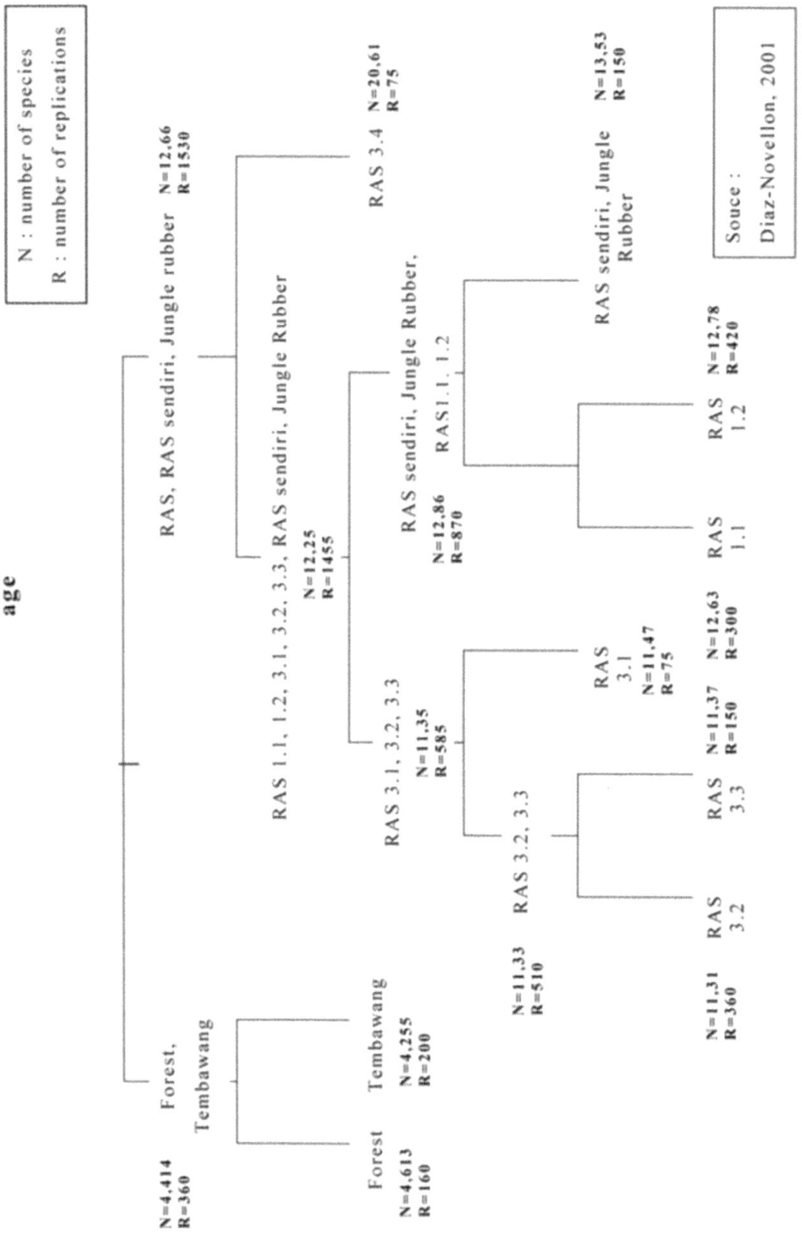

Traditional Forest Gardens in Central Sulawesi: A Sustainable Land Use System?

Frank Brodbeck, Hans-Joachim Weidelt & Ralph Mitlöhner

1 Introduction

In many parts of Indonesia, farmers are traditionally managing forest gardens, some of which have already been researched in detail, for example in the description of the "Lembos" of East Kalimantan by Sardjono (1990). The rattan gardens of the Dayak in East- and Central Kalimantan have been researched by Arifin (1995), Kraienhorst (1990) and Weidelt (1990, 1996). Sundawati (1993) has presented the "Tembawangs" of West Kalimantan and, most recently, research has been conducted in forest gardens in Central Maluku (Kaya et al. 2002).

Local people around the Lore Lindu National Park in Central Sulawesi have been managing traditional forest gardens over generations, but until now, only little is known about this land use system in this particular region.

In these forest gardens, a type of agroforestry system, a great variety of useful plants are cultivated under a canopy of forest trees and tall fruit trees. The tree cover helps to protect the soil against erosion; the multi-storey, multi-species structure of the forest gardens helps to maintain the natural soil fertility and gives these gardens a forest-like appearance. All in all, the traditional forest gardens of Central Sulawesi seem to be a very stable and sustainable land use system. Before forest gardens can be promoted as an alternative to the common cocoa and coffee plantations, the conditions and circumstances under which these forest gardens are successfully managed have to be investigated. With this aim in view, this paper presents the case studies of three different forest gardens. First, the structure of these forest gardens is analysed, using silvicultural parameters. Then, their species composition is discussed and finally, the history and genesis of the forest gardens are described.

2 Methodology

The study was conducted on Sulawesi Island (2° N – 4° S, 119°–125° E), Indonesia, in the province of Central Sulawesi. The average annual rainfall varies greatly on short distance: from 747 mm in the provincial capital Palu (Whitten et al. 2002) to 2,237 mm in Kulawi (Metzner 1981), only 70 km away.

The research sites are situated in or near the Lore Lindu National Park (218,000 ha). A major part of the National Park lies at an altitude of more than 1,000 m with Gunung Nokilalaki (2,280 m) being the highest mountain.

2.1 Selection of Study Sites

The composition of forest gardens obviously differs, depending on climate (altitude), cultural preferences for certain crops and market access. Considering these parameters, suitable forest gardens were selected in three different villages around the National Park (Table 1). Serving as a reference to compare the forest gardens, a mature undisturbed forest ("natural forest") was selected in the vicinity of each forest garden. To allow maximum comparability of the site factors, the natural forest should be as close as possible to the forest garden. On the other hand, the disturbance of a forest (harvesting of timber, rattan and other forest products) increases with its proximity to the forest gardens and villages. So, a compromise had to be found, with a greater weighting being placed on the undisturbed character of the forest. While the valleys and plains are mostly occupied by villages and cultivated land (including forest gardens), mature undisturbed forest is usually only found in remote, steep, mountainous areas, which explains the differences in altitude between forest garden and natural forest plots (Table 1).

Table 1. Research villages

Village	Rompo	Kamarora	Kulawi
Geographic position	S 01°24' E 120°20'	S 01°12' E 120°08'	S 01°27' E 119°59'
Altitude of forest garden	approx. 1,000 m asl.	approx. 700 m asl.	approx. 600 m asl.
Altitude of natural forest	approx. 1,000 – 1,100 m asl.	approx. 1,000 m asl.	approx. 1,100 – 1,200 m asl.
Ethnic group	Besoa	mixed	Kulawi-Moma
Market access	far, difficult access	close, easy access	intermediate

2.2 Structural Analyses in Forest Gardens and Natural Forests

Inventories were conducted in forest gardens (FG) as well as in natural forests (NF) to provide the basic silvicultural data for structural analyses of these systems. For the inventory, the tree-flora was divided into three compartments of different sizes, a modification of a method described by Brun (1976). The plot design with the compartments is illustrated in Figure 1. Each plot (three plots of FG and three plots of NF) is one ha in size. The NF plots have a square shape, whereas in the

FGs the plot shape had to follow the actual boundaries of the gardens. The plots were permanently marked and then divided into subplots of 20 x 20 m and recording units of 10 x 10m.

In these recording units (Figure 1: compartment A), the following parameters were measured and recorded for all **big trees** (diameter at breast height ≥ 10 cm):

- tree species (local name and scientific name)
- tree diameter at breast height (1.30 m above the ground) (dbh)
- basal area (area of the cross section of the tree(s) at breast height) for individual trees (g) and the whole stand (G)
- tree height (h)
- crown base and crown diameter
- vitality, damages, climber infestation
- coordinates of the trees (distribution in the stand)
- phenological observations (flowering, fruiting, shedding of leaves etc.)

In each subplot (20 x 20 m) one sample (compartment B; 5 x 5 m) was systematically selected to record the **small trees** (dbh < 10 cm and h ≥ 1.30 m). In compartment C (2 x 2 m, two samples per subplot), the number of **seedlings** per species (trees and rattans) was counted. In addition, all the cultivated and useful plants were counted, measured and identified, regardless of their size, as an adaptation of this silvicultural method to inventories in forest gardens.

3 Analysis of Traditional Knowledge of Forest Gardens

To obtain a deeper understanding of the forest gardens, interviews with the forest garden owners were carried out. The focus was on the following questions:

- history and genesis of the forest garden
- management practices
- knowledge of the different uses of plant species (forest plants and cultivated plants)
- knowledge of micro site preferences of plant species
- yields and prices of FG products

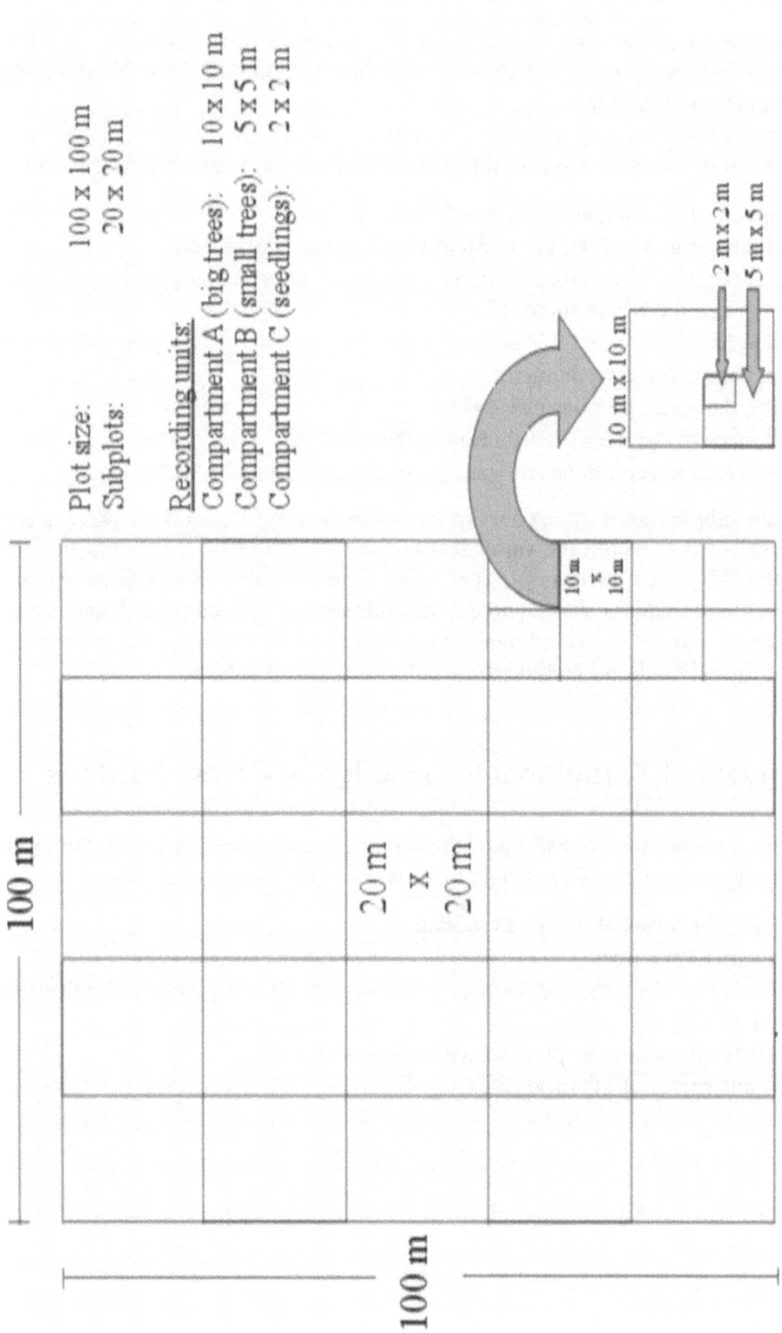

Fig. 1. Design of the sample-plots in forest gardens and natural forests

The results from the interviews about traditional knowledge were verified and completed by "participatory observation". All the research work in the forest gardens (inventory, taking of soil samples, collecting of herbarium specimens etc.) was carried out in direct collaboration with the owner, so that the above-mentioned questions could be discussed on site. The owner and his family continued their usual work in the FG (weeding, tending, harvesting, planting etc.) while the researchers were in the plots. This allowed realistic observations of the actual management practices.

4 Results

The silvicultural parameters describing the stand structure and composition of both natural forests and forest gardens (Table 2) will now be discussed in detail.

4.1 Species Diversity

The number of tree species (dbh ≥ 10 cm) per hectare is 81-92 species in the natural forest and 32-40 species in the forest gardens. The much lower number of species in the forest gardens is due to the fact that the forest garden owners only preserve and tend those species from the natural vegetation that are useful for them and remove the "useless" species. The latter are then replaced with useful species, like fruit bearing or timber producing trees. This fact is also reflected in the high diversity of cultivated and useful plants, which reaches up to 120 species per hectare in the forest garden Kulawi.

The species composition varies not only between the different forest gardens in different regions, but also over time. New species, like coffee *(Coffea canephora)*, cocoa *(Theobroma cacao)* and vanilla (*Vanilla planifolia*) have been introduced in the forest gardens while other species, like betel pepper *(Piper betle)* or betel nut *(Areca catechu)*, lose their importance.

4.2 Stand Basal Area

The stand basal area (area of the cross section at breast height of all the trees) of the forest garden Kulawi (29.4 m^2/ha) is almost as high as in the three natural forest plots (31.6 to 33.1 m^2/ha), which indicates the forest-like character of this forest garden. This high stand basal area in FG Kulawi is obtained with a relatively small number of trees (N=265), compared with the high number of trees in the natural forests (N=471 to 819).

To explain this phenomenon, the diameter distribution of the stands was analysed. The results are presented in Figure 2, which shows the number of stems (N) in 10 cm-diameter classes.

4.3 Diameter Distribution

The diameter distribution curves of FG Kulawi and FG Rompo are quite similar to the "normal" curve of the natural forest: a high number of trees in the small diameter classes, which gradually decreases in the higher diameter classes. It thus can be stated that the structure of these forest gardens is very similar to the structure of a natural forest, even though the total number of trees is slightly smaller. The relatively high basal area in these two FGs is due to an over-proportionally high number of very big trees (dbh $\geq$ 90 cm).

The structure of FG Kamarora is different: with a high diversity of useful plants (82 species) and a tree cover, it looks like a typical forest garden at first glance. But the small number of trees (N=110), the low basal area (G=17.8 m^2/ha) and the imbalance in the diameter distribution (cf. Figure 2) show that this agroforestry system does not have the forest-like structure which is typical for traditional forest gardens.

4.4 Dead Trees

The highest basal area of dead trees is found in the FG Kamarora (3.0 m^2/ha), much higher than in the other forest gardens (0.2 to 1.6 m^2/ha), and also higher than in all three natural forest plots (1.6 to 2.6 m^2/ha), so it can not be explained by natural mortality. In fact, most dead trees in FG Kamarora were girdled by the forest garden owner to kill the trees slowly, in order to open the canopy without damaging the cultivated plants (as would be the case if the big trees were felled). The very small basal area of dead trees in FG Rompo (0.2 m^2/ha), is due to the fact that dead, dry wood is regularly collected and used as firewood. In Kulawi, the most dominant species of dead trees (0.5 m^2/ha) is *Arenga pinnata*, which naturally dies 2-10 years after flowering and tapping. Its stem then remains in the FG because it cannot be used as firewood or otherwise.

Table 2. Silvicultural parameters of natural forests and forest gardens in Central Sulawesi

Parameter	Natural Forest Rompo	Natural Forest Kulawi	Natural Forest Kamarora	Forest Garden Rompo	Forest Garden Kulawi	Forest Garden Kamarora
Stand basal area (G) [m²/ha]	31.6	33.1	32.1	21.6	29.4	17.8
Basal area dead trees [m²/ha]	1.2	2.6	1.6	0.2	1.6	3.0
Number of trees[a]/ha (dbh ≥ 10 cm)	471	819	535	297	265	110
Number of stems[a]/ha (dbh ≥ 10 cm)	484	841	554	374	280	114
Tree species/ha (dbh ≥ 10 cm)	89	81	92	32	40	38
Useful plants/ha [b] (number of species)	---	---	---	58	120	82

[a] The number of stems is higher than the number of trees, because some trees are forked in two or more stems.
[b] Cultivated and wild plants (including tree species) which are used

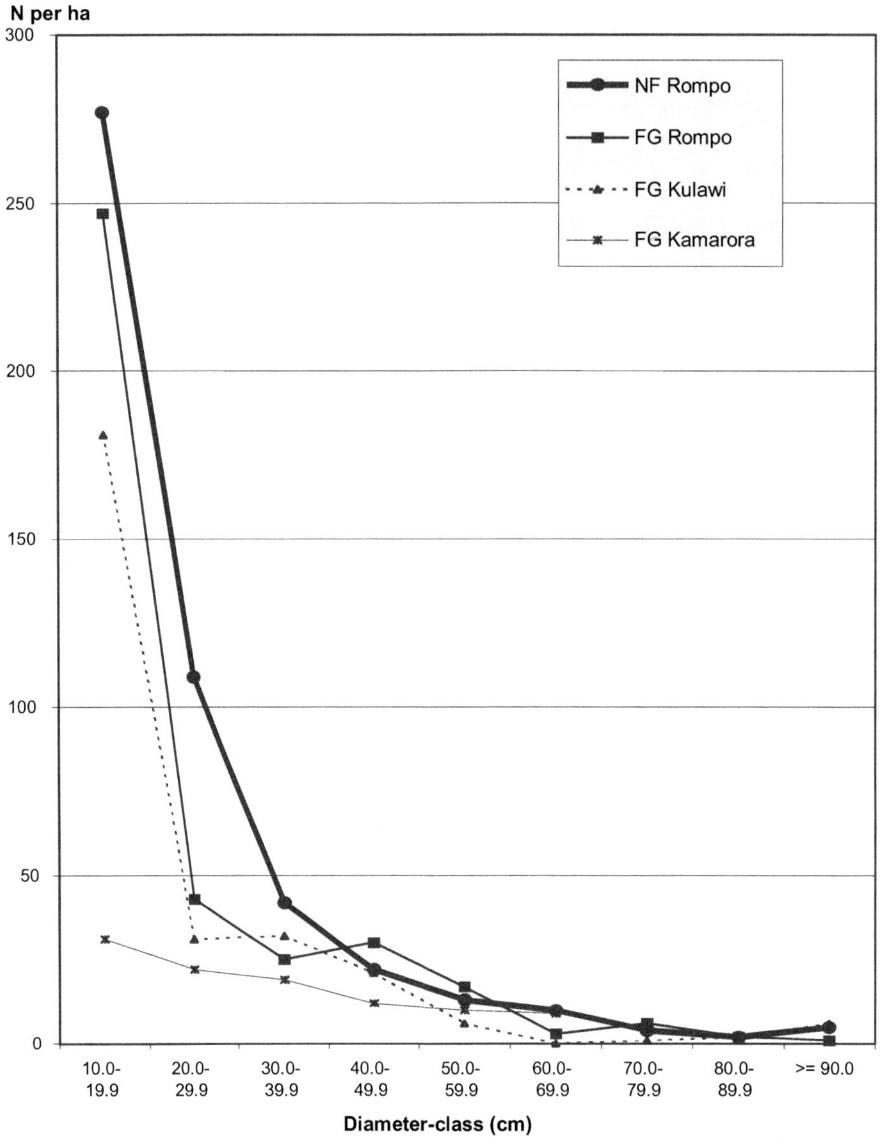

Fig. 2. Diameter distribution curves (number of stems (N) per diameter-class) in three forest gardens (FG) and a mature undisturbed forest (NF)

4.5 Species Composition and Use

So far, only the structure of the stands and the number of trees have been discussed in this paper, regardless of the botanical species. All the tree species in the forest gardens have been ranked by their Importance Value (IV). The most common species and their IVs are presented in Table 3.

Table 3. The most common tree species in three forest gardens (FG) by their Importance Value (IV)

Rank	Species	Dominance [m²/ha]	Abundance [N/ha]	Frequency [%]	IV
I. FG Rompo					
1	*Erythrina subumbrans*	11.29	50	36	89.2
2	*Coffea canephora*	1.56	84	25	49.5
3	*Eucalyptus deglupta*	1.20	12	10	15.2
4	*Acalypha caturus*	0.54	15	12	14.3
5	*Erythrina fusca*	0.38	18	9	12.9
6-32	Other species	6.64	118	87	118.9
1-32	Total	21.61	297	179	300.0
II. FG Kulawi					
1	*Ficus* sp.	14.74	5	5	55.0
2	*Theobroma cacao*	0.89	78	23	45.8
3	*Arenga pinnata*	3.39	27	20	33.3
4	*Cocos nucifera*	1.47	19	15	20.8
5	*Aleurites moluccana*	1.42	13	11	16.1
6-40	Other species	7.46	123	99	129.0
1-40	Total	29.37	265	173	300.0
III. FG Kamarora					
1	*Erythrina subumbrans*	1.35	15	12	33.1
2	*Magnolia* sp.	2.26	4	4	20.3
3	*Phoebe tenuifolia*	1.33	5	5	17.0
4	*Ficus* sp.	2.13	2	2	15.7
5	*Palaquium obovatum*	1.31	3	3	13.1
6-38	Other species	9.45	81	75	200.8
1-38	Total	17.83	110	101	300.0

4.5.1 Forest Garden Rompo

The five most common species are either planted (*Coffea canephora*, *Erythrina* spp.) or cultivated: *Eucalyptus deglupta* occurs in the natural vegetation, but wild saplings are also planted for its valuable timber; *Acalypha caturus* (Euphorbiaceae) occurs naturally but is actively tended, because its young leaves are consumed as a vegetable.

Coffee trees generate cash-income, the two *Erythrina* species provide shade and nutrient-rich litter for the coffee trees and other crops.

4.5.2 Forest Garden Kulawi

All of the five most common species are important useful plants. Cocoa tree, coconut palm *(Cocos nucifera)* and candlenut tree *(Aleurites moluccana)* are planted, while the sugar palm *(Arenga pinnata)* and Ficus tree *(Ficus* sp.*)* occur in the natural vegetation but are tended in the forest garden. With only five trees per hectare, *Ficus* sp. has the highest Importance Value, due to its immense dimensions (dbh up to 260 cm). These huge trees are important in the traditional and cultural life of Kulawi. People believe that the spirits of the dead move from their graves (in the FG) to the Ficus trees. Small branches of these trees provide the raw material for the production of bark-cloth.

Theobroma cacao and *Aleurites moluccana* provide cash income, while the fruit of *Cocos nucifera* are mostly for subsistence use. Palm wine ("saguer") and palm sugar from *Arenga pinnata* are produced partly for personal consumption and partly for sale.

4.5.3 Forest Garden Kamarora

Only the species with the highest Importance Value, *Erythrina subumbrans*, is planted; it serves as a shade tree for cocoa trees. The other four most common species are forest trees, which in some cases (*Magnolia* sp. and *Palaquium obovatum*) provide valuable timber, but apart from that are not used. The manufacture of bark cloth from *Ficus* sp. is not practiced in Kamarora.

The forest gardens in Rompo and Kulawi are old, traditional, well-established land use systems. The most common species are useful trees. Some of them, like *Arenga pinnata* or *Eucalyptus deglupta*, are left over from the natural forest. Other trees from the natural vegetation, which have no use for the local people, have long been replaced by fruit trees (e.g. *Cocos nucifera, Mangifera indica*) or cash crops (e.g. coffee, cocoa, *Aleurites moluccana*). Thus, these forest gardens offer a large variety of uses for their owners and are highly productive.

In FG Kamarora, the main focus is on cash crops, especially cocoa. The cocoa trees were planted only a couple of years ago, so in most cases their diameter (dbh) is still below 10 cm. Consequently, cocoa trees are not counted and measured in compartment A (cf. Figure 1) and do not contribute to the total number of trees or the stand basal area. The trees in the upper and middle layer of this forest garden serve mainly as shade trees for cocoa and other crops; they do not have a particular use themselves.

4.6 Genesis and Development of Forest Gardens

Taking a closer look at the genesis of forest gardens and their development in recent history, deeper insights into their stability can be obtained. The forest garden in Rompo was established some 60 years ago, then still situated at the rainforest margin. Over time, the neighbouring forest was cleared and transformed into agricultural land to grow maize (*Zea mays*), rice (*Oryza sativa*), coffee or cocoa. Only the forest garden itself was preserved in its old structure, so that today the forest garden is isolated, surrounded by agricultural land. While its structure remained more or less the same, the species composition of the forest garden was continually adapted to changing needs and conditions. Thus, this agroforestry system still provides its owner with most of his needs and there is no intention to transform the forest garden to other forms of land use.

In Kamarora, the situation is different. The village is relatively close to the provincial capital Palu and has good access to the markets, so it becomes more profitable for the farmers to concentrate on the cultivation of only few species of cash crops, a phenomenon that Salafsky (1994) also reports from mixed fruit gardens in West Kalimantan. Consequently, the forest garden in Kamarora is slowly being transformed into a cocoa plantation. Many of the big trees from the natural forest have already been removed (cf. the low number of trees and the low basal area, Table 2) to allow more light to penetrate to the ground for the cultivation of cocoa trees. The variety of useful plants is still relatively high (82 species per hectare compared to only 58 in Rompo), but this is mostly due to annual crops like vegetables, which are temporarily grown under the open canopy where a lot of light reaches the ground. Once the cocoa trees are established, the light will become insufficient for growing these vegetables.

The situation in Kulawi is quite different: the forest garden which is analyzed in this paper, and many other FGs in that village, developed from abandoned home gardens. When several floods of the Uerehe River hit the village, people decided to move to another, more elevated location a few hundred meters away to build the village in its present place. While the old houses were abandoned, the home gardens were still tended by their owners, though in a less extensive manner. So, most useful trees and shrubs from the old home gardens continued to grow while other plants from the natural vegetation or from nearby cultivated land appeared spontaneously in the former home gardens. Additional species of useful plants and trees were planted, and so these home gardens gradually developed into the forest gardens that can be found today.

This seems to be the most successful form of forest garden. With a high stand basal area and a forest-like structure, this forest garden harbours the highest variety of useful plants (120 species per hectare). When new crops like cloves (*Syzygium aromaticum*) or rambutan (*Nephelium lappaceum*) were introduced in Kulawi, the farmers rather tried to integrate these new species into the existing structures instead of transforming their forest gardens into plantations. Thus, the traditional forest garden provides not only most of the fruits, vegetables and medical plants for everyday use, but also a source of income from different types of cash crops like coffee, cocoa, vanilla, cloves, candlenut (*Aleurites moluccana*) or java almond (*Canarium indicum*). The big variety of cash crops in the forest garden gives the owner a maximum sustainability of income, quite evenly distributed over the year and mostly unaffected by price fluctuations of individual crops.

5 Discussion and Conclusions

Traditional forest gardens are complex, species-rich agroforestry systems. Even though forest gardens are quite common in Indonesia, their composition, diversity and structure vary from island to island. Data on forest gardens from different parts of Indonesia are compiled in Table 4. A comparison of the data is sometimes difficult, because different authors use a different methodology and different units of reference. In their paper on forest gardens in Central Maluku, Kaya et al. (2002) present the number of species per 200 m^2. Even though the unit of reference in this study was 1 ha, the method of dividing the plots into 10 x 10 m recording units allowed for a detailed analysis of the data for smaller areas and for comparability with other studies. Forest gardens in Central Sulawesi are similar to rattan gardens in East Kalimantan with regard to the number of stems per hectare (256 resp. 289 stems/ha) and in the stand basal area (22.9 resp. 21.7 m^2/ha). These results are far higher than the respective numbers in old dusuns in Central Maluku (150 stems/ha; 18.6 m^2/ha), where "relatively few and tall trees are widely spaced in order to increase production" (Kaya et al. 2002). On the other hand, Kraienhorst (1990) remarks that rattan gardens in East Kalimantan "have a structure which resembles that of a natural rainforest", a result that has also been found for forest gardens in Central Sulawesi.

The average tree diversity in old dusuns in Central Maluku (3.8 species per 200 m^2, dbh $\geq$ 10 cm), however, is higher than in forest gardens in Central Sulawesi (2.7 species per 200 m^2). This fact can partly be explained by a different methodology: in the study in Central Maluku, 90 disjunctive plots of 200 m^2 each were investigated. It can be assumed that only plots with full tree cover were selected. This study in Central Sulawesi, however, was carried out in three forest gardens with a size of 1 ha each, including margins and gaps with smaller numbers of trees and less tree species (cf. Table 4, column "range"). The maximum in both systems is 7 tree species per 200 m^2.

Michon (1985) found a very high stem density (525 $\pm$ 144 stems/ha; dbh $\geq$ 10 cm) in forest gardens ("kebon damar") in South Sumatra. The most important tree species are *Shorea javanica*, *Eugenia aromatica* and a variety of fruit trees. In these gardens, all the cultivated plants are trees, and thus they can form a closed canopy with a high number of stems. In other types of forest gardens, however, the number of trees has to be reduced in favour of other, more light-demanding cultivated plants (vegetables in forest gardens in Central Sulawesi and East Kalimantan, rattan species in rattan gardens in East Kalimantan).

Table 4. Silvicultural parameters of different agroforestry systems in Indonesia

AF system region author	Number and size of (sub)plots	Total area	Number of stems per ha (dbh ≥ 10 cm) [a]	Number of species per 200 m² [b]			Number of species per ha [a]	Stand basal area [m²/ha] [a]
				mean	SD	range		
Forest gardens Central Sulawesi This study	3 plots à 1 ha (150 subplots à 200 m²)	30,000 m²	256 ± 108	2.7	± 1.6	0-7	36.7 ± 3.4	22.9 ± 4.8
Old dusuns Central Maluku Kaya et al. (2002)	90 subplots à 200 m² in 4 villages	18,000 m²	150 ± 8	3.8	± 1.5	1-7	---	18.6 ± 3.9
Lembos East Kalimantan Sardjono (1990)	3 types of Lembos	---	371 ± 33.6	---	---	---	---	---
Rattan gardens East Kalimantan Kraienhorst (1990)	7 plots à 2,500 m²; 1 plot à 1,500 m²	19,000 m²	289	---	---	---	82 spp. per 1.9 ha	21.7
Kebon damar South Sumatra Michon (1985)	3 plots (600 m², 1,000 m², 400 m²)	2,000 m²	525 ± 144	---	---	---	---	---

[a] Mean and standard deviation for n = 3 gardens (Central Sulawesi) resp. 4 villages (Central Maluku) resp. 3 Lembos (East Kalimantan)
[b] Mean and standard deviation for n = 150 (Central Sulawesi) resp. 90 (Central Maluku) subplots

Traditional forest gardens in Central Sulawesi are managed so as to produce a great variety of timber and non-timber forest products in different layers. Having thus maintained their forest-like structure and a high diversity of tree species over generations, it can be stated that these forest gardens are a stable and well-adapted form of land use. In addition, this agroforestry system is both ecologically and economically sustainable: it provides steady yields and income without the use of fertilizers or pesticides. Together with permanent wet-rice cultivation, traditional forest gardens satisfy all the needs of their owners and their families, so that the present population has no need for new clearings. Thus, forest gardens are part of a system that contributes to the stability of rainforest margins in Central Sulawesi. Furthermore, it should be considered whether unproductive secondary forests could be rehabilitated by transforming them into forest gardens.

6 Acknowledgements

The data in the natural forest Kamarora were collected together with Maike Langkau (Göttingen) and Syukur Umar (Palu). The data in the forest garden Kamarora were collected together with Syukur Umar (Palu).

The collected herbarium specimens were identified by Harry Wiriadinata, Herbarium Bogoriense (Bogor) and Ramadhanil Pitopang, Herbarium Celebense (Palu).

References

Arifin YF (1995) The rattan gardens in North Barito District, Central Kalimantan. MSc thesis, University of Göttingen, Germany

Brun R (1976) Methodik und Ergebnisse zur Biomassenbestimmung eines Nebelwald-Ökosystems in den venezolanischen Anden. Proceedings of the XVIth IUFRO World Congress, Oslo

Kaya M, Kammesheidt L, Weidelt HJ (2002) The forest garden system of Saparua Island, Central Maluku, Indonesia, and its role in maintaining tree species diversity. Agroforestry Systems 54: 225-234

Kraienhorst H (1990) Die Rattangärten der Dayak-Benuaq in Ostkalimantan. MSc thesis, University of Göttingen, Germany

Metzner J (1981) Palu (Sulawesi): Problematik der Landnutzung in einem klimatischen Trockental am Äquator. Erdkunde 35: 42-54

Michon G (1985) De l'homme de la forêt au paysan de l'arbre. Agoforesteries indonésiennes. Dissertation, Université des sciences et techniques du Languedoc, France

Salafsky N (1994) Forest gardens in the Gunung Palung region of West Kalimantan, Indonesia. Agroforestry Systems 28: 237-268

Sardjono MA (1990) Die Lembo-Kultur in Ost-Kalimantan: ein Modell für die Entwicklung agroforstlicher Landnutzung in den Feuchttropen. Dissertation, University of Hamburg, Germany

Sundawati L (1993) The dayak garden systems in Sanggau district, West Kalimantan: an agroforestry system. MSc thesis, University of Göttingen, Germany
Weidelt HJ (1990) Rattan growing in SE Asia: an ecologically well adapted form of land use. Plant Research and Development 31: 26-37
Weidelt HJ (1996) Rattan: distribution, morphology, use and ecologically well adapted cultivation. In: Schulte A and Schöne D (eds): Dipterocarp Forest Ecosystems. World Scientific, Singapore, pp 627-647
Whitten T et al. (2002) The ecology of Sulawesi. The ecology of Indonesia series, vol IV. Periplus Editions, Hong Kong

Nutrient Cycling in Tropical Forest Plantations and Secondary Rainforests: The Functional Role of Biodiversity

Martina A. Langi, D. Lamb & R.J. Keenan

1 Introduction

1.1 Key Issues

The work was driven by the need to improve understanding of the structure and functioning of tropical forest ecosystems; and to develop more cost effective approaches to forest ecosystem restoration. The fundamental question was how different are contrasting forest types (forest plantations *versus* secondary rainforest) in terms of ecosystem structure and functioning? The second question was: to what extent are the ecological processes in these differing forests influenced by plant biodiversity? That is, is there any functioning redundancy in the plant diversity present in these tropical forests?

Plantations are commonly established after natural forests have been cleared and the land has been used for other purposes. Such clearing can be expected to reduce the overall fertility of these sites. In fact, the productivity of forest plantations is often high, especially in terms of commercial timber yields. But it is unclear whether nutrient cycles in these plantations have any resemblance to those in natural forests. The issue is complicated by the fact that many tropical plantations are colonized by a large number of native rainforest species. Hence, are enriched plantations more similar to natural forests than plantations with dense understorey? Or, are all systems with a similar overall productivity also similar in terms of nutrient cycling?

1.2 Establishing Plantations in the Tropics

As forest plantations have become increasingly significant to meet economic, social, or ecological demands, many tropical countries (including Indonesia, India, Brazil, and Vietnam) are practically challenged to target some substantial cleared areas for timber plantations by the end of 21st century (Brown et al. 1997, Evans 1992, FAO 2001). Relevant research, which has been extensively carried out for temperate forest plantations, is thus becoming increasingly required for the sustainable production of tropical plantations.

Timber plantations can potentially be designed to meet both economic and restoration goals. Included in such plantations are monoculture plantations of native species, plantation compartments of different species, mosaics of plantation monocultures, plantations of mixed species, and plantations with developed understorey communities (Lamb 1998). Their general aim is to restore ecosystem integrity (including productivity) and to provide for maintenance of biodiversity.

1.3 Biodiversity in Tropical Forests

The significance of biodiversity in ecosystem functioning is, in fact, still the subject of debate. Ewel (1999) has pointed out that some attempts to mimic natural systems with managed ecosystems had sometimes produced opposite effects. In questioning "how much biodiversity is enough", Main (1999) has stressed the need to establish the system goal (e.g. yield or particular ecosystem services), and time frame, in order then to determine the level of biodiversity desired.

The role of biodiversity in forest functioning depends, practically, on species' responses to environmental conditions (e.g. Naeem et al. 1998, Wright 1996). Species that tolerate over a narrow range of environmental conditions perform effectively under a limited set of environmental conditions; and increasing diversity of these species may enhance ecosystem productivity and stability (McNaughton 1993, Odum 1971). Species that tolerate over a broad environmental tolerances can be either indifferent or share similar responses to environmental gradients such as in light (Denslow et al. 1990, Welden et al. 1991), soil drainage (Hubbell and Foster 1986, Lieberman et al. 1985), or soil fertility (Baillie et al. 1987). In mature tropical forest communities with generally diverse habitats, increasing species diversity would not necessarily result in improved ecosystem productivity and stability (McNaughton 1993). The functional redundancy might be high thus ecosystem processes might have an asymptotic relationship with plant diversity (Wright 1996). As a consequence, forest productivity could be largely independent of species richness for the range normally observed in tropical forests (e.g. Berish and Ewel 1988, Brown and Ewel 1987, Jordan 1983).

1.4 Forest Restoration

Attempts on the recovery process can become highly complicated, as the original state of many degraded forests is poorly known. Consequently, the practical target of restoration is often uncertain. Nonetheless, assisting natural processes can accelerate recovery and reduce the risk of further disturbance (e.g. fire, excessive hunting or grazing). In some cases, this may depend on maintaining a given level of soil organic matter (Holt and Spain 1986), or utilising remnants of original forests to act as a source or catalyst for new successions (e.g. Parrotta et al. 1997).

Many attempts to accelerate forest recovery (e.g. Goosem and Tucker 1995, Kooyman 1996, Miyawaki 1993, Sun et al. 1995, Tucker and Murphy 1997, Yu et al. 1994) have focused on relatively small areas, but in much of the tropics ecological restoration is required over large areas of degraded forest. Furthermore, many methods of tropical forest restoration have a purely ecological goal (Goosem and Tucker 1995, Kooyman 1996, Tucker and Murphy 1997). These can also be extremely costly and may not necessarily meet the goal of many forest landowners or managers for ongoing economic production from their forest (Lamb 2001), nor meet increasing demand for wood products. Consequently, there is a need for "commercially attractive" forms of forest restoration that reinstate ecosystem services, enhance biological diversity, and produce income for forest owners.

Different forms of timber plantations coupled with well-managed secondary forests are likely to offer a more attractive forest management option in the future, with every land unit valued from both an ecological as well as a socioeconomic viewpoint. In the tropics, these largely fall into two categories: forest (timber) plantations and agroforestry.

1.5 Scope of the Study

The tropical rainforest of north Queensland contains considerable areas of secondary or regrowth rainforests adjoining timber plantations. Although the plantations were not established originally for restoration purposes, they represent the typical tropical plantation of the region where natural recovery has been taking place without any major disturbance (Keenan et al. 1997). The proximity of the plantations to the remnants of original forests has allowed the development of natural recovery processes. Furthermore, these forest plantations and secondary forests are found at a variety of locations, providing excellent opportunities for comparing ecological processes in different forests and at sites with different soils.

The fundamental aims of the whole study were: (a) to characterize and compare ecosystem structure and functioning in natural secondary rainforest and plantations of two different tree species in the wet tropics of north Queensland, Australia; and (b) to investigate the relationship between different ecosystem functionings produced by different forest types at sites with different soil types. It is hypothesized in the study that forest ecosystem functionings are independent of vegetative diversity represented by the different forest types.

2 Study Sites and Methods

This study was carried out at three sites on the Atherton Tableland, north Queensland; these are Gadgarra ($17^0$18'S, $145^0$43'E), Wongabel ($17^0$19'S, $145^0$30'E), and Danbulla ($17^0$11'S, $145^0$36'E) State Forests. At each site, three forest communities were investigated, i.e. hoop pine (*Araucaria cunninghamii*) plantation, Queensland maple (*Flindersia brayleyana*) plantation, and secondary or regrowth natural rainforest. These three forest types were adjacent to or within a maximum of 150 m from each other. The forest plantations ranged from 55-64 years old, and the adjoining natural rainforests were selectively logged at various times in the past usually following the clearing for the adjacent forest plantations. These plantations were established (within the original density of 600-1000 trees ha^{-1}) at each of the three sites following clearing and burning of the original forests. Table 1 presents the general description of the study area.

As the detailed description can be seen in Langi (2001), the summarized methods employed in this study are indicated as follow:

1. *Soil analyses for physical and chemical properties*: Soil particle size and bulk density were analyzed following those of Anderson and Ingram (1993); total organic C with a combustion analyzer (Leco CNS-2000); exchangeable cations with ICP-AES; and soil pH values (1:5 water) were calculated based on sample proton concentrations.
2. *Vegetation structure and composition*: Sampling plots of 50 x 50 m were established on each of the nine study areas for identification and measurement of all tree $\geq$10 cm dbh and subplots of 5 x 20 m for smaller trees and understorey species.
3. *(Bio)mass estimates*: Tree biomass was estimated based on stand inventory data (Brown 1997, Brown and Lugo 1992) and the published allometric regression equations; shrub and herb biomass by total harvest; root biomass by soil core extraction; forest floor mass by sampling the ground litter, coarse woody debris or fallen logs (line-intersect), and soil organic matter.
4. *Production estimates:* Biomass increment was estimated using an allometric equation on stem circumference growth measured during two years; a two-year litterfall collection was made weekly (during wet months), monthly (during dry months), and fortnightly (during the other months); the total above-ground net primary production (NPP) was estimated by summing up tree biomass increment and litter production during the period of measurement (ignoring the relatively low losses due to herbivores).
5. *Nutrient concentrations of litter*: After collection (as described earlier) the litter samples from each study plot were analyzed (by seasons) for N, P, K, Ca, and Mg concentrations.
6. *Litter decomposition and nutrient dynamics*: Freshly fallen leaf litter of hoop pine (*Araucaria cunninghamii*), Queensland maple (*Flindersia brayleyana*), mixed dominant rainforest species, and *Pinus caribaea* Mor. var. hondurensis Barr. Et Golf. (added as a standard substrate for its resistant nature) were incubated on the forest floor using the standard litterbag technique. Following collection (month 1, 3, 6, 11, and 18), the decomposing litter was oven-dried, ground, and analyzed for N, P, K, Ca, and Mg.
7. *Roots and mycorrhizae*: Root samples – extracted with a stainless steel corer – were sorted by species, size (<2 and ≥ 2 mm diameter as fine and coarse roots), and by nature (live and dead). Subsequent determination was made for root biomass, nutrient contents, root length density, and mycorrhizal infection. Fine root production was estimated from the ingrowth core measurements taken at seasonal intervals over two years using the 24-week incubation period in each season. Annual turnover was estimated as the ratio of production to standing biomass for live and total (live + dead) fine roots.
8. *Soil nutrient availability*: Soil N and P availability was determined using the ion exchange resin (IER) bag procedures collected on each seasons over a two-

year period. The procedures of nitrate and ammonium analyses followed those of Binkley and Matson (1983) and Binkley and Vitousek (1989), while the amount of phosphate was determined using the ascorbic acid-molybdate method (Olsen and Sommers 1982, Watanabe and Olsen 1965). Microbial biomass was determined using the method outlined in Anderson and Ingram (1993); and the microbial biomass C, N, and P was analyzed following those procedures detailed in Brookes et al. (1982, 1985), Hughes and Reynolds (1991), and Vance et al. (1987).

9. *Plant nutrients (nutrient concentration, nutrient retranslocation, nutrient use efficiency)*: From each of the nine study plots, five mature trees were sampled for chemical analyses. Samples of each component (i.e. leaves, branches, wood, and bark) were analyzed for N, P, K, Ca, and Mg. Nutrient concentrations were calculated based on plant dry weight, nutrient retranslocation based on the element to calcium ratios in recent leaf litter versus fresh leaves (Vitousek and Sanford 1986), and nutrient use efficiency (NUE) index as the ratio of litterfall biomass to litterfall nutrient return (Vitousek 1982, 1984).

All values from the nine study plots were calculated based on forest type, site, interaction, and season and year (where applied). Two-way ANOVA was used to compare the values in each stage, and the Tukey's HSD test was applied for further comparison when significant differences were found. The significance level used in all comparisons was 0.05.

3 Results

3.1 Vegetation

The overall density of trees (>10 cm dbh) in the 0.25 ha plots was broadly similar irrespective of forest type (Table 2). This was despite the fact that the stocking of plantation trees varied widely between sites in both *Araucaria* and *Flindersia* plantations. The similarity in overall stocking was maintained by a substantial stocking of new colonists in the two plantation forests; and the densities of these were especially high in the *Flindersia* plantations. The forest types differed rather more in basal area, which was invariably higher in the *Araucaria* plantation sites than either the *Flindersia* plantations or the natural rainforest. A consistent pattern of tree species richness was found at all sites with diversity increasing in the sequence from *Araucaria*→*Flindersia*→rainforest. Understorey diversity was usually higher under *Flindersia* plantation, although most forest types generally exhibited considerable understorey species diversity. Species diversity in rainforest at Wongabel was generally low. Despite its high density, the understorey contained only a relatively small number of species. The vegetation data indicated a clear difference between the three sites. Irrespective of forest type, the Gadgarra sites invariably had higher levels of tree density, basal area and species richness than the other two sites.

3.2 Forest Biomass

Biomass estimations of the tree, understorey, fine and coarse roots, ground litter, fallen logs, and soil organic matter are presented in Figure 1. The largest tree biomass was found in the *Araucaria* plantation, whereas the two broadleaf forest types showed similar levels. Across the sites, Gadgarra supported the greatest tree biomass, followed by Danbulla and Wongabel. The ratio of above- to belowground biomass was similar in the three forest types; however, among the sites this ratio was higher at Wongabel and Gadgarra, and lower at Danbulla. Overall, ecosystem total biomass and organic matter can be differentiated between the higher level in the *Araucaria* plantations and lower level in the *Flindersia* plantations, whereas values in the natural rainforests were between these two forest types.

3.3 Forest Production

Forest productions in term of litterfall, tree diameter increment, aboveground NPP, fine root production and turnover are presented in Figure 2. Forest productivity was high in both the *Araucaria* plantations and the natural rainforests, but invariably low in the *Flindersia* plantations. Among the sites, the highest productivity was shown at Gadgarra followed by Danbulla, and finally Wongabel.

3.4 Decomposition Processes

There were no differences in the rates of litter decomposition between the environments provided by the different forest types and sites (Table 2, selected here is the mixed rainforest litter to represent all litter types being examined). However, there were significant ($P<0.05$) differences in decomposition rates among the litter from different species (Figure 3), and this was mainly a function of initial C:N and lignin:N ratios (Figure 4). The differences between rates of litter production and decomposition throughout the three forest types at three sites have resulted in a range of nutrient fluxes among the nine study plots (Figure 5). For example, a higher rate of litterfall in the *Araucaria* plantation at Gadgarra was not compensated by an equally high rate of decomposition, resulting in lower contents of such nutrients as N in the soil.

3.5 Fine Root Dynamics

Fine root biomass was highest in the *Flindersia* plantations, and lowest in the *Araucaria* plantations (Figure 6). Across the sites, fine root biomass was highest at the least fertile site (Danbulla) and lowest at the most fertile site (Wongabel). Root length density was highest in the natural rainforests, while it was lowest in the *Araucaria* plantations (Figure 7). On the other hand, the value of specific root length was highest in *Araucaria* plantations, while it was similar between the other two forest types (Figure 8). Across the sites, root length density was highest at Danbulla and lowest at Wongabel; whereas specific root length was higher at Gadgarra than at the other two sites, where the values were similar.

Fine root production was highest in the *Araucaria* plantations and relatively similar between the other two forest types (Figure 2). Of the three sites, Gadgarra supported the highest fine root production, and Wongabel the lowest. Fine root turnover rates were highest in the *Flindersia* plantations, while they were lowest in the *Araucaria* plantations (Figure 2). Among the three sites, turnover rates were highest at Gadgarra and lowest at Danbulla.

Root nutrient contents and turnover were generally higher in the *Flindersia* plantations and natural rainforest compared with the *Araucaria* plantations (Table 3). Across the sites, the values were usually higher at Wongabel and lower at Danbulla. The degree of mycorrhizal infection was similarly greater in the *Araucaria* plantations and natural rainforests, and lower in the *Flindersia* plantations (Figure 9). The forests at Danbulla had a greater mycorrhizal infection than those at Wongabel and Gadgarra.

3.6 Nutrient Dynamics

The presence of N and P was less in the soils supporting *Araucaria* plantations than in soils still occupied by natural forests (Figure 10). By contrast, the presence of soil N in *Flindersia* plantations was comparable to that in soils supporting natural forests, even though these *Flindersia* sites had presumably also lost nutrient capital when the original forest had been cleared. Several measures including foliar nutrient concentrations (Table 4), retranslocation rates (Figure 11) and NUE (Figure 12), all suggested the *Araucaria* plantations were subject to a greater degree of nutrient stresses than either the *Flindersia* plantations or the natural (regrowth) forests. The relationship between some of these variables and the respective productivity levels of the three forest types will be discussed further.

4 Discussion

4.1 Ecosystem Functioning in Plantations and Natural Rainforests

Compared with natural forests, plantations are usually more productive because they are immature systems in terms of successional development. However, different plantation species may produce quite contrasting amounts of growth and yield. The study found that the stand biomass of the three forest types differed with higher values being found in the *Araucaria* plantations than the other two forest types; the biomass of the latter were broadly similar. These differences raise several questions; the most obvious one is - why are the *Araucaria* plantations so productive? But equally intriguing is - why is the biomass and productivity of the essentially "natural" (although logged) rainforests apparently so much higher than the *Flindersia* plantations? After all, it is a common perception that plantations are almost always more productive than natural forests. Apparently differences in functional processes between different forest types in this study are related to the functional differences between species (i.e. gymnosperms *versus* angiosperms). Some of these differences such as photosynthesis and shade tolerances, could not be explored in this study. The study did explore, however, some of the differences in nutrient use and turnover in these different forests.

Biomass was not positively related to the availability of nutrients in topsoils of the three forest types but rather to water availability. Higher values were found at the wetter Gadgarra site than at the drier Wongabel site despite the higher levels of nutrient availability at the latter. On the other hand, there were differences in the way nutrients were used and cycled in the three forest types. In acquiring nutrients, for example, there was a greater degree of mycorrhizal infection (Figure 9) in the *Araucaria* plantations than in the *Flindersia* plantations. By contrast, there was a greater fine root biomass in the *Flindersia* plantations and rainforests than in the *Araucaria* plantations. More N retranslocation occurred in *Araucaria* plantations than the other two forest types, and there was a positive relationship between stand biomass and N retranslocation (r^2=0.44). This can be viewed as a form of nutrient use "efficiency".

There are several other indications of the high N use efficiency in the *Araucaria* plantations, and these include a comparatively small amount of nutrients returned in litterfall and a high NUE associated with this litterfall. In general it seems the *Araucaria* plantations relied more heavily on internal recycling of nutrients because of these "efficiencies". Additionally, larger amounts of nutrient were also conserved in litter layers of this forest type compared with the other two forest types. Overall the N cycle in *Araucaria* forests was very "tight" as also reflected by the lower foliar N concentrations. As noted, the fraction of N retranslocated upon senescence in this forest type was the greatest compared with the other forest types. As one measure of nutrient use efficiency, nutrient retranslocation among the three forest types follows a similar pattern to tree biomass. Although some of the site nutrient capital would also have been lost when the *Flindersia* plantation was established, the plantation never achieved the same biomass as the *Araucaria* plantations or indeed the rainforest. Again some of the differences might be due to slower growth, and physiological attributes such as photosynthetic capabilities not assessed here.

In terms of nutrient cycling, however, the differences between the *Flindersia* plantations and the natural forests seem relatively modest. Available soil P was less in the *Flindersia* plantations than in the natural rainforests, but P retranslocation was much greater. Similarly, litter NUE for P was much higher in the *Flindersia* plantations than the rainforest, and the rate of nutrient release from decomposing litter was also slower. These data suggest P was more limiting in the *Flindersia* plantations than in the natural rainforest (and that N was more limiting in the *Araucaria* forest). The greater plant biomass produced in *Araucaria* plantations was for the cost of less nutrient demand than that in the natural rainforest. *Araucaria* gained more of its nutrient requirements through retranslocation, and was less dependent on the decomposition pathway. This is also explained by the much lower fine root biomass but greater mycorrhizal association in *Araucaria* plantation for the activity of nutrient uptake. On the contrary, the rainforest and *Flindersia* retranslocated less nutrients, but the litter decomposed faster thus released more nutrients. The abundant fine roots in the rainforest and *Flindersia* not only reflect their potential to "recapture" released nutrients, but also indicate the forests' dependence on those available nutrients for their nutrient cycling.

High amounts of available nutrients in the rainforest will be taken up by roots, react with soil minerals, caught by soil microbes, or lost through leaching or volatilization. This explains the "necessity" or dependence of rainforest production on an undisturbed nutrient cycle in order to minimize the amount of nutrients lost from the system, and hence the susceptibility of this forest type to disruption such as in clear-felling. On the other hand, *Araucaria* plantations that depend more on internal nutrient pathways might be expected to be more resilient to disturbances. The patterns in the *Flindersia* plantations followed those of the rainforests.

Of the nutrient elements being retranslocated upon senescence, the fraction of P was the largest. This suggests that P might be the most limiting nutrient in these forest ecosystems of north Queensland. Results from mycorrhizal study confirm the low P availability, that is, with the dominance of VA mycorrhizal rather than ectomycorrhizal associations commonly found in environments where soils are low in N (Allen 1991, Harley and Smith 1983, Vogt et al. 1997). Isbell (1994) pointed out the frequent occurrence of P and Fe-Al oxy-hydroxide complexes leading to a reduction in the availability of P in these basaltic krasnozems soils.

All forests at each site experienced disturbances (such as logging and cyclones) practically at the same time, and recovered their own course either as monoculture plantation or secondary rainforest. After more than fifty years, the three forest types shared a similar level of soil organic matter (SOM), indicated partly by the similar microbial C values within all the three forests. The recovery processes appeared to be comparable among the three forests at each site, most probably attributed to the different strategies in adjusting to differing nutrient status of the sites. All forest types in this study generally accumulated a similar amount of nutrients, but natural rainforests tended to cycle nutrients faster than *Araucaria* plantations, which in turn accumulated more biomass.

4.2 The Role of Diversity in Ecosystem Functioning

Many species have colonised the plantations, presumably because all study sites are within 150 m of one another. Thus, plantations that started as monocultures are now more diverse both in numbers of species and in numbers of life forms present, particularly the *Flindersia* plantations (Table 2). While tree density was comparable across the three forest types, the plant biodiversity varied between the more diverse broadleaf forests and the less diverse conifer forest. Some of the ecosystem functional processes measured in this study have been plotted against species richness (for trees $\geq$10 cm dbh) and are presented in Figure 21.

The tree diversity was related most strongly to nitrogen dynamics (i.e. N availability, retranslocation, NUE, and foliar content) followed by decomposition rates, P availability, understorey biomass, tree biomass, and tree biomass. In spite of the difference in degree of relationships (shown by various r^2), those are significant at $P < 0.05$ for the factors referred to above. Meanwhile, the effect of tree diversity is not significant ($P < 0.05$) on fine root production and phosphorus dynamics (i.e. P retranslocation and foliar P).

There are indications that some of the relationships between tree diversity and the various attributes displayed were masked by site differences. Since there were only three forests represented at each site it is difficult to explore these patterns in great detail. However the following variables had positive relationships with tree diversity at particular sites:

- Nutrient dynamics (especially N, and also decomposition processes and available P) - this suggests that a diverse range of litter materials promoted better organic decomposition and mineralisation in topsoils.
- Understorey biomass - this probably reflects the fact that a higher biomass is simply a consequence of a greater degree of understorey colonisation.
- Fine root biomass - probably because a higher tree stocking (irrespective of diversity) yields a larger fine root biomass.
- Soil organic matter - perhaps another reflection of the enhanced decomposition noted earlier with respect to available N.
- Foliar N and P – may be caused by improvements in the availability of topsoil nutrients.

The other general trend across the three forest types suggests that the differences were stronger *between Flindersia* plantations and natural rainforests as one side *and* the *Araucaria* plantation as the other than *between* plantation forests *and* natural forests. This suggests that the number of species may not be as important as the role of particular species in ecosystem functioning within the mature tropical forest ecosystems (*see* also Berish and Ewel 1988, Cuevas *et al.* 1991, Jordan 1983, Lugo 1992, Michelsen *et al.* 1996, Silver *et al.* 1996).

There were, in fact, also some insignificant relationships at each site between overall tree diversity and fine root production and some phosphorus dynamics. These are probably a consequence of more intense competition for resources, such as light and nutrients, due to the greater tree densities resulting from colonisation rather than from biodiversity *per se.* There are many cases in tropical forest ecosystems where species' tolerances, responses, and ecological niches broadly overlap (e.g., Baillie *et al.* 1987, Denslow *et al.* 1990, Hubbell and Foster 1986, Lieberman *et al.* 1985, Welden *et al.* 1991). As a consequence, numerous species become either indifferent *or* highly competitive to environment gradients. In some cases of this study certain ecosystem functional processes are within a close range across a spectrum, including the plantations of different species and the adjacent natural rainforest.

5 Conclusion

The major responses to the original questions posed in the beginning of this study are therefore as follow.

1. The different forests in this study had contrasting patterns of biomass accumulation, productivity, and nutrient cycling, but the differences were not based simply on plantation forests *versus* natural forest. Rather there was a contrast between the conifer and the broadleaf forests. The cycling of nutrients was much faster in the natural and broadleaf plantation forests than in the conifer forests. Despite all being established after clearing and burning an original rainforest cover, the *Araucaria* plantations appeared to be more limited by N resources while the *Flindersia* plantations were limited by P resources.
2. Ecosystem functional processes that are directly affected by soil nutrient status - such as litter decomposition, nutrient availability and efficiency – are more affected by species diversity. On the other hand, ecosystem attributes such as forest biomass and productivity were affected more by age and physical environment, such as water availability. These processes are less affected by species diversity. This may confirm that the issue of biodiversity is more significant for forest plots where land rehabilitation is the primary objective, but less significant where production is the main concern.

6 Acknowledgements

The authors would like to acknowledge the roles of the CRC for Tropical Rainforest Ecology and Management, the DPI Forestry (QFRI) and the CSIRO (TFRC) at Atherton, Australia for financially supporting and facilitating the whole research.

References

Allen MF (1991) The ecology of mycorrhizae. Cambridge University Press, Cambridge

Anderson JM, Ingram JSI (eds) (1993) Tropical soil biology and fertility: a handbook of methods. Commonwealth Agricultural Bureau International, Aberystwyth

Baillie IC, Ashton PS, Court MN, Anderson JAR, Fitzpatrick EA, Tinsley J (1987) Site characteristics and the distribution of tree species in mixed dipterocarp forest on Tertiary sediments in central Sarawak, Malaysia. Journal of Tropical Ecology 3: 210-220

Berish CW, Ewel JJ (1988) Root development in simple and complex tropical successional ecosystems. Plant and Soil 106: 73-84

Binkley D, Matson P (1983) Ion exchange resin bag method for assessing forest soil nitrogen availability. Soil Science Society of America Journal 47: 1050-1052

Binkley D, Vitousek P (1989) Soil nutrient availability. In: Pearcy RW (ed) Plant Physiological Ecology: field methods and instrumentation. Chapman & Hall, London, pp 75-96

Brookes PC, Landman A, Pruden G, and Jenkinson DS (1985) Chloroform fumigation and the release of soil nitrogen: a rapid direst extraction method to measure microbial biomass nitrogen in soil. *Soil Biology and Biochemistry* 17: 837-842.

Brookes PC, Powlsan DS, Jenkinson DS (1982) Measurement of microbial biomass phosphorus in soil. Soil Biology and Biochemistry 14: 319-329

Brown A, Lugo AE (1992) Aboveground biomass estimates for tropical moist forests of the Brazilian Amazon. Interciencia 17: 8-18

Brown AG, Nambiar EKS, Cossalter C (1997) Plantation for the tropics. In: Nambiar EKS, Brown AG (eds) Management of soil, nutrients, and water in tropical plantation forests. ACIAR Monograph No 43

Brown BJ, Ewel JJ (1987) Herbivory in complex and simple tropical successional systems. Ecology 68: 108-116

Brown S (1997) Estimating biomass and biomass change of tropical forests. A Primer. FAO-UN, Rome

Cuevas E, Brown S, Lugo AE (1991) Above and belowground organic matter storage and production in a tropical pine plantation and a paired broadleaf secondary forest. Plant and Soil 135: 257-268

Denslow JS, Schultz JC, Vitousek PM, Strain BR (1990) Growth responses of tropical shrubs to treefall gap environments. Ecology 71: 165-179

Evans J (1992) Plantations forestry in the tropics. Clarendon Press, Oxford

Ewel JJ (1999) Natural systems as a model for the design of sustainable systems of land use. Agroforestry Systems 45: 1-21

FAO (2001) Global forest resources assessment 2000. Main report, FAO Forestry Paper No 140. FAO, Rome

Goosem S, Tucker N (1995) Repairing the rainforest: theory and pratice of rainforest re-establishment in North Queensland's wet tropics. Wet tropics management authority, Cairns

Harley JL, Smith SE (1983) Mycorrhizal symbiosis. Academic Press, London

Holt JA, Spain AV (1986) Some biological and chemical changes in a north Queensland soil following replacement of rainforest with *Araucaria cunninghamii* (Coniferae: Araucariaceaea). Journal of Applied Ecology 23: 227-237

Hubbell SP, Foster RB (1986) Commoness and rarity in a neotropical forest: implications for tropical tree conservation. In: Soule M (ed) Conservation biology: Science of scarcity and diversity. Sinauer, Sunderland (MA)

Hughes S, Reynolds B (1991) Effects of clearfelling on microbial biomass phosphorus in the Oh horizon of an afforested podsol in Mid-Wales. Soil Use and Management 4: 183-188

Isbell RF (1994) Krasnozems: a profile. Australian Journal of Soil Research 32: 915-929

Jordan CF (1983) Productivity of tropical rain forest ecosystems and the implications for their use as future wood and energy sources. In: Golley FB (ed) Ecosystems of the world. Elsevier, Amsterdam

Keenan R, Lamb D, Woldring O, Irvine T, Jensen R (1997) Restoration of plant biodiversity beneath tropical tree plantations in Northern Australia. Forest Ecology and Management 99: 117-131

Kooyman R (1996) Growing rainforest: rainforest restoration and generation. Greening Australia and state forests of new South Wales, Australia
Laffan MD (1988) Soils and land use on the Atherton tableland, north Queensland. CSIRO Division of Soils, Soils and Land Use Series 61. CSIRO, Canberra.
Lamb D (1998) Large-scale ecological restoration of degraded tropical forest lands: the potential role of timber plantations. Restoration Ecology 6: 271-279
Lamb D (2001) Reforestation: ways of retrieving biodiversity at degraded sites. Encyclopedia of Biodiversity 5: 291-1 to 291-12
Langi MA (2002) Nutrient cycling in tropical plantations and secondary rainforests. Ph.D. thesis, University of Queensland (Australia)
Lieberman D, Lieberman M, Peralta R, Hartshorn GS (1985) Mortality patterns and stand turnover rates in a wet tropical forest in Costa Rixa. Journal of Ecology 73: 915-924
Lugo AE (1992) Comparison of tropical tree plantations with secondary forests of similar age. Ecological Monographs 62: 1-41
Main AR (1999) How much biodiversity is enough? Agroforestry Systems 45: 23-41
McNaughton SJ (1993) Biodiversity and function of grazing ecosystems. In: Schulze ED, Mooney HA (eds) Biodiversity and ecosystem function. Springer, New York
Michelsen A, Lisanework N, Friis I, Holst N (1996) Comparison of understorey vegetation and soil fertility in plantations and adjacent natural forests in the Ethiopian highlands. Journal of Applied Ecology 33: 627-642
Miyawaki A (1993) Restoration of native forest from Japan to Malaysia. In: Leith H, Lohman M (eds) Restoration of tropical forest ecosystems. Kluwer Academic Publ., Dordrecht (The Netherlands)
Naeem S, Kawabata Z, Loreau M (1998) Transcending boundaries in biodiversity research. Trends in Ecology and Evolution 13: 134-135
Northcote KH (1971) A Factual key for the recognition of Australian soils, 3rd edn, Rell Technical Publication, Glenside
Odum EP (1971) Fundamentals of ecology. Saunders, Philadelphia
Olsen SR, Sommers LW (1982) Phosphorus. In: Page AL et al. (eds) Methods of soil analysis: chemical and microbiological properties. American Society of Agronomy, Madison, pp 403-448
Orians GH, Dirzo R, Cushman JH (1996) Impact of biodiversity on tropical forest ecosystem processes. In: Mooney HA, Cushman JH, Medina E, Sala OE, Schulze ED (eds) Functional roles of biodiversity. John Wiley & Sons
Parrotta JA, Knowles OH, Wunderlie JM (1997) Development of floristic diversity in 10 year old restoration forests on a bauxite mined site in Amazonia. Forest Ecology and Management 99: 21-42
Silver WL, Brown S, Lugo AE (1996) Biodiversity and biogeochemical cycles. In: Orians GH, Dirzp R, Cushman JH (eds): Biodiversity and ecosystem processes in tropical forests. Springer, Berlin
Sun D, Dickinson GR, Bragg AL (1995) Direct seeding of *Alphitonia petrei* (Rhamnaceae) for gulley revegetation in tropical northern Australia. Forest Ecology and Management 73: 249-257
Tucker NJ, Murphy TM (1997) The effect of ecology rehabilitation on biodiversity recruitment: some observations from the wet tropics of north Queensland. Forest Ecology and Management 99: 133-152

Vance ED, Brookes PC, Jenkinson DS (1987) An extraction method for measuring soil microbial biomass C. Soil Biology and Biochemistry 19: 703-707

Vitousek PM (1982) Nutrient cycling and nutrient use efficiency. American Naturalist 119: 553-572

Vitousek PM (1984) Litterfall, nutrient cycling, and nutrient limitation in tropical forests. Ecology 65: 285-298

Vitousek PM, Sanford Jr RL (1986) Nutrient cycling in moist forest. Annual Review of Ecology and Systematics 17: 137-168

Vogt K, Asbjornsen H, Ercelawn A, Montagnini F, Valdés M (1997) Roots and mycorrhizas in plantation ecosystems. In: Nambiar EKS, Brown AG Management of soil, nutrients, and water in tropical plantation forests. ACIAR Monograph No 43

Watanabe FS, Olsen SR (1965) Test of an ascorbic acid method for determining phosphorus in water and $NaHCO_3$ extracts from soil. Soil Science Society of America Proceedings 29: 677-678

Welden CW, Hewett SW, Hubbell SP, Foster RB (1991) Sapling survival, growth, and recruitment: relationship to canopy height in a neotropical forest. Ecology 72: 35-50

Wright SJ (1996) Plant species diversity and ecosystem functioning in tropical forests. In: Orians GH, Dirzp R, Cushman JH (eds) Biodiversity and ecosystem processes in tropical forests. Springer, Berlin

Yu ZY, Wang SH, He SY (1994) Rehabilitation of eroded tropical coastal land in Guangdong, China. Journal of Tropical Forest Science 7: 28-38

The Use of Models to Assess the Impact of Land Use Change on Ecological Processes: Case-Studies of Deforestation in South-East Asia

Peter H. Verburg, A. Veldkamp, W. Engelsman, R. van Zalinge, M.E.F. van Mensvoort & K.P. Overmars

1 Introduction

Landscape ecology is based on the analysis of the linkages between spatial pattern and ecological processes within a landscape (Turner, 1989). Three landscape characteristics are important to the linkage between pattern and process: structure, function and change (Forman and Godron 1986). "Structure" refers to the spatial relationships between components of a landscape, that is, the distribution of energy, materials, and species in relation to the size, shape, numbers, kinds and configuration of components. "Function" refers to the interactions between the spatial elements, such as the flow of energy, materials, and organisms among the components of the landscape. "Change" refers to alteration in the structure and function of the landscape mosaic through time. Understanding the structure, function and change of landscapes helps to determine the relationship between landscape pattern and ecological process and has been the main research item in many studies of ecological systems.

Human activities have large impacts on the structure, function and change of landscapes (Kasperson *et al.* 1995). Most directly, land use change (i.e., the modification of the way land is used, such as clearing of forest for agricultural use) has immediate consequences for the spatial structure of the landscape. Land use change, therefore, influences many ecological processes through its effect on habitat fragmentation, diversity, patch shape and size etc. One of the most important land use change processes, deforestation, is seen as the main threat to global biodiversity through its effects on the habitat of species.

The importance of land use change as a determinant of landscape pattern makes it essential to have reliable methods to project land use change. Such methods are an important element of the assessment of potential future changes in the functioning of landscapes.

An increasing number of methods and models are available to simulate changes in the spatial pattern of land use at multiple scales (e.g., Veldkamp and Fresco 1996; Pontius *et al.* 2001; Veldkamp and Lambin 2001a; Pijanowski *et al.* 2002). In this paper we will describe how such spatially explicit land use change models can contribute to landscape ecological assessments. One land use change model, CLUE-s (Verburg *et al.* 2002a), is used for two applications in South-East Asia to illustrate the potential value of land use change simulations for ecological impact assessments.

2 Land Use Change Modelling

Reviews of land use change models are provided by Lambin (1997) and Kaimowitz and Angelsen (1998) for deforestation, Miller *et al.* (1999) for integrated urban models, Lambin *et al.* (2000) for agricultural intensification models, Bockstael and Irwin (2000) for land use models based on economic theory and by Parker *et al.* (2003) for multi-agent models. More general reviews are provided by Agarwal *et al.* (2001), Briassoulis (2000) and Verburg *et al.* (2002b).

The CLUE-S model, used in this study, is a land use change model that has already been used and validated in a wide range of applications (Verburg *et al.* 2000; Kok and Winograd 2001; Veldkamp *et al.* 2001b). The model is especially useful for assessments of changes in ecological pattern because of the explicit attention that is given to the temporal and spatial dynamics that lead to changes in land use pattern.

The CLUE-s model (CLUE-s: the Conversion of Land Use and its Effects at Small regional extent) is a special version of the CLUE modelling framework (Veldkamp and Fresco 1996) for the simulation of changes in land use pattern at the landscape scale. In this paper we will give a short overview of the functionality of the CLUE-s model, a more detailed description of the model is provided by Verburg *et al.* (2002a). Figure 1 shows the components of the CLUE-s model. Together, these components create the set of conditions on which the calculation of the changes in land use pattern is based.

Land use requirements (demand) are calculated at the aggregate level (the level of the case-study as a whole) as part of a specific scenario. The land use requirements constrain the simulation by defining the totally required change in land use, i.e. all land requirements defined at the aggregate level are allocated at the pixel level. All changes in individual pixels should add up to these requirements. In our approach, land use requirements are calculated independently from the CLUE-s model itself and can be based on a range of methods, depending on the case study and the scenario, e.g. the extrapolation of trends in land use change of the recent past into the near future is a common technique to calculate land use requirements.

Location characteristics determine the relative suitability of a location for the different land use types. The relative probability of finding a land use type at a particular location is related to the biophysical and socio-economic conditions of that location. The relation between the relative probabilities and the biophysical and socio-economic location characteristics is defined in a logit model. The coefficients are estimated through logistic regression using the actual land use pattern as dependent variable. This method is similar to econometric analysis of land use change, which is common in deforestation studies (e.g. (Chomitz *et al.* 1996; Nelson *et al.* 1997; Geoghegan *et al.* 2001; Nelson *et al.* 2001). In our case studies we assume that locations are devoted to the land use type with the highest 'suitability'. 'Suitability' includes the monetary profit, but can also include cultural and other factors that lead to deviations from (economic) rational behaviour in land allocation. This assumption makes it possible to include a wide variety of location

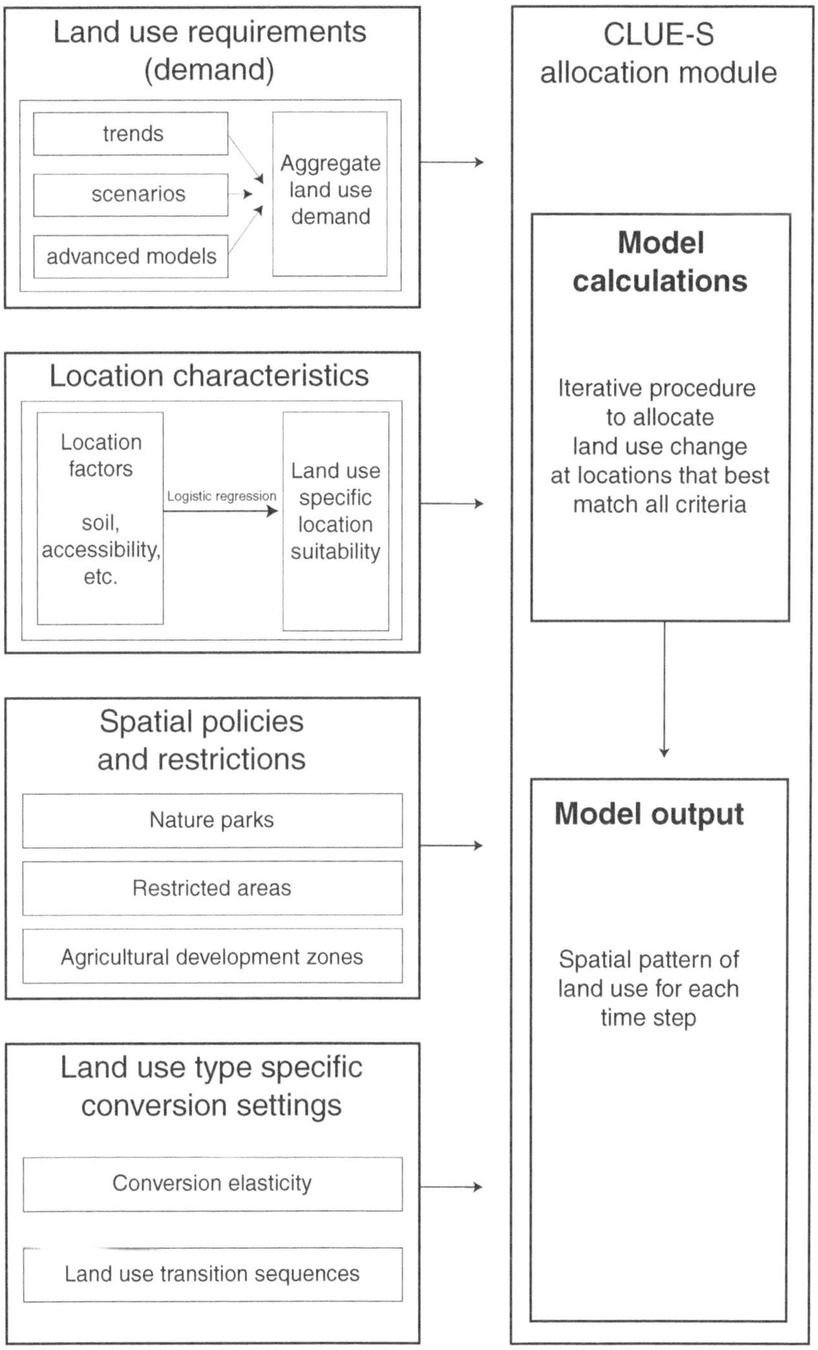

Fig. 1. Structure of the CLUE-s model

characteristics to estimate the logit function that defines the relative probabilities for the different land use types.

Spatial policies and restrictions indicate areas where land use changes are restricted through spatial policy or tenure status. Case studies have shown that spatial policies and land tenure can influence the pattern of land use change (McMorrow *et al.* 2001). Maps that delineate the area for which the spatial policy is implemented must be supplied. Some spatial policies restrict all land use change in a certain area, e.g., a log-ban within a forest reserve. Other land use policies restrict a set of specific land use conversions, e.g., residential construction in designated agricultural areas or permanent agriculture in the buffer zone of a nature reserve. The conversions that are restricted by a certain spatial policy can be indicated in a land use conversion matrix: for all possible land use conversions it is indicated if the spatial policy applies.

Land use type specific conversion settings determine the temporal dynamics of the simulations. Two sets of parameters are needed to characterize the individual land use types: conversion elasticities and land use transition sequences. The first parameter set, the conversion elasticities, is related to the reversibility of land use change. Land use types with high capital investment will not be easily converted into other uses as long as there is sufficient demand. Examples are residential locations but also plantations with permanent crops (e.g., fruit trees). Other land use types easily shift location when the location becomes more suitable for other land use types. Arable land often makes place for urban development while expansion of arable land occurs simultaneously at the forest frontier. An extreme example is shifting cultivation: for this land use system the same location is mostly not used for periods exceeding two seasons as a consequence of nutrient depletion of the soil. These differences in behaviour towards conversion can be approximated by conversion costs. However, costs cannot represent all factors that influence the decisions towards conversion such as nutrient depletion, aesthetic values etc. Therefore, for each land use type a value needs to be specified that represents the relative elasticity to change, ranging from 0 (easy conversion) to 1 (irreversible change). The user should decide on this factor based on expert knowledge or observed behaviour in the recent past.

The second set of land use type characteristics that needs to be specified is the allowed land use transition sequence. Not all land use changes are possible or realistic (e.g., arable land cannot be converted into primary forest directly) and many land use conversions follow a certain sequence. Examples include fallow land and forest regrowth following shifting cultivation. These sequences are specified in a land use conversion matrix. For each land use type it is indicated in what other land use types it can change during the next time step.

When all input is provided the CLUE-s model calculates, with discrete time steps, the most likely changes in land use given the above described restrictions and suitabilities. Some changes are irreversible, while others are dependent on the changes in earlier time steps. Therefore, the simulations tend to result in complex, non-linear changes in land use pattern, characteristic for complex systems.

3 Applications

3.1 Location of the Case Studies

The use of the CLUE-s model is illustrated with two case-studies in South-East Asia: Selangor river basin in Malaysia and Sibuyan Island in the Philippines. The location of both case studies is shown in Figure 2.

Fig. 2. Location of the case studies in South-East Asia

Selangor river basin is located in the state of Selangor of Peninsular Malaysia just north of Kuala Lumpur, the national capital. Selangor state is the most densely populated state of Malaysia facing fast expansion of urban land uses. The study area covers an area of 1,617 square kilometres and has a diverse landscape. The eastern part is hilly with large areas of forest remaining on the hill slopes. The central and western part of the area is flatter with large areas of land covered by oil palm and rubber estates. Currently many of these estates are being converted into urban, residential, recreational and industrial areas. The area has a long history of tin mining. The English colonists established the tin mines in the end of the 19th century, which spurred Malaysian economic development. Large areas without vegetation remain after these mines are abandoned. The remaining forest areas are under large pressure due to urban expansion. These dynamics make the area relevant for an analysis of possible future pathways of land use change.
The second case study, Sibuyan island, is located in Romblon Province of the Philippines. The island measures 28 km east to west at its widest point and 24 km north to south, with a land area of approximately 456 km^2 surrounded by deep water. Steep mountain slopes covered with forest canopy characterize the island. The land surrounding the high mountain slopes gently to the sea and is mainly used for agricultural, mining and residential activities. The island was selected as a case-study because of its very rich biodiversity. There are an estimated 700 vascular plant species including 54 species endemic to Sibuyan Island and 180 species that are endemic to the Philippine archipelago. Fauna diversity is low, but endemism is high (Velasco et al. 2000). This makes the island a 'hot spot' for nature conservation and relevant for a detailed study of land use change.

3.2 Land Use Change Simulations for the Selangor River Basin, Malaysia

3.2.1 Model Configuration

All analysis for this case-study is made at a spatial resolution of 750x750 meter. This resolution was chosen based on the scale of the available data. Eight land use types were distinguished and dynamically modelled in simulations from 1999 to 2014. These land use types include forest, oil palm plantations, rubber plantations, mixed cultures, scrub-grassland, mining area, urban/residential area and water.

Location factors. The suitability of locations for the different land use types is determined by relating the different land use types to a set of location factors. For this case-study we have included a set of accessibility measures (distance to road, river, center of residence, forest), indicators of the socio-economic conditions (population pressure and labor force), altitude and the suitability of the different soils in the area. Based on logistic regression relations were established between these location factors and land use. These empirical relations are used as input for the dynamic simulation model.

Allowed land use change sequences. A number of unrealistic land use conversions are excluded to limit the possible range of conversions made during the model simulations. In the model it is assumed that urban/residential area cannot be converted into other land uses. Furthermore the water surfaces are assumed to be constant and the former mining areas are assumed to be unsuitable for agricultural use.

Conversion elasticities. Different conversion elasticities are assigned to the different land use types based on the specific characteristics of the land use types. Oil palm and rubber plantations are not easily converted into other land uses due to the investments made in the plantations and have therefore been given relatively high conversion elasticity. Scrub-grassland has been assigned a low conversion elasticity and mixed culture an intermediate value.

Scenarios. Two different scenarios of plausible developments in land requirements during the period between 1999 and 2014 have been generated and used as input to simulate the changes in the spatial pattern of land use. The first scenario is based on a linear extrapolation of the changes in land requirements between 1990 and 1999. During this period large increases in urban/residential area occurred at the cost of forest and agricultural land use types. The continuation of these trends is most likely, since the population projections for the area indicate population growth rates for the next 20 years that equal the growth rates between 1990 and 1999. In the second scenario it is assumed that forest within the official forest reserves (see Figure 3) is well protected. Land use change is assumed to occur only outside these protected areas. The land requirements are therefore somewhat different from the first scenario: only slight decreases are expected in forest area while urban expansion occurs at the same rate as above. Instead of the forest, the areas for agricultural use and scrub-grassland are expected to decrease.

3.2.2 Simulation Results

Figure 3 shows the extent of the two main land use types, forest and urban/residential area, for different years. The 1990 and 1999 data are based on respectively a land use map and a satellite image interpretation. The two maps for 2014 are simulated by the CLUE-s model for the scenarios based on respectively trend-extrapolation and the protection of forest reserves. The results illustrate that forest protection in the center of the study area is very difficult due to the large pressure on land use in this part of the area. Already in 1999 significant areas designated as forest reserve are no longer forest. From the simulations it is clear that in the central and western part of the study area only very small fragments of forest will remain if protection is not enforced. The forest on the hill slopes in the eastern part of the study area suffers less loss except for the area along the highway in the north-eastern part of the area. The hill slopes are much less suitable for urbanization and agricultural use. Furthermore, this area is not located along the main transportation corridor. Further urbanization will, most likely, give rise to a more intensive use of these forest areas for recreational purposes. Under both scenarios all forest will disappear in the western part of the study area due to expansion of the large-scale oil palm plantations in this area. In the 'forest protection' scenario large areas that were formerly used for tin mining are converted to urban area. Although this conversion is rather costly it is a feasible option if land pressure increases. This conversion is also seen in other parts of the area surrounding Kuala Lumpur, where urbanization has already increased the pressure on remaining forestlands considerably.

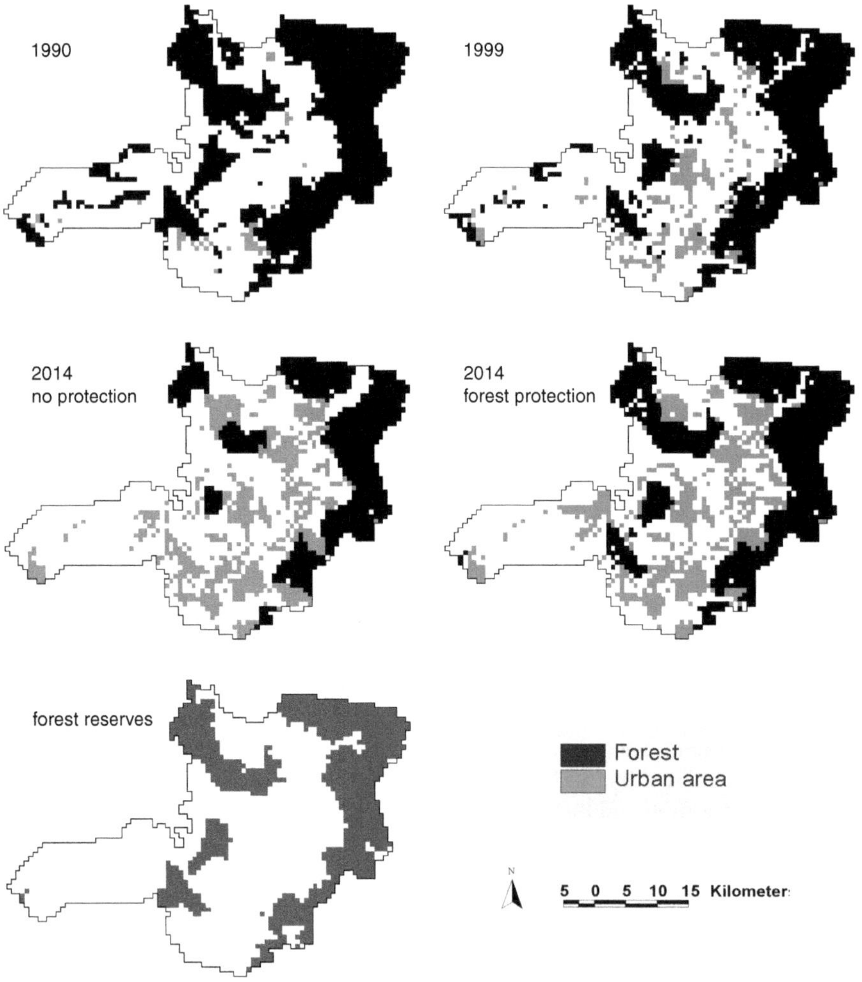

Fig. 3. Location of forest and urban area for different years: 1990 and 1999 are based on observed data and the maps for 2014 are based on the simulation of two scenarios.

3.3 Land Use Change Simulations for Sibuyan Island, the Philippines

3.3.1 Model Configuration

For this application all simulations are made at a high spatial resolution, namely 100 x 100 meter, based on a land use map derived from a participatory 3-D Model of the island (Rambaldi *et al.* 2000a; Rambaldi *et al.* 2000b). This land use map was reclassified in 8 different land use classes that were dynamically modelled: mangroves, coconut, rice, grassland, secondary forest, shifting cultivation, primary forest and a class of miscellaneous land uses.

Location factors. Data were collected for factors that were supposed to influence the allocation decisions of land users based on general knowledge derived from case-studies in literature (e.g., (Geist *et al.* 2002), a rapid rural appraisal in the study area (Velasco *et al.* 2000) and a number of informal interviews. The biophysical factors are related to landform. The rugged terrain of the island makes slope and altitude important factors influencing the suitability of the land for agricultural use (Figure 4). Other biophysical factors included are the geomorphology and geology of the island. Distance to rivers was included because agricultural land is often located in river valleys, where water is available for wetland rice cultivation and access is easy. The socio-economic conditions influencing land allocation decisions are represented by accessibility and population pressure (two indices based on population numbers and village territories). Logit models were used to link the location factors to the land use distribution. The high data quality and the selection of a large number of relevant variables, based on knowledge of the island, resulted in logit models that are highly capable to explain the land use pattern. These logit models were used as input for the dynamic simulation model.

Allowed land use change sequences. Land use types considered in this case study are related to each other in fixed sequences. Secondary forest can develop from grasslands or fallow land as long as burning and grazing are restricted. New areas of shifting cultivation are most likely to develop through the clearance of secondary or primary forest. After a short period of cultivation they are abandoned or converted into permanent agriculture. Based on this type of knowledge we classified each possible conversion either as realistic or unrealistic. This information was used as model input.

Conversion elasticity. High conversion elasticities were assigned to rice fields and coconut plantations based on the observation that these land use types are not easily converted into other land use types due to the investments made in land preparation. Primary forest is assumed be highly inelastic as a consequence of the time involved in regrowth from secondary vegetation. Grassland and shifting cultivation are assumed to be highly dynamic land use types and have been assigned a relatively low elasticity.

Scenarios. In this study we have evaluated three different scenarios representing different degrees of implementation of nature conservation policies.

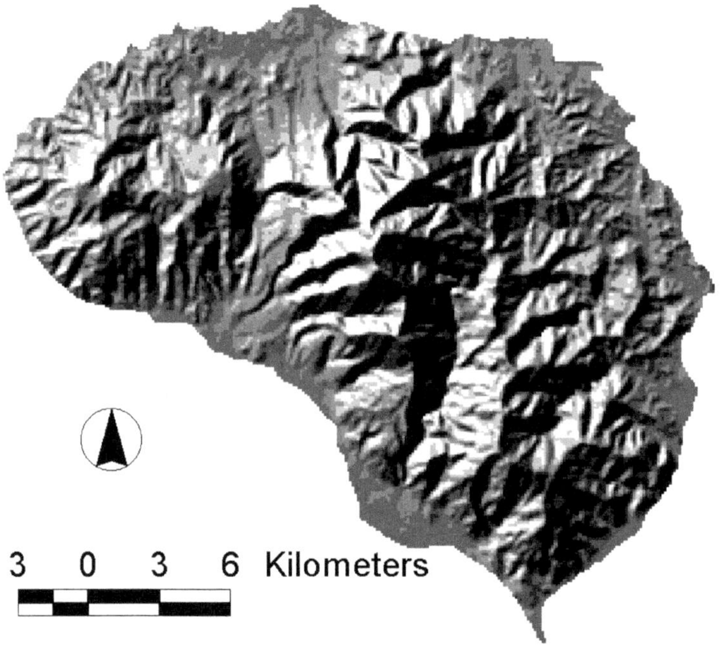

Fig. 4. Hillshade view of Sibuyan Island

Scenario 1: Baseline. The land requirements are assumed to follow the current trend of land use change: agricultural expansion and deforestation, both through land clearance for shifting cultivation and (illegal) wood extraction. The land requirements are calculated by linear extrapolations of the developments of the past twenty years, leading to increases in rice area, coconut plantations, shifting cultivation and grassland while decreases occur in mangrove, primary and secondary forests. In this scenario no spatial policies or restrictions are included. All land use changes are, therefore, allocated based on the suitability of locations as determined by the location factors, the existing land use pattern and competition between land use types.

Scenario 2: Spatial policies. For this scenario the same land requirements are assumed as in the baseline scenario. The spatial allocation of these requirements is, however, restricted by a number of spatial policies. The major restriction to land use change is the establishment of Mt. Guiting-Guiting natural park in the central part of the island. This area covers about 15,000 hectares, approximately one third of the island, and was proclaimed a natural park in 1996. The protected area is surrounded by a buffer zone, another 10,000 hectares. In the buffer zone of the natural park we have assumed that the establishment of permanent agricultural practices (plantations and wetland rice) is no longer possible. Although officially not allowed within the buffer zone, we assumed that deforestation still occurs for wood extraction and shifting cultivation.

Scenario 3: Sustainable agricultural development. This scenario simulates the potential spatial configuration of land use in case of successful implementation of the protected area and associated livelihood programs organised by NGOs. These livelihood programs aim at sustained use of forest resources and income generation through intensification of permanent agriculture outside the park and buffer zone area. If successful, land requirements will deviate from current trends. The assumptions for this scenario include protection of the remaining mangrove area, expansion of the coconut and rice areas (similar to scenarios 1 and 2), a 50% decrease in area of shifting cultivation during the 20 year simulation period, a small increase in primary forest area as a consequence of the protection and regrowth of secondary forest, a two-third decrease in grassland as a result of less logging and shifting cultivation, and the protection of secondary forest areas. It is assumed that new agricultural activities, including shifting cultivation, are now completely banned from the buffer zone.

3.3.2 Simulation Results

Figure 5 shows the spatial distribution of the remaining forest cover (including primary and secondary forest and rock outcrops within the forest area) at the start of simulation and after 10 and 20 years of simulation for scenario 2 (spatial policies scenario). Different parts of the island show different trajectories of land use change. The upland valleys in the southern part of the island show the conversion of secondary and primary forest into shifting cultivation. After abandonment grassland remains the most important land cover. Regrowth of secondary vegetation hardly occurs at these locations. Small areas of secondary regrowth are found in the north-western part of the island, where population pressure is much less. Large-scale development of wetland rice fields is found on the south-western coast. Most developments of coconut plantations are found in the sloping area on the eastern side of the island where all remaining fragments of secondary forest are disappearing.

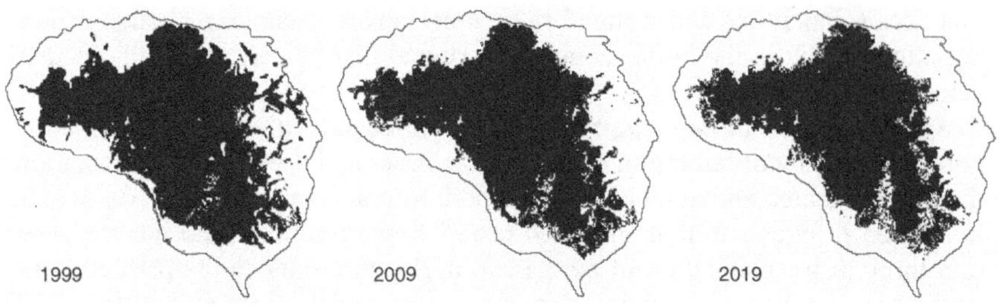

Fig. 5. Forest cover (black) for Sibuyan Island at the start of simulation (1999) and simulation results of 2009 and 2019 for the spatial policies scenario (scenario 2)

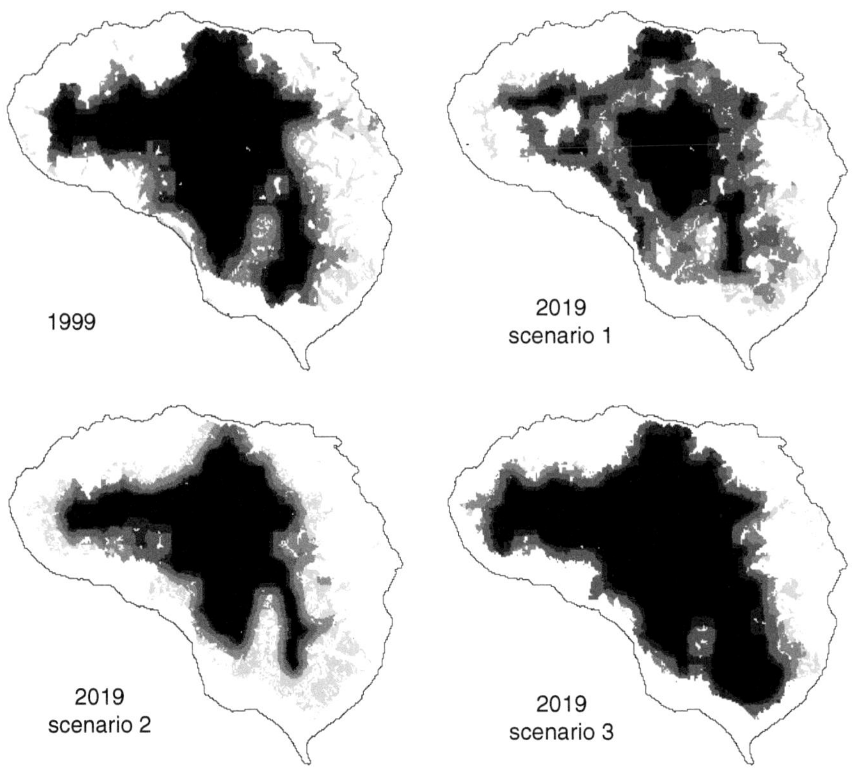

Fig. 6. Forest cover (all gray-shades) for Sibuyan island at the start and end of three different scenarios; gray-shades indicate the aggregation index (black: high; gray: low). Scenarios: 1. baseline; 2. spatial policies; 3. sustainable agricultural development.

The other scenarios result in different patterns of land use change. For all scenarios, the model predicts a further development of agricultural use along the footslopes of the mountains, but the spatial pattern is clearly different (Figure 6). Scenario 1 (baseline), without protection of the Natural Park and buffer zone causes a lot of forest fragmentation. At different locations pieces of secondary and primary forest are cut for shifting cultivation or for wood extraction. In the other scenarios this fragmentation is much less severe, due to the assumed strict protection of the natural park. However, the buffer zone is, in scenario 2, still heavily influenced by deforestation. For scenario 1 and 2 the total remaining forest area is the same. Therefore, the differences in spatial pattern are entirely due to the spatial policies. Good protection of the natural park clearly makes sense to keep a larger, undisturbed core area with forest. Scenario 3 (sustainable agricultural development) assumes less demand for wood extraction and shifting cultivation, leading to a larger remaining forest area. The concentration of all activities outside the natural park and buffer zone leads to a more intensive use of the lowlands, where almost all forest is expected to disappear.

3.4 Effects of Changes in Land Use Pattern on Forest Fragmentation for Sibuyan Island

The three scenarios simulated for Sibuyan Island clearly have a different impact on forest fragmentation (Figure 6). The high resolution of the results allows us to analyse the effects of land use change on forest fragmentation in more detail. It is possible to quantify the change in spatial pattern of the forest resources by spatial indices, which allow a quantitative comparison of the simulation results. We selected two different indices to characterize the spatial pattern of the remaining forest based on the importance of disturbance and patch cohesion for biodiversity. The Aggregation Index (AI) is a measure used to quantify the shape of the forest patches (He *et al.* 2000). AI equals 1 when the forest is completely aggregated into a single, square, patch. It reports numbers closer to 0 when each patch is narrow in one direction and long in another. The aggregation index can also be calculated for every location on the island based on the surrounding forest pixels. Such a spatial representation of the aggregation index is a local measure of the fragmentation of the surrounding forest area. High values of the spatially explicit aggregation index indicate a non-fragmented forest area while low values indicate highly fragmented patches. A second index for spatial pattern was calculated to characterize the influence of edge effects on the forest ecosystem. The distance over which edge effects are important depends on the specific organism or process studied. Since we do not address specific species in this study, we have used a more general index, GISFrag (Ripple *et al.* 1991), to assess the influence of edge effects. GISFrag is calculated by finding the average of the distance to the nearest edge of all the pixels classified as forest. The larger this average distance, the larger the core area of forest. Low values of the GISFrag index indicate that the forest area is very small or highly fragmented. We used the FragStats package (McGarigal *et al.* 1995) to calculate the Aggregation Index and a simple GIS procedure to calculate the GIS-Frag index.

Figure 7, 8 and 9 give the temporal change in respectively the forest area, aggregation index and GISFrag index for the three scenarios. Although the forest area shows a constant rate of change, this does not translate into linear changes in the spatial indices of forest pattern. Clearly, the pattern of land use change is not linearly related to the rate of land use change. The aggregation index decreases with time for scenario 1 and 2, due to fragmentation of the forest in the bufferzone and Natural Park. In scenario 3 (sustainable agricultural development) the aggregation index increases relative to 1999, due to the concentration of agricultural activities in the lowlands and secondary regrowth on grasslands and fallow fields within the buffer zone and Natural Park. For scenario 1 and 2 the aggregation index slightly increases during the first few years of the simulation. This is due to the loss of small forest patches in the lowlands of the island. These small patches contribute to a lower aggregation index at the start of the simulation, as can be seen from the spatially explicit representation of the aggregation index in Figure 6. These patches are preferentially cleared due to their easy access and suitability for agricultural purposes. Also in scenario 3 these patches are cleared, leading to the strong increase in aggregation index at the start of the simulation. Protection of the forest area within the park and the buffer zone maintain a high aggregation index during the rest of the simulation. The decrease in aggregation index indicates the large impact scenario 1 (baseline) has on the remaining area of undisturbed forest. In Figure 6 it can be seen that the major part of the forest area faces a decrease in aggregation index as compared to the situation in 1999. For scenario 2 (spatial policies) a much smaller area of forest experiences a decrease in aggregation index. However, the aggregation index for the island as a whole is, at the end of the simulation, lower than for scenario 1. This is due to the relatively high fragmentation of forest in the buffer zone and at the edges of the nature reserve.

Edge effects have become more important for the forest area remaining at the end of scenario 1 (Figure 9). The average distance to a forest edge decreases from 972 meters to 551 meters during the 20 years of the simulation. For scenario 3 a substantial increase in the GISFrag index occurs, partly due to the deforestation of remaining forest patches in the lowlands and regrowth of secondary forest inside the buffer zone. The temporal change of the GISFrag index for scenario 2 is non-linear. The loss of small lowland patches of forest first increases the average distance to forest edges. However, as soon as deforestation shifts to the main forest area the GISFrag index starts to decrease due to the diminishing size of the main forest area.

For scenario 3 the spatial indices tend to lead to more favourable values for biodiversity conservation as compared to the 1999 situation. However, it should be realized that the remaining patches of lowland forest suffer considerable losses in this scenario as well. The few remaining stands of lowland forest are small, isolated patches, with limited value for biodiversity conservation.

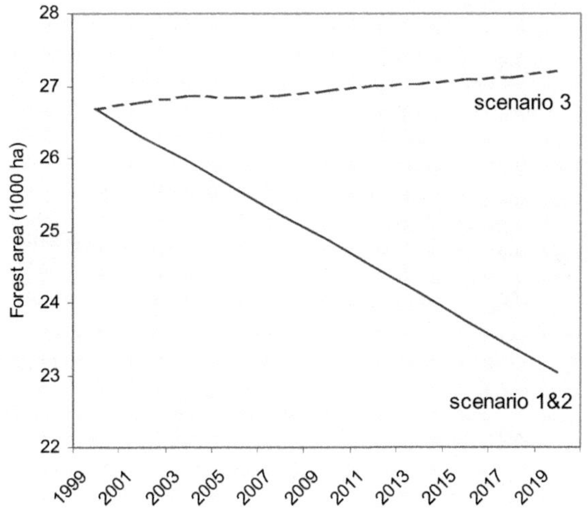

Fig. 7. Change in forest area assumed in the different scenarios; scenarios: 1. baseline; 2. spatial policies; 3. sustainable agricultural development.

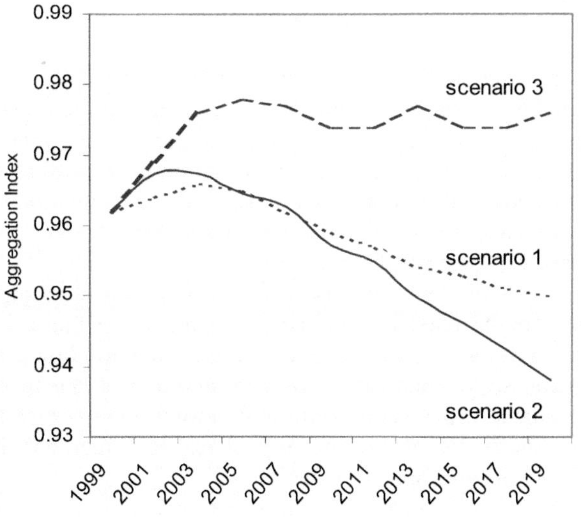

Fig. 8. Change in aggregation index based on the simulated changes in spatial distribution of forest; scenarios: 1. baseline; 2. spatial policies; 3. sustainable agricultural development.

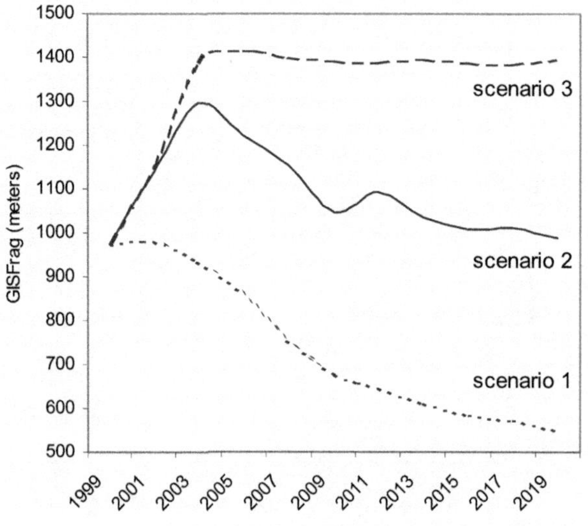

Fig. 9. Change in GISFrag index based on the simulated changes in spatial distribution of forest; scenarios: 1. baseline; 2. spatial policies; 3. sustainable agricultural development.

4 Discussion

The two model applications for the simulation of land use change scenarios have shown the potential use of land use models in assessments of the impacts of deforestation. The application in the Selangor river basin shows in which parts of the region the forests are most susceptible to deforestation under different implementations of a forest conservation policy The results of the simulation visualize the potential result of different strengths of political authority in forest conservation. Furthermore, based on these results it is possible to make qualitative assessments of the potential effects of these land use changes on biodiversity and other ecological functions. The decrease in size and fragmentation of these forest patches can be identified from the simulation results. The results are, however, at a relatively coarse spatial scale, inherent to the configuration of the model. A more quantitative assessment of the spatial pattern of deforestation is therefore not feasible, since most spatial variability in patch shape and fragmentation is, most likely, found at more detailed spatial scales.

The high spatial resolution of the simulation for Sibuyan Island is more suitable for a detailed analysis of spatial patterns. The results clearly show that the spatial pattern is not linearly related to the rate of land use change. The effects of land use change on ecological processes can, therefore, not be assessed without specific attention to the spatial dynamics. The study of spatial patterns is an essential link in research that aims at relating human activities to ecological processes. Non-linear behavior arises through the multiple feedbacks between the spatial and temporal dimensions of land use change (Turner II, 1997), including spatial interactions, path dependence and other forms of system complexity (Janssen, 1998; Manson, 2001). Land use models aim at capturing this type of complexities to translate land use change processes to spatial patterns. Landscape ecologists are faced with the task of assessing the impact of changes in spatial pattern on the relevant ecological processes. Spatial indices as used in this paper can be a tool to help to translate spatial patterns to ecological processes. However, there is no straightforward relation between spatial indices and ecological processes. Spatial indices calculated from simulated land use patterns are valid only at the scale of simulation. Often a mismatch exists between the scale of the ecological processes of interest and the spatial and thematic scale of the simulation results (Gustafson, 1998). The scale dependency of the landscape pattern is therefore an important consideration when translating the simulation results of land use change models to impacts on ecological processes (Wu *et al.* 2000; Wu *et al.* 2002). Furthermore, landscape change does not affect all species similarly, different species have different home ranges, dispersal distances and are differently affected by edges and landscape diversity (White *et al.* 1997; Tischendorf, 2001). A more detailed analysis should be based on the specific requirements of the species found in the study area, taking into account the differences between primary and secondary forest, the temporal changes in landscape pattern and the impact of the land use types at the edge of the forest on the species and processes of interest.

Most land use change models simulate the transition of land use from one distinct land use type into another. In reality, most land use changes happen more gradually and involve modification rather than change of land use. Forest degeneration through selective logging can have important ecological effects but does not appear in the simulation results. Different intensities of agricultural land use can have differential impacts on ecological processes. Therefore, changes in land management and land use intensity are important issues to be given attention in land use change modelling in order to enhance the assessment of the ecological impact of land use change (Lambin *et al.* 2000).

5 Conclusions

Spatially explicit land use models are able to link human activities leading to land use conversion to spatial patterns of landscape change. The spatio-temporal complexity of land use change processes causes non-linear relations between the rate and spatial pattern of land use change. The simulated spatial patterns of landscape change provide essential information to assess the impact on ecological processes.

The case studies in this paper have shown that land use change models can simulate different pathways of land use change based on scenarios of socio-economic developments and spatial policies. The translation of these scenario conditions into spatial patterns of landscape change and the relation between landscape pattern and ecological function makes it possible to evaluate the consequences of different development trajectories and policies. As such, this type of analysis can be a powerful tool to identify conflicts between development objectives, localize 'hot-spots' for nature conservation and recommend alternative options to policy makers and conservation programs.

6 Acknowledgements

The authors would like to acknowledge the Foundation for the Advancement of Tropical Research (WOTRO) of the Netherlands Organization for Scientific Research (NWO) for funding of the research on which this paper is based.

References

Agarwal C, Green GM, Grove, JM, Evans TP, Schweik CM (2001) A Review and Assessment of Land-Use Change Models. Dynamics of Space, Time, and Human Choice. CIPEC Collaborative Report Series 1. Center for the Study of Institutions, Population, and Environmental Change, Indiana University and USDA Forest Service Bloomington and South Burlington

Bockstael NE, Irwin EG (2000) Economics and the Land Use-Environment Link. In: Folmer H, Tietenberg T (eds) The International Yearbook of Environmental and Resource Economics 1999/2000. Edward Elgar Publishing

Briassoulis H (2000) Analysis of Land Use Change: Theoretical and Modeling Approaches. In: Loveridge S (ed) The Web Book of Regional Science. West Virginia University, Morgantown

Chomitz KM, Gray DA (1996) Roads, Land Use, and Deforestation: a spatial model applied to Belize. The World Bank Economic Review 10: 487-512

Forman RTT, Godron M (1986) Landscape Ecology. Wiley, New York

Geist HJ, Lambin EF (2002) Proximate Causes and Underlying Driving Forces of Tropical Deforestation. Bioscience 52: 143-150

Geoghegan J, Villar SC, Klepeis P, Mendoza PM, Ogneva-Himmelberger Y, Chowdhury RR, Turner II BL, Vance C (2001) Modeling Tropical Deforestation in the Southern Yucatán Peninsular Region: Comparing Survey and Satellite Data. Agriculture, Ecosystems and Environment 85: 25-46

Gustafson EJ (1998) Quantifying Landscape Spatial Pattern: What Is the State of the Art? Ecosystems 1: 143-156

He HS, DeZonia BE, Mladenoff DJ (2000) An aggregation index (AI) to quantify spatial patterns of landscapes. Landscape Ecology 15 591-601

Janssen MA (1998) Use of Complex Adaptive Systems for Modeling Global Change. Ecosystems 1 457-463

Kaimowitz D, Angelsen A. (1998) Economic Models of Tropical Deforestation - A Review. Center for International Forestry Research, Bogor (Indonesia)

Kasperson JX, Kasperson RE, Turner II BL (1995) Regions at Risk: Comparisons of Threatened Environments. United Nations University, Tokyo

Kok K, Winograd M (2002) Spatially explicit modelling of the Central American land-use system. Ecological Modelling 149: 53-69

Lambin EF (1997) Modelling and monitoring land-cover change processes in tropical regions. Progress in Physical Geography 21: 375-393

Lambin EF, Rounsevell MDA, Geist HJ (2000) Are agricultural land-use models able to predict changes in land-use intensity? Agriculture, Ecosystems and Environment 82 321-331

Manson SM (2001) Simplifying complexity: A review of complexity theory. Geoforum 32: 405-414

McGarigal K, Marks BJ (1995) FRAGSTATS: spatial pattern analysis program for quantifying landscape structure. USDA For. Serv. Gen. Tech. Rep. PNW-351

McMorrow J, Talip MA (2001) Decline of forest area in Sabah, Malaysia: Relationship to state policies, land code and land capability. Global Environmental Change 11: 217-230

Miller EJ, Kriger DS, Hunt JD (1999) TCRP Web Document 9: Integrated Urban Models for Simulation of Transit and Land-Use Policies: Final Report. University of Toronto Joint Program in Transportation and DELCAN Corporation, Toronto

Nelson GC, Harris V, Stone SW (2001) Deforestation, Land Use, and Property Rights: Empirical Evidence from Darien, Panama. Land Economics 77: 187-205

Nelson GC, Hellerstein D (1997) Do Roads Cause Deforestation? Using Satellite Images in Econometric Analysis of Land Use. Agricultural Economics 79: 80-88

Parker DC, Manson SM, Janssen MA, Hoffmann MJ, Deadman P (2003) Multi-Agent Systems for the Simulation of Land-Use and Land-Cover Change: A Review. Annals of the Association of American Geographers 93(2) forthcoming

Pijanowski BC, Brown DG, Shellito BA, Manik GA (2002) Using neural networks and GIS to forecast land use changes: a Land Transformation Model. Computers, Environment and Urban Systems 26: 553-575

Pontius RG, Cornell JD, Hall CAS (2001) Modeling the spatial pattern of land-use change with GEOMOD2: application and validation for Costa Rica. Agriculture, Ecosystems and Environment 85: 191-203

Rambaldi G, Callosa-Tarr J (2000) Manual on Participatory 3-D Modeling for Natural Resource Management. Essentials of Protected Area Management 7. National Integrated Protected Areas Programme, Quezon City (Philippines). European Commission, DENR

Rambaldi G, Mendoza M, Ramirez F (2000a) Adding the fourth dimension to Participatory 3-D Modelling. PLA Notes 39: 19-24

Ripple WJ, Bradshaw GA, Spies TA (1991) Measuring landscape pattern in the Cascade Range of Oregon, USA. Biological Conservation 57: 73-88

Tischendorf L (2001) Can landscape indices predict ecological processis consistently? Landscape Ecology 16: 235-254

Turner II BL (1997) Spirals, Bridges and Tunnels: Engaging Human-Environment Perspectives in Geography. Ecumene 4: 196-217

Turner MG (1989) Landscape ecology: the effect of pattern on process. Annual Review of Ecology and Systematics 20: 171-198

Velasco E, Miclat A, Garana J, Lucas-Fernan ML, Ladra WF, Blastique T, Velarde A, Torres MJA, Sespene P, Casilin AC, Ronion D, Rambaldi G (2000) Socio-economic and Cultural Profile of the Island of Sibuyan, Romblon: Findings and Recommendations for PA Planning resulting from a Rapid Rural Appraisal conducted within the communities of Barangays Taclobo, Tampayan and Lumang Weste. National Integrated Protected Area Programme (NIPAP), Manilla

Veldkamp A, Fresco LO (1996) CLUE-CR: an integrated multi-scale model to simulate land use change scenarios in Costa Rica. Ecological Modelling 91: 231-248

Veldkamp A, Lambin EF (2001a) Editorial: Predicting land-use change. Agriculture, Ecosystems and Environment 85: 1-6

Veldkamp A, Verburg PH, Kok K, de Koning GHJ, Priess J, Bergsma AR (2001b) The need for scale sensitive approaches in spatially explicit land use change modeling. Environmental Modeling and Assessment 6: 111-121

Verburg PH, Chen YQ, Veldkamp A (2000) Spatial explorations of land-use change and grain production in China. Agriculture, Ecosystems and Environment 82: 333-354

Verburg PH, Schot P, Dijst M, Veldkamp A (2002b) Land-Use Change Modelling: Current Practice and Research Priorities. Geojournal in press

Verburg PH, Soepboer W, Limpiada R, Espaldon MVO, Sharifa M, Veldkamp A (2002a) Land use change modelling at the regional scale: the CLUE-S model. Environmental Management 30: 391-405

White D, Minotti PG, Barczak MJ, Sifneos JC, Freemark KE, Santelmann MV, Steinitz CF, Kiester AR, Preston EM (1997) Assessing Risks to Biodiversity from Future Landscape Change. Conservation Biology 11: 349-360

Wu J, Jelinski DE, Luck M, Tueller PT (2000) Multiscale Analysis of Landscape Heterogeneity: Scale Variance and Pattern Metrics. Geogrphic Information Sciences 6: 6-19

Wu J, Shen W, Sun W, Tueller PT (2002) Empirical patterns of the effects of changing scale on landscape matrics. Landscape Ecology in press

Agricultural Intensification, Population Growth and Forest Cover Change: Evidence from Spatially Explicit Land Use Modeling in the Central Highlands of Vietnam

Daniel Müller & Manfred Zeller

1 Introduction

Land use change, the physical change in land cover caused by human activities such as agriculture and silviculture, is a common phenomenon associated with population growth, market development, technical and institutional innovation, and related rural development policy. Changes in land use can have various consequences on economic growth, the level and distribution of income, and on natural resources such as biodiversity, ecosystems, water, and soils. Land use change leads to change in - and is influenced by - socio-economic indicators such as agricultural productivity, wealth, and human capital. A better understanding of the complex interactions of these changes over time can enable decision makers at various policy levels to design and implement regionally adapted policy interventions which stimulate benefits and counteract negative consequences of land use change by considering the trade-offs among economic, environmental, and social objectives in the process of sustainable rural development.

The paper emanated from a larger research program that attempts to assess the impact of policy, technology, socio-economic, and geo-physical conditions on land use change and its related outcomes on economic growth, welfare, and the protection of forests. The research program is taking place in two districts of Dak Lak province in the Central Highlands of Vietnam. Dak Lak exhibits an interesting case in the study of land use dynamics with its abundant forest resources, ethnic diversity, high immigration rates, and dynamic agricultural and socio-economic development, particularly during the last decade.

Spatial data analysis gives valuable insights into processes of land use change and their underlying causes. The paper employs recently developed methods for combining data from a village-level survey with remote sensing data derived from LANDSAT images. Our objective is to analyze the influence of socio-economic, agro-ecological, geo-physical, and policy variables on land use, using an ex-post research design and a reduced-form, spatially referenced multinomial logit model.

After a brief description of major trends in agricultural and socio-economic development in the research area, we present the conceptual framework and methodology, including issues and related innovative methods employed in combining geo-referencing data from various sources, and present the hypotheses concerning determinants of land use change. Section three presents results of the econometric model. We conclude with a discussion of results and their implications for policy and future research.

2 Trends in Population, Agricultural Growth, and Land Use in the Research Area

The research area is situated in Dak Lak province, the largest of a total of 63 provinces in Vietnam. The province has a forest cover of 52 % compared to 33 % in the overall country (General Statistical Office 2001). Dak Lak has experienced high immigration rates since Liberation Day in 1976. During the last decade, the economy has grown rapidly. Major factors of growth include the rising population supported by in-migration from other provinces, increases in the agricultural and irrigated area, improved agricultural productivity, and rising production and marketed surpluses in higher-value crops and animals. Apart from rice, the rapidly growing cash crop production of mainly coffee and rubber, and more recently pepper, is the major factor for the growth in agricultural exports from Vietnam (Minot and Goletti 2000). Despite annual growth rates in the population of about 7.7 % during 75-90 and 6.4 % during 91-2000, the massive agricultural growth enabled large increases in per-capita income during the 1990s. According to official data, forest cover in Dak Lak province has declined steadily, particularly where soil conditions allow large-scale coffee production. The basaltic soils suitable for coffee cultivation cover 40% of Dak Lak province (Dak Lak Statistical Office 2001).

The research area comprises two districts (Lak and Krong Bong) in the south-eastern part of Dak Lak province. In 2000 the population in the research area was around 124,000 inhabitants (Dak Lak Statistical Office 2001). Population has doubled since 1981. Roughly 20% of the population growth was due to government-controlled resettlement programs in areas delineated as New Economic Zones (NEZ) following the Fixed Cultivation and Resettlement Program (FCRP) implemented mainly during the 1980s. Uncontrolled immigration has been strongest during the last decade with high coffee prices stimulating migration from Central and North Vietnam to the Central Highlands. However, the research area has not benefitted the 'coffee boom' as much as other districts of the Central Highlands because only small areas of the two selected districts have the basaltic soils suitable for coffee production.

Agriculture is the main source of income. Paddy is the most important crop with respect to the factors income share and area cultivated. Degrees of market integration differ significantly between villages and ethnic groups. Ethnic Vietnamese (*kinh*)[1] have a long tradition of intensively growing wetland rice and market-

[1] People of ethnic Vietnamese origin are called *kinh* people, which we will use hereinafter to distinguish them from people belonging to ethnic minority groups (referred to as *ethnic people*). Politically, both groups are Vietnamese citizens. Also note that we use the

ing surplus production to local traders. Their villages have a tendency to be located close to roads, and they usually market a large share of their agricultural production. On average, they own more paddy area per head, which they cultivate more intensively than villages inhabited by ethnic minorities. Before 1990, very little fertilizer and pesticides were used in the research area. During the nineties, the adoption of new crops (maize and coffee), fertilizers (Urea and NPK), and pesticides was promoted by governmental extension programs and related investments in road and market infrastructures. Today, the second most important crop is maize, with hybrid varieties cultivated by 65% of households in the research area. Maize has turned into a major source of cash income during the last five years, especially for households which are not able to produce rice above their subsistence needs and lack suitable soil for coffee production. Livestock is another important production activity; however, it will not be considered further in this article because of its minor importance for land use change in the area. For possible effects of livestock on land use, see for example Cattaneo (2001).

According to our analysis of the LANDSAT images, forest cover in 1992 accounted for 69 % percent of the total research area. In 2000, forested area increased to 77 %, only two absolute percentage points below the level observed in 1975. At lower elevations and in locations close to villages and roads, bush land, open canopy, and bamboo forests prevail. Mountainous evergreen forests dominate areas with altitudes above 600 meters. Bush and bamboo forests as well as young open canopy forests are found as secondary vegetation on areas formerly used for upland cultivation. About 7 % of the research area belongs to nature reserves and protected areas designated for forest rehabilitation and biodiversity conservation. Strict enforcement of forest protection in recent years might have facilitated the restoration of forest cover in these areas to a fairly homogenous, closed canopy cover.

3 Methodology

3.1 Conceptual Framework

Our conceptual framework draws heavily from the theory of induced innovation (Boserup 1965; Ruthenberg 1980; Ruttan and Hayami 1984). We distinguish two key processes of land use change as a source for increasing agricultural production: first, agricultural expansion into previously uncultivated areas, which usually takes place at an extensive and constant technological level; and second, agricultural intensification on already cultivated land. Intensification involves the substitution of land with labor or capital-intensive technology such as irrigation, improved seeds, and fertilizer. Boserup (1965) and Ruthenberg (1980) emphasize the responses of households, communities, and societies to pressures and opportuni-

expression *'minority'* in a purely quantitative sense indicating that people belonging to ethnic *minority* groups are outnumbered by the *kinh* in Vietnam.

ties induced by exogenous shift factors such as population growth, improvement of rural road networks, and investments in markets and the agribusiness sector enabling the introduction of higher-value perennial crops. In the research area, additional pressures on land use change include the introduction of protected areas during the 1990s, and a more effective enforcement of state policies discouraging shifting cultivation.

These policies and driving forces at the macro and regional levels induce changes in land use undertaken by local agents (Pender et al. 1999; Scherr et al. 1996; Templeton and Scherr 1997). With the switch from a planned rural economy to a household responsibility system in the early 1990s, changes in prices, transaction costs, and availability of technology induced private farmers and entrepreneurs to adjust their land use patterns quite rapidly, as can be witnessed in Dak Lak Province. Moreover, the Vietnamese policy transformation towards more market orientation was supported by significant public investments in the rural infrastructure and agricultural extension services, which promoted the adoption of new technologies, crops, and animal husbandry systems. As an indication for the scale of improvement in the rural infrastructure, the average distance from a village to an all-year road in the research area decreased from 17 km to six between the two periods under consideration. Apart from policy and socio-economic factors, changes in the use of land by agriculture or forestry are determined by a range of natural conditions (Ruthenberg 1980) such as rainfall, topography, soil quality, and other geo-physical and agro-ecological variables.

Chomitz and Gray (1996) have developed a widely cited spatially explicit model based on survey and satellite data from Belize. They assess the effect of roads on land use and employed a land rent model based on theories developed by Ricardo and von Thünen. Nelson and Hellerstein (1997) have estimated a similar model for Mexico, including corrections for spatial dependence. Both models rely on satellite data to derive land cover and employ geophysical data and limited secondary information for socio-economic characteristics as independent variables.

Following the conceptual framework of induced innovation and taking into account recent advances in the literature using spatial analysis in combining geophysical and socio-economic data, we explain land use change as a function of geophysical variables (level and variance of rainfall, soil suitability, the altitude and slope of land), endogenous (and therefore lagged) population growth, socio-economic village characteristics, and operational variables for measuring the spatial placement of policy-induced investments in road and market infrastructures and the introduction of new agricultural technology and crops. We will explore the causal relationships between these variables in a spatially explicit framework and attempt to quantify their respective direction and magnitude.

3.2 Sources and Methods Used for Collecting Data

This section briefly describes the specification of the left-hand-side variable land cover[2] and distinguishes various categories for the underlying hypothesized determinants as right-hand-side variables. The right-hand variables are separated into geophysical variables mainly determining the natural agricultural potential, variables describing the socio-economic village characteristics influencing land use, and policy-induced variables.

Land Cover Data

Changes in land cover were measured using time series of satellite data. LANDSAT images for the research area were visually interpreted for the years 1975 (Multi-Spectral Scanner, MSS), 1992 (Thematic Mapper, TM), and 2000 (Enhanced Thematic Mapper Plus, ETM+).[3] All images were taken in the spring season between March 4th and April 7th. Hence, they stem from the same cropping season and from comparable climatic conditions. This enhances the interpretation as the spectral reflection of land cover is easier to compare. In addition, recently updated topographic maps with a scale of 1:50,000 were scanned, geo-referenced, and merged to obtain a consistent set of base information. These maps allow the verification of land cover delineation using additional point information and linear features such as contour lines and roads. The interpretation resulted in land cover maps, which indicate physical land cover of the respective years for the major land use classes. Overlays produce quantitative estimates of changes in land cover over time.

For the econometric analysis, the land cover data was aggregated into five classes, which were used for subsequent analysis:

1. mixed agricultural land (cash and food crops) including cultivated upland plots;
2. paddy fields, both with one and two crops a year;
3. closed canopy, dense forest;
4. open canopy, mixed and secondary forest;
5. mixed grassland (bamboo, bush, and shrub) including re-growing upland plots.[4]

---

[2] Precisely defined, land cover is the (bio-) physical cover of the land surface and immediate subsurface. Land use, on the other hand, describes human employment of that land. In that sense, e.g., hunting would be a kind of land use, while shrub could be the corresponding cover. However, in literature the two are used almost synonymously. For the purposes of this paper, we will use 'cover' to denote the categories of the dependent variables and refer to land 'use' when interpreting the results of the analysis.

[3] To match remote sensing data with recall information, cloud-free satellite images from 1980 and 1990 should have been used. They were, however, not available.

[4] Field verification reveals that most shifting cultivation plots, which were in fallow for one or two years, were still classified as agricultural land. Longer fallow periods result in the category 'mixed grassland.'

Geophysical Data

To a considerable extent, data on geophysical variables determine the agricultural potential inherent in a specific plot. Geophysical variables include rainfall-time series, soil types, elevation, and slope values. Rainfall data stems from five rainfall stations in, and several others around the research area. Daily time series were available starting from Liberation Day in 1976. We calculated the yearly mean and the variance of daily rainfall amounts for both time periods considered in the data analysis.[5] Continuous surfaces for these layers were created by interpolating the point coverage with a regularized spline interpolator.

A digital soil map (DSM) on a scale of 1:250,000 has been reclassified into nine suitability classes for paddy and mixed agricultural production with 'one' exhibiting the most suitable soils and 'nine' the least suitable soils. The suitability classes are based on interviews with soil experts from the region. The Digital Elevation Model (DEM) is based on contour lines from American topographic UTM maps, scaled at 1:50,000. The DEM has a spatial resolution of 50 by 50 meters. Slope degrees were calculated from this DEM.

Socio-Economic Village Characteristics

A village survey was conducted in 101 randomly selected villages out of a total of 194 villages in the research area. The survey provides primary data on the current status and changes in land use practices since Liberation Day in 1975, and on the causes of these changes and their effects on economic, socio-cultural, and ecological variables. Recall techniques were employed to obtain estimates from villagers for Liberation Day in 1976, 1980, and 1990.[6] Additional socio-economic indicators were collected from interviews with communal people's committees and from secondary sources such as official government statistics.

A metric variable was created for the age of the villages in years since their foundation. Villages in the research area are usually inhabited by *kinh* or ethnic minority people. Mixed villages do not exist, apart from one or two Vietnamese families, who usually own a shop in an ethnic village. For that reason a dummy variable coded one for an *ethnic minority village*, and zero otherwise, was included to capture the effect of ethnic composition on patterns of land use. Another

---

[5] The rain-fed cropping season is almost all-year-round, and only four months of the year, December to March, are considered to be the dry season. During this dry season, however there is considerable variation with respect to the commencement and end of rainfall due to regional diversity in altitude and micro-climate (proximity to forests). Rainfall during the dry season influences soil moisture for the planting season and has an effect on water availability in semi-technical irrigations systems that use collected rainfall water as a source. Therefore, we chose to calculate the mean and variance of rainfall data for the entire year.

[6] We related the questions for 1980 to a large flood that occurred in 1981, which nearly all of the villagers remembered. The recall information from 1990 is related to a period characterized by economic renovation policy (*doi moi*), which also provided a suitable point of reference.

dummy was created for villages awarded the status of a '*hero village*'. Villages, which assisted the Northern Vietnamese government in the fight against the South and the U.S.A, were awarded this title after the war. For *population* data, an inverse distance weighted interpolation technique, which calculates a linearly weighted combination from the set of sample points, was used to derive a continuous population surface. Thus, we account for the fact that the population is highest at actual village locations and pressure decreases as one moves away from population centers. The population surface aims to represent population distribution more realistically with a distribution of the data values as a property of location rather than an entire area (Bracken 1994). The lagged value of the continuous population map was included in the data analysis as a pre-determined variable and, therefore, treated as an exogenous factor in our reduced-form equations.

There is possibly a range of other important variables, which can have a substantial effect on land use. For instance, land tenure and property rights shifted recently from a system based on customary rights to an official recognition of property rights. The first land use titles for agricultural land were issued in 1996. Until today, 65% of the households received these certificates – 45% of them in the past two years. Therefore, it is too early to estimate the effects of secure property rights on land use, given our land cover data.

Policy Variables

Policy-induced variables reflect investments from external sources such as governments and donors in infrastructure and technologies as well as restrictions on land use imposed by the government.

Surface vectors for roads obtained on a scale of 1:250,000 were corrected for the research area to a scale of 1:50,000 using the digitized topographic base maps and the pan-chromatic band from the LANDSAT ETM+. Market access is measured by the Euclidean *distance to the nearest all-year roads* for trucks. Based on recall information, we approximated a second road network, designed to reflect access for the year 1990. The number of *years since the establishment of a primary school* in a village was included as an indicator of investment in primary education.

Technology variables are incorporated in the analysis with regard to the time of the introduction of a new technology for agricultural production: (a) the number of *years since the introduction of NPK* as a yield-increasing technology, (b) the number *of years since the construction of a reservoir* for irrigation of agricultural land, and (c) *the increase of irrigated area per village multiplied by the number of years already irrigated* as a measure for the number of irrigated hectare-years.

The boundaries for National Parks and Nature Reserves were included as a policy variable (Chomitz and Gray 1996; Deininger and Minten 2001). We considered the *time of protection* as the number of years that a pixel belongs to a protected area. If a pixel does *not* belong to a protected forest area, this variable assumes the value of zero.

3.3 Integration of Village and Remotely Sensed Data

Farmers, and other land users, make their land management decisions based on objectives, constraints, and the potentials of their households and individual members. At the village level, we observed aggregate behavior of all land users and other stakeholders in land allocation and management. Ideally, to integrate spatially explicit data with survey data, the scale of the analysis should match the unit of decision-making. Yet, in Vietnam as in most developing countries, plot maps and village boundaries are not available. This renders spatial modeling a time-consuming and costly task due to the necessary delineation of the spatial extent of plots or villages, e.g., using Global Positioning Systems (GPS).

The spatial extent of village boundaries could not be delineated with the time and budget available.[7] To demarcate the spatial base unit for the integration of socio-economic variables, the geographic positions of all villages were recorded with GPS and point coverages created. Village boundaries for all surveyed villages were then approximated by applying a cost-distance algorithm to define a set of explicitly defined 'accessibility catchments', generated around each village location and based on estimated transport costs (Bigman and Deichmann 2000). Spatial accessibility is similar to Euclidean distance functions, but instead of calculating the actual distance from one point to another, the shortest cost distance (or accumulated travel cost) from each cell to the nearest source cells is determined.[8] Cost distance functions apply distance not in geographic units but in cost units (Environmental Systems Research Institute 2000). The resulting polygons were then used as a base unit for village-level data in subsequent analyses. Hence, survey data – apart from population – take the value of the interviewed village for each point, which has lower transportation costs to the geographic location of that village than to any other village location (see figure 1).[9] The vast areas of virtually unused and remote forestland are implicitly integrated via geophysical indicators such as topography, soil quality, and slope as well as distance to major roads, which all render land use unprofitable on these marginal lands and should therefore not distort results very much.

In this study, the unit of analysis is a pixel. The research area was spatially referenced to pixels with a size of 50 by 50 meters. Each pixel therefore represents an area of a quarter of a hectare. After resampling all spatially available information

[7] Delineating plot boundaries as the base unit of analysis would match the unit of decision-making of an individual land user much more accurate, thereby being considerably closer to economic reality. However, this is obviously not feasible for an area of 2,400 km^2 with approximately 30,000 households involved in agriculture, each typically having several dispersed plots. Examples of analysis at the plot level linking geophysical, agro-ecological data on plots with socio-economic data of farms or villages include Fox et al. (1994) and Mertens et al. (2000).

[8] For an example of spatial data integration using purely Euclidean distance measures see Müller and Zeller (2002).

[9] This method implies that surveyed and (non-sampled) neighbor villages are similar with respect to land use data. Apart from numerous field visits into the research area, however, we cannot validate this assumption because of lack of data on non-sampled villages.

to above grid size, data was stored in a geographical information system as a raster GIS with 964,000 grids cells covering the entire research area. To compensate for potential spatial dependence in the dependent variables, Besag's coding scheme was used (Besag 1974), also employed by Nelson and Hellerstein (1997), Munroe et al. (2002) and Nelson et al. (2001) in similar studies. The regular spatial sample was drawn by selecting every sixth cell in the X and Y directions so that no selected cells are physical neighbors. This sampling procedure allows us to apply standard estimation techniques (Anselin 1999). To control for potentially inaccurate borders, an inside buffer of 75 meters was cut off the outer boundary of the research area. After merging data from the socio-economic survey with geophysical and land use data, 50 % of the grid cells were randomly sampled. These sampling procedures resulted in a dataset of 15,000 observations used for subsequent econometric modeling.

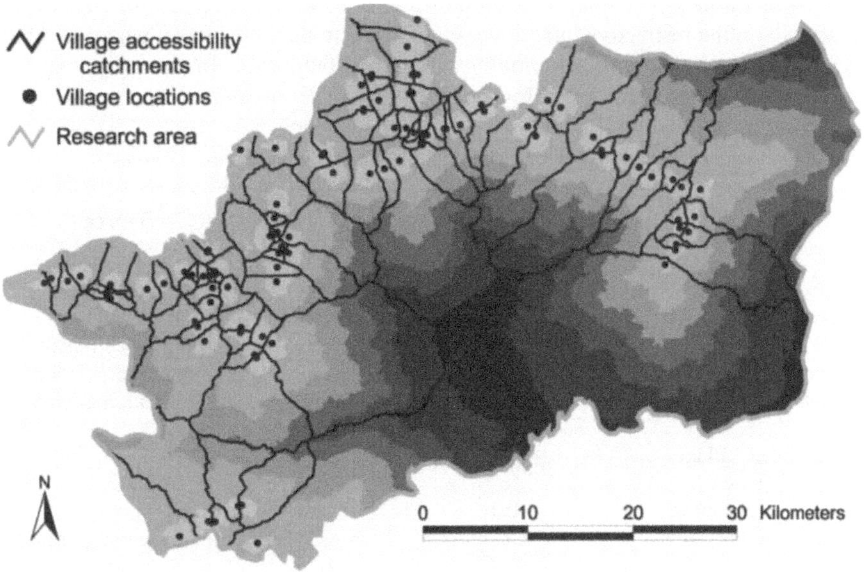

Fig. 1. Cost distance surface, village positions and accessibility catchments

3.4 Expected Effects of Regressors

We ran regressions on four types of regressors. First, we used time-invariant geophysical variables, with the exception of rainfall, which we observe in both periods, and second, historical variables, which describe socio-economic characteristics of the villages. Here, we include population as a time-variant, but lagged variable referring to the state in the previous period. Third, exogenous policy variables to describe government investments and macro-level policies are included in the model.[10]

At present, there are no models and test statistics available which account for substantive spatial interaction in a qualitative dependent variable framework due to the difficulties of incorporating spatial effects into limited dependent variables models like spatial models of land use change (Anselin 2002). To compensate for potential spatial effects, we used the coding scheme described in previous sections. In addition, we included two exogenous variables indicating the X and Y coordinates of each cell in the grid as a spatial filter (Fox et al. 1994; Nelson et al. 1999; Munroe et al. 2002).

Hypothesized effects of regressors on land use categories can be anticipated based on previously outlined theoretical considerations, research findings, and field experience. **Table 1** summarizes the variables including their sources and level of measurement. The expected effects of the exogenous variables are depicted in **Table 2** where a plus sign indicates a positive expected influence of that variable on the respective land cover class and a minus indicates a negative effect. Question marks indicate potentially ambiguous outcomes. In the following section, the hypotheses are stated for the four types of regressors.

Table 1. Level of measurement, description and source of exogenous variables

Variables	Level of measurement	Pixel/ village	Source
	Geophysical		
Slope degrees	degrees	pixel	digital elevation model
Elevation	100 meters	pixel	digital elevation model
Soil suitability classes*	9 classes	pixel	digital soil map
Mean rainfall amount	100 millimeters	pixel	point interpolation
Mean variance of rainfall	100 millimeters	pixel	point interpolation
	Socio-economic history of village		
Years since village establishment	2000 - year	village	survey
Ethnic composition	0/1	village	survey
Lagged population	households	pixel	survey (interpolation)
Hero village	0/1	village	survey
	Policy Variables		
Distance to all-year road	km	pixel	road network/survey
Years since primary school	2000 - year	village	survey
	Technology		
Years since construction of reservoir	2000 - year	village	survey
Increase in irrigated area	hectare	village	survey
Years since introduction of NPK	2000 - year	village	survey

[10] The former we assume to be exogenous, but later may deal with sample selection problems.

Table 1. (cont.)

	Nature conservancy		
Nature reserve areas	0/1	pixel	official park boundaries
	Spatial Effects		
X and Y coordinates	every 5th	pixel	sampled from grid

* One variable for soil suitability for paddy and one for mixed agricultural production with 1=highest suitability and 9=lowest suitability.

Table 2. Predicted signs of exogenous variables

Variables	**Mixed agriculture**	**Paddy**	**Closed forest**	**Open forest**	**Mixed grass**
		Geophysical			
Slope degrees	-	-	+	+	?
Elevation	-	-	+	+	?
Soil suitability classes	-	-	+	+	+
Mean rainfall amount	+	+	+	-	-
Mean variance of rainfall	-	+	+	+	+
		Socio-economic history of village			
Years since village establishment	+	+	-	-	-
Ethnic composition	+	-	-	-	+
Lagged population	+	+	-	-	?
Hero village	+	+	+	-	-
		Policy Variables			
Distance to all-year road	-	-	+	+	-
Years since primary school	+	+	+	-	-
		Technology			
Years since construction of reservoir	-	+	+	+	-
Increase in irrigated area	-	+	+	+	-
Years since introduction of NPK	?	?	?	?	?
		Nature conservancy			
Nature reserve areas	-	-	+	+	-
		Spatial Effects			
X and Y coordinates	?	?	?	?	?

Geophysical Variables

On higher *elevations*, agricultural land use is less feasible and less likely due to more difficult climatic conditions such as lower temperatures and more difficult terrain. Equally, agriculture on steeper *slopes* and poorer *soils* is usually less profitable and more difficult. We hypothesize that the coefficients for elevation, slope, and soil are negative for agricultural land use and positive for forest land.

The *amount and variance of rainfall* critically affect agricultural activities in several ways. In areas with high absolute rainfall, more water-demanding crops can be cultivated. To decrease risks from highly variable and, therefore, more insecure rainfall, more drought-resistant crops and a higher crop diversification should be expected. However, since the land cover data at hand does not enable us to explain crop choice, we can only deduce the following hypotheses: more absolute amounts of rainfall in the moderate climate prevailing in the research area[11] can be expected to increase the natural potential for agriculture, whereas, due to increasing risks, one would expect higher rainfall variation to reduce land under agriculture in the absence of irrigation facilities and increase the likelihood of other uses.

Variables Describing the Socio-Economic History of the Villages

Rural development is essentially path-dependent. Hence, initial states and historical events influence current and future socio-economic situation of villages, including land use. In this context, several variables will be considered and explained in the following.

A positive relation is expected between ethnic minority villages and former shifting cultivation areas, whether fallow or cultivated. Fallow fields are either interpreted as grass land or as agricultural land, depending on the length of fallow. Some fields might have already recovered to become secondary, open canopy forest. Due to these traditional land intensive cultivation techniques and presumably less political power, ethnic villages might be associated with more agricultural area (cultivated less intensively), less irrigated paddy area, and more grass land. Common belief also attributes deforestation to ethnic people; therefore, we would expect less open and closed canopy forest.

Higher lagged *population* should be associated with increasing demand for agricultural land uses and is often cited as a major factor influencing deforestation (Tachibana et al. 2001; Kaimowitz and Angelsen 2000).

The status of 'hero villages' is used as a proxy operational variable to measure the *political capital* of a village. We hypothesize that higher political capital (Birner and Wittmer 2002) is expected to improve the access to government programs and outside assistance. This may well lead to an improved infrastructure such as irrigation facilities and better access to extension services and credit supplies. Subsequently, the intensification effect might be associated with a reduction of land expansion, therefore reducing the pressure on forest land and allowing grass land to recover to become forest again.

---

[11] In more tropical settings with higher amounts of rainfall other authors found the opposite effect (Chomitz & Thomas 2001). In the research area rainfall averaged 1,800 mm between 1976 and 1999. In drier climates where absolute rainfall is a restricting factor at least part of the year, we do expect an increase in agricultural production for higher amounts of rainfall.

Policy Variables

With respect to *infrastructure investments* by the government, we employed several operational variables derived from the village survey. Distance from each cell to all-year roads for both periods serve as proxies for market access and access to political centers. Lower values of these variables might be associated with lower forest cover because of higher producer prices, lower input prices, and lower transaction costs compared to remote locations. Yet, Deininger and Minten (2001) find that these variables are highly correlated with population density, soil fertility, and physical capital such as irrigation facilities. However, that was not the case in our dataset where these variables all showed correlation coefficients below 0.5. Since no farm-gate prices for inputs or outputs are available with spatial variation, Euclidean distance to roads serve as a proxy for prices as well as transaction costs (Deininger and Minten 2001). Agricultural land use is expected to be more likely to occur closer to all-year roads, while forest cover should be less likely (Chomitz and Gray 1996; Pfaff 1997; Nelson and Hellerstein 1997). Again, the effect on mixed grass land cannot be hypothesized a priori.

Access to education is proxied by the availability of primary schools in years since their opening. The effect might be uncertain. Common sense tells us that farmers with better access to education should be expected to use land more intensively and adopt new technologies earlier, thus saving land to achieve a given income or farm output. However, there is also empirical evidence that educated households might engage more in land clearing (Pichon 1997).

Technological change, which has been promoted by a variety of government agencies, is incorporated via the introduction of NPK, one example for a yield-increasing technology. NPK increases first and foremost the productivity of land, and one might expect land expansion to become more attractive with NPK, provided farmers have sufficient labor and the ability to finance the purchase of agricultural inputs. However, given current labor constraints and incomplete rural financial markets, we rather expect increasing yields and more input in production on existing fields. The longer since the introduction of NPK, the more likely the possibility of forest cover. On the other hand, more profitable agriculture could increase the incentive to expand area under cultivation and, therefore, lead to a lower forest cover and a positive coefficient for agriculture (Angelsen and Kaimowitz 2001). Therefore, the overall expected sign of the effect is ambiguous. Another variable related to technological change are years since the construction of a reservoir for agricultural *irrigation*. We anticipate that this labor-intensive intensification on existing agricultural areas will induce households to allocate less time to land clearing and land expansion into upland areas; accordingly, forested pixels will be more likely (Shively and Martinez 2001; Pender et al. 2001).

Inside a *protected area*, it is impossible to establish property rights for long-term secure cultivation, thereby decreasing the profitability of agriculture (Deininger and Minten 2001). We expect a significant positive coefficient for non-agricultural land uses and a regeneration of grass to secondary forest. Protected areas may be located on more marginal lands with steeper slopes and on less suitable soils. For that reason, we "interact" the protection dummy with slope and soil suitability for mixed agriculture. If our hypothesis holds, the introduction of these interaction terms should reduce the significance and the effect of the protection dummy on preserving forest land.

4 Models and Results

4.1 Descriptive Statistics

Agricultural land increased by nine percent from 1975 to 1992, and has stagnated since then (**Table 3**). The agricultural area mainly developed from open canopy forest and grass. Grassland decreased in the second period from eleven to four percent. It is interesting to look at total forest cover, which declined by ten percent in the first period, but has grown by eight percent since 1992. Thus, no severe deforestation has taken place over the last 25 years, with total forest cover decreasing by about two percent.

Table 3. Land cover in km^2 and percentages of total

	2000		1992		1975	
Cover class	**km^2**	**%**	**km^2**	**%**	**km^2**	**%**
Mixed agriculture	339.8	14.2	352.0	14.6	146.7	6.1
Paddy	111.0	4.6	117.0	4.9	67.2	2.8
Closed forest	866.6	36.4	828.2	34.5	836.2	34.9
Open forest	962.6	40.5	827.9	34.6	1052.3	44.0
Grass	101.5	4.3	273.5	11.4	295.7	12.2
Total	**2381.6***	**100**	**2398.6**	**100**	**2398.2**	**100**

* 18 km^2 missing in 2000, because part of the research area was not covered by the scene.

Another interesting feature are the changes in the spatial patterns of land fragmentation, measured as the number of patches under a specific land cover (**Table 4**). Land fragmentation in the last eight years decreased dramatically from 746 patches of land cover in 1992, to 432 in 2000. Most of that change was due to the high fragmentation of agriculture and grass in 1992, compared to 2000, and it reflects the virtual disappearance of shifting cultivation in the research area, which until today has often been made responsible for large-scale deforestation.

Detailed descriptive statistics of the independent variables are reported in **Table 6** of the appendix. The range of values in socio-economic and policy variables demonstrates the diverse characteristics of villages. The mean population increased from thirty households in the first period to more than fifty in the second period, indicating higher population pressure in the last decade. The Euclidean distance to all-year roads averages 6 km for all villages, albeit with a large standard deviation indicating the considerable variation in village locations. Construction of reservoirs and more land under irrigation stimulated rice production by allowing two harvests of wetland rice per year.

Table 4. Number of patches*

Cover class	2000	1992
Mixed agriculture	141	339
Paddy	62	57
Closed forest	14	17
Open forest	56	74
Grass	159	259
Total	**432**	**746**

* Number of patches only counted for the images from 1992 and 2000, which have the same spatial resolution of 30 by 30 meters.

4.2 Multinomial Logit

To explore relationships between exogenous and predetermined variables and the land cover categories as left-hand side (LHS) variables, a multinomial logit specification (MNL) was employed. MNL models estimate the direction and intensity of the explanatory variables on the categorical dependent variable by predicting a probability outcome associated with each category of the dependent variable. The probability that $Y = h$ can be stated as (Greene 1997):

$$PROB(Y = h) = \frac{e^{\beta' x_{lh}}}{\sum_{m=1}^{M} e^{\beta' x_{lm}}} \tag{1}$$

In (1), m denotes the land cover classes used for analysis, β is a vector of estimation parameters and x_l are the exogenous variables for all Y and at all locations l. Equation (1) holds, if the error terms are independently and identically distributed as log Weibull (McFadden 1973). If model assumptions hold, the maximum likelihood estimators are asymptotically normally distributed, with a mean of zero and a variance of one for large samples. Significance of estimators is tested with z-statistics, which are reported in the output tables. To obtain robust variance estimates we employ the Huber and White sandwich estimator.

4.3 Empirical Results

The MNL has five land cover classes as categorical, unordered dependent variables. One multinomial logit model was performed for each of the two time periods. Hereafter, we will refer to the 'model 1990' for the first period (1975 to 2000) and 'model 2000' for the second period (1992 to 2000). Survey recall data for Liberation Day and 1980 was used as 'best estimate' for predetermined variables in 1975 and, similarly, recall data from 1990 was employed for 1992. It would have been more suitable to arrive at equally long time periods, but this was not possible.[3] To control potential endogeneity problems, only lagged values for time-variant independent variables such as population growth and placement of schools are considered in the empirical applications. In addition, all variables were tested for multicollinearity. Model results are reported as raw coefficients in **Table 7** for the model 2000 and **Table 8** for the model 1990 of the appendix. Closed forest is chosen as the comparison group in both models. In the discussion, we will also mention the signs of other interesting coefficients not included in **Table 7** and **8**.

The model does not include lagged values of land use due to relatively little variation in the overall land cover over time. About 70 % of the pixels from 1992 to 2000 and 60% in the period 1975 to 1992 did not change over time. Therefore, the lagged values predict present land use almost perfectly, making model estimation problematic. Since the first nature reserve in the research area was established in 1991, we included the protection variable and its interaction terms only in the 'model 2000'.

Overall, the model for the second period performs better than the model for the first period due to fewer missing values and more exact data on land cover and recall information. Likelihood-ratio tests for combining dependent categories let us reject the hypothesis that categories are indistinguishable. Overall predictive power as measured by the locations predicted correctly is high for both models, with 80% of the pixels predicted correctly in 'model 2000' and 71% in the 'model 1990' (**Table 5**). The majority of the false predictions in both models are pixels covered with grass, which were predicted to be covered by mixed agriculture in 30% and open forest in 50% of the cases in the 'models 2000'.

Coefficients for most topographic variables are significant at the 1% level and show the expected signs.[12] Agriculture is more likely at lower altitudes, on flatter land with better soils and in areas with higher amount and lower variance of rainfall. Surprisingly, better soils increase the odds of closed canopy forest compared with all other land use categories; however, it also makes agricultural land uses more probable compared with grass and open forests.

Older villages increase the likelihood of closed forest in the 'model 1990'. All coefficients have the opposite direction relative to closed forest in the 'model 2000', however with low levels of significance. This result mirrors the location of younger villages closer to roads and populated areas, while older villages are more evenly scattered across the area. Ethnic villages have the lowest odds for closed forest in both models. In the 'model 2000', open forest and grass are significantly more likely closer to ethnic villages than other land uses, which points to a regeneration of forest on formerly used shifting cultivation plots and to a decrease in these cultivation techniques over time.

[12] Note that the multinomial logit model with five outcomes yields 10 possible combinations of outcomes to be compared for each independent variable, hence 200 coefficients for 20 variables. Here, we will focus on closed forest as the comparison group (base category), if not indicated otherwise.

Table 5. Predictive power of land use

Observed land use	Predicted land use 2000						
2000	**Mixed agriculture**	**Paddy**	**Closed forest**	**Open forest**	**Grass**	**Total**	**Correct [%]**
Mixed agriculture	**1,893**	119	4	165	40	**2,221**	85.2
Paddy	280	**377**	0	2	3	**662**	56.9
Closed forest	41	3	**4,450**	1,000	10	**5,504**	80.9
Open forest	190	10	697	**5,644**	31	**6,572**	85.9
Grass	183	3	34	312	**65**	**597**	10.9
Total	**2,587**	**512**	**5,185**	**7,123**	**149**	**15,556**	**79.9**
1990				**1990**			
Mixed agriculture	**1,366**	190	34	505	118	**2,213**	61.7
Paddy	225	**488**	0	18	3	**734**	66.5
Closed forest	11	0	**4,553**	650	64	**5,278**	86.3
Open forest	173	1	781	**4,335**	160	**5,450**	79.5
Grass	477	35	141	913	**313**	**1,879**	16.7
Total	**2,252**	**714**	**5,509**	**6,421**	**658**	**15,554**	**71.1**

Note: Each entry corresponds to one pixel, i.e. to 0.25 hectares. Land use with highest predicted probability is assumed to be the predicted land use. Bold entries along the diagonals indicate correct predictions. For example, 280 pixels of paddy are predicted to be mixed agricultural land in 2000.

Lagged population increases the probability of other land use classes relative to closed forest in 'model 2000', as we expected. A comparison between the two periods reveals some interesting changes. Greater population density significantly increases the probability of both agricultural land uses compared to open forest and grassland in both models. In the 'model 1990' closed canopy forest was however the most likely land cover category in areas with higher population density. This might illustrate the more even distribution of population as well as of plots of dense forest in the first period. It is as well an indication that - at lower levels of population density during the first period - population induced less pressure on forest cover, but growing population density generated a higher demand for paddy and agricultural land.

The coefficients for distance from all-year roads have the expected signs and higher values increase the odds for closed canopy forest. The availability of primary schools increases in both models the probability of closed forest and of paddy land. Irrigation has a two-fold effect. Older reservoirs decrease while more irrigated area increase the likelihood of pixels to be covered by primary forest in the first period. More irrigated area might, thus, have a positive effect on forest protection through an increase in land use intensity. The age of protected areas increases the likelihood of open and closed forest cover compared to all other categories and, therefore, contributed to forest regeneration. On more suitable soils, forest protection increases the probability of a pixel covered with closed forest relative to the other categories while the interaction term with slope does not affect land use classes.

5 Discussion and Policy Implications

The findings from the satellite-image interpretation suggest a decrease in open canopy forest in the first period due to conversion into rain-fed and irrigated agricultural land. Forest cover increased during the 1990s by 8, mainly due to the natural regeneration of mixed grassland. Fallowed fields formerly used for shifting cultivation were largely abandoned during the last decade and regenerated to become bamboo and open canopy secondary forest. Overall, rainfed mixed agriculture remained constant in the 90s, as the reduction in shifting cultivation was compensated by an expansion of cropping area. Over time, agricultural production became more locationally concentrated (less patches) with potential environmental benefits for preserving the integrity of ecosystems and endangered species populations. These changes in land use show that shifting cultivation as the traditional farming system practiced by the indigenous population in the research area almost entirely disappeared in its traditional form during the last decade.

The effect of ethnic minority villages on land use shows strong signs in both models. Ethnic villages are on the average closer to forested and grass areas and further away from all-year roads and political centers. Forest regeneration at the expense of agricultural area predominantly occurs closer to ethnic villages. The increase in forest cover over the last decade, despite the observed population growth, does not correspond to the widely stated positive correlation between higher population density and lower forest cover. Instead, it seems to support the findings from Pender et al. (2001). Permanent conversion of primary forest into agricultural land was hardly observed and, if so, it took place between Liberation Day and 1990, the key period of the government resettlement programs.

Access to all-year roads improved substantially in the last decade, thereby facilitating market integration, access to the infrastructure, agricultural inputs, and public services. The investments in irrigation and infrastructure, combined with improved access to roads, markets, and services, were successful in intensifying agricultural production. Higher agricultural productivity on existing land reduced the need for shifting cultivation, thus preserving forest cover while sustaining a much greater population on virtually the same agricultural land area. Zeller et al. (2000) find similar linkages between agricultural intensification and rural income diversification on the one hand and reduced pressure on forests and soils on the other hand in one of their research areas in Madagascar. This was possible through increases in the productivity of food and cash crops (mainly rice), as well as an increase in cash crop production (mainly maize and to a lesser extent coffee). Because of this and on-going urbanization, the upland arable area as well as the paddy area reduced slightly, whereas much of the grassland of the earlier period (due to shifting cultivation) was reforested to become open canopy forest and, to a lesser extent, closed canopy forest. Protected areas contributed to this development by decreasing the likelihood of non-forest land uses. Thus, the policy measures have contributed to a reduction in shifting cultivation and facilitated agricultural intensification.

During the first decade, we can observe a pathway of land expansion into previously uncultivated areas. The low population density facilitated extensive land cultivation based on forest fallow. Few farmers applied intensive cultivation techniques and production was subsistence-oriented. The last decade brought about significant changes in the policy environment such as investments in road upgrades, irrigation facilities, and the introduction of new technologies. This period was characterized by intensification without further land expansion and allowed forests to regenerate. Today, forests cover nearly the same area as in 1975. The land saving effect of technological progress prevailed over the labor saving effect. In other words, labor-intensive technical progress diverted resources away from land intensive farming systems such as the various forms of shifting cultivation by augmenting the return on labor. At present, more people produce more per capita on less land than ten years ago.

Investments in the road network did not significantly reduce forest cover as it was outweighed by intensification on existing land. Road development seems to contribute significantly to improvements in productivity and welfare, in that way reducing land expansion and removing pressure from forests. At the same time, the shift to a market-driven economy in 1989 (*doi moi*), state policies on sedentarisation and forest protection encouraged production and made traditional farming techniques more difficult to pursue, thus further contributing to the transformation from traditional subsistence-based to a market-oriented production. Moreover, government's expansion and improvement of irrigation systems enabled and facilitated agricultural development. Rice yields and production almost tripled in the research area in the last decade. Intensification from improved irrigation systems may have greatly reduced the pressure to expand cultivated area at the expense of forest.

As we showed, the intensification of agriculture – combined with the (enforced) protection of forested areas - can reduce the pressure on forested land, and slow down or even halt the expansion of agricultural land if coupled with the right policy instruments. The analysis shows that investments in the infrastructure can facilitate this much-needed intensification of agriculture without detrimental consequences for forest resources.

The two districts in our study were purposively selected and are not representative of the entire province. Our secondary data and satellite images of other districts in Dak Lak Province not presented in this paper show a less positive picture as forest cover continued to decline there during the second decade. This is presumably due to less mountainous topography and the widespread existence of the basaltic soil in these other districts. For that reason, the plantation of coffee and pepper, much of it by state-owned enterprises, expanded rapidly in the 1990s to the detriment of forest cover. Therefore, our findings relate to one part of Dak Lak province and cannot be generalized to all of the Central Highlands. Nonetheless, the policy implications of this study call for a renewed emphasis on rural and agricultural development – combined with effective enforcement of nature conservancy measures – that can address the subsistence and income needs through improved technology and institutions.

With respect to implications for further research, we conclude that econometric modeling with larger samples no longer causes serious computational problems. Remote sensing scanners of resolution as used here are available at low costs for the public and continue to increase in quality. Problems for spatially explicit modeling and spatial statistics are found more frequently in the combination of data on natural resources with socio-economic information at an acceptable scale and under reasonable assumptions and simplifications. The integration of socio-economic data with continuous spatial layers poses several challenges. Our attempt to combine land use data with interpolation techniques and accessibility catchments at the village level addresses this issue. It could be a reasonable approach considering the size of our research area. However, this aggregation masks decision-making processes at the farm and plot levels. A combination with household level data and more GPS measures for field boundaries would improve the strength of the overall model.

Our analysis suggests a variety of possible extensions for future research. The dynamics of forest cover change could be more deeply examined by employing binary outcome models and panel estimation procedures. With regard to the assumptions of the independence of irrelevant alternatives, a nested logit approach could yield more consistent estimates, but adds complexity to the interpretation. Inferences from underlying causalities are still difficult to draw given the reduced-form model used in our analysis. Structural models would yield further insights, but would also be more data demanding. An additional analysis could explore the potential endogeneity of the placement of government programs to improve the rural infrastructure (Bigman and Fofack 2000). It is our hope that the analysis presented here may have highlighted the usefulness of spatially differentiated models of land use change for agricultural and rural development policy analysis.

6 Acknowledgements

The research was initiated by and is affiliated with the Sustainable Management of Resources in the Lower Mekong Basin Project (SMRP). SMRP is implemented by the Mekong River Commission (MRC) with support from the German Agency for Technical Cooperation (GTZ) and the German Ministry of Economic Cooperation and Development (BMZ). The research project is supported by the Tropical Ecology Support Program with funding from the BMZ. We are grateful to Regina Birner, Silke Boger, Gerald Nelson, Thomas Sikor and John Pender for helpful comments on previous versions.

References

Angelsen A, Kaimowitz D (2001) Agricultural Technology and Forests: a Recapitulation. In: Agricultural Technologies and Tropical Deforestation. Oxon, pp 383-402

Anselin L (1999) Spatial Econometrics. http://geog55.gis.uiuc.edu/~luc/ [accessed Nov. 2001]. Richardson, Texas, Bruton Center, School of Social Sciences, University of Texas at Dallas

Anselin L (2002) Under the hood: Issues in the specification and interpretation of spatial regression models. Agricultural Economics 27(3): 247-267

Besag J (1974) Spatial interaction and the statistical analysis of lattice systems. Journal of the Royal Statistical Society, Series B 36: 192 - 236

Bigman D, Fofack H (2000) Geographical targeting for poverty alleviation: Methodology and applications. World Bank, Washington D.C.

Bigman D, Deichmann U (2000) Spatial Indicators of Access and Fairness for the Location of Public Facilities. In: Geographical Targeting for Poverty Alleviation: Methodology and Applications. The World Bank, Washington D.C., pp 181-206

Birner R, Wittmer H (2002) Using Social Capital to Create Political Capital, How Do Local Communities Gain Political Influence? A Theoretical Approach and Empirical Evidence from Thailand. In: The Commons in the New Millennium: Challenges and Adaptation. MIT Press

Boserup E (1965) The Conditions of Agricultural Growth. Aldine, Chicago

Bracken I (1994) A surface model approach to the representation of population-related social indicators. In: Spatial Analysis and GIS. Taylor & Francis, London, pp 247-259.

Cattaneo A (2001) Deforestation in the Brazilian Amazon: comparing the impacts of macroeconomic shocks, land tenure, and technological change. Land Economics 77: 219-240.

Chomitz KM, Gray D (1996) Roads, Lands Use, and Deforestation: a Spatial Model Applied to Belize. World Bank Economic Review 10(3): 487-512

Chomitz KM, Thomas TS (2001) Geographic Patterns of Land Use and Land Intensity in the Brazilian Amazon. Development Research Group. http://econ.worldbank.org/ [accessed Nov. 2001]. The World Bank, Washington D.C.

Dak Lak Statistical Office (2001) Statistical Yearbook 2000. Buon Ma Thuot, Vietnam

Deininger K, Minten B (2001). Determinants of Deforestation and the Economics of Protection: an Application to Mexico. The World Bank, Washington D.C.

Environmental Systems Research Institute (ESRI) (2000) ArcView 3.2 On-line Help

Fox J, Kanter R, Yarnasarn S, Ekasingh M, Jones R (1994). Farmer Decision Making and Spatial Variables in Northern Thailand. Environmental Management 18: 391-399

General Statistical Office (2001) Statistical Yearbook 2000. Statistical Publishing House, Hanoi

Greene WH (1997) Econometric Analysis. Prentice-Hall International, Inc

Kaimowitz D, Angelsen A (2000) Economic Models of Tropical Deforestation: a Review. Center for International Forestry Research, Bogor (Indonesia)

McFadden D (1973) Conditional Logit Analysis of Qualitative Choice Behavior. In: Frontiers in Econometrics. Academic Press, New York

Mertens B, Sunderlin WD, Undoye O, Lambin EF (2000) Impact of Macroeconomic Change on Deforestation in South Cameroon: Integration of Household Survey and Remotely-Sensed Data. World Development 28: 983-999

Minot N, Goletti F (2000) Rice Market Liberalization and Poverty in Vietnam. International Food Policy Research Institute, Washington D.C.
Müller D, Zeller M (2002) Land use dynamics in the central highlands of Vietnam: a spatial model combining village survey data and satellite imagery interpretation. Agricultural Economics 27(3): 333-354
Munroe D, Southworth J, Tucker CM (2002) The dynamics of land-cover change in western Honduras: exploring spatial and temporal complexity. Agricultural Economics 27(3): 355-369
Nelson GC, Harris V, Stone S (1999) Spatial Econometric Analysis and Project Evaluation: Modeling Land Use Change in the Darien. http://www.iadb.org/sds/ [accessed May 2001]. Report for the Interamerican Development Bank, Washington D.C.
Nelson GC, Harris V, Stone S, DePinto A (2001) Land Use and Road Improvements: a Spatial Econometric Analysis. http://www.ace.uiuc.edu/faculty/gnelson/ [accessed Nov. 2001]. Agricultural and Consumer Economics Staff Paper AE-4741. University of Illinois Urbana-Champaign, Illinois
Nelson GC, Hellerstein D (1997) Do Roads Cause Deforestation? Using Satellite Images in Econometric Analysis of Land Use. American Journal of Agricultural Economics 79: 80-88
Pender JL, Jagger P, Nkonya E, Sserunkuuma D (2001) Development Pathways and Land Management in Uganda: Causes and Implications. International Food Policy Research Institute, Washington D.C.
Pender JL, Scherr SJ, Durón G (1999) Pathways of Development in the Hillsides of Honduras: Causes and Implications for Agricultural Production, Poverty and Sustainable Resource Use. International Food Policy Research Institute, Washington D.C.
Pfaff ASP (1997) What Drives Deforestation in the Brazilian Amazon? Evidence from Satellite and Socioeconomic Data. The World Bank, Washington D.C.
Pichon FJ (1997) Settler Households and Land-Use Patterns in the Amazon Frontier: Farm-Level Evidence from Ecuador. World Development 25: 67-91
Ruthenberg H (1980) Farming Systems in the Tropics, 3rd edn. Oxford
Ruttan VW, Hayami Y (1984) Towards a Theory of Induced Institutional Innovation. Journal of Development Studies 20: 203-223
Scherr SJ, Bergeron G, Pender JL, Barbier B (1996) Policies for Sustainable Development in Fragile Lands: Methodology Overview. International Food Policy Research Institute, Washington D.C.
Shively GE, Martinez E (2001) Deforestation, Irrigation, Employment and Cautious Optimism in Southern Palawan, the Philippines. In: Agricultural Technologies and Tropical Deforestation. Oxon, pp 335-346
Tachibana T, Nguyen TM, Otsuka K (2001) From Deforestation to Reforestation through Tenure Reforms: The Case of the Northern Hill Region of Vietnam. Journal of Environmental Economics and Management 41: 44-69
Templeton SR and Scherr SJ (1997) Population Pressure and the Microeconomy of Land Management in Hills and Mountains of Developing Countries. International Food Policy Research Institute, Washington D.C.
Zeller M, Lapenu C, Minten B, Ralison E, Randrianaivo D, Randrianarisoa C (2000) The critical triangle between environmental sustainability, economic growth, and poverty alleviation. In: Minten B, Zeller M (eds) Beyond market liberalization: Income generation, welfare and environmental sustainability in Madagascar. Ashgate Publishing Company, Aldershot (GB), pp 209-236

Appendix

Table 6. Descriptive statistics for randomly sampled research area

Variable	Obs.	Mean	Std.Dev	Min.	Max.
Mixed agriculture, 2000	15556	0.14	0.35	0	1
Paddy, 2000	15556	0.04	0.20	0	1
Closed forest, 2000	15556	0.35	0.48	0	1
Open forest, 2000	15556	0.42	0.49	0	1
Grass, 2000	15556	0.04	0.19	0	1
Mixed agriculture, 1992	15554	0.14	0.35	0	1
Paddy, 1992	15554	0.05	0.21	0	1
Closed forest, 1992	15554	0.34	0.47	0	1
Open forest, 1992	15554	0.35	0.48	0	1
Grass, 1992	15554	0.12	0.33	0	1
Slope (degrees)	15548	13.60	9.49	0	50
Elevation (100m)	15548	8.30	3.67	4.2	23.9
Soil suitability for mixed agriculture	15557	2.30	0.69	1	6
Soil suitability for paddy	15557	6.59	1.29	1	7
Rainfall amount (100mm), 1992-1999	15539	20.43	0.79	18.9	22.7
Rainfall variance ("), 1992-1999	15539	29.27	6.27	16.7	39.5
Rainfall amount ("), 1976-1991	15539	15.97	2.61	11.6	20.1
Rainfall variance ("), 1976-1991	15539	20.51	2.52	14.2	25.3
Age of village (years), 2000	15557	33.05	16.80	1	55
Ethnic minority village (0/1)	15557	0.75	0.43	0	1
Population (households), 1990	15545	53.10	32.58	0	292.9
Population (households), 1980	15545	29.20	21.84	0	338.9
Hero village (0/1)	15557	0.36	0.48	0	1
Km to all-year road, 1998	15557	5.83	4.94	0	20.2
Km to all-year road, 1990	15557	16.76	10.05	0	43.2
Years since primary school, 2000	15557	11.49	9.12	0	25
Years since reservoir, 2000	15557	8.63	9.59	0	68
Age of reservoir*irrigated area, 2000	15557	8.32	12.44	0	62.5
Years since application of NPK, 2000	15557	3.53	3.20	0	20
Age of protected area (years), 2000	15557	1.13	2.65	0	9
Coordinate in x-direction	15557	828.05	348.58	4	1529
Coordinate in y-direction	15557	542.17	223.07	4	1029

Source: Primary data on village level collected by authors through village survey and land cover data through interpretation of LANDSAT images; secondary data on geo-physical and agro-ecological variables from the Mekong River Commission (Digital Elevation Model) and provided by the Department for Agriculture and Rural Development, Dak Lak (Digital Soil Map and protected areas); rainfall data from own interpolation of data from nine meteorological stations, classification of soil suitability classes from expert opinion.

Table 7. Multinomial logit model for 2000 with closed forest as comparison group

	Mixed agriculture	**Paddy**	**Open forest**	**Grass**
Slope (degrees)	**-0.163** (12.95)**	**-0.308** (7.86)**	**0.013** (3.43)**	**-0.062** (7.08)**
Elevation (100m)	**-3.711** (16.54)**	**-9.271** (13.54)**	**-0.745** (41.28)**	**-1.166** (11.11)**
Soil suitability for paddy	-0.099 -1.13	**-0.34** (3.65)**	**0.205** (2.30)*	**0.203** (1.98)*
Soil suitability for mixed agriculture	**-1.103** (6.63)**	**-1.029** (6.01)**	-0.231 -1.56	**-0.806** (4.68)**
Rainfall amount (100mm)	0.147 -0.58	**1.704** (3.67)**	**-0.474** (2.85)**	-0.271 -1.08
Rainfall variance (100mm)	**-0.54** (6.98)**	**-0.477** (3.77)**	**0.139** (4.23)**	0.008 -0.11
Age of village (years)	0.006 -0.88	0.011 -1.39	**0.014** (2.97)**	0.008 -1.26
Ethnic minority village (0/1)	0.318 -1.24	0.31 -1.03	**1.013** (5.46)**	**1.134** (4.35)**
Population (households)	**0.016** (4.17)**	**0.024** (4.61)**	0.004 -1.58	0.004 -1.18
Hero village (0/1)	0.294 -1.49	**-0.904** (3.28)**	**-0.733** (6.21)**	**-0.63** (3.70)**
Km to all-year road	**-0.667** (13.46)**	**-0.647** (7.34)**	**-0.12** (8.91)**	**-0.273** (6.42)**
Years since primary school	**-0.021** (2.15)*	-0.008 -0.7	**-0.029** (3.80)**	**-0.051** (5.02)**
Years since introduction of NPK	**-0.1** (4.15)**	0.003 -0.11	**-0.084** (4.93)**	**-0.07** (2.79)**
Years since reservoir	**0.053** (3.82)**	**0.079** (5.35)**	**0.08** (6.36)**	**0.042** (2.91)**
Age of reservoir * irrigated area	**-0.031** (5.09)**	**-0.022** (2.98)**	**-0.025** (5.34)**	**-0.014** (2.27)*
Age of protected area (years)	**-0.612** (3.78)**	**-1.108** (5.24)**	**-0.35** (2.75)**	**-0.687** (3.97)**
Interaction protection - slope	-0.016 -0.39	-0.006 -0.03	0.003 -0.34	-0.08 -1.62
Interaction protection - soil	**1.129** (2.46)*	**2.013** (3.77)**	**1.198** (3.30)**	**1.4** (2.91)**
Coordinate in x-direction	**0.007** (4.71)**	0 -0.02	**-0.003** (3.78)**	**-0.004** (2.89)**
Coordinate in y-direction	**-0.003** (3.65)**	-0.001 -1.12	**0.004** (9.18)**	**-0.002** (2.71)**
Constant	**34.244** (7.18)**	**28.184** (3.46)**	**11.346** (3.39)**	**18.139** (3.82)**
Observations	15525	15525	15525	15525

Notes: Absolute value of z statistics in parentheses below coefficients; * significant at 5%; ** significant at 1% (coefficients with 5% or higher significance level in bold).

Table 8. Multinomial logit model for 1990 with closed forest as comparison group

	Mixed agriculture	Paddy	Open forest	Grass
Slope (degrees)	**-0.081**	**-0.222**	**0.012**	0.008
	(12.22)**	(6.14)**	(3.28)**	-1.49
Elevation (100m)	**-1.318**	**-8.655**	**-0.674**	**-1.223**
	(27.70)**	(11.25)**	(35.38)**	(35.73)**
Soil suitability for paddy	-1.709	-1.823	-0.916	-1.404
	-1.2	-1.28	-0.64	-0.98
Soil suitability for mixed agriculture	**0.703**	**0.975**	**1.054**	**1.08**
	(2.92)**	(4.00)**	(4.44)**	(4.55)**
Rainfall amount (100mm)	**2.34**	**3.314**	**1.84**	**2.494**
	(12.68)**	(11.86)**	(11.39)**	(13.73)**
Rainfall variance (100mm)	**-1.096**	**-1.663**	**-0.549**	**-0.999**
	(12.28)**	(12.39)**	(7.09)**	(11.37)**
Age of village (years)	**-0.059**	**-0.044**	**-0.056**	**-0.058**
	(13.77)**	(5.85)**	(17.50)**	(14.82)**
Ethnic minority village (0/1)	**2.492**	**1.503**	**2.424**	**2.135**
	(11.23)**	(5.35)**	(12.85)**	(9.75)**
Population (households)	**-0.013**	**-0.009**	**-0.016**	**-0.018**
	(3.74)**	(2.20)*	(4.93)**	(5.06)**
Hero village (0/1)	**0.384**	**-1.230**	0.24	0.163
	(2.35)*	(4.78)**	-1.71	-1.03
Km to all-year road	**-0.1**	**-0.163**	**0.092**	-0.024
	(6.29)**	(4.68)**	(7.04)**	-1.66
Years since primary school	**-0.081**	-0.006	**-0.107**	**-0.054**
	(7.12)**	-0.46	(11.38)**	(5.11)**
Years since introduction of NPK	0.057	0.04	**0.159**	**0.089**
	-1.73	-0.91	(5.47)**	(2.84)**
Years since reservoir	**0.097**	**0.099**	**0.111**	**0.113**
	(5.23)**	(3.35)**	(7.14)**	(6.23)**
Age of reservoir * irrigated area	**0.094**	**0.11**	**0.112**	**0.085**
	(8.50)**	(9.28)**	(10.45)**	(7.76)**
Coordinate in x-direction	**0.007**	**0.01**	**0.002**	**0.01**
	(7.92)**	(7.23)**	(2.08)*	(8.32)**
Coordinate in y-direction	**-0.005**	-0.001	**-0.009**	**-0.007**
	(5.15)**	-0.38	(11.27)**	(8.20)**
Constant	3.339	**26.531**	-6.284	-4.811
	-0.34	(2.55)*	-0.63	-0.48
Observations	15523	15523	15523	15523

Postscript

Between Difference and Synergy: Cultural Issues in an International Research Scheme

Michael Fremerey & Hana Panggabean

1 Introduction

The contributions in this reader document the result of two years of research in an intercultural context. It has been managed by joint German-Indonesian planning and steering committees, and it has been conducted in close co-operation between more than 100 German (including a few other European nationals) and Indonesian scholars and researchers, both parts being equally represented in numbers.

Conventional conceptions may suggest that in the field of scientific research, which is assumed to be universally characterized by "rationality" and "objectivity", there is only little room for cultural variations. Yet, two years of scientific co-operation, which aimed at determining crucial factors of stabilization and destabilization in rainforest margins, have demonstrated the profound impact of cultural values, views and behaviour on the research process. This refers in particular to modes of communication, co-operation and management, but it also includes basic approaches towards scientific work.

The research scheme on "Stability of Rainforest Margins" (STORMA) has been designed as a joint German-Indonesian project. In this capacity it has to recognize culture as a crucial factor, if it wants to fully develop its potentials. This includes both preventing destructive emotional and procedural disturbances and utilizing cultural diversity as an asset which eventually may produce synergies. However, such effects do not emerge by themselves. They rather result from conscious processes of cultural learning, reflexion, understanding and adaptation.

Considering the given variety of culturally determined orientations, the joint management and research process went on remarkably well. This is not to say that frictions and misunderstandings have not occurred - many of which may even have escaped the awareness of the actors as being cultural in nature. However, intercultural communication and co-operation has made considerable progress, particularly in the context of field research, though it may be premature to refer to synergetic effects already at this stage. In such a situation, which is characterized by a kind of self-steering mutual acculturation process, it would be counterproductive to expose the cultural factor too much to extensive deliberation. One can literally drown cultural awareness by assuming that everything is related to culture and by constantly trying to surface the cultural factor. Yet, as will be commented upon later, we emphasize that culture is an omnipresent (though often hidden) "orientation system" (Thomas 1999c). At the same time, we rather desist from Hofstede's definition, who once referred to culture as a "collective programming of mind" (Hofstede 1997). It is precisely the experience of two years of intensive intercultural contact which tells us that, on the one hand, culture is a significant variable in co-operation but, on the other hand, the individual person is more than his or her culture. If it weren't so, what could we expect from cross-cultural learning?

Acknowledging individuality as a prime source of behaviour and scientific mode of work and, at the same time, recognizing and appreciating that a good part of our individual profile is culturally determined, is basic to STORMA's approach towards research and research management in an intercultural context. It is this attitude which guided the focus- group discussions of an intercultural workgroup, on which this article is based. It was held during the STORMA "International Symposium on the Stability of Rainforest Margins" in Bogor, September 29 – October 2, 2002, and it aimed at defining, sharing and documenting the experience of cultural difference in the co-operation of the past two years.

2 Culture as Orientation System

Among the various formulations and definitions of culture, this article finds the one from Alexander Thomas (1999c, p 3) as the most comprehensive one. He defines culture as :

> "..a universal orientation system, also typical of a society, nation, organisation or group. This orientation is formed by specific symbols, such as language, meaningful signs, and typical behavior, and is handed down within the specific society or group, etc. It influences perception, thinking, evaluation and the behavior of all members and thus defines their membership of their society. The orientation system enables society members to master their environment in complex social fields and promotes smooth and effective co-operation".

The definition views culture in its broad sense and focuses on its dynamic nature. That is, the term 'culture' comprises an integrated human function in experiencing, comprehending and responding to the surroundings. Furthermore, culture is perceived as a result of socialization, i.e. it is learned and continuously adapted to its social context. Thus, the above definition provides an appropriate basis for understanding the intercultural issues discussed below.

The first issue is the urge to define the scope of cultural problems at work, to draw a line separating work problems and cultural problems. Following the logic of culture as an orientation system means to affirm that such a line does not exist. It means, culture influences individuals' thinking and behaviour in working contexts. Hence, a large number of work-related problems can be traced back to their cultural origins.

Another important issue is the classic nature-nurture dichotomy. The debate focuses on whether being culturally adaptive is attributed to personality tendencies or to learned skills. According to the definition, culture most likely covers both polars. Individuals acquire culture through socialization, an indication of the existence of social learning experiences. However, at the point when it has been acquired, the orientation system becomes a part of the individual's identity. Culture, then, is an aspect of personality. It is in this sense that we can understand the degree of flexibility in adapting to different cultures. The two-fold influence of culture is clear, as culture "..offers facilities and stimuli towards behavior, but also imposes conditions and limits" (Boesch 1980 in Thomas 1999c, p 3). Consequently, there are people who are able to adjust easily to cultural differences and there are those who fail to adapt without proper cultural preparation.

Most important, pointing out learning aspects of culture means setting the focus on its developmental nature rather than its deterministic one. To perceive culture as an endstate is a major constraint in developing models for intercultural communication. This view neglects the changing and dynamic process of culture and therefore is less effective to meet the demand of human interaction processes.

3 Intercultural Problems: The Challenge of Culture Contacts

Intercultural situations are characterised by encounters of different orientation systems. Since culture guides individuals in their social interactions, including the working contexts, it is reasonable to assume that communication and work in such situations are more difficult and less efficient. The interaction becomes more complex because "..customary behavior, thinking and emotions overlap with what seems unusual and strange" (Thomas 1999c, p 3). Concepts and behaviour are either completely different so that they mutually appear as irrational, or they are seemingly similar yet carrying different meanings. Confusions, doubts, and distrust may be the results, leading in turn to friction and conflict.

Studies on intercultural groups highlight conflict sources which have emerged as consequences of cultural misunderstandings or incompatibilities. Some of these areas are: interpersonal relationship, task accomplishment process, stereotypes and prejudices (Brislin 1993); the role of national culture, leadership style, national management style (Shuter & Wiseman 1994); communication style, environmental and economic factors (Teboul et al. 1994). Specifically, Tjitra (2001) reveals main conflicting sources of Indonesia-German workgroups. Some of them are: asymmetrical power structure, communication style, problem solving style and working style. That is, the difference between the Indonesian and the German national culture is so distinct that it leads to different patterns in accomplishing group tasks. The differences eventually lead to communication failures and interpersonal misunderstandings. For instance, the Indonesians refer to their indirect communication style and subordinate structural positions as obstacles to the optimal expression of their initiatives, and therefore perceive unequal opportunities between Germans and Indonesians in the group.

Within STORMA, three major issues can be identified in this context:

- *Communication*
  Basically, the problem is rooted in conflicting communication styles. The Indonesian pattern of indirect communication is incompatible with the German direct communication style. A typical example of the communication problem is the high rate of failure to keep appointments on the Indonesian side.
- *Work-ethic*
  Problems in this area cover issues of work commitment and working style. The Germans (and a few Indonesians) mentioned that Indonesians often pursue activities and tasks irrelevant to the project. Accordingly, that conduct leads them to perceive that Indonesians put project works at lower priority than those sideline activities. This situation has its impact in the form of slower work pace. Another point contributing to emotional stress in daily working situations is the unsystematic pattern in doing the task.
- *Power*
  The group mentioned unbalanced power structure in the research scheme as a source of conflict. Realistically, it is admitted that the structure of the project is, to a certain extent, pre-set. Accordingly, the focus is on how to manage the given structure rather than to change it. The asymmetrical structure creates different levels of autonomy and, up to a certain point, gives way to possessive attitudes towards the project and its facilities. As an example, different levels of user rights in the laboratory were mentioned.

Looking into the above problems, it becomes clear that cultural challenges in Indonesian-German work groups are real. In order to put these challenges into their contextual meanings, we try to look at them from an Indonesian and a German cultural perspective.

4 Indonesia and Germany: Basic Cultural Values

Knowing Indonesian and German cultural values, one cannot help concluding that both cultures are very different. Empirical studies evidently convey affirmation to that conclusion. The two cultures are located at opposite ends of almost every cultural dimension (see Hofstede 1997; Adler1991). These facts are either discouraging or challenging for those involved in Indonesian-German work groups. One thing is certain, though: knowledge of and familiarity with values of both cultures become a vital necessity for the members. Accordingly, this section discusses German and Indonesian core values in order to promote mutual understanding.

German Work Values

Sources on German cultural values agree that work is regarded as an important part of one's identity. German society takes work as a serious matter and, therefore, work-related values are pronounced strongly as major guidance for the nation. Max Weber has referred to this value complex as "protestant ethics", which still plays a significant role in the present (Craig 1991). It comprises, among others, industrious attitude (e.g. hard work, discipline, rule-orientation) and personal achievement, which are rated as particularly high in the scale of values (compare : Craig 1991; Hall & Hall 1990; Peabody 1985).

Another signified value in German culture refers to rationality and objective thinking (Hall & Hall 1990; Peabody 1985). Like the work ethics, the rational and objective thinking patterns are rooted in the society as the legacy of a major historical idea, i.e. the period of enlightenment (Craig 1991; see also Panggabean 2001).

Indonesian Social Values

Though Indonesia is a multi-cultural society, we shall concentrate here on the Javanese culture, due to its majority status and widespread influence across the nation (see Panggabean 2001 for more details).

As it is common in many Asian cultures such as, for instance, the Japanese and Chinese, Indonesia's primary cultural value is the idea of harmony (Magnis-Suseno 1996; Mulder 2001). However, even more than in most other Asian countries, the principle of harmony has a very strong impact on social life in the Indonesian context, expressing itself in a "harmonious social appearance" (Geertz in Magnis-Suseno 1996, p 40). Basically, harmony is rooted in the Javanese spiritual concept that the universe is arranged in harmonious order and the main human task is to preserve this harmony. Hence, a harmonious condition is something preserved, not something achieved. In its socially transformed existence, the idea of preserving harmony results in various conflict avoidance mechanisms and indirect ways to convey negative messages (Brandt 1997). In accordance with this principle, Indonesian society makes a clear distinction between open and discreet conflict. Open conflicts should be avoided since they are harmful to harmony, but it does not mean that the culture forbids conflict completely. It is rather a matter of how to perform conflicts.

Another important value is the concept of *rasa.* According to Geertz (1964), the concept has a dual meaning. The first one is 'feeling' and the second one is 'meaning'. That is, one should use one's refined feelings to acquire the true meaning. A refined *rasa* is an important indicator of maturity and significant for harmony preservation. *Rasa* is manifested in sensitivity characteristics of Indonesians (see Panggabean 2001). Discussions on harmony and *rasa* as Indonesian central values clearly demonstrate that Indonesian society highly values smooth and subtle conducts, even to express power (Anderson 1990). The Indonesian indirect communication style should be viewed particularly from this perspective.

Cultural Perspectives of STORMA Problems

Considering both frameworks discussed, it becomes evident that the two cultures imply diverging standpoints as they enter the working context. Indonesians focus strongly on the social-harmonious aspect, whereas Germans concentrate on the work-related aspects. It is then understandable that the meeting of the two cultures under working conditions is prone to misunderstandings and conflict.

The differential *communication style* is the first problematic area. The German style is direct and straightforward, striving after a speedy common perspective and objective, a matter-of-fact pattern in dealing with things. On the other hand, Indonesians prefer an indirect style for completely different reasons. The indirect style is a suitable way to deal with the existing socio-cultural heterogeneity, attended by a high level of uncertainty and a high probability of intergroup problems. Moreover, in German culture honesty is highly rated as an indicator of individual integrity. Indonesian indirectness, *de facto* a pattern which serves to maintain harmony and to avoid insult, is easily misinterpreted from this perspective as a sign of insincerity.

Regarding the *work values*, German work ethics are firmly established: discipline, commitment and responsibility are values which enjoy a high appreciation in working contexts. On the other hand, the most important working condition for Indonesians is the existence of a comfortable working atmosphere. This is not to say that Indonesians are not prepared to work in a less comfortable work climate, but they would rather sacrifice their motivation and commitment for the sake of a more comfortable working atmosphere.

The different emphasis put on intercultural sensitivity illustrates how both cultures vary in the rating of social aspects in the working context. However, the empirical findings also disclose considerable learning capacities within German-Indonesian teams as to the mutual understanding and appreciation of such work related values. By revealing common characteristics of intercultural sensitivity in mixed work groups, Panggabean (2001) indicates that both cultures can learn to draw close to each other while maintaining culture-specific characteristics.

Discussing *power issues* in Indonesian-German work groups presupposes the consideration how Indonesian and German cultures perceive hierarchy. Hierarchy is important in both cultures. Even though, according to Hofstede's seminal work (1997), the German culture is less hierarchical than the Indonesian, it is considered as more hierarchy oriented than most western cultures (Hall & Hall 1990). The crucial point in both cultures is how they manage hierarchy. The German performance could be classified as a "direct" one, though individuals tend to perceive and accept the power structure in a quite relaxed way. Interactions are characterised by sufficient autonomy and space for the development or maintenance of an independent attitude. Hierarchy is not supposed to prevent independent initiatives or critical discussions. On the other hand, in accordance with the highly valued rasa, Indonesians manage hierarchy in a much more indirect and smoother way. Persons in power are expected to exercise it without overly displaying it (Anderson 1990; Magnis-Suseno 1996). However, among Indonesians the strong hierarchical order is clearly sensed. Hence, the crucial point for Indonesian-German workgroups on this point is more how power is conveyed rather than whether it is recognized.

Summing up, it may be stated that German-Indonesian work groups need to manage their cultural differences, if they want to maximise their efficiency and productivity. Though the management of cultural differences is, as mentioned above, anything but an easy task, the discussions in the STORMA intercultural work group revealed some positive signs of mutual cultural learning. These indications will be discussed below.

5 Optimizing the Advantages of Intercultural Work Groups

Though research regarding the effects of cultural diversity in work groups is not totally conclusive, a large number of studies confirm the benefits of cultural diversity in promoting creativity and synergetic achievements (see Cox 1994). Empirical evidence of synergetic potentials in German-Indonesian work groups has been provided, among others, by Zeutschel (1999) and Tjitra (2001). Such findings conform to increasing realizations of the advantages of culturally diverse work groups (see e.g. Thomas 1999a).

In its discussions, the STORMA work group agreed on the fact that the intercultural profile of the research program constitutes a major challenge to all researchers and administrators involved. Though difficulties in cross-cultural adaptation were not denied, there was a unanimous feeling that intercultural learning has occurred on both sides, and that this learning process is about to open up hidden potentials. Below are some conclusions and perceptions which have been mentioned in this regard.

Moving Towards Power Balance

The first 2 years of intercultural co-operation in the STORMA context have witnessed a gradual shift from an asymetric distribution of power towards a greater power balance. This has been highlighted by the group as a major (though still fragmentary) breakthrough in the intercultural design of the research program. In particular, the increasing involvement of Indonesian research supervisors was rated as an indicator for change in the distribution of power and authority. Both studies which have dealt in depth with the problems and potentials of German-Indonesian workgroups (Panggabean 2001; Tjitra 2001) emphasize the crucial role of power distribution in intercultural ventures. It largely determines the climate of work, the determination to make the joint project a success, and the willingness to adopt a learning attitude with regard to characteristics of the other culture.

Willingness to Learn

Irrespective of genuine difficulties in understanding and accepting some peculiarities of the other culture, an overall readiness towards intercultural learning and to adjust to new ways of thinking and acting was emphasized. This should be valued as an important precondition of promoting cultural sensitivity. Basically, the willingness to learn and to approach another culture with an open mind and flexibility are identified as essential indicators for intercultural effectiveness. These qualities repeatedly occur as competences needed to overcome distress and frustrations in processes of cultural adjustment (see Panggabean 2001).

Developing Synergies

Creating intercultural synergies in joint research efforts is a complex and difficult process. Yet, the group agreed in identifying complementary competencies and skills in the interculturally composed research schemes. For instance, the Indonesian flexibility and ability to cope with unexpected situations or results may productively complement the German tendency towards advance planning and prognoses (see Stumpf & Zeutschel 1999). Thus, the seemingly oppositional cultural tendencies of Indonesians and Germans, which come near the methodological differentiation between "inductive" and "deductive", could both benefit the cooperation when they are seen and treated as complementary. This could be reinforced by a mutual learning process which includes the supplementation of one's own cultural habits by some practices of the other culture.

Such experiences, which could be acknowledged as initial synergetic effects, have been gained in the process of STORMA research. It is believed that they will help to move towards true synergy by combining different views of reality into a shared one with new and peculiar qualities. Actually, some of the papers presented in this volume, which have been co-authored by German and Indonesian scientists, clearly point into this direction.

The Role of Intercultural Mediators

Intercultural learning, establishing a power balance and moving towards synergetic effects are processes in intercultural teams which rarely grow by themselves. Nor can they be "prescribed" in a top-down approach. Initiating and sustaining such processes rather depend on the influence of individuals who are familiar with both cultural frameworks and therefore have the potential to develop culturally-adequate behavior for both parties. The concept of the 'mediating person' has been around for decades (Bochner 1981; Adler 1974), and Panggabean (2001) has emphasized the role of mediating persons as intercultural facilitators in Indonesian-German work groups.

It can be regarded as fortunate that STORMA has at its disposal a few of such individuals in various functions, both with Indonesian and German cultural background. They act as communication catalysts, enabling effective communication between cultures. They perform their roles by customizing the message and the way it is conveyed in order to adjust it to the given cultural framework. They have the capability to actively perform cultural empathy, and to transform cultural knowledge into constructive action. In short, the existence of these individuals is valuable since they are highly appreciated by members from both cultures and thus are able to prevent or mediate group conflicts.

6 Conclusions

It remains uncontested that the joint endeavour to find sustainable solutions for the interaction between man and natural environment in the case of rainforests constitutes an outstanding and world-wide challenge, which demands not only interdisciplinary but also intercultural approaches. As increasing heterogeneity is a concomitant phenomenon of work groups, systematic efforts are needed to manage cultural differences. In the context of Indonesian-German work groups, as in the case of STORMA, the marked distance between the two cultures affects the work accomplishment and potentially leads into conflict situations. However difficult it seems, it is a common conviction that these challenges are manageable, and that a proper management of cultural differences may even lead the group into some sort of synergy.

While it is confirmed that STORMA has some advantageous characteristics, systematic and deliberate efforts are still needed in the field of intercultural learning. To learn culture in a systematic way is frequently referred to as a beginning to promote synergy (Thomas 1999c; Brislin et al. 1983; Albert 1994). Though the benefits of intercultural training are not totally uncontested (Ward et al. 2001), accumulated works over decades suggest that individuals involved in intercultural assignments would benefit from some kind of systematic preparation and training, enabling them to cope with cultural challenges.

Based on these considerations, the STORMA discussion group brought forward the following recommendations to increase the intercultural performance in the STORMA research scheme:

- *Individual cultural preparation prior to the assignments*. This preparation should take the form of information-gathering, targeted at raising the level of cultural knowledge. This first stage of culture learning normally implies the cognitive acquisition of culture-related information, like norms, values, habits, symbols etc.. The learning materials are books, films, or any relevant documents. This kind of learning is the most practical and easy to access, but it also has many limitations, since the information is often too general and does not include the behavioral dimension (Ward et al. 2001).
- *Exposure to supervised real or simulated second-culture experiences* (Ward et al. 2001). This learning process is based on situational cases as well as on concrete experiences from work practices in the host country. Experiential learning is evidently more effective than the pure cognitive approach, due to the greater personal involvement of the learner (Ward et al. 2001). A typical technique is role-playing with members from the host culture, providing a chance for the learner to experience the local interaction patterns.
- *Intercultural learning in mixed teams*. Positive learning effects can be particularly expected from learning situations, where both cultures are represented and, if possible, incidents of immediate and common interest are shared and reflected. This concerns first of all project members who are interacting in the field, having experienced actual situations of cultural diversity. The major po-

tential of the mixed team approach lies in the mutuality of learning. It follows the principle that, irrespective of host and guest roles, intercultural communication and management needs awareness and understanding among all actors involved.

- *Intercultural coaching*. This learning approach provides opportunities for discussion or sharing with individuals, who can provide cultural insights or practical solutions in immediate field problems. The coach thus assumes the role of a project-attendant and *ad hoc* available consultant, who is accepted by all sides concerned as a neutral person with adequate intercultural knowledge and communicative skills. In this role, the coach may take over crucial functions not only in facilitating intercultural understanding and preventing or mitigating conflicts, but also in creating a mutual basis on which synergies can develop (Zeutschel 1999).

The above-mentioned steps towards the improvement of joint Indonesian-German research indicate the efforts which have to be invested in order to optimize the intercultural composition of the research scheme. On the whole, these efforts comprise three major steps: (1) to develop an awareness about cultural differences, (2) to avoid the threats (e.g. in terms of misunderstandings and offenses), which different systems of orientation may have for a beneficial cooperation, and (3) to exploit the huge opportunities embodied in the intercultural composition of the research scheme.

The latter refers to the creation of synergies in the research process which somehow implies the vision of a joint "STORMA culture" with its own system of orientation. Though a vision is something which one may never totally achieve, it provides guidance for efforts in intercultural learning. And these efforts deserve high priority and attention if synergies are to be attained – or, as U. Zeutschel has appropriately put it: "synergy is not for free" (Zeutschel 1999, p 6).

References

Adler N (1991) International dimensions of organizational behaviour. Kent, Boston

Adler P (1974) Beyond cultural identity: Reflections on cultural and multicultural man. In: Brislin R (ed) Topics in culture learning. East-West Centre, Honolulu, pp 23-40

Albert RD (1994) Cultural diversity and intercultural training in multinational organizations. In: Wiseman RL, Shuter R (eds) Communicating in multinational organizations. International and Intercultural Communication Annual 18: 153-165

Anderson B (1990) Language and power: Exploring political cultures in Indonesia. Cornell University Press, Ithaca

Bochner S (ed) (1981) The mediating person: Bridges between cultures. Schenkman, Cambridge (Mass.)

Brandt T (1997) 'Kunci Budaya'. Business in Indonesia: The cultural key to success. Goasia, Bad Oldeslohe

Brislin R (1993) Understanding culture's influence on behavior. Harcourt Brace College Publ., Orlando

Brislin R, Landis D, Brandt ME (1983) Conceptualizations of intercultural behaviour and training. In: Landis D, Brislin R (eds) Handbook of intercultural training I: Issues, theory and design. Pergamon, New York, pp 1-35

Cox Jr T (1994) Cultural diversity in organizations. Theory, research, and practices. Berrett-Koehler, San Francisco

Craig GA (1991) The Germans. Penguin Books, New York

Geertz C (1964) The religion of Java. The Free Press of Glencoe, New York

Hall ET, Hall MR (1990) Understanding cultural differences: Germans, French, and Americans. Intercultural Press, New York

Hofstede G (1997) Cultures and organizations: Software of the mind. Mc Graw-Hill, New York

Magnis-Suseno F (1996) Etika Jawa: Sebuah analisa falsafi tentang kebijaksanaan hidup Jawa. Gramedia Pustaka Utama, Jakarta

Mulder N (2001) Mistisisme Jawa. Ideologi di Indonesia. LKIS, Yogyakarta

Panggabean H (2001). Characteristics of intercultural sensitivity in Indonesian-German work groups. Dissertation, Universität Regensburg.

Peabody D (1985) National characteristics. Cambridge University Press, Cambridge

Pedersen P (1994) A handbook for developing multicultural awareness. American Counseling Association, Alexandria

Shuter R, Wiseman R (1994) Communication in multinational organizations: concepts, theoretical, and practical issues. In: Wiseman R, Shuter R (eds) Communicating in multinational organizations. Intercultural Communication Annual 18: 3-11

Stumpf S, Zeutschel U (1999) Synergy effects in multinational work groups: what we know and what we don't know. In: Thomas A (ed) Contributions to intercultural psychology. Unpublished papers. Universität Regensburg

Teboul J, Ling C, Fritz L (1994) Intercultural organizational communication research in multinational organizations. In: Wiseman RL, Shuter R (eds) Communicating in multinational organizations. Intercultural Communication Annual 18: 12-29

Thomas A (ed) (1999a). Contributions to intercultural psychology. Unpublished papers. Universität Regensburg

Thomas A (1999b) Divergences, convergences and synergy potential of German-Japanese teams. In: Thomas A (ed): Contributions to intercultural psychology. Unpublished papers. Universität Regensburg

Thomas A (1999c) Intercultural orientation and action. In: Thomas A (ed) Contributions to intercultural psychology. Unpublished papers. Universität Regensburg

Tjitra H (2001) Synergiepotenziale und interkulturelle Probleme. Chancen und Herausforderungen am Beispiel deutsch-indonesischer Arbeitsgruppen. Deutscher Universitäts-Verlag, Wiesbaden

Ward C, Bochner S, Furnham A (2001) The psychology of culture shock. Routledge, East Sussex

Zeutschel U (1999). Intercultural synergy in professional teams. Views from the field of practice. In: Thomas A (ed) Contributions to intercultural psychology. Unpublished papers. Universität Regensburg

Printing: Mercedes-Druck, Berlin
Binding: Stein+Lehmann, Berlin

Batch number: 10762625

Printed by Printforce, the Netherlands